Manual of Water Supply Practices

M68

Water Quality in Distribution Systems

American Water Works Association

Manual of Water Supply Practices—M68

Water Quality in Distribution Systems

Disclaimer

The authors, contributors, editors, and publisher do not assume responsibility for the validity of the content or any consequences of its use. In no event will AWWA be liable for direct, indirect, special, incidental, or consequential damages arising out of the use of information presented in this book. In particular, AWWA will not be responsible for any costs, including, but not limited to, those incurred as a result of lost revenue. In no event shall AWWA's liability exceed the amount paid for the purchase of this book.

If you find errors in this manual, please email books@awwa.org. Possible errata will be posted at www.awwa.org/resources-tools/resource.development.groups/manuals-program.aspx.

Senior Managing Editor/Project Manager: Melissa Valentine
Cover art: Melanie Yamamoto
Production: Janice Benight
Manuals Specialist: Sue Bach

Library of Congress Cataloging-in-Publication Data

Names: Smith, Kira S., author. | Slabaugh, Rebecca, author. | American Water Works Association, issuing body.
Title: Water quality in distribution systems / by Kira S. Smith, Rebecca Slabaugh.
Other titles: AWWA manual ; M68.
Description: First edition. | Denver, CO : American Water Works Association, [2017] | Series: AWWA ; M68 | Includes bibliographical references and index.
Identifiers: LCCN 2017028391 | ISBN 9781625762269
Subjects: LCSH: Water quality management. | Water--Distribution.
Classification: LCC TD365 .S62 2017 | DDC 628.1/44--dc23 LC record available at https://lccn.loc.gov/2017028391

Printed in the United States of America

ISBN 978-1-62576-226-9 eISBN-13 978-1-61300-423-4

American Water Works Association

American Water Works Association
6666 West Quincy Avenue
Denver, CO 80235-3098
awwa.org

Contents

This page intentionally blank.

Figures

This page intentionally blank.

Tables

This page intentionally blank.

Acknowledgments

The AWWA Manual M68 was written through the persistent and dedicated work of the following authors:

Kira S. Smith, *Manual Chair,*
City of Houston Drinking Water Operations, Houston, Texas

Nick Ashbolt, University of Alberta, Edmonton, Alb., Canada
Hélène Baribeau, AQUAlity Engineering, Arcadia, Calif.
Patrick Cole, H2M Architects and Engineers, Brielle, N.J.
Ricard Devesa, Aigues de Barcelona (Agbar), Barcelona, Spain
Andrea Dietrich, Virginia Tech, Blacksburg, Vir.
John Dyksen, SUEZ Water NA, North Haledon, N.J.
Kevin Fisher, Southern Nevada Water District, Las Vegas, Nev.
Peter Fiske, Lawrence Berkeley National Laboratory, Berkeley, Calif.
Melinda Friedman, Confluence Engineering Group LLC, Seattle, Wash.
Rich Giani, CH2M, Lees Summit, Mo.
Amlan Ghosh, Corona Environmental Consulting, Austin, Texas
Alex Gorzalski, US Army Corp of Engineers, Washington, D.C.
John Graham, California Water Service Co, Chico, Calif.
Ellen Hall, Hazen and Sawyer, Fairfax, Vir.
Amie Hanson, Confluence Engineering Group LLC, Seattle, Wash.
Adam Jacoby, Hatch Mott MacDonald, New York, N.Y.
Aneta King, Halton Regional Municipality, Oakville, Ont., Canada
Mark LeChevallier, American Water, Voorhees, N.J.
Dave MacNevin, Tetra Tech, Tampa, Fla.
Yvonne Mazza, CH2M, Lakewood, Colo.
Laura Meteer, Regional Municipality of York, Newmarket, Ont., Canada
Jack W. Moyer, AECOM Technical Services, Morrisville, N.C.
Andrew Ohrt, Arcadis, Minneapolis, Minn.
Susan Rivera, Central Washington University, Ellensburg, Wash.
Meg Roberts, Hazen and Sawyer, Greensboro, N.C.
Caroline Russell, Carollo Engineers, Austin, Texas
Janice Skadsen, Ann Arbor, Mich. (formerly with CDM Smith)
Rebecca Slabaugh, Arcadis, Indianapolis, Ind.
Justin Sutherland, Carollo Engineers, Inc., Austin, Texas
Alejandro Ureta, Asesoramiento a la Gestión de los Sistemas de Agua (AGeSA), Montevideo, Uruguay
Raj Vaidya, Chastain Skillman, Lakeland, Fla.
Francesc Ventura, IDAEA-CSIC, Barcelona, Spain
Jennifer Wade, Wachs Water Services, Buffalo Grove, Ill.

David Wahman, US Environmental Protection Agency, Office of Research and Development, Cincinnati, Ohio

Carol Walczyk, Hatch Mott MacDonald, Iselin, N.J.

Thomas M. Walski, Bentley Systems, Nanticoke, Pa.

Jian Yang, American Water–Engineering, Voorhees, N.J.

The following individuals provided peer review of this manual. Their knowledge and efforts are gratefully appreciated:

Colleen M. Arnold, Aqua America, Inc., Bryn Mawr, Pa.

Alicia Diehl, Texas Commission on Environmental Quality, Austin, Texas

Michelle De Haan, Park City MCWD, Park City, Utah

Tarrah Henrie, Corona Environmental, Fremont, Calif.

Randy Moore, Tnemic Company, Saint Louis, Mo.

Lauren Wasserstrom, AWWA, Denver, Colo.

The authors would also like to acknowledge the following individuals who provided editorial and technical comments:

Mike Duer, Tideflex, Carnegie, Penn.

Laith Furation, University of British Columbia, Vancouver, B.C., Canada

Mark Graves, HDR, Austin, Texas

Paul Handke, TideFlex, Venetia, Penn.

Katherine D. Martel, The Cadmus Group, Inc., Steep Falls, Me.

Dennis O'Connor, Philadelphia Water Department, Philadelphia, Pa.

The following individuals provided peer review of this manual. Their knowledge and efforts are greatly appreciated:

Tim Brown, Albermarle County Service Authority, Charlottesville, Vir.

Gary Burlingame, City of Philadelphia, Philadelphia, Pa.

Anne Camper, Montana State University, Bozeman, Mont.

Alan Degnan, Wisconsin State Lab of Hygiene, Sun Prairie, Wisc.

Paul Handke, TideFlex, Venetia, Penn.

Tarrah Henrie, Corona Environmental, Fremont, Calif.

Simon Horsley, Stantec, York, Ont., Canada

Kathy Martel, Cadmus Group, Steep Falls, Me.

Mark Pinkney, Ontario WaterMarks, Midland, Ont., Canada

Ravi Ravisangar, Brown and Caldwell, Atlanta, Ga.

Camille George Rubeiz, Plastics Pipe Institute, Irving, Texas

Robert Ryder, Kennedy/Jenks, San Francisco, Calif.

Charlotte Smith, Charlotte Smith and Associates, Berkley, Calif.

Tom M. Walski, Bentley Systems, Nanticoke, Penn.

Linda Wojcicka, Associated Engineering, Burnaby, B.C., Canada

Don Wood, KYPipe, Carrollton, Texas

This new manual was reviewed and approved by the Distribution System Water Quality Committee. Special thanks to the following committee members during the time of development and for review comments:

Rich Giani, Chair, CH2M, Kansas City, Mo.
Paul R. Easley, Central Arkansas Water, Little Rock, Ark.
Peter Fiske, Lawrence Berkeley National Laboratory, Berkeley, Calif.
Stewart Husband, Sheffield Water Centre, Western Bank, Sheffield, U.K.
Quirien Muylwyk, Toronto, Ont., Canada
Meg Roberts, former Chair, Hazen and Sawyer, Greensboro, N.C.
Susan Teefy, East Bay Municipal Utility District, Oakland, Calif.
Thomas M. Walski, Bentley Systems, Nanticoke, Pa.

This manual was also reviewed and approved by the Water Quality Technology Division under the Technical and Educational Council and included the following personnel at the time of approval:

Brent Alspach, Arcadis, Oceanside, Calif.
Glen Boyd, The Cadmus Group, Seattle, Wash.
George Di Giovanni, Metropolitan Water District of Southern California, La Verne, Calif.
Kimberly Kunihiro, Orange County Utilities-Water Division, Orlando, Fla.
Kerry Meyer, CH2M, Englewood, Colo.
Meg Roberts, Hazen and Sawyer, Greensboro, N.C.
Richard Sakaji, El Cerrito, Calif.
Jeff Swertfeger, Greater Cincinnati Water Works, Cincinnati, Ohio
David Wahman, US Environmental Protection Agency, Office of Research and Development, Cincinnati, Ohio

This page intentionally blank.

Chapter 1

Introduction

Kira S. Smith, City of Houston Drinking Water Operations, Houston, Texas

Rebecca Slabaugh, Arcadis, Indianapolis, Ind.

A drinking water distribution system is a system of pipes that carry potable water from treatment plants or water sources to consumers. It is also the last barrier available to water systems to maintain safe and high-quality water. This manual presents typical distribution system water quality challenges, providing summaries of typical responses and best practices as a "first stop" for drinking water system professionals.

Since each distribution system is unique, this manual is not intended to be all inclusive. Rather, it is a guide that summarizes the issues and actions to be taken when distribution system issues arise and provides references to other industry standards and publications that provide more detail. Additionally, this manual does not delve into treatment process or source water changes that can affect the quality of water that enters the distribution system. Readers are encouraged to familiarize themselves with the American Water Works Association (AWWA) standards and manuals of practice related to water source and treatment. These manuals are available for purchase online at http://www.awwa.org/publications/manuals-of-practice.aspx (accessed May 16, 2016).

For purposes of this manual, distribution systems include pump stations, ground and elevated storage tanks, potable water mains, potable water service lines, and all associated valves, fittings, and meters. Potable water customer service lines are excluded (Texas Commission on Environmental Quality 2015).

The chapters in this manual are organized based on the most common distribution system challenges that water systems face today. These are listed in Table 1-1.

Each chapter provides an introduction and description of a Distribution System Water Quality (DSWQ) Challenge, followed by:

- Discussion and description of the factors associated with each challenge. See text box(es) in each chapter for a summary of Characterizing the DSWQ Challenge;

Table 1-1 M68 chapters and their focus

Chapter	Title
2	Capacity and Water Age
3	Understanding and Managing Biofilm, Coliform Occurrence, and the Microbial Community
4	Infrastructure Integrity and Water Quality
5	Taste, Odor, and Appearance
6	Nitrification
7	Corrosion Control
8	Disinfectants and Disinfection By-products
9	Management of Low Pressure
10	Cross-Connection Control and Backflow Prevention
11	Security Issues

- Summaries of available response actions. See text box(es) in each chapter for a summary of Responding to the DSWQ Challenge; and
- Recommended best practices for optimizing DSWQ. See the table in each chapter that lists DSWQ Best Practices, which are rated as "basic" or "advanced." Some chapters include specific definitions for these terms. Generally, basic practices are the actions all water systems should perform if they have experienced the DSWQ challenge noted. Advanced practices are for water systems that wish to optimize and/or develop a comprehensive program related to the specific DSWQ challenge. Some chapters include an intermediate, or "moderate," rating that is clearly defined within that chapter.

GETTING STARTED—HOW TO IDENTIFY A PROBLEM OR CHALLENGE

Understand Baseline Water Quality Conditions

Many water systems have developed acceptable ranges for various water quality parameters in their distribution systems, for example, pH, disinfectant residual, iron, and manganese. These ranges are based on regulatory requirements and specific utility goals. Some have also developed target levels that trigger various responses, such as notifying the water quality manager and taking action (e.g., flushing) to restore water quality to acceptable levels.

AWWA Standard G200-09, *Distribution Systems Operations and Maintenance* (AWWA 2010), provides guidance to help water systems establish the documented programs and practices needed to understand water quality problems in their distribution systems. In addition to meeting regulatory requirements and preparing written sampling plans, the utility should annually review sampling plans and make adjustments based on historical data trends (AWWA 2010). All AWWA standards are available at http://www.awwa.org/publications/standards.aspx (accessed May 16, 2016).

Customer complaint data can also help a utility identify the source and extent of a water quality problem if the complaint information is managed properly and is accessible. Standard G200 recommends that a utility have a system for documenting customer inquiries that includes the date, inquiry type, customer location, sampling results, and resolution (AWWA 2010).

Water systems that wish to take a holistic approach to the improvement of overall distribution system water quality should consider joining AWWA's Partnership for Safe Water. This is a voluntary optimization program that provides various tools and guidance for tracking disinfectant residual, pressure, and main breaks, as well as a self-assessment guide for systems that wish to optimize distribution system operations.

Some water systems have also implemented proactive quality assurance practices by using hazard analysis and critical control points (HACCP) programs in order to ensure higher levels of tap water safety. HACCP is a systematic preventive approach, designed in the 1960s to protect food and beverages from microbiological hazards. For drinking water systems, the HACCP system promotes a "source to tap" philosophy in order to identify the critical points throughout the entire system (i.e., those points within the system or its operation whose disruption or failure would result in a greater public health risk compared to other points) and then to focus resources on these critical locations and processes. This systematic approach helps water systems identify potential hazards that are closer to their source, thus minimizing the occurrence and effects of incidents that degrade water quality and potentially threaten public health (USEPA 2006).

Consider Regulatory and Policy Frameworks

Legislative frameworks should support effective management of drinking water quality. These frameworks vary throughout the world based on constitutional settings. Legislation should define the roles and responsibilities of the primary agencies and entities associated with ensuring the delivery of safe drinking water.

Drinking water quality regulation requires public health and treatment technology expertise, which typically resides within governmental health and/or environmental protection agencies. Local government environmental health officers may also assist in the administration of legislation. Irrespective of the identities of regulators, they need to be given sufficient power to monitor and enforce legislative requirements. In the United States, for instance, the Safe Drinking Water Act (SDWA) provides authority to the US Environmental Protection Agency (USEPA) to set national drinking water regulations, with enforcement potentially including national, regional, and local authorities. Summaries of US and international regulations are provided at the end of this chapter. It is important to note that in the United States, each primacy agency (USEPA region, state, or territory) may have rules and regulations that local public water systems must follow.

In addition to legislation, provision of safe drinking water may be supported by standards and policies that apply to matters such as certification of water operators, water treatment chemicals, and distribution system materials and plumbing. These can take the form of mandatory requirements or recommended practices (Cunliffe 2014).

A summary of current international and US regulatory and standard frameworks associated with distribution systems is included at the end of this chapter. Most have common themes, which include most or all of the following five categories of distribution water quality indicators:

- Microbial activity,
- Disinfectant residuals,
- Disinfection by-products (DBPs),
- Corrosion, and
- Aesthetics.

Microbial activity and disinfectant residuals. Many potential factors associated with microbial activity and loss of disinfectant residual are similar and, therefore, characterized together. Microbial growth in the distribution system can cause changes to drinking water quality that can adversely affect public health. Direct microbiological indicators are generally based on the presence or absence of bacteria, the most common include total coliforms, fecal coliforms, *Escherichia coli*, and heterotrophic plate count bacteria. The presence of these bacteria may indicate the potential existence of a pathogen or issues associated with the treatment process, distribution system integrity, and/or distribution system operations and maintenance. Chapter 3, Managing Biofilm, Coliform Occurrence, and the Microbial Community, provides a summary of the regulatory frameworks associated with these indicators, as well as information on each indicator.

Disinfectants in drinking water are used to control microbial growth and the potential hazards associated with that growth. Chlorine-based disinfectant residuals have both minimum and maximum regulatory limits. Chapter 8, Disinfectants and Disinfection By-products, discusses these types of disinfectants and summarizes the regulatory frameworks associated with them.

Chapters 3 and 8 provide information on the indicators and possible hazards associated with these indicators. These chapters should be used as starting points for water systems in the detection of microbial indicators and/or loss of disinfectant residuals. However, the contributing factors or root causes for the presence of these indicators can be found in other chapters of this manual. Table 1-2 summarizes potential contributing factors for the presence of microbiological indicators and loss of disinfectant residual in the distribution system, as well as other chapters that provide more information on these factors.

Disinfection by-products. DBPs are chemicals that form when disinfectants/oxidants react with organic and inorganic substances in the water. Several DBPs are currently regulated based on potential health effects, and water systems should focus on controlling the levels of these regulated compounds. Chapter 8, Disinfectants and Disinfection By-products, provides information on these currently regulated DBPs, as well as emerging compounds, in terms of the regulatory framework and health effects. Chapter 8 also provides information on the DBPs that are associated with different types of disinfectants used by water systems. Table 1-3 lists the chapters that include information on changing DBP levels in distribution systems.

Lead, copper, and other corrosion indicators. USEPA promulgated the Lead and Copper Rule (LCR) in 1991 to reduce consumer exposures to lead and copper in drinking water (56 FR 26460; *Federal Register* 1991). The purpose of the LCR framework is to protect public health by minimizing lead and copper levels in drinking water, primarily by reducing water corrosivity (USEPA 2004). Health Canada's Water Quality and Health Bureau also has developed guidelines for lead and copper. Table 1-4 lists the chapters that include information on internal corrosion challenges.

Aesthetics. In the United States, the USEPA sets secondary standards related to constituents that cause unpleasant aesthetic conditions in distribution systems. Customer complaints are an initial indicator for aesthetic issues. These are typically related to color, taste, odor, particles, pressure, and sometimes perceived or actual sickness. AWWA Standard G200 recommends use of action plans to address color and staining, as well as taste and odor inquiries (AWWA 2010). Many customer complaints can be attributed to the factors listed in Table 1-5.

Table 1-2 Chapters on microbial activity and disinfectant residual challenges

Potential Contributing Factor	Relevant Chapter(s)
Microbial growth, pathogen presence	Chapter 3, Understanding and Managing Biofilm, Coliform Occurrence, and the Microbial Community
Disinfectant residual	Chapter 8, Disinfectants and Disinfection By-products
Aged water	Chapter 2, Capacity and Water Age
Main breaks	Chapter 4, Infrastructure Integrity and Water Quality
	Chapter 9, Management of Low Pressure
Nitrification	Chapter 6, Nitrification
Unprotected cross-connections	Chapter 10, Cross-Connection Control and Backflow Prevention
Contamination	Chapter 11, Security Issues

Table 1-3 Chapters on disinfection by-products challenges

Potential Contributing Factor	Relevant Chapter(s)
Changes in disinfectant and/or source water organic content	Chapter 8, Disinfectants and Disinfection By-products
Aged water	Chapter 2, Capacity and Water Age

Table 1-4 Chapters on internal corrosion challenges

Potential Contributing Factor	Relevant Chapter(s)
Inadequate corrosion control	Chapter 7, Corrosion Control
Contamination event evidenced by pH/alkalinity variations	Chapter 10, Cross-Connection Control and Backflow Prevention
	Chapter 11, Security Issues
Change in disinfectant or treatment	Chapter 8, Disinfectant and Disinfection By-products
Microbial-induced corrosion	Chapter 3, Understanding and Managing Biofilm, Coliform Occurrence, and the Microbial Community
	Chapter 6, Nitrification

Table 1-5 Customer complaint issues

Potential Contributing Factor	Relevant Chapter(s)
Corrosion	Chapter 7, Corrosion Control
Secondary and other constituents that cause aesthetic issues	Chapter 5, Taste, Odor, and Appearance
Main breaks	Chapter 4, Infrastructure Integrity and Water Quality
	Chapter 9, Management of Low Pressure
Microbial growth	Chapter 3, Understanding and Managing Biofilm, Coliform Occurrence, and the Microbial Community
	Chapter 6, Nitrification
Changes in disinfectants	Chapter 8, Disinfectants and Disinfection By-products

SUMMARY OF STANDARDS AND REGULATIONS

US Regulations

In the United States, Congress established the SDWA to protect the quality of drinking water throughout the nation. This law focuses on all waters actually or potentially designed for drinking use, whether from above ground or underground sources. SDWA authorizes USEPA to establish minimum standards to protect tap water and requires all owners or operators of public water systems to comply with these primary (health-related) standards (USEPA 2015). The 1996 amendments to the SDWA require that USEPA consider a detailed risk and cost assessment and best available peer-reviewed science when these standards are developed (USEPA 2015). USEPA also encourages attainment of secondary standards (nuisance-related; USEPA 2015).

The USEPA regulations that are based on the SDWA and its amendments are listed in Title 40 of the Code of Federal Regulations (CFR), which can be found at www.ecfr.gov (accessed April 26, 2016).

The SDWA, like most environmental regulations, is typically implemented through primacy agencies, where individual states or territories establish and enforce requirements that are at least as stringent as the federal requirements. Prior to the SDWA, states adopted US Public Health Service standards as regulations or guidelines (USEPA 1999). However, a 1970 survey of community water supplies revealed that 79 percent of systems were not being surveyed by state or local health departments (US Department of Health, Education, and Welfare 1970). The SDWA establishes and authorizes USEPA to allow states to assume primary oversight and enforcement, or primacy, for public water systems (Tiemann 2014).

According to Title 40 CFR regulations, primacy agencies have special requirements beyond implementation of the SDWA, which include ensuring that water systems are appropriately designed, constructed, and have appropriately certified operators based on water system size and water type.

Several SDWA requirements apply to the quality of water that enters distribution systems. There are also rules that apply to distribution systems, including the Total Coliform Rule, Surface Water Treatment Rules, Disinfectants/Disinfection Byproduct Rules, and Lead and Copper Rule. USEPA develops these regulations by setting acceptable public health standards based on the best available peer-reviewed science and what is feasible for water systems. To ensure these levels are met, USEPA regulations also set minimum monitoring requirements. Table 1-6 summarizes the current USEPA regulations that are associated with distribution systems, the public health requirement associated with distribution system levels, and required monitoring. Some chapters in this manual contain more detailed information on regulations as related to specific DSWQ challenges.

Additionally, it is important to note that the USEPA released its third Six-Year Review (Six-Year Review 3) in December 2016. The SDWA requires the USEPA to review each national primary drinking water regulation at least once every six years and to revise them, if appropriate. As part of the Six-Year Review 3, USEPA identified the following eight contaminants as candidates for regulatory revision: chlorite, cryptosporidium, haloacetic acids, heterotrophic bacteria, giardia lamblia, legionella, total trihalomethanes, and viruses. Many of these contaminants are related to distribution system challenges and are reflected in chapters of this manual. The reader is encouraged to reference the USEPA website on the Six-Year Review 3 for the most current information (USEPA 2016).

Table 1-6 Summary of USEPA distribution system regulatory requirements and monitoring

Regulation	Distribution System Requirement	Monitoring
TCR and RTCR (USEPA 2014)	No more than 5 percent of samples positive for total coliform and no fecal contamination or *Escherichia coli*; RTCR requires assessments for exceedances	Collect monthly samples at sites representative of the distribution system and population served
Surface Water Treatment Rules (USEPA 2004a)	At least 95 percent of distribution samples in water originating from a surface water source must have detectable residual*	Check residual disinfectant concentration at entry points to the distribution system and at TCR/RTCR routine sites
Disinfectant/Disinfection Byproduct Rules (USEPA 2006a)	Chlorine maximum residual disinfectant level (MRDL) = 4.0 mg/L; TTHMs and HAA5 locational running annual average maximum contaminant levels = 80 and 60 μg/L, respectively	Check residual disinfectant concentration at TCR/RTCR routine sites; measure TTHM and HAA5 quarterly in distribution system
Lead and Copper Rule (USEPA 1999a)	Lead levels must be <15 μg/L and copper levels must be <1.3 mg/L in 90 percent of samples†	First-draw tap samples for lead and copper every 6 months, annually, or every 3 years; alkalinity, pH, and corrosion inhibitor (if appropriate) monitoring every six months or annually

Abbreviations: HAA5, group of five haloacetic acids; TCR, Total Coliform Rule; TTHM, total trihalomethanes; RTCR, Revised Total Coliform Rule.

* Several states have rules regarding disinfectant residual requirements in distribution systems. Check with local primacy agencies for specific requirements.

† At the time this chapter was written, USEPA's National Drinking Water Advisory Council work group had finalized and made recommendations to revise the Lead and Copper Rule (LCR). These recommendations include proactive lead service line replacement, stronger public education, improved corrosion control treatment, tap sampling modifications, expanded water quality parameter monitoring, a household action level for lead, and separate requirements for copper. For more information on these recommendations and any subsequent proposed LCR changes, visit http://www.epa.gov/dwreginfo/lead-and-copper-rule (accessed April 26, 2016).

World Health Organization

The World Health Organization (WHO), an agency of the United Nations responsible for public health, developed a comprehensive set of guidelines related to drinking water quality that are intended to assist policymakers around the world to ensure sound drinking water practice. These nonenforceable guidelines have been established for various microbial, chemical, radiological, and aesthetic-based contaminants (WHO 2011). The WHO has also formalized a HACCP management approach, known as Water Safety Plans (WSPs; WHO 2009).

Australia

The National Health and Medical Research Council in collaboration with the National Resource Management Ministerial Council developed the Australian Drinking Water Guidelines in order to provide individual communities and regulatory agencies with a structure for proper treatment and management of drinking water supplies. The guidelines propose limits for microbial, physical, and chemical parameters in drinking water to ensure supplies are safe for human consumption. In addition, the guidelines include recommendations for selecting and monitoring the effectiveness of various treatment processes. These are nonenforceable standards. Each state or territory may choose to implement any or all of the guidelines. Readers are advised to consult the individual state or territory for additional information on drinking water requirements (NHMRC, NRMMC 2011).

Canada

Health Canada's Water Quality and Health Bureau, on behalf of the Federal–Provincial–Territorial Committee on Drinking Water, is the federal organization responsible for developing the Guidelines to Canadian Drinking Water Quality (Health Canada 2014). Similar to maximum contaminant levels established in the United States, these guidelines are designed to protect public health by establishing maximum acceptable concentrations (MACs) for microbiological, physical, chemical, and radiological parameters in drinking water (Health Canada 2014). These are nonenforceable federal standards. In Canada, the establishment of drinking water legislation is the responsibility of the individual provinces and territories (Health Canada 2014).

Each province or territory may choose to adopt a portion or all of the MACs established under the Guidelines to Canadian Drinking Water Quality or to establish new levels. To date, only 2 of the 12 provinces and territories (Alberta and Nova Scotia) have adopted the Guidelines to Canadian Drinking Water Quality in their entirety as enforceable drinking water regulations (Health Canada 2014). In addition, each province or territory is responsible for developing a set of drinking water requirements related to treatment, monitoring, and reporting (Health Canada 2014). Alberta has also regulated the need for drinking water safety plans for each system, which follows the WHO's WSP management approach (Government of Alberta 1995–2015).

While the individual requirements vary across Canada, the overarching nationwide goal is to use a multibarrier approach to protect public drinking water by reducing or preventing contamination at the source, during treatment, and throughout the distribution system (Health Canada 2014). Table 1-7 summarizes some of the Health Canada guidelines that apply to distribution systems (Health Canada 2014).

Table 1-7 Summary of Health Canada distribution system guidelines and monitoring

Guideline	Maximum Acceptable Concentration (MAC)	Recommended Monitoring
Total coliform/ *Escherichia coli*	None detectable in a 100-mL sample leaving the treatment plant and in nondisinfected water leaving the well	For systems that serve up to and including 5,000 people: monitor water leaving pump house or treatment plant and at representative locations in the distribution system four times per month; no consecutive samples from the same site or not more than 10 percent of samples should show the presence of total coliform bacteria*
Chlorine	None required	Free and/or total chlorine residuals should be tested when bacteriological samples are taken*
Monochloramine	3.0 mg/L	Free and/or total chlorine residuals should be tested when bacteriological samples are taken*
Lead	0.010 mg/L	Option 1: two-tier protocol; tier 1, annual first draw, if 10 percent >0.015 mg/L, then tier 2: four consecutive 1-L sampling in highest 10 percent locations Option 2: lead service line residences; four consecutive 1-L sampling; site average <MAC in >90 percent of locations†
Trihalomethanes	0.10 mg/L for the sum of chloroform, chlorodibromomethane, bromodichloromethane, and bromoform	For systems that use chemical disinfection, monitor every three months to compute a locational running annual average of quarterly samples*
Haloacetic acids	0.08 mg/L as low as reasonably achievable; for the sum of monochloroacetic acid, dichloroacetic acid, trichloroacetic acid, monobromoacetic acid, and dibromoacetic acid	For systems that use chemical disinfection, monitor every three months to compute a locational running annual average of quarterly samples*

* Health Canada Guidance for Providing Safe Drinking Water in Areas of Federal Jurisdiction, Version 2. Accessed May 16, 2016. http://healthycanadians.gc.ca/publications/healthy-living-vie-saine/water-federal-eau/index-eng.php.

† Health Canada Guidance on Controlling Corrosion in Drinking Water Distribution Systems. Accessed May 16, 2016. http://healthycanadians.gc.ca/publications/healthy-living-vie-saine/water-corrosion-eau/index- eng.php.

Emirate of Abu Dhabi

In 2009, the Regulation and Supervision Bureau for the Water, Wastewater, and Electricity Sector in the Emirate of Abu Dhabi developed the Water Quality Regulations to ensure that safe drinking water is provided to all customers throughout the Emirate of Abu Dhabi. The regulations establish maximum limits for select microbial, inorganic, and organic contaminants in drinking water based on WHO guidelines. The regulations include requirements for monitoring and recordkeeping for water suppliers. While the regulations do not directly discuss treatment or disinfection, they do state that "the water does not contain any element, organism or substance at a concentration or value which would be detrimental to public health" and contain limits for various DBPs (Regulation and Supervision Bureau 2009).

European Union

In 1998, the European Union (EU) issued Directive 98/83/EC, which established minimum standards for water intended for human consumption for EU Member States. Although disinfection is not explicitly required, the directive includes limits for microbiological parameters and DBPs that are similar to the WHO guidelines. Member States are also encouraged to minimize DBP formation, but not at the expense of disinfection. Member States may adopt stricter standards as well as limits for additional parameters. In general, most Member States require disinfection of surface waters using chlorine, chlorine dioxide, ozone, or ultraviolet light but do not require disinfection of groundwater supplies. Only some Member States require that a minimum disinfectant residual be maintained at all times in the distribution system. Other Member States, such as the Netherlands, apply strategies that include the use of only high-quality water supplies and biologically stable distribution system materials in addition to distribution system operational strategies such as flushing to minimize the potential for bacterial regrowth (EU Council 1998).

In October 2015, the European Commission adopted a directive to revise Directive 98/83/EC to include an option to perform monitoring in a more flexible way, provided a risk assessment is performed to ensure full protection of public health. It follows the principle HACCP and the WHO's WSP approach (European Commission 2015).

New Zealand

The New Zealand Water Quality Guidelines were developed based on WHO guidelines and include recommended maximum values for microbial, radiological, and chemical parameters, including DBPs. The guidelines are enforceable and apply to all drinking water providers regardless of source, treatment type, or distribution system size. The guidelines require that some form of disinfection be provided for all surface or spring water sources. The guidelines also include requirements for monitoring disinfection efficiency at treatment plants and in the distribution system in addition to reporting and corrective action requirements (Ministry of Health 2008).

Republic of South Africa

The South African Water Quality Guidelines were established in 1996 and provide target water quality ranges for various microbial, physical, chemical, and aesthetic parameters. The guidelines recommend disinfection via chlorination or chloramination and that an adequate residual be maintained to minimize any bacterial regrowth or biological corrosion (Department of Water Affairs and Forestry 1996).

Drinking water must also comply with the requirements set forth in the South African National Standard (SANS) 241 Drinking Water Specification. The specification establishes maximum allowable concentrations for select parameters monitored at the "pointe of delivery." Values have been established based on WHO guidelines (Department of Water Affairs and Forestry 1996).

Singapore

Recently, the Singapore Government promulgated the Environmental Public Health Act, which established standards for the quality of drinking water in piped systems. The act also includes information on development of sampling and safety plans. Although the act does not plainly require disinfection, the Public Utilities Board, which is the national water agency in Singapore, provides disinfection of its water supplies prior to distribution (Government of Singapore 2008).

REFERENCES

AWWA (American Water Works Association). Standard G200, *Distribution Systems Operations and Maintenance*. Denver, CO: AWWA.

Cunliffe, D. et al. 2014. *Water Safety in Distribution Systems*. Geneva: World Health Organization.

Department of Water Affairs and Forestry. 1996. South African Water Quality Guidelines. Pretoria, Republic of South Africa: CSIR Environmental Services.

European Commission. 2015. *Commission Directive (EU) 2015/1787 of 6 October 2015 amending Annexes II and III to Council Directive 98/83/EC on the quality of water intended for human consumption*. Accessed May 16, 2016. http://ec.europa.eu/environment/water/water-drink/review_en.html

European Union Council. 1998. *Council Directive 98/83/EC of 3 November 1998 on the quality of water intended for human consumption*. Accessed May 16, 2016. http://ec.europa.eu/environment/water/water-drink/legislation_en.html

Federal Register. 1991. *National Primary Drinking Water Regulations; Lead and Copper; Final Rule*. 56(110):26460. https://nepis.epa.gov/Exe/tiff2png.cgi/P100DP2P.PNG?-r+75+-g+7+D%3A%5CZYFILES%5CINDEX%20DATA%5C06THRU10%5CTIFF%5C00001389%5CP100DP2P.TIF

Government of Alberta. 1995–2015. Drinking Water Safety Plan. Accessed May 16, 2016. http://environment.alberta.ca/apps/regulateddwq/DWSP.aspx

Government of Singapore, Environmental Public Health. 2008. Quality of Piped Drinking Water Regulations. Singapore.

Health Canada. 2014. Guidelines for Canadian Drinking Water Quality–Summary Table. Accessed May 16, 2016. http://www.hc-sc.gc.ca/ewh-semt/pubs/water-eau/sum_guide-res_recom/index-eng.php

Ministry of Health. 2008. Drinking Water Standards for New Zealand 2005, revised 2008. Wellington, New Zealand.

NHMRC, NRMMC. 2011. Australian Drinking Water Guidelines. Paper 6, National Water Quality Management Strategy. Commonwealth of Australia, Canberra: National Health and Medical Research Council and National Resource Management Ministerial Council.

TCEQ (Texas Commission on Environmental Quality). July 30, 2015. Title 30 Texas Administrative Code, Chapter 290–Public Drinking Water. Austin, TX: TCEQ.

Regulation and Supervision Bureau for the Water, Wastewater, and Electricity Sector in the Emirate of Abu Dhabi. 2009. The Water Quality Regulations. Abu Dhabi, U.A.E.

Tiemann, M. 2014. *Safe Drinking Water Act (SDWA): A Summary of the Act and Its Major Requirements*. Congressional Research Service. 7-5700. RL31243. Accessed May 16, 2016. https://www.fas.org/sgp/crs/misc/RL31243.pdf

US Department of Health, Education, and Welfare. 1970. *Community Water Supply Study—Significance of National Findings.* Washington, DC: DHEW. Accessed May 16, 2016. nepis.epa.gov/Exe/ZyPURL.cgi?Dockey=9100W0XR.TXT

USEPA (US Environmental Protection Agency). 1999. *25 Years of the Safe Drinking Water Act—History and Trends*. EPA 816-R-99-007. Accessed May 16, 2016.nepis.epa.gov/Exe/ZyPURL.cgi?Dockey=200027R1.TXT

USEPA. 1999a. Lead and Copper Rule (LCR). Accessed January 29, 2016. 1/29/16. http://water.epa.gov/lawsregs/rulesregs/sdwa/lcr/index.cfm

USEPA. 2004. Lead and Copper Rule: Quick Reference Guide. EPA 816-F-04-009. Washington, D.C.: USEPA, Office of Water.

USEPA. 2004a. Surface Water Treatment Rules: Comprehensive Surface Water Treatment Rules Quick Reference Guides. Accessed May 22, 2017. https://www.epa.gov/dwreginfo/surface-water-treatment-rules

USEPA. 2006. Hazard Analysis Critical Control Point (HACCP) Strategies for Distribution System Monitoring, Hazard Assessment and Control. Washington, DC: USEPA.

USEPA. 2006a. Stage 1 and Stage 2 Disinfectants and Disinfection Byproducts Rules (Stage 2 DBPR). Accessed January 29, 2016. http://water.epa.gov/lawsregs/rulesregs/sdwa/stage2/index.cfm

USEPA. 2014. Revised Total Coliform Rule and Total Coliform Rule. Accessed January 29, 2016. http://water.epa.gov/lawsregs/rulesregs/sdwa/tcr/index.cfm

USEPA. 2015. Summary of the Safe Drinking Water Act. Accessed January 29, 2016. http://www.epa.gov/laws-regulations/summary-safe-drinking-water-act

USEPA. 2016. Six-Year Review 3 of Drinking Water Standards. Accessed March 10, 2017. https://www.epa.gov/dwsixyearreview/six-year-review-3-drinking-water-standards

WHO (World Health Organization). 2009. Water Safety Plan Manual: Step-by-Step Risk Management for Drinking Water Suppliers. Geneva: WHO.

WHO. 2011. Guidelines for Drinking Water Quality, fourth edition. Accessed May 16, 2016. http://www.who.int/water_sanitation_health/publications/2011/dwq_guidelines/en/

Chapter 2

Capacity and Water Age

Meg Roberts, Hazen and Sawyer, Greensboro, N.C.

Ellen Hall, Hazen and Sawyer, Fairfax, Vir.

Historically, water distribution systems have been sized using a capacity-based approach, with future demand scenarios and economies of scale often driving sizing decisions. The main focus was on providing hydraulic reliability and maintaining positive pressures throughout the system (Kelley 1994). While these considerations are still critical to the function of a water distribution system, there is a general understanding now that the negative impacts to water quality when systems are oversized should also be considered in sizing system components. More recent trends include selecting the right-sized facility in the right location (Mays 2000).

Distribution system water quality may be characterized in a variety of ways, many of which are discussed throughout this manual. One general indicator of water quality in distribution systems is water age. Water age is defined as the average time it takes for water to travel from the treatment facility to any given point within the distribution system. In general, water quality tends to deteriorate with increasing water age. For example, disinfectant residuals decrease, microbial activity increases, and disinfection by-products (DBPs) increase as water increases.

With this increased awareness of the relationship between water age and water quality, water quality issues are no longer thought of as something to be addressed only at the treatment facility, nor can they be separated from the infrastructure and hydraulic issues that distribution systems face today.

DSWQ Challenge:

As water age increases, disinfectant residuals decrease, microbial activity increases, and disinfection by-products increase.

DETERMINING CAPACITY

The first step in evaluating a distribution system's water quality is to understand the existing system capacity. This includes setting appropriate design criteria and water age criteria for the system and then comparing existing conditions to the selected criteria. Design criteria vary from system to system depending on a variety of factors. This section reviews the basic steps and considerations in setting design criteria for the capacity of different distribution system components. Additionally, water age goals vary from system to system as a function of water quality goals, source water quality, treatment types, and other factors. This section also reviews water age goals considering these factors.

Design criteria for many systems have changed over the decades; the design criteria used for existing infrastructure is likely to differ from that to be used for new infrastructure. Changes in design criteria or changes in system demand over time result in facilities that are either under- or overdesigned with respect to current needs. This adds to the challenges many utilities face with respect to infrastructure investment.

In 2010, investment needs for buried drinking water infrastructure totaled more than $1 trillion just within the United States over the succeeding 25 years (AWWA 2010). The World Bank estimated global infrastructure investment needs at $35 trillion in the succeeding 20 years (HSNW 2010).

Today, prioritizing infrastructure renewal based on water age and water quality goals is more imperative than in previous decades in order to not only maintain water quality but make the best possible use of limited available funds for infrastructure renewal. Before making significant new investments, water quality in the distribution system must be considered. Use of objective methods to determine the capacity of existing infrastructure and water age in existing distribution systems is therefore an important first step toward evaluating design criteria. The following section reviews common methods to measure and/or estimate the capacity and water age of distribution systems.

Determining Needed Capacity

The major components of a water distribution system include pumps, pipes, valves, and storage tanks (or reservoirs). Methods of determining and evaluating existing and required capacity for each type of component are discussed below.

Pumps. Pump capacity criteria are based primarily on water demand. Pumps are sized to provide the maximum day demand in cases where storage is available to meet peak hour, fire, and emergency demands. If storage is not available, the latter three considerations will impact pump sizing.

Designers must consider the full range of possible demands, including minimum as well as average and maximum demands, and select a pump or pumps that will operate within the manufacturer's recommended operating range under the range of anticipated demand conditions. Pump sizing typically starts with maximum day demand conditions (design flow) and pressure design criteria (design head). A pump curve describes the relationship between flow and head for a given pump, and pumps operate at different efficiencies along this curve. A representation of a typical centrifugal pump curve is shown in Figure 2-1.

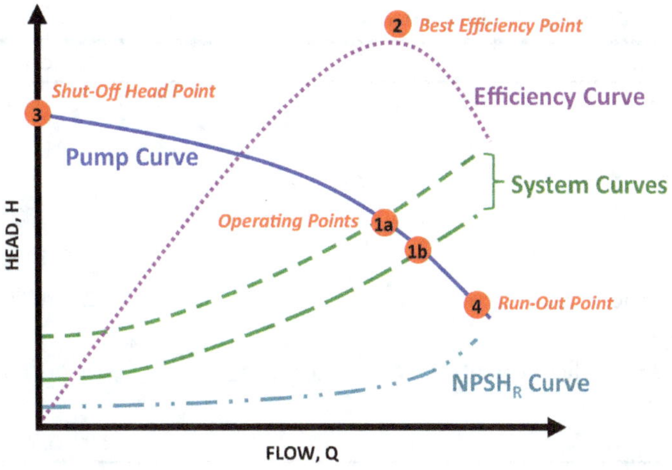

Courtesy of Ellen Hall, Hazen and Sawyer.

Figure 2-1 Pump and system head curves

Abbreviation: NPHS$_R$, net positive suction head required.

The operating point for a given centrifugal pump in a given system is determined as the point at which the pump curve intersects the system head curve. System head curves are graphical representations of the first law of thermodynamics (Bernoulli equation for water) and are calculated as a function of the elevation (static head) and head loss in the system. System curves can be calculated for different design conditions. For example, the two system curves and resulting operating points shown in Figure 2-1 (points 1a and 1b) might represent operating conditions at high and low tank levels. System curves can also be developed for different demand conditions, different Hazen-Williams coefficients (C-factors), multiple pumps running in parallel, and similar conditions. To properly size a pump, multiple system curves should be developed to represent the full range of operating conditions that the pump or pumps are expected to experience in the system. The most efficient pump selection is one with a curve that crosses the average day system curve (i.e., the curve on which the pump will operate for the majority of the time) at or near its best efficiency point (point 2) and otherwise falls within the envelope of all the system curves.

Continuous operation near the shutoff head of the pump (point 3) or the run-out point (point 4) should be avoided due to low efficiencies, potential for vibration, and, in the case of point 4, potential for insufficient suction head. The required net positive suction head (NPSH$_R$ in Figure 2-1) may exceed the net positive suction head available, causing cavitation. Failure to consider the full range of possible system curves can result in inadequate supply to the system and/or poorly operating pumps. For more information on the pump selection process, refer to Mays (2000), Jones et al. (2008), and the Hydraulic Institute Standards (latest edition).

The desired head is related to system pressure criteria (Table 2-1). Most utilities seek to maintain system pressures between 40 and 150 psi at all times of day. Some utilities aim for an even higher minimum of 50 psi to accommodate items such as taller structures, residential sprinkler systems, or head loss through building backflow preventers (AWWA Manual M35); others may select lower values in order to reduce leakage through pressure control. The higher end of this range (150 psi) is a typical value based on common pressure ratings for pipe (e.g., the thickness classes given in ANSI A21.50 for ductile iron pipe). However, utilities in mountainous terrain may require pipes that have significantly higher pressure ratings. While pump capacity can be largely impacted by system pressures in certain situations, such as a closed pressure zone without elevated storage, pipe size is often the more critical factor for maintaining adequate pressures, as discussed below.

Table 2-1 Typical pipe capacity design criteria

Criteria	Typical Range
Minimum pressure under average day, maximum day, and peak hour conditions	30–50 psi*
Minimum pressure under fire flow conditions	20–30 psi*
Maximum pressure	80–150 psi†
Maximum velocity	5–10 fps*
Maximum head loss under peak hour conditions in pipes smaller than 12 in.	2–3 ft/1,000 ft
Maximum head loss under peak hour conditions in pipes ≥12 in.	1–2 ft/1,000 ft
Required fire flow	500–3,500 gpm‡

* See AWWA Manual M32 (2005).

† Many plumbing codes require pressure-reducing valves (PRVs) on services if pressure is greater than a set value, such as 80 psi. However, distribution system piping may be designed for greater pressures if needed due to topography or other factors.

‡ Gallons per minute varies depending on the type of structure; Insurance Services Office, Inc. ratings; local fire officials' requirements; and similar guidelines. The low end of the range shown represents the typical capacity of a single hydrant; the high end of the range is discussed in AWWA Manual M31 (1998).

Utilities must also size pump stations for the proper firm capacity. Firm capacity is defined as the capacity of the pump station when the largest pump is out of service. For systems with storage, pump stations are typically designed with a firm capacity equal to the maximum day demand. Utilities often design pump stations to ensure that existing and future demand conditions can be met with the installed pumps. However, it is important from both hydraulic and water quality standpoints to not oversize individual pumps for future demand conditions. Multiple smaller pumps are often a better choice, so that they function more efficiently under existing demands. Multiple pumps of different sizes or variable-speed pumps can also be considered in order to accommodate a range of demands. Providing space for future pumps allows flexibility while avoiding excessive installed capacity under existing conditions.

Pump capacity should always be considered in conjunction with pipe and storage capacity. A capacity restriction in any given area could potentially be resolved using a pump upgrade, a pipe improvement, a storage improvement, or a combination of these.

Pipes. The primary design criteria for pipe capacity include velocity and pressure. The goal is to select pipe sizes that are sufficient to prevent excessive head losses (and thus minimize pumping energy required) and, for systems designed to provide fire protection, adequate pressure at needed fire flows. Fire flows impact both pipe and storage sizing and are often the dominant factor in pipe sizing.

Table 2-1 lists some typical ranges of values for pipe design criteria in North America. Design criteria vary widely around the world. While this table presents typical ranges for comparison purposes, individual systems may have circumstances that call for criteria outside these ranges. For example, a need to deliver flow over varying elevations may result in a transmission main being designed for pressures that are much higher than typical pressures in some areas, or the need to provide fire protection to a specific type of industrial customer may result in a higher required fire flow based on the International Organization for Standardization guideline (AWWA 1998). Surge pressures can be substantially higher than normal operating pressures and must also be considered in determining design pressures for pipes. With respect to pipe size and fire flows, 6-in. pipe is the minimum diameter recommended in North America on which to install a hydrant (AWWA 1998).

Determining system pressures in pipes under existing or future conditions can be accomplished in different ways. To investigate an existing pressure issue, pressure gauges are installed in accessible locations such as hydrants and pumping stations. Several options exist for online distribution pressure monitors that can detect low pressures during various demand conditions and over long periods of time. Chapter 9 discusses online pressure monitoring for pressure management in distribution systems in more detail. However, pressure measurements cannot predict pressures in a network under future conditions when demands may be higher, lower, and/or distributed differently than under existing conditions; when operations might be significantly different; or when future facilities may be in place. Hydraulic models can be used for this purpose. They can also help identify higher elevation areas where low-pressure issues are most likely to occur. These higher elevation areas are often the limiting factor in determining the desired hydraulic grade for a given zone within the distribution system.

In order to design new pipes for an existing distribution system, the designer must know the head losses in the existing system. This is done with loss-of-head tests. Head loss measurements are equivalent to the drop in the hydraulic grade line between the inlet and outlet of a test section. Pressure measurements are taken at hydrants, air valves, or other taps. Measurement techniques depend on field conditions and the amount of head loss measured (AWWA 2005). Similar to other pressure measurements, the results are valid only for the flow conditions created for the test. However, results can be used to calculate the roughness coefficient for the particular segment of pipe tested. This information can be used for hydraulic model calibration, and models can be used to examine head loss under future demands and/or operations. The hydraulic model can also be used to identify any current and new pipes above the maximum velocity. Refer to AWWA Manual M32 (2005) for further information on conducting loss-of-head and other types of hydraulic tests, and see Use of GIS and Hydraulic Models, later in this chapter, for further discussion on the use of hydraulic models.

Valves. Isolation valves are nominally sized based on pipe diameter and are located based on the need to shut down distribution system segments. Selecting the location, size, and settings of control valves in a distribution system involves detailed analyses.

Control valves include PRVs, pressure-sustaining valves, throttling valves, and check valves. Control valves are often used at pressure zone boundaries. Size, type, and settings are very system specific and depend on the valve's intended purpose. PRVs, for example, are often used in low-elevation areas. Use of PRVs to avoid overpressurization can help reduce water loss; however, this can also lead to higher water age in the PRV zone due to the isolation from surrounding demands. PRVs can also be used for fire protection between zone boundaries where the boundary needs to remain closed unless a certain low pressure is reached during a fire flow condition. The pressure setting allows the valve to fully open, providing flow across the zone boundary as needed during a fire.

Valve sizing is outlined in AWWA Manual M44 (2015b). However, valve settings and controls should be set with both pressure criteria and water quality goals in mind. For example, allowing more flow through PRVs by having a small bypass line (approximately 1-in. diameter) open at zone boundaries can relieve both sides of potential dead ends and improve water age (Cruikshank 2010).

Storage. The design of new storage or the evaluation of existing storage is dependent on two basic factors: volume and hydraulic grade. Storage volume is typically determined using either standards-driven or functional methods (Mays 2000). Standards-driven methods are often general requirements for total storage in which a total storage volume that equals some percentage of average or maximum day demand is required. One-half the average day demand or an equivalent definition based on gallons per connection is an approach used in several US states. This approach tends to produce more conservative results than a functional approach (AWWA 1998; Ten States Standards 2012).

The functional method considers the following three categories of storage:

- *Equalizing storage* provides water when the demand exceeds the rate of supply. Providing flow equalization from storage allows water production facilities to be sized for maximum day (rather than peak hour) demand and allows them to operate at a relatively constant rate, which is advantageous for process control. Equalizing storage is most commonly expressed as a percentage of the maximum day demand. Typical values range from 10 percent to 25 percent of the maximum day demand. Larger systems or pressure zones tend to be at the lower end of this range, while smaller systems tend to be at the higher end due to the demand averaging effect that occurs in larger systems.
- *Fire storage* is the volume of water required for fire fighting, usually expressed as a fire flow multiplied by a fire duration (e.g., 3,500 gpm for 3 hours [AWWA 1998]). The utility determines the fire flow and duration based on building codes, input from local fire officials, the desire to maintain the community's insurance rating from the Insurance Services Office, or other criteria. The flow and duration can vary significantly depending on land use (residential, commercial, institutional, or industrial), building construction materials, and spacing of structures. AWWA Manual M31 (1998) outlines many of the recommended practices.
- *Emergency storage* is the storage volume required to maintain service during an emergency, such as loss of a supply or a main break. Emergency storage requirements vary from utility to utility and are typically based on an assessment of the level of reliability and redundancy present in the system, the typical frequency and duration of emergencies, and other factors. For example, if a system has more than one source and has standby power, there is less need for emergency storage, whereas with a single source and no standby power, more emergency storage may be warranted. Values are typically expressed as a percentage of demand or time at the average day demand or maximum day demand (e.g., 25 percent of the average day demand or 4 hours of supply at the maximum day demand).

Along with the functional categories, storage may also be categorized as effective or ineffective as follows:

- *Effective storage,* also known as usable storage, is the volume of stored water at or above the level at which the minimum pressure is attained (in tanks that float on the system) or above which the water can be pumped (in pumped storage).
- *Ineffective storage,* also referred to as nonusable storage or dead storage, is the volume of stored water below the level at which the minimum pressure is attained (in tanks that float on the system) or below which the water cannot be pumped (e.g., due to suction head limitations in pumped storage).

Figure 2-2 is a conceptual schematic of these storage volumes and elevations, representing the boundaries of effective storage for each category. Methods for determining the volumes and elevations (or hydraulic grades) for each category are discussed below.

Determining the equalizing storage volume. To determine the needed equalizing storage volume, diurnal demand patterns can be calculated using measurements of flow and storage tank volume changes and/or historical production, pumping, and tank level records (Bates et al. 1996). An example calculation is shown in Table 2-2 for a system with one water treatment plant (assuming no purchased water supplies) and two storage tanks using hourly data.

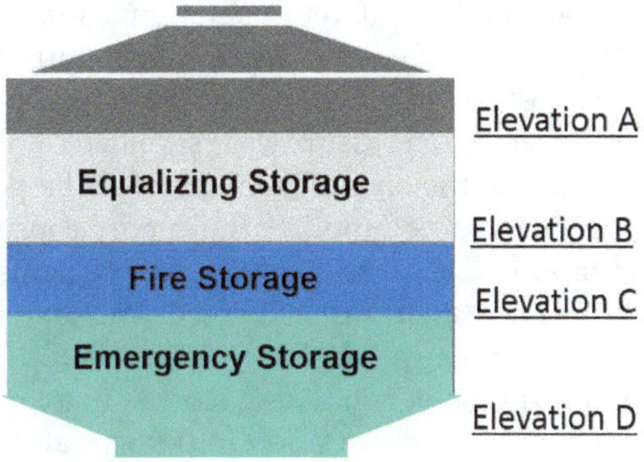

Courtesy of Meg Roberts, Hazen and Sawyer.

Figure 2-2 Distribution system storage volume design concept

Table 2-2 Example diurnal pattern and equalizing storage volume calculation

Time of Day	WTP Supply (mgd)	Tank 1 Level (feet)	Tank 2 Level (feet)	Tank 1 Flow* (mgd)	Tank 2 Flow* (mgd)	System Demand (mgd)	Fraction of Average	Volume Over Average (million gal)
0 - 1	1.99	21.1	26.2	-0.2	0.5	2.23	1.04	
1 - 2	1.00	22.1	24.3	-0.2	-0.5	0.34	0.16	
2 - 3	1.01	22.8	26.2	-0.4	-0.3	0.28	0.13	
3 - 4	2.35	24.3	27.6	-0.2	0.3	2.51	1.17	0.02
4 - 5	1.99	25.0	26.2	0.3	0.5	2.74	1.28	0.03
5 - 6	2.00	23.9	24.4	0.5	0.1	2.61	1.22	0.02
6 - 7	2.37	21.9	23.9	0.5	0.3	3.12	1.46	0.04
7 - 8	2.36	20.1	22.7	0.1	0.0	2.48	1.16	0.01
8 - 9	2.35	19.6	22.7	-0.1	-0.2	2.06	0.96	0.00
9 - 10	2.49	20.1	23.4	-0.2	-0.5	1.80	0.84	0.00
10 - 11	2.49	20.9	25.4	0.0	-0.6	1.87	0.87	0.00
11 - 12	2.47	21.0	27.7	-0.2	0.3	2.59	1.21	0.02
12 - 13	2.49	21.7	26.6	0.0	0.3	2.77	1.30	0.03
13 - 14	2.35	21.9	25.3	-0.1	0.1	2.38	1.11	0.01
14 - 15	2.01	22.1	25.0	0.0	0.2	2.27	1.06	
15 - 16	2.10	21.9	24.1	0.1	-0.6	1.63	0.76	
16 - 17	2.19	21.5	26.4	0.2	-0.4	2.00	0.94	0.00
17 - 18	2.36	20.9	27.8	0.0	0.7	3.09	1.45	0.04
18 - 19	2.35	20.8	25.0	0.2	0.0	2.52	1.18	0.02
19 - 20	2.36	20.1	25.0	0.2	-0.2	2.42	1.13	0.01
20 - 21	2.05	19.1	25.7	0.0	-0.3	1.81	0.85	
21 - 22	2.02	19.1	26.7	-0.3	-0.1	1.64	0.77	
22 - 23	2.01	20.1	27.2	-0.4	0.3	1.96	0.92	
23 - 24	2.02	21.7	25.8	0.1	0.1	2.18	1.02	
0 - 1		21.4	25.5					
					Average (mgd):	2.14	Sum (mgd):	0.2

Equalizing Storage =

$$\frac{0.2}{2.14} = 11\%$$

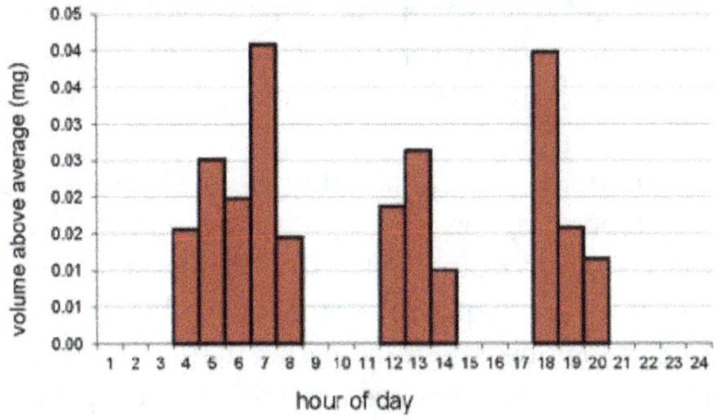

*assuming cross sectional area of each tank is 1400 square feet

For sizing new tanks, different operational strategies might be considered in order to determine the amount of equalizing storage required. For example, a utility may want to reduce energy costs by shifting booster pumping operations to times of the day when power costs are lower. For a pressure zone supplied by a booster pump within this system, a calculation similar to the one in Table 2-2 could be performed to determine the equalizing storage needed to accommodate the energy-conserving pumping strategy. Enough volume in the tank would be required to supply demand when the booster pumps are off. The water age impacts of this storage volume should be considered in evaluating the energy conservation strategy.

The hydraulic grade of equalizing storage in a tank depends on whether or not the tank is equipped with a pump station to pump water out of the tank into the distribution system. Most elevated storage tanks are not equipped with this; therefore, the entire equalizing storage volume is only effective if it is located above a hydraulic grade that provides adequate pressure to all customers served by the tank at all times of day (elevations between A and B in Figure 2-2). Whether or not pressure is adequate depends on the distribution system's pressure criteria. As discussed above, many utilities have a lower daily pressure requirement of between 30 and 50 psi. However, this design criterion can vary significantly, even within a single distribution system, depending on specific customer needs.

Determining fire storage volume. Fire storage volume requirements are often calculated by multiplying the maximum fire flow demand by the required duration. However, some utilities do not include a fire storage volume in their storage volume criteria. Instead, they supply the required fire flow and duration using other system capacity, such as fire pumps, additional treatment plant capacity, and/or interconnections with other systems. Hydraulic models can be used to test the effectiveness of storage or other methods in providing the required fire flow and duration at various locations in the system.

For most distribution systems, the minimum required pressure for fighting fires is 20 psi. At this pressure, fire pumper trucks can adequately supply water and contamination from vacuum pressures in the piping network is prevented (AWWA 1998). If the bottom of the fire storage volume (elevation C in Figure 2-2) is not at a hydraulic grade to provide 20 psi, the volume is ineffective. Fire pumps would be required to meet fire storage design criteria. Other systems may require a higher distribution system pressure for fighting fires. Some facilities are equipped with their own fire pumps that require a higher suction side pressure. If these facilities are critical (e.g., hospitals), the utility may have higher pressure criteria at certain locations.

Determining emergency storage volume. Different utilities or regulatory agencies have differing requirements for emergency storage volume. Some do not require any emergency storage volume, as they may have adequate volume at adequate pressure from emergency interconnections with other systems. Others require enough to supply as much as one to two days of average demand. Some utilities require an emergency volume in addition to equalizing and fire storage. Others assume that a fire and emergency condition would not occur simultaneously and so either include only a fire volume or include the larger of the fire or emergency volume.

The required hydraulic grade for this volume (elevation D in Figure 2-2), such as equalizing and fire volumes, varies depending on the topography of the area served, the pressure design criteria, and the ability to pump out of the storage tank. For example, many utilities have clear wells at water treatment plants in order to attain sufficient disinfectant contact time. Utilities may choose to consider their clear well volume as water available during an emergency if generators are available to pump the volume into the distribution system during a power outage. Acceptable system pressures during an emergency generally vary from 0 to 20 psi. However, some utilities with sensitive or critical facilities require higher pressures under emergency conditions to ensure these facilities remain functional.

Summary. The goal in sizing storage tanks is to store the minimum volume that will meet a utility's capacity needs and to make that volume effective. For example, standpipes with a great deal of ineffective storage add to water age and typically provide only limited capacity for equalizing or fire storage. Excess storage in the distribution system can contribute significantly to higher water age (Snyder et al. 2002).

DETERMINING WATER AGE

The spatial and temporal variation of water quality in a distribution system reflects very complex phenomena, including hydraulic mixing patterns and biological and chemical water quality transformations that can occur within a piping network. Age is a water quality parameter that can serve as a general indicator of water quality conditions that result from this complex set of interactions.

There are several ways to determine water age:

- Hydraulic modeling,
- Tracer studies, and
- Estimate from measured hydraulic and/or water quality data.

Before concerns about water age are addressed, it must first be understood that where treatment processes and source quality are excellent, it is still possible that very old water will have good quality. Conversely, less than optimal treatment and/or a challenging water source with respect to quality can result in faster water quality degradation at lower water age. The distribution system is the third step in water quality optimization; source water and treatment are first and second, respectively.

Hydraulic Modeling

Changes in the age of water throughout a distribution system can be simulated with hydraulic models or indirectly measured using tracer studies (AWWA 2005). New water that enters the network from treatment plants or supplies has an age of zero. Age is treated as a reactive constituent whose growth follows zero-order kinetics with a rate constant equal to 1 (meaning each second the water becomes 1 second older).

Mathematical determination of water age with a hydraulic model or tracer studies is based on conservation of constituent mass. Advective transport of mass within pipes is assumed (i.e., the constituent mass moves with the bulk liquid), and longitudinal dispersion is not typically considered an important transport mechanism under most operating conditions. This means that there is an assumption of no intermixing of mass between adjacent parcels of water traveling through a pipe.

Distribution system water quality models represent water transport within pipes, mixing at junctions and storage tanks, and appropriate reaction kinetics (zero-order for water age) with a set of mathematical equations. These equations are solved according to an appropriate set of boundary and initial conditions in order to predict the variation of water age through the distribution system (AWWA 2005).

The advantage of using hydraulic models to estimate water age is that the age calculation is purely hydraulic. If a model is calibrated and extended-period simulations are used, water age can be readily predicted at any location in the distribution system under both existing and alternative operating and/or demand conditions. Extended-period simulations are multiple days in length; water age simulations are run for long periods of time in order to reach a stable, long-term water age in the tanks. Note that the exact length varies for each system and must be determined iteratively using plots similar to those shown in Figure 2-3. Water age is highly dependent on water demand in the network as well as piping configuration and storage tanks. Changes to those inputs can aid in the prediction of improvements to or deterioration of water quality.

The mixing of mass within storage tanks can be simulated by most models using the following four types of mixing regimes: complete mixing, plug flow (first in/first out, or FIFO), short-circuiting (last in/first out, or LIFO), or compartment (two zone). Figure 2-3 shows an example of how the four types of assumed mixing affect water age within storage tanks. In reality, specific storage tank configurations often result in mixing characteristics that are not fully captured by any one of these model representations. Computational fluid dynamics models can be used to represent mixing characteristics of specific storage tanks more effectively. However, for purposes of using hydraulic models to estimate water age, selection of one or more of the available mixing options serves as a basis for comparing possible improvements. Different mixing options are tested in the model in order to determine the sensitivity of the water age results in the choice of mixing regime. Paired temperature or chlorine residual measurements taken at different depths within an existing tank provide an indication of the type of mixing regime present. Calculating the Reynolds number for the flow entering and/or exiting the tank provides an indication of the flow regime (laminar, transitional, or turbulent) at the tank entrance, which indicates how well mixed a tank might be based on its inlet and/or outlet configuration.

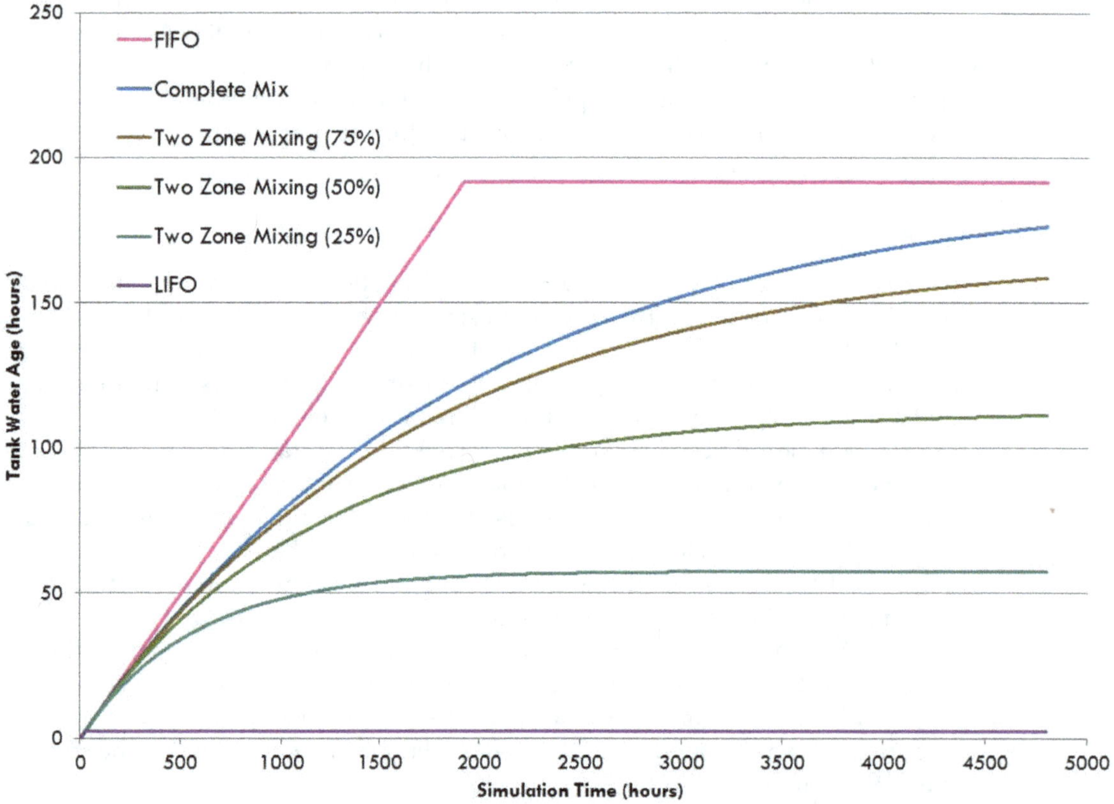

Courtesy of Meg Roberts, Hazen and Sawyer.

Figure 2-3 Example of how storage tank mixing characteristics affect tank water age effluent

Using hydraulic models to determine water age, however, does have drawbacks. Models are fairly input intensive and, in addition to software costs, time and information are required to build, calibrate, and effectively use them. Many utilities, especially smaller systems, do not have sufficient staff to create and maintain a model in-house. While calibration can improve the assumptions used, models do require a certain number of assumptions and have limitations in accuracy (hydraulic grade lines within 5 ft and flows within 10 percent are good goals for calibration [AWWA 2005]). If a model is not available, the residence time contributions from the tanks in a system can be calculated and trended over time using supervisory control and data acquisition data (e.g., tank level, FIFO, tank dimensions), with FIFO inferred from change in the tank level over time.

The important thing to consider with these mixing regimes is that the LIFO and zone mixing regimes shown in Figure 2-3 have water age that is lower than the other regimes at the outlet of the tank, but this is due to short-circuiting. This means that the water age and water quality at higher levels in the tank can be significantly worse and may be detrimental to water quality in the distribution system at times when that water is released (higher demands or operations that draw the tank levels down more than usual).

To address this issue and achieve more uniform water quality in tanks, many utilities use active or passive mixers in an effort to achieve a mixing regime that is closer to the complete mix regime. To be effectively sized and placed, such mixers need to be evaluated based on the geometry and inlet/outlet configuration of the particular tank before installation.

Tracer Studies

Tracer studies are another method for estimating water age. A tracer is a conservative compound, that is, it does not grow or decay. Conducting a tracer study to determine water age involves injecting a tracer at the entry point to the distribution system, shutting off or reducing a continuous chemical feed at the entry points, or using a substance that occurs naturally in the source water. The concentration of the tracer is measured over time at various places in the distribution system, and the tracer data is used to calibrate the water quality model. A tracer study requires careful planning and implementation but may still be a more economical alternative for some utilities compared to developing a hydraulic model (Clark 2012). However, the results of a tracer study reflect only the specific locations sampled and the operating conditions at the time of the study. They cannot be used to predict water age under future conditions or for what-if scenarios.

Estimating from Hydraulic and Water Quality Data

Real-time monitoring of constituents such as disinfectant residual or a naturally occurring tracer provides results similar to those gained from a tracer study but over a longer period of time. Average age in a particular storage tank can also be estimated directly from operating data (water level and tank geometry), and paired measurements of disinfectant residual at different depths within a tank can provide an indicator of stratification within the tank. Improvements in automated monitoring equipment continue to be made, and software tools to integrate real-time data with hydraulic models are becoming available and will continue to evolve. As these tools are developed, they will become more reliable and accessible to more utilities and will help to provide a better understanding of existing system conditions over time.

Summary

It is important to note, however, that each distribution system is a unique biological and chemical reactor with different source waters, treatment processes, and facilities. As such, the absolute value of water age has limitations as a water quality indicator. Water quality data should be assessed in conjunction with water age to determine age values that are appropriate for each system. However, what-if scenarios that involve predicted water age are useful for the purpose of evaluating potential changes in system configuration and/or operations.

Characterizing the DSWQ Challenge:

Determine Needed Capacity

- Pumps: demand and pressure conditions
- Pipes: pressure, velocity, and head loss
- Valves: sizing and hydraulic functionality
- Storage: storage volume criteria for equalizing, fire, and emergency storage

Determine Water Age

- Hydraulic modeling
- Tracer studies
- Estimate from measured hydraulic and/or water quality data

WAYS TO BALANCE CAPACITY AND WATER AGE

Every water distribution system has many of the same basic components: pipes, pumps, valves, storage facilities, and so forth. However, the particular combination of source water quality, geography, and infrastructure in any given system makes it unique, and finding solutions to any water quality issue requires careful examination of all the factors involved. Water age can be a surrogate for a number of different parameters and can be determined using a calibrated hydraulic model or tracer study or be estimated from measured hydraulic and/or water quality data. Consequently, water age is a good place to start when analyzing a water quality issue. This section describes ways in which capacity criteria and operational practices can be modified to better balance capacity and water age.

Water Age Considerations for Capacity

The design criteria selected for use in sizing distribution system components can have a significant impact on water age. Some criteria are related to the essential function of the distribution system—to deliver water at sufficient flow rates and pressures; any significant deviation from these criteria would have the potential to compromise the system's basic functionality. However, there are areas of criteria selection and application that can be more individualized to a particular system, giving more flexibility to consider water quality impacts when selecting the criteria. Storage volume criteria, in particular, fall into this latter category and are discussed in detail later in this section.

Pipes. A consequence of sizing pipes to provide fire flows is higher water age under normal demand conditions (Snyder et al. 2002), especially for dead-end mains with limited demand. While fire flow criteria may be predetermined and some dead ends are inevitable due to topography, land use limitations, and cost, utilities can evaluate solutions such as looping pipes and providing small-diameter bypasses on zone boundary valves in order to eliminate dead ends with water age issues. For cases in which a branched rather than looped design is necessary or preferred, utilities may consider flushing to improve water age (see Flushing Programs later in this chapter). When analyzing alternatives, it is beneficial to consider the trade-offs between increased fire flow and reliability and the potential for water quality degradation in a looped system as compared with a branched system (see Chapter 5 for more discussion of branched and grid/loop designs).

When planning for future demands, installation of very large transmission mains can add a significant amount of water age and degrade water quality, especially in the early years of development before an area is built out. Phased parallel piping may be preferable. Although higher in cost, the near-term improvement in water quality is often favorable, and, in the long term, parallel pipes provide added redundancy. Alternatively, phasing capital improvement projects by demand triggers instead of planning years can help reduce overcapacity issues. Demand forecasts, while helpful for capital improvement planning, are often based on population projections that vary over time and are based on assumptions that may change. For major hydraulic upgrades that are required to meet future demand scenarios, tracking water demand in specific areas of the system is useful as a trigger to schedule construction on large transmission mains. In this way, the full capacity that is needed in the future does not cause water age problems in the present. For example, a utility may track demand by district or pressure zone and initiate a new pipe project when demand in that district or zone reaches 80 percent of the demand at which the pipe is needed. Water age modeling may be used to assess the impacts of pipe improvements under different demand conditions, and the results may be used to aid in the phasing of capital improvements.

Pumps. Sizing multiple pumps so that there is continuous pumping from a pump station helps to minimize water age by preventing stagnant water in transmission mains. The energy costs of different sizing options and operational strategies should be considered when evaluating pumping alternatives. Energy costs can vary significantly depending on the rate structure in effect for a particular facility. Utilities need to balance water age improvement with energy costs or find opportunities to improve both. Similar to pipeline capital improvement projects, pumps are upsized based on demand triggers in lieu of planning years.

Storage. There is an increasing awareness that the standards-driven storage volume criteria of the past do not adequately address the potential water quality impacts of this quantity of storage. Excessive storage should be avoided (Kennedy et al. 1993). An updated approach to storage volume criteria is to focus more on effective rather than total storage and consider functional criteria such as equalizing, provision of fire flows, and emergency supply and system reliability. Hydraulic modeling can be used to check the true functionality of the criteria selected, consider the interconnectedness of systems, and assess system reliability under different scenarios such as fires, major pipe breaks, and use of alternate sources.

Of the three functional categories, equalizing storage is the most basic hydraulic requirement. If the supply is designed for the maximum day demand, equalizing storage must be provided to handle peak demands in excess of the daily average. Fire and emergency storage can be more flexible in terms of the criteria that a utility selects and the ways in which the utility applies the criteria. For example, the fire volume of a given tank could potentially be shared with a lower pressure zone if sufficient pipe capacity exists and the second zone is at the same or lower hydraulic grade or if reliable pumping is provided (AWWA 1998). This type of option can be tested using a hydraulic model or field measurements to determine whether required fire flows can be met in both zones. Water age rises sharply in cases where fire and emergency storage exceed 80 percent of the total storage volume (Snyder et al. 2002), so fire and emergency storage criteria are key places to look in evaluating (or reevaluating) storage capacity.

Ineffective storage has limited usefulness and is one of the largest contributors to increased water age. While ineffective storage can serve a structural function of supporting the water above it, it has the negative impacts of increased water age and stratification within a tank. One way to help meet capacity needs while minimizing water age is to minimize ineffective storage in the system. One method of doing this is to choose elevated tanks rather than standpipes for floating storage. Standpipes have a greater amount of ineffective storage than other tank types. Also, stratification is more likely to occur in them, and it is harder to break up the stratification with the inflow jet entering the tank (Grayman et al. 2000). A ground storage tank with an operating range comparable to that of an elevated tank may also be a good option for floating storage if a site with a sufficiently high ground elevation is available. For pumped storage, the ground storage tank should be designed such that the pumps can draw a large fraction of the tank's volume.

Designing storage tanks for conditions that rarely, if ever, occur is a challenge for many utilities. This is especially true for systems with wide seasonal fluctuations in demand; sizing storage for the maximum day demand results in long residence times in the off-season. It may be possible to take some storage off line during low-demand periods. Utilities also look for opportunities to develop emergency interconnections or other redundancy plans that would provide emergency supply in lieu of excessive emergency storage.

Water age criteria. Water age goals should be determined by examining existing water quality measurements and comparing them to water age. A statistical analysis of chlorine residual at locations with known water age, for example, can help set a water age goal. Quarterly DBP correlations are another option.

Figure 2-4 shows an example of a system that uses free chlorine with a water age goal of five days to minimize total trihalomethanes based on historical data review and predicted water age calculations from a calibrated hydraulic model. In systems where temperature fluctuations are high, seasonal trends should be evaluated with different water age goals.

Water age design criteria are very system specific, and there is no ideal water age for all systems (Friedman et al. 2010). The extent to which water age impacts water quality depends on numerous factors, including source water quality, treatment process efficiency, the water's microbial and chemical stability, disinfectant type and dose, and distribution system configuration, condition, and operation (Friedman et al. 2010). However, water age can be estimated within a given system and the results used to help guide design and operational decisions.

Simulated distribution system testing can estimate a bulk disinfectant decay curve or DBP formation curve to predict and manage disinfectant and DBP levels based on estimated water age in the distribution system. (see Figure 2-5 for an example and refer to *Standards Methods for the Examination of Water and Wastewater*, 22nd ed., for procedure). Measuring decay or formation due to pipe wall interactions is more complex, and a standard

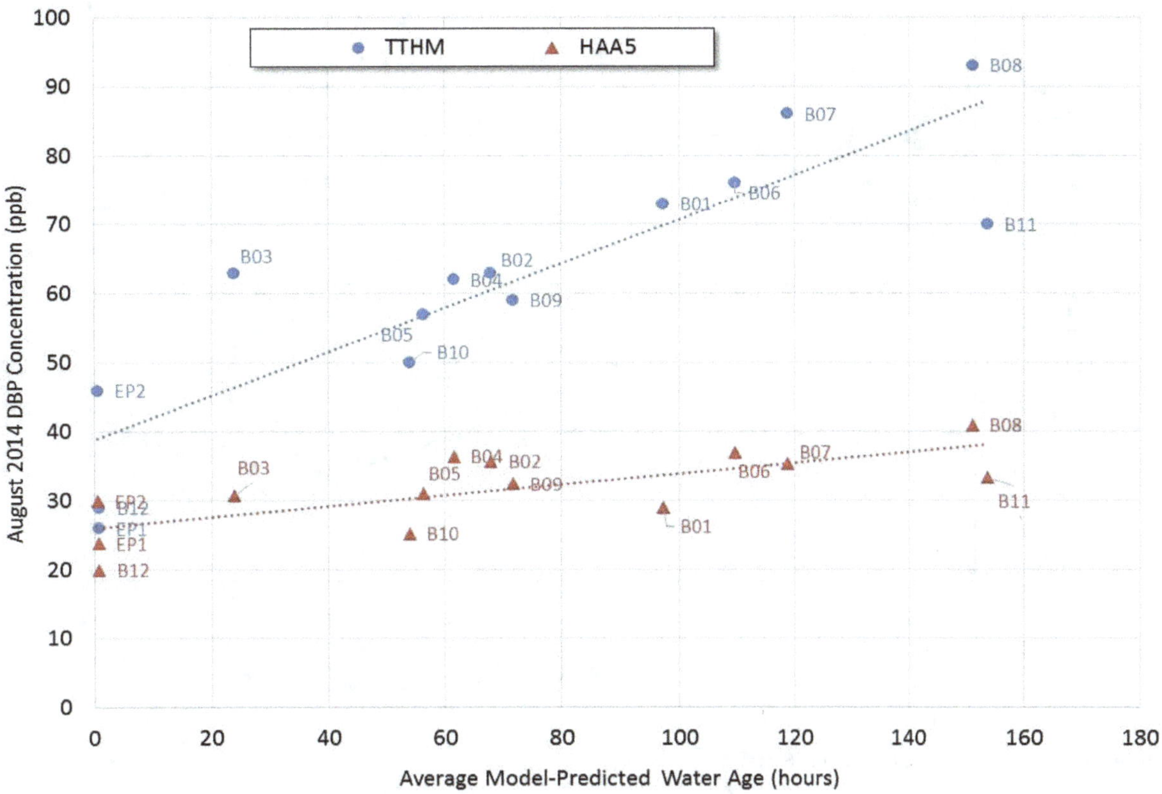

Courtesy of Meg Roberts, Hazen and Sawyer.

Figure 2-4 Increasing disinfection by-products with water age in a free-chlorinated system

procedure has yet to be developed. Even between constituents within a single water, there is no ideal age; however, measuring several parameters will give an idea of the most desirable range (Cruikshank 2010). Alternatives designed to bring the system within the desired range can then be developed. If modeling or other studies indicate that this range is not practically attainable with changes to distribution system capacity or operations alone, changes in treatment may be needed (Kirmeyer et al. 2000). For more information on disinfection and disinfectants, refer to Chapter 8.

Operations

Capacity and water age depend not only on the physical characteristics of a system but also on how the system is operated. Many operational changes reduce water age at relatively little cost, making operations a good place to start in identifying solutions to capacity and water age issues.

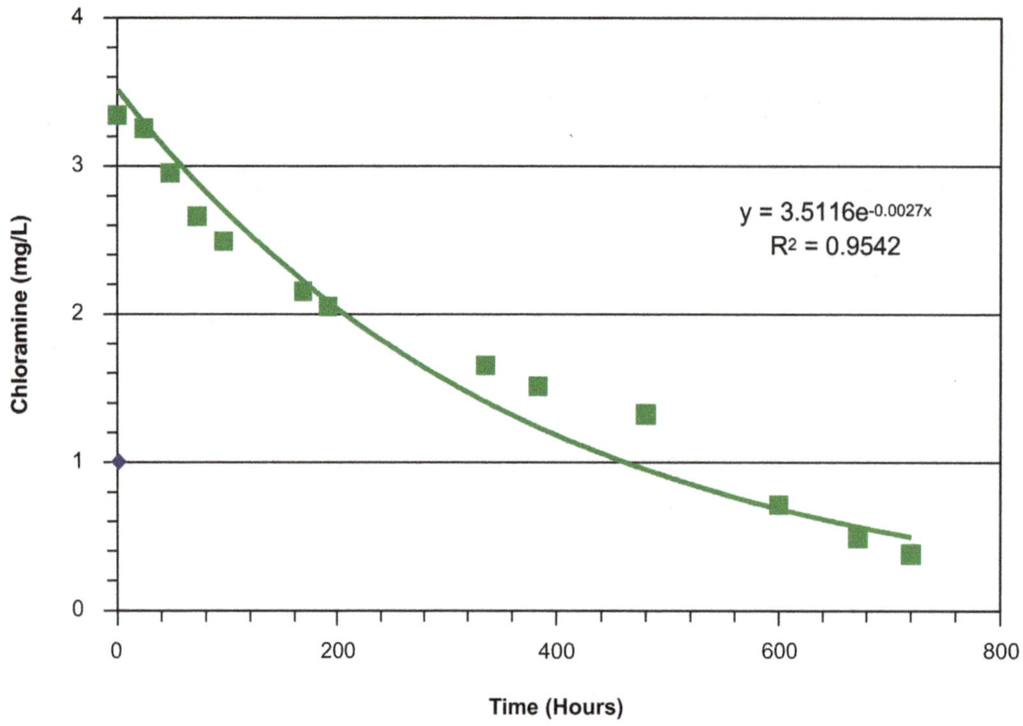

Courtesy of Jeff Cruickshank, Hazen and Sawyer.

Figure 2-5 Example of chloramine bulk decay curve

Pipes. Adjustments to piping configurations to improve water age may be made operationally by changing valve positions or settings. Examples of changes that can be made to improve water age include the following:

- Allow flow through zone boundary valves: Closed valves at pressure zone boundaries often function as dead ends in a distribution system. Allowing these valves to bleed back a small amount of flow from a higher pressure zone to a lower pressure zone by opening a bypass around a gate valve, throttling a valve, or changing the opening setting of a PRV can improve water age in the area of the valve without consuming any water (Cruikshank 2010). Care should be taken when throttling valves to ensure integrity is not compromised.
- Reconfigure tank connections: The configuration of the piping entering and leaving a tank can have a significant impact on water age in the tank (Grayman et al. 2000). A tank configured with a separate inlet and outlet and control valves to prevent reverse flow can help improve the flow pattern, provide some mixing, and reduce stratification in the tank. Keeping bypass piping closed also helps to maintain a higher rate of flow into the tank. This allows a higher velocity from the inflowing jet, which improves mixing.
- Close valves to change flow patterns: One way to improve circulation through an area with high water age is to strategically close valves to increase velocity through the area (Cruikshank 2010) and possibly route the water to areas with higher consumption within the system. Closure options should be tested using a hydraulic model or water quality sampling to ensure that new areas of high water age near the closed valves will not be an issue. An increase in velocity could cause

a release of accumulated sediments. Flushing to remove sediment prior to the change can help provide a controlled release and removal of accumulated sediments (see Appendix B for more information on flushing).

- Encourage water use at system extremities: The utility can consider encouraging large water users at extremities of the piping network to use more water (e.g., irrigation of golf courses or athletic fields, industrial uses) through education or rate incentives. Additionally, spot flushing can reduce water age at selected locations (see Flushing Programs later in the chapter).
- Consider the effect of multiple supply sources: If there is a choice between multiple supply locations, the supply to use can be chosen in order to minimize water age in the distribution network of pipes. For example, two sources on opposite sides of a distribution system tend to produce lower water age results than two sources that are close together, as both sides have initial water ages of zero at each entry point (Cruikshank 2010). This should also be a consideration in evaluating future supply alternatives. The quality of the source water and treated water from each supply should also be taken into consideration, depending on the system's water quality challenges.

Pumps. The way that pumps are operated can have a significant impact on water age. Some examples of how pump operation can be modified to improve water age include the following:

- Change pump settings to improve tank turnover: Adjust pump on and off settings to produce the maximum amount of volumetric turnover allowable without compromising pressure or fire flow design criteria.
- Pump across zone boundaries: Pumping at multiple locations along a zone boundary rather than pumping at one boundary and using only PRVs at other boundaries is another method for eliminating dead ends at zone boundaries.
- Control variable-speed pumps advantageously: Many variable-speed pumps are controlled based on maintaining a pressure setting. This produces consistent system pressures but can result in low turnover in tanks. Controlling variable-speed pumps based on a flow setting can improve tank turnover by allowing the pumps to shut off and the tank to draw (Cruikshank 2010). Variable-speed pumps can also be controlled based on tank level, which can improve water age even (Figure 2-6). In this case, running the pump at a high speed when the tank drops to a certain level and at a low speed as it approaches full can produce optimal results.

Storage. Areas of high water age are often centered around storage tanks. Following are some operational strategies for improving water age in tanks:

- Reduce storage volume: Use less volume within an existing tank by adjusting both pump-on and pump-off settings. Do this to the extent feasible without compromising pressure or fire flow design criteria (Cruikshank 2010).
- Increase turnover: Use a larger fraction of the volume in a tank during regular operations by balancing pump settings and customer draw down (Cruikshank 2010).
- Use mixing strategies to improve flow patterns and produce a consistent water age: While mixing cannot reduce the average water age within a tank, it can help provide homogeneity and prevent stratification and zones of high water age from forming within the tank.

- Take a tank off line seasonally or permanently: If a tank with high water age causes seasonal issues, consider temporarily removing the tank from service during the period when the issues typically occur. If the tank volume exceeds capacity requirements or represents emergency storage that could be provided in other ways, a utility should consider taking the tank off line permanently. The pressure and fire flow impacts should be evaluated before implementing either alternative, and the tank should be drained and cleaned before placing it back into service.
- Direct bulk water haulers to draw water from high-age tanks: A hydrant near a tank can be designated for this purpose to increase tank turnover.
- Use ground storage with pumps and/or control valves to circulate water: Many distribution systems have ground storage tanks, the hydraulic grade of which requires pumps to draw down water and/or control valves for filling. Turning over the tank volume on a regular basis with valve and pump controls can help circulate water in the distribution system and reduce aging inside the ground storage tank.

Summary. There are only two ways to truly achieve a net reduction in water age in a system—increase demand (e.g., by flushing) or reduce the volume of water stored in the system. Other operations such as tank mixing do not cause a net reduction in water age but rather make the age more uniform. Unless demand is increased or storage is reduced, when localized reduction in water age is created in one place, a commensurate increase is occurring somewhere else (though potentially spread over a larger area). The goal is to ensure that changes do not inadvertently create additional high water age areas. Such areas can occur at locations completely removed from the initial area of concern, which is one reason why hydraulic models of the complete system are useful in evaluating water age reduction alternatives.

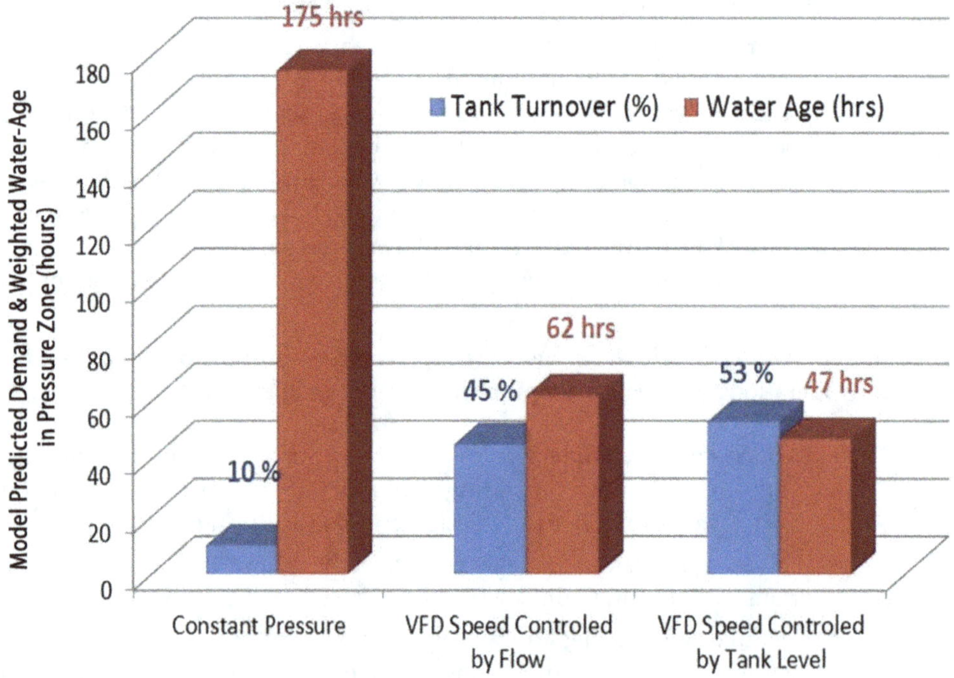

Courtesy of Jeff Cruickshank, Hazen and Sawyer.

Figure 2-6 Example improvements in water age by controlling pump speed by flow or tank level

Maintenance Programs

While every distribution system requires some reactive maintenance, proactive maintenance programs or asset management programs can function to reduce reactive maintenance needs, manage costs, and, in general, produce a better functioning system.

Pipes. Pipe condition and maintenance impact many aspects of distribution system water quality, including chemical and microbial interactions with the pipe as well as capacity and water age issues.

Tuberculation or other deposits on the interior pipe wall can result in greatly reduced capacity. For example, Figure 2-7 shows a sample from a pipe that was determined to have significant head loss, as reflected by a C-factor of 22. The head loss in this pipe could create a capacity problem, particularly for fire flows, and also exacerbate water age–related issues such as increased microbial activity (see Chapter 3 for more information on managing the microbial community). Utilities can use fire flow capacity needs and water age as criteria when prioritizing a capital improvement program. For example, a flow path to move water directly to an area with high water age or low fire flow may be created by rehabilitating or replacing certain sections of pipe first. Refer to Chapter 4 for more information on aging infrastructure and pipe rehabilitation and replacement programs.

Valves. Valves are an often overlooked system component that, if not maintained, can lead to an increased risk of property damage and service disruption during main breaks, as well as hydraulic restrictions that diminish the quality of service. There are two basic components to a valve maintenance program (Bloetscher 2011):

- Locating and accessing the valves, and
- Operating the valves on a regular schedule (annually is a good goal). Any valves that are found to be not operational should be repaired or replaced.

Figure 2-7 Tuberculated pipe with reduced capacity

Additionally, all utilities find valves that are closed but should be open. More closed valves can be identified if all valves are located and operated on a regular basis. Closed valves create hydraulic constrictions that can impede fire flows and reduce hydraulic capacity in an area. Areas isolated by closed valves can also exhibit increased water age (and deteriorated water quality) due to the lack of flow. Depending on their location, closed valves have the potential to impact water age and quality in large areas of the distribution system.

Pumps. Pump capacity and operations have a direct impact on water age and water quality. Pump capacity and operations are directly impacted by pump maintenance. As pumps age, impellers wear, leading to loss of capacity. Pumps also have a number of components that need periodic checks and replacement, such as seals and bearings. A pump that is vibrating excessively due to worn bearings or improper balancing will wear out more quickly than one that is maintained. A pump maintenance program should include scheduled maintenance based on the typical useful life of these components.

Pump capacity can be checked using a pump test. The pump is operated using a minimum of three discharge valve settings while the following parameters are measured:

- Suction pressure (for dry pit pumps) or wet well level (for submersible pumps);
- Discharge pressure;
- Flow, measured using an installed meter if available and calibrated or using a pitometer rod;
- Speed, if variable speed (multiple speed points can be used to develop a system curve); and
- Current (electrical draw), can use to calculate efficiency.

The results are then plotted against the original pump characteristic curve from the manufacturer, and the system curve is developed either from the test data or hydraulic modeling. The results show how much the pump has changed, whether it still meets the design intent, and whether its efficiency has decreased to the point where it needs to be repaired or replaced. Replacing inefficient pumps with more efficient units provides operational cost savings. The results of the pump test can also be used when calibrating the hydraulic model. See AWWA Manual M32 (2005) for more information on pump tests and model calibration.

Even if the pump has not changed significantly, the system may have, and the pump test results will show how the system is actually operating and whether the pump is still an appropriate choice for the system. Well-maintained pumps that operate as intended can reduce water age, allowing utilities to implement operational strategies discussed earlier in the section on Operations.

Storage. Routine tank inspection and maintenance is an important part of distribution system maintenance. The typical inspection includes draining, cleaning, and removing sediment; chlorination; and total coliform testing. Coating repair and mixing system inspection and repair can also be performed as part of each inspection. AWWA Manual M42 (2013) and (Deb et al. 2000) both recommend periodic inspections. The appropriate intervals between inspections and cleanings vary depending on state regulatory agency guidelines and site-specific needs. Routine inspections allow utilities to observe the amount of sediment deposited in a given tank over time and to adjust cleaning frequency accordingly.

Tank maintenance is important in balancing capacity and water age because tank operations have such a large impact on water age. If the tank mixing system is in disrepair, for example, water age (and quality) in parts of the tank can increase. Excessive sediment in a tank not only has the potential to contribute to water quality problems such as nitrification, loss of residual, and biofilm development but may reduce storage capacity as well.

Flushing Programs

Flushing, or opening, a distribution system connection to the atmosphere and allowing water to be discharged from the system is a common method for reducing water age. Flushing has been a long-standing practice for periodic removal of sediment from pipes.

More recently, flushing has been recognized as a way to address areas with high water age and/or poor water quality. If customer demand alone is insufficient to remove enough water from a particular area of the system, flushing can be used to achieve the same effect, essentially creating an artificial demand. While flushing does consume water, the water is not wasted if it is used to improve water quality (Cruikshank 2010). The cost of the water consumed is low in comparison with the cost of more capital-intensive methods of reducing water age. Flushing is a strategy that is often used in a reactive mode (e.g., a low residual is found during routine sampling and that area is then flushed), but it can also be used in a more proactive way to help control water age. For example, conventional spot flushing or dead-end flushing can be done to reduce water age at specific locations, while quasi-unidirectional or strict unidirectional flushing provides bulk water turnover and a temporary reduction in water age over a larger area (see Appendix B for more information on these different types of flushing). Conventional spot flushing or dead-end flushing is manual or automated:

- Manual (intermittent): Operators perform intermittent, manual flushing at locations with known high water age or water quality, such as dead ends, and areas with seasonal low demands. This technique is also used in response to a customer complaint or a low residual measurement. Manual flushing becomes labor intensive if numerous locations need localized flushing. However, in some cases, it may be the only option if site constraints do not permit installation of an automated flushing unit.
- Automated (intermittent or continuous): Automated flushing units are used to periodically or continuously flush a given location. They are available in a variety of shapes, sizes, and configurations. The units consist of electrically actuated valves, a controller, a battery, a piping connection to the water main, and a means of discharging to the atmosphere with appropriate backflow prevention (air gap or double check valve), all housed in a compact enclosure. Dechlorination using a chemical tablet feeder may also be included. Maintenance for these units consists of checking operation, refilling tablets, and replacing batteries. They are available in a variety of models that have timers, the ability to set flow rates, and even incorporate disinfectant residual sensors and instrumentation to trigger flushing based on a residual set point. Space requirements, access, power supply options, adequate drainage, and community impacts should be evaluated in order to determine site feasibility and select the appropriate equipment for each site.

Flushing for water age reduction, whether manual or automated, can be optimized by flow rate (Cruikshank 2010). At any given location in the distribution system, there are diminishing returns on the flow rate of the flusher. Flushing locations can be evaluated using a hydraulic model to determine the minimum flow rate needed to achieve the desired water age improvement. Figure 2-8 shows an example of model-predicted water age with various continuous flushing rates. In selecting a flow rate, the resulting flushing velocity should also be considered with respect to the water quality objective, for example, water age improvement only, sediment flushing (Friedman et al. 2010).

Use of Geographic Information Systems and Hydraulic Models

A geographic information systems (GIS) are very powerful and important tools for utilities. A GIS can serve as a central repository for all types of spatially referenced information about the system, from pipe location, size, material, age, and condition to sampling data, billing addresses, and modeling results. Because distribution systems are geographically widespread and the distance from water treatment plants, storage tanks, pump stations, and similar components can dramatically impact water age and water quality, the ability to visually display distribution system components helps utilities better understand the operations and challenges of their systems.

Today, hydraulic and water quality models are often integrated with a GIS platform, such that GIS data can be easily used as model , and model output can be displayed using the map-making capabilities of GIS. This makes it possible to overlay model results, geography, asset management information, and similar information in order to better understand what is happening in relation to location in the distribution system.

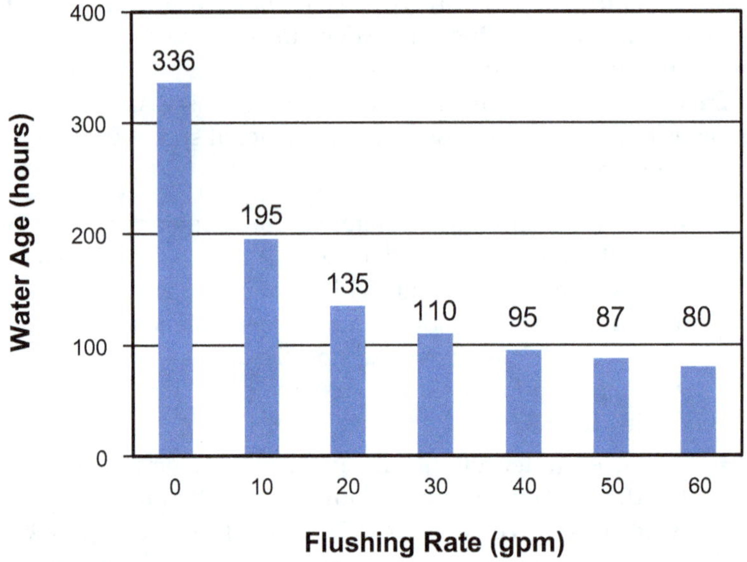

Courtesy of Jeff Cruickshank, Hazen and Sawyer.

Figure 2-8 Example of diminishing returns on water age from flushing rates

GIS and hydraulic models have numerous applications in balancing capacity and water age. They can be used to perform numerous hydraulic and water age analyses including, but not limited to, the following:

- Test alternative scenarios for meeting capacity requirements—consider pump upgrades, pipe improvements, storage improvements, or a combination; evaluate water age for each alternative; and, use the modeled water age as part of the alternative selection criteria.
- Perform a source trace analysis to understand the impact of storage and/or supply on water age and/or quality and to understand areas influenced in the distribution system by each facility.
- Test functionality of storage under different scenarios—find ways to minimize storage volume, such as sharing storage between zones, replacing aging standpipes with elevated storage, changing pump settings to improve turnover, all while meeting equalizing, fire, and emergency needs.
- Identify effective and ineffective storage and develop alternatives to minimize ineffective storage.
- Identify zones of influence in tanks and places in the system where blending issues can occur with multiple water sources using the source trace function in the model.

- Overlay multiple parameters on maps in order to troubleshoot capacity and water quality issues. For example, low fire flow could be caused by an undersized pipe, a low C-factor pipe, a closed valve, or low pressure due to high elevation. Mapping the potential factors is a good way to identify which factors can be ruled out and which merit further investigation.
- Find opportunities for pipe looping in order to reduce dead ends.
- Store pipe asset management information such as pipe size and material, date installed, and date cleaned/lined.
- Record valve maintenance information such as location and date last operated.
- Optimize flushing rates to achieve water quality goals while minimizing water usage.
- Use add-on modules to design a flushing program more efficiently than with manual planning.
- Animate operational sequences over specified periods of time to observe the impact of tank fill–drain cycles and diurnal demand patterns in comparison to water age and/or quality data.

In addition to these examples, most of the strategies mentioned in the previous sections can be tested using a hydraulic model in order to identify the most effective alternatives and optimize parameters prior to implementing changes in the system. The actual concentrations of interest, such as disinfectant residual and DBPs, can also be modeled; however, building and calibrating this type of model takes greater effort. The use of modeling for assessment of water quality is discussed in greater detail by Powell et al. (2004) and Walski et al. (2003).

DSWQ Response Actions:

Reevaluate design criteria to incorporate water age considerations.

Develop and implement operational adjustments that improve water age by moving water and/or changing supply and demand conditions.

Implement maintenance programs for pipes, valves, pumps, and tanks.

Conduct localized and/or system-wide flushing programs.

Use GIS and hydraulic models to investigate issues and test alternative solutions.

BEST PRACTICES FOR OPTIMIZING DISTRIBUTION SYSTEM CAPACITY AND WATER AGE

All distribution systems are unique. This makes development of the formula for providing adequate capacity while reducing water age even more challenging. Table 2-3 includes some recommended metrics that utilities can use as a starting point to evaluate their performance with respect to best practices.

Table 2-3 Water quality evaluation criteria for balancing system capacity and water age

Best Practice	Element Importance
Determining needed capacity	
• Evaluate storage volume criteria	Basic
• Conduct master planning every 5–10 years	Basic
• Minimize pressure	Advanced
Determining water age	
• Estimate potential impacts based on historical water quality measurements	Basic
• Use supervisory control and data acquisition and data analysis	Basic
• Incorporate a hydraulic model with limited or infrequent calibration	Basic
• Perform tracer studies	Advanced
• Calibrate hydraulic model regularly	Advanced
Water age considerations for capacity	
• Reduce dead ends	Basic
• Reduce ineffective storage	Basic
• Minimize pipe diameters	Advanced
• Upsize pumps based on demand triggers	Advanced
• Evaluate ways to reduce fire and/or emergency storage volume requirements	Advanced
• Determine water age criteria	Advanced
Operations	
• Evaluate tank turnover and optimize based on results of evaluation	Basic
• Use less volume within existing tanks	Advanced
• Evaluate stratification and mixing	Advanced
• Optimize pressure zone boundaries	Advanced
• Control variable-speed pumps advantageously	Advanced
• Evaluate use of valve closures to change flow patterns	Advanced
• Evaluate operating methods to increase low flows	Advanced
• Use multiple supply sources	Advanced
Maintenance programs (see also Chapter 4)	
• Replace distribution components (i.e., pipes, pumps, and valves) at failure	Basic
• Assess valves and hydrants regularly	Basic
• Perform regular master meter testing	Basic
• Perform regular pump tests	Basic
• Clean and inspect tanks regularly	Basic
• Prioritize rehabilitation/replacement plan	Advanced
• Ensure an asset management system is in place	Advanced
Flushing programs (see also Appendix B)	
• Perform spot/dead-end flushing	Basic
• Perform automated flushing	Advanced
• Perform unidirectional flushing	Advanced
• Track flushing volume	Advanced
• Develop a written flushing plan	Advanced

(continued)

Table 2-3 Water quality evaluation criteria for balancing system capacity and water age *(continued)*

Best Practice	Element Importance
Use of GIS and hydraulic models	
• Store asset management information in GIS	Basic
• Overlay multiple parameters on GIS maps to troubleshoot issues	Advanced
• Use hydraulic model to test alternative scenarios and/or design a flushing program	Advanced
• Perform a source trace analysis	Advanced
• Calibrate a water quality model	Advanced

Abbreviation: GIS, geographic information system.

REFERENCES

AWWA (American Water Works Association). 1998. M31, *Distribution System Requirements for Fire Protection*. Denver, CO: AWWA.

AWWA. 2005 (2nd ed.). M32, *Computer Modeling of Water Distribution Systems*. 2nd ed. Denver, CO: AWWA.

AWWA. 2013. M42, *Steel Water Storage Tanks*. Denver, CO: AWWA.

The American National Standards Institute (ANSI) and AWWA. 2014. Standard C150/A.21.50-14, *Thickness Design of Ductile Iron Pipe*. Denver, CO: AWWA.

AWWA. 2015a. Standard G200-15, *Distribution Systems Operation and Management*. Denver, CO: AWWA.

AWWA. 2015b (3rd ed.). M44. *Distribution Valves: Selection, Installation, Field Testing, and Maintenance.* 2015. Denver, CO: AWWA.

AWWA. 2016. *Buried No Longer: Confronting America's Water Infrastructure Challenge*. Denver, CO: AWWA.

AWWA. 2016 (5th ed.). M17, *Fire Hydrants: Installation, Field Testing, and Maintenance*. Denver, CO: AWWA.

Bates, R.T., Cesario, L., & Ranger, M. 1996. Sizing of Treated Water Storage Reservoirs. Proceedings 1996 AWWA Engineering and Construction Conference. Denver, CO: AWWA.

Bloetscher, F. 2011. *Utility Management for Water and Wastewater Operators*. Denver, CO: AWWA.

Clark, R. M. 2012 (2nd edition). *Modeling Water Quality in Distribution Systems*. Denver, CO: AWWA.

Cruikshank, J. 2010. Hydraulic Models Shed Light on Water Age. *Opflow*, 36:6.

Deb, A.K.; Momberger, K.A.; Hasit, Y.J.; & Grablutz, F.M. 2000. *Guidance for Management of Distribution System Operations and Maintenance*. Denver, CO: AWWA Research Foundation.

Friedman, M.; Kirmeyer, G.; Lemieux, J.; LeChevallier, M.; Seidl, S.; & Routt, J. 2010. *Criteria for Optimized Distribution Systems*. Denver, CO: Water Research Foundation.

Grayman, W.M.; Rossman, L.A.; Arnold, C.; Deininger, R.A.; Smith, C; Smith, J.F.; & Schnipke, R. 2000. *Water Quality Modeling of Distribution Storage Facilities*. Denver, CO: Water Research Foundation.

HSNW (Homeland Security News Wire). 2010. Aging Infrastructure Poses Economic, Security Risks. Feb. 9, 2010. http://www.homelandsecuritynewswire.com/aging-infrastructure-poses-economic-security-risks (accessed Sept. 19, 2016).

Jones, G.M.; Sanks, R.L.; Tchobanoglous, G.; & Bosserman II, B.E. 2008 (3rd ed.). *Pumping Station Design*, revised. Burlington, MA: Elsevier, Inc.

Kelley, W.A. 1994. Hydraulic Reliability—What Drives the Size/Location of Reservoirs? Proceedings 1994 AWWA Annual Conference. Denver, CO: AWWA.

Kennedy, M.S.; Moegling, S.; Sarikell, S.; & Suravallop, K. 1993. Assessing the Effects of Storage Tank Design on Water Quality. *Jour. AWWA*, 85:7:78.

Kirmeyer, G.J.; Friedman, M.; Clement, J.; Sandvig, A.; Noran, P.F.; Martel, K.D.; Smith, D.; LeChevallier, M.; Volk, C.; Antoun, E.; Hiltebrand, D.; Dyksen, J.; & Cushing, R. 2000. *Guidance Manual for Maintaining Distribution System Water Quality*. Denver, CO: Water Research Foundation.

Mays, L.W. 2000. *Water Distribution Systems Handbook*. New York: McGraw Hill.

Powell, J., Clement, J.; Brandt, M.; Casey, R,; Holt, D.; Grayman, W.; & LeChevallier, M. 2004. *Predictive Models for Water Quality in Distribution Systems*. Denver, CO: Water Research Foundation.

Rice, E.W., Baird, R.B., Eaton, A.D., & Clesceri, L.S. (editors). 2012. *Standard Methods for the Examination of Water and Wastewater.* 22nd ed. ISBN: 9780875530130. American Public Health Association, American Water Works Association, & Water Environment Federation: Washington, D.C.

Snyder, J.K.; Deb, A.K.; Grablutz, F.M; McCammon, S.B.; Grayman, W.M.; Clark, R.M.; Okun, D.A.; Tyler, S.M.; & Savic, D. 2002. *Impacts of Fire Flow on Distribution System Water Quality, Design and Operation*. Denver, CO: Water Research Foundation.

Ten State Standards. 2012. *Recommended Standards for Water Works*. Great Lakes–Upper Mississippi River Board of State and Provincial Public Health and Environmental Managers. http://10statesstandards.com/index.html (accessed May 23).

Walski, T.M.; Chase, D.V.; Savic, D.A.; Grayman, W.; Beckwith, S.; & Koelle, E. 2003. *Advanced Water Distribution System Modeling and Management*. Exton, PA: Bentley Systems.

Chapter 3

Understanding and Managing Biofilm, Coliform Occurrence, and the Microbial Community

Melinda Friedman, Confluence Engineering Group LLC, Seattle, Wash.

Nick Ashbolt, University of Alberta, Edmonton, Alberta, Canada

Amie Hanson, Confluence Engineering Group LLC, Seattle, Wash.

Laura Meteer, Regional Municipality of York, Newmarket, Ontario, Canada

Alejandro Ureta, Asesoramiento a la Gestión de los Sistemas de Agua (AGeSA), Montevideo, Uruguay

The occurrence of microbes in distribution systems can lead to the presence of coliform bacteria, nitrification, corrosion, waterborne pathogens, and other unfavorable water quality conditions. This chapter describes microbial growth challenges that water utilities may experience in the distribution systems; microbial indicators of water quality; techniques to characterize microbial communities; factors that contribute to microbial growth as well as mitigation and corrective actions recommended for addressing microbial problems; and emerging waterborne pathogens.

MICROBIAL GROWTH CHALLENGES IN THE DISTRIBUTION SYSTEM

Even after treatment, including disinfection, neither drinking water nor the distribution system are free of microbes. Every drinking water distribution system contains free-floating (planktonic) microorganisms and surface-associated (such as biofilm and those attached to particles) microorganisms, which are important with respect to active biomass. Surface-associated microorganisms thrive on all surfaces, including sediments that accumulate in pipes and within storage facilities. Thus, a key objective of secondary disinfection and distribution system operations and maintenance (O&M) practices is to constrain microbial growth, especially that of surface-associated microbes.

Microbial growth in the distribution system can change drinking water quality, increase corrosion, and potentially affect public health. Microbial biofilms and planktonic cells may exert a disinfectant demand (destroying residual), cause taste and odor episodes, and increase the potential for disinfection by-products (DBPs) within the distribution system. Secondary effects include changes in pH and redox chemistry that can increase corrosion and metals release within piping and premise plumbing materials. As discussed in Chapter 6, in chloraminated distribution systems, nitrification can result in the following:

- Loss of disinfectant residual,
- pH decrease,
- Alkalinity decrease,
- Dissolved oxygen decrease, and
- Nitrite and nitrate increase (AWWA 2013).

This section describes important aspects of distribution system microbial growth, including total coliforms, source water stability, and other issues; waterborne pathogens and emerging concerns; waterborne disease outbreaks caused by distribution system deficiencies; regulations and guidelines that apply to distribution system microbes; and pathways through which microorganisms reach water in the distribution system.

Total Coliforms

Total coliform bacteria are ubiquitous to aquatic environments, soil, and vegetation and also occur in the feces of warm- and cold-blooded animals (Ashbolt, Grabow, and Snozzi 2001). Therefore, they are not specific fecal indicators. Coliform bacteria, with a few exceptions, are not harmful to humans.

Total coliform monitoring has a significant role in distribution system water quality monitoring. For example, the US Environmental Protection Agency (USEPA) adopted the Revised Total Coliform Rule (RTCR) in 2013, which formally recognizes that total coliforms are a possible indicator of the need for system maintenance (USEPA 2013). However, in the RTCR, a subset of total coliforms (i.e., *Escherichia coli*) is used as the indicator of possible fecal contamination.

The presence of total coliforms can indicate the following:

- Source water events or short-term conditions that affect water quality (e.g., rain/snowmelt events that result in significantly increased turbidity and changes to organic matter nutrients);
- The efficacy of water treatment processes (pretreatment including coagulation and flocculation, oxidation, filtration, nutrient removal, and disinfection);
- Compromised integrity of the distribution system (e.g., pressure issues, cross-connections, main breaks) that could signal intrusion points for fecal or soil contamination; and/or

- Inadequate distribution system operation and/or maintenance that have the potential to impact water quality (e.g., water age, disinfectant residual maintenance, flushing programs, storage facility management).

Factors that affect regrowth of coliforms in distribution systems include temperature, availability of nutrients (carbon/energy sources), unlined cast iron, and insufficient residual disinfectant (LeChevallier, Welch, and Smith 1996). The presence of total coliforms in drinking waters could be caused by mobilization of sediments, corrosion products, and biofilms or by detachment events caused by changes in pressures or flow regimes. Therefore, in distribution systems, coliforms can be a natural part of pipe sediments, corrosion products, or biofilms and not always an indicator of a loss in distribution system integrity. In the United States, the USEPA, under the RTCR, no longer has a maximum contaminant level (MCL) for total coliforms; instead, their presence is used to trigger an assessment of distribution system conditions and integrity.

Bulk Water and Particle-Associated Microorganisms

Microorganisms are present in bulk water and can associate with (or attach to) particulate matter or sediments within distribution systems. Liu et al. (2014) investigated the composition of the bacterial community with respect to microbial presence in bulk water, suspended solids, loose deposits, and biofilm in an unchlorinated distribution system. Results showed that bulk water bacteria (including the contribution of suspended solids) contributed <2 percent of the active biomass measured by the presence of adenosine triphosphate (ATP). The microorganisms associated with loose deposits and pipe wall biofilm that accumulated in the drinking water distribution system accounted for more than 98 percent of the equivalent active bacterial biomass (Figure 3-1).

Ormeci and Linden (2002) showed that particle- and biofilm-associated microorganisms were some three-fold more resistant to disinfection than planktonic cells (i.e., those free-floating in the bulk water). Miller et al. (2015) reported that the free-living pathogenic protozoan *Naegleria fowleri* survived more than 30 times the recommended concentration of chlorine for drinking water systems when associated with distribution system biofilm. The investigators associated the increased resistance of the pathogen to biofilm consumption of the disinfectant and reduced disinfectant penetration into the biofilm, which was further reduced by the presence of the other microorganisms in the biofilm.

Biofilms

Microbial growth is concentrated within biofilms that form on all moist surfaces. A biofilm is largely comprised of extracellular material excreted by bacteria, in which other organic and inorganic material accumulate. Most of a biofilm's structure is a hydrated matrix of extracellular polymeric substances that enable microorganisms to persist in nutrient-poor, hydraulically dynamic, disinfectant-containing distribution systems (Doutelero, Husband, and Boxall 2014b; Costerton et al. 1987; Simoes, Simoes, and Vieira 2010; Flemming, Percival, and Walker 2002; Flemming and Wingender 2010; USEPA 2002). In drinking water distribution systems, extracellular polymeric substances provide mechanical stability; decrease loss of lysed cell nutrients to the broader environment; provide tolerance against biocides, metals, and toxins; protect microbes within the biofilm against predatory species; and allow for the accumulation of drinking water nutrients, despite their low concentration (Flemming, Percival, and Walker 2002).

Microorganisms present in biofilms can reach customers through changes in water quality or flow patterns. These changes result in sloughing or erosion of the biofilm from pipe surfaces and introduction of microorganisms into the potable water that moves through the distribution system pipes.

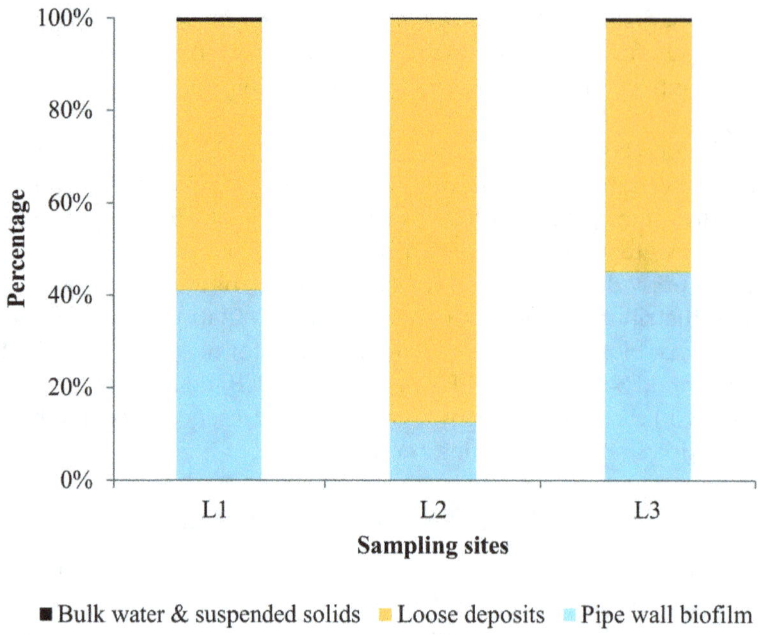

Reprinted with permission from Liu et al. 2014. Copyright 2014 American Chemical Society.

Figure 3-1 Comparison of bacterial abundance (biomass as inferred from adenosine triphosphate content of living cells) of different phases within a 1-m water main (polyvinyl chloride, 110 mm)

Biofilms can affect distribution systems in the following ways:

- Increase pipe corrosion (Miller et al. 2013);
- Adversely affect pipe hydraulics (Miller et al. 2013);
- Reduce the usefulness of total coliforms as process indicator organisms (Miller et al. 2013);
- Act as a source of water-based pathogens (e.g., *Legionella pneumophila, N. fowleri,* and nontuberculous mycobacteria [NTM]), accumulating fecal pathogens and facilitating increased disinfectant resistance/virulence transfer (Wingender and Flemming 2011; Ashbolt 2015a, 2015b; Miller 2015);
- Exert a disinfectant demand (Miller et al. 2013);
- Contribute to bad taste and odor (Miller et al. 2013);
- Increase DBPs (Wang, Choi, and Seo 2013);
- Contribute to the proliferation of macroinvertebrates (Buse et al. 2014); and
- Accumulate contaminants and other materials that enter the distribution system (Wingender and Flemming 2011).

Drinking water biofilms are dynamic and highly competitive microbial niches that develop into self-protecting environments for functional gene activity (e.g., nitrification [Pinto et al. 2016]) and to some degree maintain taxonomic groups (Bautista-de Los Santos et al. 2016; Revetta et al. 2016; Li et al. 2016). While culture-based methods have been superseded by 16S ribosomal DNA gene-based identification methods, currently available sequencing platforms may only identify down to the genus level and miss minor members. Until "very deep sequencing" can be cost-effectively achieved (Singer et al. 2016), our understanding of biofilms and drinking water bacteria is limited to the major players (i.e, coliform and heterotrophic bacteria) and may miss less numerous members (such as opportunistic strands of *Pseudomonas* discussed later in this chapter) that are of health or water quality concern. Furthermore, while laboratory-scale studies shed light on some of the basic principles (Culotti and Packman 2014), biofilms are multispecies environments influenced by a complex and dynamic array of factors. For example, recent work undertaken on the microbial diversity of biofilm versus plankton indicates that there are many different members in the two zones that differ depending on the nature of the source water (Gomez-Alvarez et al. 2016). We have much to learn about the minor but key species in drinking water systems and how the more numerous, benign members interact with microeukaryotic members to protect water quality in drinking water systems (Prest et al. 2016).

Drinking Water Stability

In several European countries such as Switzerland, Austria, Germany, and the Netherlands, many drinking water systems meet public health requirements while providing finished water without a disinfectant residual. These systems focus on providing biologically stable water by limiting nutrient levels present in finished water. This is done to minimize microbial growth as water moves through the distribution system (Chowdhury 2012; Hammes et al. 2010; van der Kooij 1999).

Smeets, Medema, and van Dijk (2009) described the following Dutch approach for providing a safe supply without using chlorine for primary or secondary disinfection:

- Use the best source available, with a preference for microbiologically safe groundwater.
- Use a treatment process such as sedimentation, filtration, or ultraviolet disinfection. Oxidation with ozone or peroxide can be used, if absolutely necessary, but chlorine is avoided to limit DBPs.
- Prevent contaminant entry into the distribution system through practices such as maintaining adequate pressure throughout the system, following maintenance protocols, and preventing cross-connections and backflow.
- Prevent microbial growth in the distribution system by producing and distributing biologically stable water and using biostable materials. Biostability of materials is tested using the biofilm formation potential test developed by van der Kooij and Veenendaal (1993); the majority of distribution system materials consist of relatively biostable asbestos-cement or polyvinyl chloride (PVC), compared to unlined cast iron.
- Monitor for timely detection of any system failure to prevent significant health consequences, noting that in the absence of a residual, *Escherichia coli* is a sensitive indicator of integrity loss in the drinking water distribution system.

Chowdhury (2012) reviewed research that described maximum drinking water nutrient levels that may not support microbial growth. Results varied due to the complexity of factors influencing growth in drinking water distribution systems and differences in measuring available nutrients. For waters with low levels of carbon, nitrogen, and phosphorus, biologically stable water may be assessed based on the growth of two bacterial strains in assimilable organic carbon (AOC) or by use of native bacteria from the system's drinking water (Hammes et al. 2010). van der Kooij (2003) recommended maintaining an AOC level of 10 μg/L acetate carbon (C) equivalents to achieve biostability for Dutch groundwaters.

With free chlorine, Liu et al. (2002) recommend up to 50–100 μg/L acetate C equivalents and 10–20 μg/L acetate C equivalents in the absence of residual chlorine. In the United

States, LeChevallier, Schulz, and Lee (1991) demonstrated that as little as 50 μg/L of acetate C equivalents of AOC is required to encourage coliform regrowth in drinking water from surface waters with a disinfectant residual. In contrast, Boe-Hansen et al. (2002) observed significantly higher bacterial regrowth at very low nutrient conditions in the distribution system (4.6–5.3 μg/L acetate C equivalents of AOC).

It is important to recognize that an assay such as AOC, which was developed for groundwaters with very low nutrients (generally at low temperatures), may not be appropriate for use as a measure of biostability. As an alternative, Niquette, Servais, and Savoir (2001) suggest that biodegradable dissolved organic carbon (BDOC) is a more inclusive assay for bacterial growth. These researchers demonstrated that bacterial regrowth could be controlled by lowering BDOC to <1.25 mg/L of carbon at the outlet of a water treatment plant and by ensuring sufficient free chlorine residual in the distribution system. Also, with the advent of near online flow cytometry, biostability of distribution system waters can be readily assayed (Hammes et al. 2012). Nonetheless, enumeration of total active cells may be less informative as to regrowth than other measures of microbial activity (such as ATP) or the slow growth of fungi and NTM (van der Wielen and van der Kooij 2010).

Biofouling and Microbially Induced Corrosion

In addition to water quality issues associated with biofilm growth, biofouling can lead to issues with pipe flow and clogging of meters and other appurtenances. Biofouling, which is caused by the excessive growth of biofilms, can be defined operationally as the point at which biofilm presence exerts a level of interference that is unacceptable to performance, quality, and/or economic considerations (Flemming 2011). Many aquatic bacteria have traits that can lead to development of biofouling (Coetser and Cloete 2005). Most members of distribution system biofilms are thought to be nonculturable and include members of the phyla α-, β-, and γ-*Proteobacteria, Actinobacteria, Verrocomicrobia,* and *Bacteroidetes* and what appear to be *Cyanobacteria* (Gomez-Alvarez et al. 2015; Ling et al. 2015) but could be misclassified (Bautista-de Los Santos et al. 2016).

The patchy nature of biofilm growth leads to areas of high, low, and sometimes no oxygen at the pipe–water interface, creating a localized oxygen concentration cell (Coetser and Cloete 2005). Thus, biofouling can also lead to microbially induced corrosion (MIC). Generally, MIC is more prevalent in water systems where the source water has sufficient concentrations of organic carbon, organic nitrogen, sulfate, and phosphate to support microbial growth. Sulfate-reducing bacteria (SRB) are considered to be significant contributors to MIC (Sun et al. 2014; Yang et al. 2014). Sulfate is an electron acceptor for SRB and derives energy from organic carbon in the absence of oxygen; this has been demonstrated to be a key factor in MIC of copper piping. Under oxygen-deficient conditions, SRB reduce sulfate to sulfide, which is a critical anion in the pitting corrosion of copper pipe (Boulay and Edwards 2011). Chapter 7 discusses MIC and corrosion control in more detail.

Waterborne Pathogens and Emerging Concerns

In the United States, drinking water managers can track contaminants that may be regulated in the future by reviewing the Contaminant Candidate List (CCL). The drinking water CCL, developed by the USEPA every five years, is a list of chemical and microbiological contaminants currently not subject to proposed or promulgated national primary drinking water regulations. The CCL contaminants are known or suspected to be present at water sources, entry points to the distribution system, and in distribution systems and may require future regulation. The USEPA published the fourth CCL in 2016. The following microorganisms are included in the 2016 list (USEPA 2016):

- Bacteria: *Campylobacter jejuni, E. coli* (O157), *Helicobacter pylori, Legionella pneumophila, Mycobacterium avium, Salmonella enterica, Shigella sonnei;*
- Viruses: adenovirus, calciviruses, hepatitis A virus, enterovirus; and
- Protozoan: *N. fowleri.*

According to the World Health Organization (2011), waterborne pathogens have the following properties that distinguish them from chemical contaminants in drinking water:

- They can cause acute and/or chronic health effects.
- They are often aggregated or can adhere to or become enmeshed in particulate matter in distribution systems.
- Exposure that results in disease depends on the pathogen's dose, invasiveness, and virulence, as well as the immune status of the individual.
- If infection is established, they multiply in their host.
- Certain waterborne pathogens are also able to multiply in food, beverages, and warm water systems, perpetuating or even increasing the likelihood of infection.
- They do not, typically, result in a cumulative health effect.
- Some can grow in the environment; these are so-called water-based (saprozoic) pathogens (Ashbolt 2015b).

A distinction between the two groups of drinking water pathogens can also be drawn according to their origins. These differences are important because they can also describe the pathogen's ability to thrive in the distribution system. Pathogens can have enteric origins, caused by fecal contamination of the source or distribution system water, or they can originate from the environment and grow within the drinking water distribution system.

Ashbolt (2015a) defines waterborne pathogens as enteric viruses, bacteria, parasitic protozoa, and fungi that are transmitted from excreta and ingested via water. Waterborne pathogens depend on the presence of a human or zoonotic host. Once they reach the distribution system, waterborne pathogens will gradually lose the ability to infect and will die off in the drinking water environment. These pathogens become undetectable yet may be present at levels that can cause infection and undergo secondary spread by person-to-person contact or poor hygiene (WHO 2011). In contrast, water-based pathogens are environmental viruses, bacteria, free-living protozoa, and fungi that naturally grow in aquatic and soil environments. Water-based pathogens such as *Legionella* may reach high concentrations in the distribution system as part of biofilms and on sediment (Ashbolt 2015a; Lu et al. 2015). However, warm, stagnant conditions that favor their amplification are generally most difficult to control in premise plumbing and in hot climates.

As summarized by Ashbolt (2015a), the classic waterborne enteric pathogens include *Vibrio cholerae* (serogroups O1 and O139, causing cholera), *Salmonella enterica* (subsp. *enterica* ser. Typhi, causing typhoid), and *Shigella* spp. (four species causing dysentery), which have largely been controlled by water treatment and disinfection and are therefore very rarely a drinking water issue in developed regions.

However, person-to-person contact and foodborne spread maintains *Shigella sonnei* within sewage in developed regions. This could impact drinking water systems, along with closely related shiga toxin and verotoxin-producing *E. coli* and pathogenic species of *Campylobacter, Salmonella, Arcobacter, Helicobacter,* and *Yersinia.* Table 3-1 provides general information on pathogens that are of relevance to drinking water supply management.

Naegleria fowleri is an "emerging" saprozoic pathogen that few utilities have direct experience with. *Naegleria fowleri,* also referred to as brain-eating amoeba, causes primary amebic meningoencephalitis, which can be acquired when contaminated water enters the nose (Bartrand, Causey, and Clancy 2014). *Naegleria fowleri* can grow in warm potable water with little to no disinfectant residual. In the 1970s and 1980s, Australia experienced multiple deaths associated with swimming and other nasal exposure to *N. fowleri* and concerns were raised regarding detections in drinking water (CDC 2015a). The detections were linked to drinking water supplies that were piped overland, sometimes for hundreds of miles, resulting in warm drinking water with little to no disinfectant present. In the United States, at least one death has been attributed to *N. fowleri* contamination of a disinfected drinking water supply; this occurred in Louisiana in 2013 (CDC 2015a). As a result, the Louisiana Department of Health issued an emergency rule that took effect in 2014. The rule requires water systems to maintain a higher minimum disinfectant residual of 0.5 mg/L (measured as free or total chlorine) in water being delivered in the distribution system. Prior to the Emergency Rule, Louisiana's regulations, which were implemented in 1995 in accordance with federal guidance, stipulated that drinking water systems were required to have a "trace" or "detectable" level of free chlorine residual at all points in their systems at all times (State of Louisiana 2013).

Waterborne Disease Outbreaks and the Distribution System

Waterborne disease outbreaks in the United States are reported to the Waterborne Disease and Outbreak Surveillance System at the Centers for Disease Control and Prevention (CDC). The CDC identifies an event as a waterborne disease outbreak based on the following: two or more persons must be linked epidemiologically by time, location of water exposure, and illness characteristics and the epidemiologic evidence must implicate water as the probable source of illness (CDC 2013). For each outbreak, the CDC gathers the following data: the number of illnesses, hospitalizations, and deaths; the confirmed or suspected etiologic agent; the implicated water system; deficiencies that contributed to the outbreak; and the setting of exposure. Figure 3-2 presents data for waterborne disease outbreaks that occurred between 1971 and 2010. Figure 3-3 presents the deficiency sources associated with drinking water outbreaks and outbreak-related cases reported between 2011 and 2012.

Distribution system deficiencies are of significant concern, particularly for enteric virus risks (Borchardt et al. 2012). As shown in Figure 3-3 (a), distribution system deficiencies accounted for more than 3 percent of outbreak cases and, as shown in Figure 3-3 (b), 6 percent of reported illnesses between 2011 and 2012. The CDC reports an outbreak as being related to the distribution system if a deficiency has been found relating to storage, cross-connection, backflow, or contamination of water mains during construction or repair. This reporting includes illnesses due to distribution deficiencies and those that were due to multiple deficiencies, with distribution deficiencies being one of the causes.

With respect to distribution system deficiencies, most cases of gastrointestinal illness result from Noroviruses from just a few outbreaks within noncommunity systems that supply undisinfected water (Beer et al. 2015). The presence of a disinfectant residual might have provided protection to the consumers and prevented these outbreaks.

Regarding the causes and types of waterborne illness, the drinking water industry has seen a change recently, as shown in Figure 3-2. Craun et al. (2010) reviewed waterborne disease outbreak information available for 1971 through 2006, noting that the illness profile of outbreaks began to change in 2001 when *Legionella* became part of outbreak reporting. Before 2001, acute gastrointestinal illness was the dominant illness type among reported outbreaks, and no outbreaks associated with acute respiratory illness were reported. Therefore, the apparent increase in Legionella outbreaks, as shown in Figure 3-2, may be a result of increased awareness and reporting and may not indicate a rise in actual infection.

Table 3-1 Recognized and potential enteric and water-based microbial pathogens to manage community drinking water risks

Microbial Group	Enteric Waterborne		Water-based (Opportunistic)	
	Recognized	Potential	Recognized	Potential
Viral	*Adenovirus* 40 and 41 *Avipolyomavirus*[*] *Enterovirus* A to D *Hepatitis* A and E *Norovirus* G1 and G2 *Rotavirus* A *Sapovirus* G1	*Mamastrovirus* 1 *Orthoreovirus* C	None	*Mimivirus*[†] *Mamavirus*[†]
Bacterial	*Aeromonas hydrophila*[‡] *Campylobacter coli* *Campylobacter jejuni* *Salmonella enterica* (non-typhoid) *Shigella sonnei* *Vibrio cholerae*[¶]	*Acinetobacter baumannii*[‡] *Arcobacter butzleri* *Helicobacter pylori* *Clostridium difficile*[‡] *Listeria monocytogenes* *Pseudomonas aeruginosa*[‡] *Staphylococcus aureus*[‡] *Yersinia enterocolitica*	*Legionella longbeacheae*[||] *Legionella micdadei*[||] *Legionella pneumophila*[||] *E. coli* O157:H7[**] Nontuberculous mycobacteria[||,‡,††] *Pseudomonas aeruginosa*[‡]	*Acinetobacter baumannii*[‡] *Aeromonas hydrophila*[‡] Amoeba-resisting bacteria (*Afipia, Bosea, Parachlamydia* spp., *Coxiella burnetii*)[||] *E. coli* (toxigenic strains) *Listeria monocytogenes*[||] *Staphylococcus aureus*[‡] *Stenotrophomonas maltophilia*[‡,||]
Protozoan	*Cryptosporidium hominis* and *parvum* *Cyclospora cayetanensis* *Giardia intestinalis* assemblages A and B *Toxoplasma gondii*	*Blastocystis hominis*	*Acanthamoeba T4* *Balamuthia mandrillaris* *Naegleria fowleri*	*Acanthamoeba* spp.[||] *Vahlkampfia* spp.[||] *Vannella* spp.[||] *Vermamoeba vermiformis*
Fungal	None	*Candida albicans* Microsporidia (e.g., *Encephalitozoon bieneusi, E. intestinalis*)	None	*Aspergillus fumigatus* *Aspergillus terreus* *Candida albicans* *Candida parapsilosis* *Exophiala dermatitidis*

Source: Ashbolt 2015a. Reprinted with permission from Springer Science + Business Media LLC.

* Main species being *JC polyomavirus,* which is largely excreted in urine, as it infects the kidneys along with the respiratory system or brain.

† *Acanthamoeba polyphaga* mimivirus may cause respiratory disease. The health effects from mamavirus are unknown (Dornas, et al 2014).

‡ Most strains of the species from the environment may be nonpathogenic; however, there is future potential for antimicrobial-resistaning strains. For *Pseudomonas Aeruginosa,* most clinical disease is identified with otitis media, a type of ear infection. with less severe disease spread via drinking water aerosols, leading to diffuse bronchopneumonia and more severe disease in high-risk children with cystic fibrosis. Folliculitis (infected hair follicle) is spread directly via drinking water used in pools and spas.

|| Largely nonpathogenic amoeba hosts contain amoeba-resisting bacteria that are of key concern as they relate to drinking water (Thomas V, et al 2014), except possibly *Coxiella burnetii,* which is of low theoretical risk estimated by quantitative microbiological risk assessment (QMRA) (Sales-Ortells H and Medema G 2014).

¶ Cholera may reemerge if a major event interferes with drinking water disinfection and an epidemic strain is introduced from an endemic region of the world.

** Shiga toxin and verotoxin-producing *E. coli* (and various *Salmonella* and *Listeria* intracellular) strains may grow within free-living protozoa (Gourabathini P, et al 2008) and nonpathogenic VBNC *E. coli* in distribution system biofilms (Juhna T, et al 2007).

†† Various nontuberculous mycobacteria, including *Mycobacterium avium* (*Mycobacterium intracellulare*) complex, *Mycobacterium chelonae Mycobacterium fortuitum, Mycobacterium gordonae,* and *Mycobacterium kansasii.*

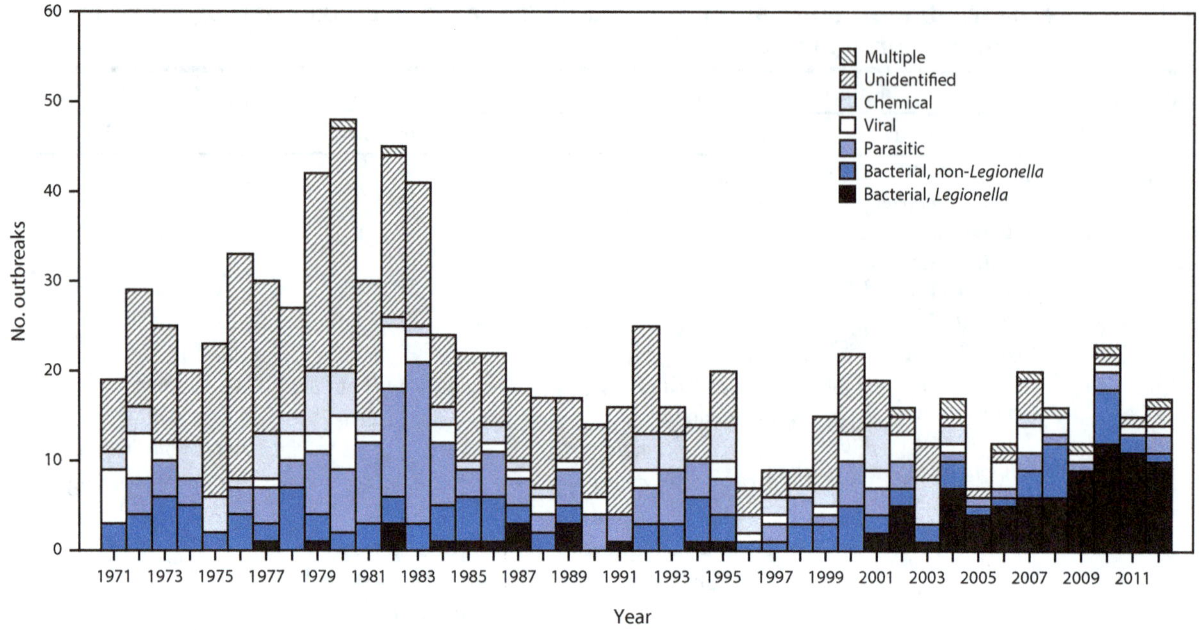

Source: Beer et al. 2015.

Figure 3-2 Number of waterborne disease outbreaks associated with drinking water (N = 851), by year and etiology—United States, 1971–2012

* Legionnaires' disease outbreaks were first reported to the Waterborne Disease and Outbreak Surveillance System in 2001; Legionnaires' disease outbreaks before 2001 were added retrospectively during the 2007–2008 reporting period.

† Includes all bacteria except *Legionella*.

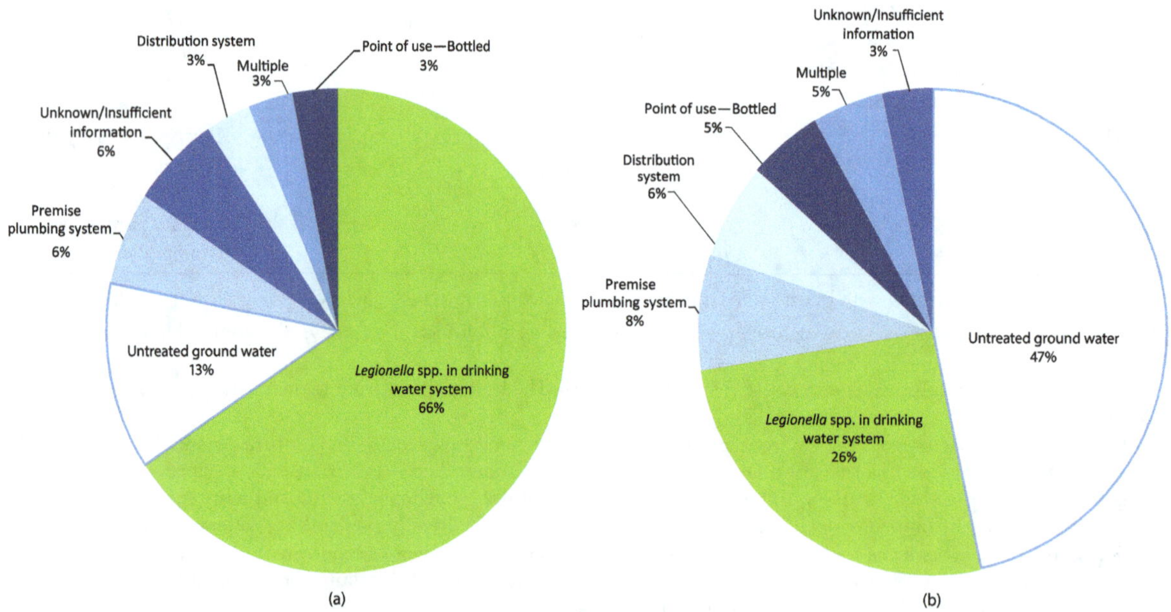

Data Source: CDC 2015b.

Figure 3-3 Deficiencies assigned to (a) drinking water outbreaks (N = 32) and (b) outbreak-related cases (N= 431) from the Waterborne Disease and Outbreak Surveillance System, 2011–2012

Note: Multiple deficiencies were assigned to one giardiasis outbreak that contributed 21 cases: untreated surface water and untreated groundwater.

Recent CDC assessments (CDC 2013; Beer et al. 2015) of waterborne disease outbreaks indicate that *Legionella* was the most frequently reported etiologic agent among drinking water and nonrecreational outbreaks. Eight of nine deaths related to drinking water during 2009 and 2010 were due to *Legionella pneumophila.* During 2011 and 2012, *L. pneumophila* was responsible for 66 percent of the 32 reported outbreaks, underscoring the need for improved *Legionella* control and mitigation. Since data on legionellosis (severe respiratory illness known as Legionnaires' disease) were recorded by CDC under waterborne illness in 2001, *Legionella* has been identified as the major agent to cause waterborne illness. Nearly all reported deaths related to public surface water supplies were due to Legionnaires' disease, with most cases reported from growth in premise plumbing of healthcare facilities (Beer et al. 2015); these facilities are typically not under the direct control of a water system. These findings underscore the new importance of *Legionella* when managing drinking water and public health.

It is important to note that while the most severe form of legionellosis is now recognized in the reporting of drinking water–related outbreaks, many milder lung infections can occur. These are caused by various legionellae (e.g., Pontiac fever), possibly via their endotoxin cell lipopolysaccharides (Neumeister et al. 1998), along with various NTM such as the *Mycobacterium avium* complex (MAC), which are thought to be spread via water, not through person-to-person contact (Falkinham 2011).

After investigating drinking water–associated pathogens using hospitalization medical claims, CDC revealed that NTM and *Legionella* accounted for approximately 90 percent of Medicare and Medicaid hospitalization claims associated with drinking waters in the United States (Collier et al. 2012). NTM have been found in drinking water distribution systems, hospital water systems, domestic tap water (Thompson et al. 2013), and even in certain types of blood heaters/coolers used in surgery (Sommerstein et al. 2016). Thompson et al. (2013) discussed the factors associated with NTM presence, which are distance of a sampling point from the main treatment point, pipe diameter at the sampling point, and piping material. Whiley et al. (2012) described the types of drinking water that were contaminated by MAC; they include warm water distribution systems, showers, faucets, household drinking water, swimming pools, and hot tub spas.

Deficiencies at points that are not under the jurisdiction of water systems or the federal government (e.g., private wells and plumbing systems), continue to cause illness, mainly from enteric viruses. Many states have developed minimum chlorine residual limits, and some systems verify residuals at the meter to show that they are delivering water with appropriate chlorine residuals.

In December 2016 as part of its Six-Year Review 3, the USEPA identified *Legionella* as a candidate for revision, along with *Cryptosporidium,* heterotrophic bacteria, *Giardia lamblia,* and viruses. The Safe Drinking Water Act requires the USEPA to review all national primary drinking water regulations at least once every six years and to revise them, if appropriate (USEPA 2016).

Canada does not have a nationally standardized surveillance system for waterborne disease outbreaks (Public Health Agency of Canada 2013). Nonetheless, national estimates have been made (Murphy et al. 2016a, 2016b). For the surveys of drinking water outbreaks that do exist, distribution systems are listed as one of the contributing factors (National Collaborating Centre for Environmental Health 2011). The majority of outbreaks are related to microbial contaminants, with *Giardia* appearing to be the most common pathogen related to individual outbreak events.

SUMMARY OF REGULATORY FRAMEWORKS APPLICABLE TO MICROBES IN DISTRIBUTION SYSTEMS

Chapter 1 of this manual provided an overview of regulations and standards associated with distribution system water quality. This section focuses on and provides more details associated with those standards that apply to microbes.

US Drinking Water Distribution System Standards

In the United States, microbial monitoring is the most frequent regulatory-driven distribution system monitoring conducted by water systems. The federal regulations are summarized in Table 3-2. It is important to note that primacy agencies (states, territories, and tribes) may have a variety of definitions of detectable residual under the Surface Water Treatment Rule (SWTR), with some establishing a numeric definition. The USEPA reviewed this provision as part of the Six-Year Review 3 and determined that there is an opportunity to revise the treatment technique provisions of the SWTR. As noted, the USEPA identified five microbial contaminants as candidates for revision under the Six-Year Review 3 process. These are included in the SWTR, with *Cryptosporidium* also included in the Interim and Long-Term 1 Enhanced Surface Water Treatment Rules (USEPA 2016).

International Standards

Health Canada provides guidance on distribution system microbial quality. The guidance recommends that total coliform, *E. coli*, chlorine residual, and turbidity be regularly measured (Health Canada 2013a). The frequency of measurement is not strictly stated. Instead, monitoring is based on the characteristics of the drinking water system. If *E. coli* presence is confirmed in a distribution system, a boil-water advisory should be issued. If total coliform is detected in more than 10 percent of samples collected in a given sampling period or at the same location in two consecutive samples, an investigation into the source of the detections should be conducted. Actions should be taken to resolve the source of the total coliforms. Chloraminated systems should conduct free ammonia, nitrate, nitrite, and heterotrophic plate count (HPC) monitoring.

The Ministry of Environment and Climate Change in Ontario, Canada, oversees drinking water requirements. Ontario has established secondary disinfectant requirements for drinking water systems; chlorinated systems must meet at least a residual level of 0.05 mg/L free chlorine, and chloraminated systems must provide 0.25 mg/L as combined chlorine at all points in the distribution system. Systems that rely on chlorine dioxide must provide a residual of 0.05 mg/L. The agency recommends an optimum target for free chlorine residual of 0.2 mg/L in the distribution system for chlorinating systems and a combined residual of 1.0 mg/L for chloraminating systems. Systems are expected to take grab samples or use continuous monitoring throughout the distribution system, with results below the requirement considered to be "adverse." The regulations outline corrective actions required for systems that measure adverse results; the residual level must be adjusted in order to quickly meet the standard or a notice to customers indicating the need for alternate sources or boiling water should be issued (Safe Drinking Water Act [Ontario] 2002).

The World Health Organization (WHO) guidelines also emphasize more frequent testing for indicator organisms rather than less frequent testing with more complicated and expensive methods (WHO 2011). WHO recommends that no *E. coli* be detected in treated water or in the distribution system. Total coliform should not be present in more than 5 percent of samples collected in the distribution system. The WHO guidelines recognize the challenges of widespread fecal contamination in developing countries. A community should use the least contaminated water source available, and governing agencies should apply the guidelines in a thoughtful manner. A free chlorine residual of 0.2 mg/L should be maintained in the distribution system (WHO 2011).

Table 3-2 US Safe Drinking Water Act regulations related to microorganisms in the distribution system

Regulation	Microbial Parameter	Treatment Technique/MCL/MCLG/Other
Revised Total Coliform Rule	TC*	Treatment technique Level 1 assessment and corrective actions to be taken if: • A PWS collecting fewer than 40 monthly routine samples has 2 or more TC+ samples in one month* • A PWS collecting 40 or more monthly routine samples has >5.0 percent positive coliform samples (of routine plus repeat samples) in the same month • A PWS fails to take every required repeat sample after any single positive total coliform sample *And no samples are positive for *E. coli* MCLG = 0 Boil water notice, level 2 assessment, and corrective actions must be taken if an MCL violation occurs when:
	*E. coli**	Routine sample is. . . / . . . and repeat sample is . . . *E. coli*+ / TC + *E. coli*+ / Missing sample *E. coli*+ / *E. coli*+ TC+ / *E. coli*+ TC+ / TC + (no *E. coli* analysis) *Known as an acute MCL under the previous Total Coliform Rule
Surface Water Treatment Rule	Heterotrophic bacteria (measured as HPC)	Treatment technique • Disinfectant residual must be present in at least 95 percent of monthly samples in 2 consecutive months • HPC <500 cfu/mL is equivalent to a detectable residual†
Ground Water Rule	TC	Other: A TC+ sample triggers source water monitoring for groundwater systems that do not conduct compliance monitoring for disinfection

Abbreviations: HPC, heterotrophic plate count; MCL, maximum contaminant level; MCLG, maximum contaminant level goal; PWS, public water system; TC, total coliform.

* In a 100-mL sample.

† USEPA Six Year Review 3 identified the Surface Water Treatment Rule treatment technique provisions as candidates for revision (USEPA 2016).

Other international microbial guidelines and standards for the distribution system include the following:

Australia: Australia's guidelines, while not mandatory, are specific and detailed. For coliform, guidelines include measuring total coliform after disinfection and in the distribution system, following with total coliform detection with *E. coli* determination, and monitoring source water for *E. coli*. If coliphages, intestinal enterococci, *E. coli*, or thermotolerant coliform are detected in any 100-mL sample of drinking water, the guidelines state that immediate action should be taken. Numerous other microorganisms, for example, total coliform, HPC, pathogenic *E. coli*, and *Legionella*, are listed with no guideline level set.

Israel: Israel's People's Health Regulations provide enforceable standards regarding the sanitary quality of drinking water, including regular monitoring of diverse microbial indicators. Indicator organisms mentioned in the standards include coliform bacteria, fecal coliform bacteria, *E. coli*, and fecal streptococci.

Philippines: The Philippine National Standards for Drinking Water are essentially the same as the USEPA rules with respect to distribution system microbial requirements.

Japan: In Japan, the Waterworks Act, first adopted in 1890, is similar to the WHO guidelines for microbes. The act goes beyond guidance by requiring that tap water contain more than 0.1 mg/L of residual chlorine and be measured daily at the end of a distribution pipe network.

Several European Union countries have recently enacted *Legionella pneumophila* culture-based standards for distribution systems (<100/100 mL or <1000/1000 mL), including the Netherlands, Germany, France, and Spain.

MICROBIAL OCCURRENCE PATHWAYS

As discussed by Friedman et al. (2009), there are three primary pathways that hazardous microbes can follow in the distribution system: treatment breakthrough or inadequate treatment, accumulation and regrowth in biofilms and sediments, and direct contamination (e.g., air and water intrusion, backflow, cross-connection) of the distribution system. Also, microbial indicators can be detected due to sampling or analytical errors. Such detection provides results that are not representative of actual distribution system conditions. The USEPA defines a potential pathway for pathogens as a "sanitary defect." This section describes these pathways.

Treatment Breakthrough or Inadequate Treatment

Source conditions or treatment problems can result in total coliform/*E. coli* presence in the distribution system. This could happen if:

- The source is contaminated by significantly higher loads of fecal microorganisms compared to typical conditions;
- A treatment failure occurs, allowing microbial contamination or nutrient loading into the distribution system; or
- An unusual turbidity event occurs, overwhelming the water system's treatment processes.

Efficacy of drinking water treatment processes can also affect distribution system coliform levels. In terms of public health, it is the detection of any *E. coli* that is of highest concern and requires immediate response (Table 3-2 provides a summary of regulations).

Survival and Regrowth in Biofilms and Sediments

Biofilms can cause a reduction or loss of disinfectant residual, reduction of dissolved oxygen, and problems with taste and odor. Depending on the biofilm's microbial ecology, public health issues associated with water-based pathogens include *Legionella* and NTM (Ashbolt 2015b; Lu et al. 2015), and water quality issues include red or black water caused by iron or SRB.

In addition to water quality problems, biofilms have been associated with MIC, hydraulic roughness, and materials having a shorter life (Characklis and Marshall 1990). Coliform bacteria and other microorganisms unrelated to *L. pneumophila* and other water-based pathogens can flourish within distribution system biofilms, sediments, and scales if water quality conditions become optimum for their growth.

Bulk water quality conditions that can cause microbial regrowth include the following (LeChevallier, Welch, and Smith 1996):

- Warmer water temperatures (typically >15°C);
- Low to no disinfectant residual in the distribution system (including stagnation zones); and
- Increased nutrients (such as AOC, iron, and ammonia).

Direct Contamination of the Distribution System

As discussed by Friedman et al. (2009), direct contamination of the distribution system can occur due to:

- Physical breaches in infrastructure, such as a main break, hole in a storage facility vent screen, or improperly vented air-vacuum relief valves;
- Low pressure events that result in outside hydrostatic pressure that is higher than the pressure within the distribution piping, allowing contaminated water to enter through infrastructure breaches;
- Unsanitary main installation/repair practices; and
- Back-pressure events that allow contaminants to enter the pressurized portion of the distribution system through unprotected cross-connections.

Sampling and Analytical Issues

A sample can be positive for total coliform due to contamination of the sample tap, faulty sample collection procedures, or issues with laboratory analysis. The review of problematic microbial data should include determining whether the samples were collected at the same site and by the same sample collector. Wherever possible, dedicated sampling stations that are protected from adverse conditions should be installed and maintained for routine microbial monitoring. The following should be noted regarding sampling and analysis:

- Sample site: Systems should review and report the condition of the sample site while collecting samples, noting items such as nearby vegetation, presence of animal feces, and site conditions, such as whether sample taps are threaded correctly or whether hot and cold taps are separated.

- Sample technique: Systems should have detailed sampling procedures in the form of written standard operating procedures. These procedures should be disseminated and used by all operators. Samplers should document site conditions, such as those listed above, and weather conditions that could affect sample collection.
- Laboratory analyses: Quality assurance procedures should be in place to ensure that laboratory analyses provide reliable results. Laboratories should be certified or accredited.

DSWQ Challenge:

The occurrence of microbes in distribution systems can lead to the presence of coliforms, biofilm, waterborne and water-based pathogens, nitrification, corrosion, loss of residual, and other unfavorable water quality conditions.

MICROBIAL INDICATORS OF WATER QUALITY

The main goal of drinking water treatment is to remove contaminants and, in turn, distribute and deliver high-quality water for which the consumer's risk from microbial pathogens is low and properly controlled through correct distribution system management. From this perspective and because it is virtually impossible to determine every known microbial hazard, microbial water quality is assessed based on the presence of indicator microorganisms and other water quality characteristics such as chlorine residual, rapid changes in pressure, turbidity, and organic carbon. Distribution systems are monitored for microbial indicators in order to determine the presence of the following types of hazards or risks:

- Process issues: inadequate water treatment processes (e.g., pretreatment, oxidation, filtration, nutrient removal, disinfection) or source water anomalies (e.g., significantly increased turbidity, disinfectant demand, or presence of nutrients);
- Distribution system integrity issues: distribution system vulnerabilities (e.g., pressure changes, cross-connections, main breaks) that could provide pathways for fecal contamination;
- Distribution system O&M issues: inadequate distribution system O&M (e.g., excessive water age, low disinfectant residuals, not enough flushing, storage facility stagnation); and
- Potential growth of opportunistic pathogens: inadequate O&M practices, lack of disinfectant residual, excessive water age, and high temperatures that can stimulate growth of *Legionella, N. fowleri, E. coli,* and other pathogens.

When assessing distribution system water quality, the drinking water industry has traditionally relied on the detection and/or enumeration of indicator bacteria. Use of indicator bacteria relies on the understanding that their presence in a water sample could be associated with the presence of human microbial pathogens. Cabral (2010) outlined the traits of bacteria that indicator fecal contamination of drinking water. They include:

- Exists in high numbers in human or animal intestines and feces;
- Is not pathogenic to humans;

- Is easily, reliably, and cheaply detectable;
- Does not multiply outside the enteric environment;
- Is present in numbers greater than the pathogenic bacteria; and
- Has similar survival, die-off, and transport behavior as the pathogenic bacteria.

Table 3-3 summarizes distribution system microbial indicators and potentials to identify hazards or risks.

The value of certain indicators has been disputed and consequently revised due to the physiological, environmental distribution, and persistence differences between many groups of pathogens and their assigned indicators (Ashbolt, Grabow, and Snozzi 2001). Most bacteria do not grow on artificial media, which is used to assess coliforms and HPCs. Prest et al. (2014) point out that <0.1 percent of drinking water bacteria can be cultured. Therefore, alternative methods such as direct vital staining and enumeration by flow cytometry are emerging, for example, in Switzerland, as the preferred method to measure disinfection efficacy (Hammes et al. 2012) and maintenance of biostability in distribution systems (Hammes et al. 2010).

The microorganisms most commonly used by distribution system water quality managers to gauge distribution system microbial conditions—total coliforms, fecal coliforms, *E. coli,* and HPC bacteria—are described here.

Total Coliforms

Regulators and drinking water system operators sample for total coliforms/*E. coli* in order to evaluate water quality conditions in the distribution system and to determine whether a pathway for drinking water contamination exists. In many countries, detection of total coliform requires follow-up investigation for *E. coli* to assess potential public health concerns associated with fecal contamination in the distribution system. While most coliform bacteria are nonpathogenic, the presence of *E. coli* indicates possible fecal contamination and therefore the possible presence of waterborne enteric pathogens.

Water-based pathogens such as *Legionella* and NTM that may be growing in the distribution system, primarily within biofilms and sediments (Donohue et al. 2015; Lu et al. 2015), are not indicated by coliforms (Ashbolt 2015a, 2015b). Therefore, general bacterial growth conditions, as measured by HPC and new indicators, are currently being investigated to clarify the presence of these CCL3 and CCL4 listed water-based pathogens.

Fecal Coliforms

Fecal coliform bacteria are a subgroup of coliforms, traditionally distinguished by their ability to grow at 44°C, referred to as "thermotolerant." Their thermotolerance led to their association with fecal contamination; however, this group of bacteria are also widely distributed in environmental waters, soil, and vegetation and therefore are no longer included in the USEPA's RTCR as fecal indicators.

Escherichia coli

Escherichia coli is a normal inhabitant of the digestive tract of humans and other warm-blooded animals but can persist, for varying periods of time, in terrestrial and aquatic environments (Ashbolt 2015a; Dirk van Elsas et al. 2010). It can be considered an indicator of fecal contamination and, potentially, the presence of human-infectious pathogens. In fact, USEPA-approved standard methods for *E. coli* detection do not specifically identify pathogenic *E. coli* (e.g., *E. coli* O157:H7). Specialized assays and methods are required for the identification of these groups of pathogens. Although the *E. coli* O157:H7 serotype is the most well known, the 2011 outbreak of *E. coli* O104:H4 in Europe highlights the importance of monitoring and controlling the range of pathogenic strains of *E. coli.*

Table 3-3 Microbial parameters and use as indicators

Microbial Parameter	Indicator of... Process Issue	Distribution System Integrity	Distribution System Operations and Maintenance	Pathogen/Fecal Contamination Presence	Comments
Total coliforms	x	x	x		• Used as an indicator by USEPA to evaluate treatment breakthrough and lack of disinfection efficacy and to determine whether a pathway for drinking water contamination or sanitary defect may exist (USEPA 2013) • Can colonize biofilms with no health association • Are nonspecific with respect to contaminant pathway
Fecal/thermotolerant coliforms	x	x		x	• No longer included as an indicator organism under the RTCR • More specific indicator of fecal contamination than total coliform but with easier methods now available to detect *E. coli*, which is a more specific fecal indicator; fecal/thermotolerant coliform no longer recommended
E. coli	x	x	x	x	• Indicator of recent fecal contamination of drinking water supply but not necessarily indicator of human pathogens (present from warm- and cold-blooded animals) that are more treatment resistant • Retained as fecal indicator in RTCR • Can survive/grow in biofilms (Juhna et al. 2007) • Sensitive to disinfection (USEPA 2006)
Enterococci and fecal streptococci	x	x	x	x	• Potentially more sensitive indicators of fecal contamination than *E. coli* due to higher environmental and treatment resistance (Schwartz, Hoffmann, and Obst 2003; Sorlini et al. 2013) • As an indicator of treatment effectiveness and distribution system integrity, enterococci and fecal streptococci are present in sewage at lower numbers than total coliforms (USEPA 2006); however, because they are more persistent, may still outnumber *E. coli* in aged fecal sources • As with *E. coli*, not as good a fecal indicator when pathogenic protozoa are present since *Giardia* cysts and *Cryptosporidium* oocysts are more resistant to environmental stress and disinfection than enterococci and fecal streptococci (USEPA 2006)

(continued)

Table 3-3 Microbial parameters and use as indicators ***(continued)***

Microbial Parameter	Indicator of... Process Issue	Distribution System Integrity	Distribution System Operations and Maintenance	Pathogen/Fecal Contamination Presence	Comments
Heterotrophs (measured as HPC)	x	x	x		• Significant increase above typical levels could indicate biofilm problems, increased nutrients in the distribution system, intrusion events, inadequate disinfectant residual • Not directly related to health risks or pathogen occurrence (Bartram et al. 2003) • Indicator for effectiveness of disinfection of bacteria (WHO 2011) • Indicator for effectiveness of cleanliness and integrity of distribution systems (WHO 2011)
Ammonia-oxidizing bacteria and nitrite-oxidizing bacteria	x		x		• Used as indicators of nitrification within the distribution system • Not commonly measured directly; instead, parameters such as loss of total chlorine, decrease in ammonium, and increases in nitrite-N and nitrate-N • Could be indicator of excess ammonium leaving the plant
Total bacteriological counts and total viable bacterial counts	x	x	x		• Provides basic information on bacterial presence in drinking water, amenable to rapid measurement versus HPC by using flow cytometry (Helmi et al. 2015) • Nonspecific and does not provide indication of fecal contamination or predictor of waterborne disease outbreaks (USEPA 2006)
Pseudomonas and *Aeromonas*			x	x	• Indicates presence of inadequate chlorine residual and/or bacterial growth or potential for biofilms • Neither are of direct health concern, although strains within certain species are opportunistic pathogens
Clostridium perfringens	x	x		x	• Indicator for effectiveness of disinfection and physical removal processes for protozoa and viruses (WHO 2011) • Spores are mainly of fecal origin (human, dog, cattle) and are more resistant to disinfection compared to *E. coli* and thus a better indicator of persistent organisms such as viruses and protozoa (Ashbolt, Grabow, and Snozzi 2001)

(continued)

Table 3-3 Microbial parameters and use as indicators *(continued)*

Microbial Parameter	Indicator of... Process Issue	Distribution System Integrity	Distribution System Operations and Maintenance	Pathogen/Fecal Contamination Presence	Comments
Coliphages	x	x	x	x	• Indicator for effectiveness of disinfection and physical removal processes for viruses (WHO 2011) • Based on the premise that these viruses will behave more like human enteric viruses than do bacterial indicators (USEPA 2006) • May be particularly effective for indicating contamination during main breaks and repairs (USEPA 2006)
Human-targeted *Bacteroides* (e.g. *HF183*, *HumM2*)		x		x	• Indicator of sewage contamination • Quantitative polymerase chain reaction assays targeting *HF183* provide greater sensitivity and *HumM2* greater specificity for sewage contamination of waters compared to direct assays for human enteric viruses (Boehm, Soller, and Shanks 2015)

Abbreviations: HPC, heterotrophic plate count; RTCR, revised Total Coliform Rule.

Escherichia coli has been retained as a fecal indicator with an MCL in the RTCR in order to trigger immediate follow-up actions (replacing the Total Coliform Rule's [TCR's] "acute MCL"). If the water has been exposed to potential *E. coli* contamination, the drinking water regulatory agency should be contacted immediately. It may be necessary to issue a boil-water advisory. This type of contamination can be due to a water main and a sewer line both broken in the same hole or to an animal in a storage tank.

Heterotrophs

Measurement of heterotrophic bacteria using the HPC method is one way to determine the microbiological quality and biostability of drinking water, as well as to test the efficacy of drinking water treatment processes (e.g., disinfection). The main factors for regrowth of heterotrophic microorganisms in distribution systems are temperature, availability of nutrients (carbon, nitrogen, phosphorus/energy sources), and lack of residual disinfectant (Wolfe et al. 1990; LeChevallier, Schulz, and Lee 1991; Norton and LeChevallier 2000; Escobar, Randall, and Taylor 2001; Niquette, Servais, and Savoir 2001; Zhang and DiGiano 2002; Lehtola, Miettinen, and Martikainen 2002; Chowdhury 2012).

Based on the carbon source organisms use for growth, they are classified as autotrophs, heterotrophs, or mixotrophs. Heterotrophs require organic carbon for growth and include bacteria and fungi (yeasts and molds). Allen, Edberg, and Reasoner (2004) point out that both indicator bacteria and opportunistic bacterial pathogens are heterotrophs. Although the heterotrophic group includes pathogens, there is no conclusive epidemiological evidence that HPC values alone are directly related to health risks or correlated with the occurrence of water-based pathogens (Bartram et al. 2003).

Only some heterotrophic species may be able to grow on HPC media. Selective and/or differential media are necessary to determine whether pathogenic bacteria or fungi are present. Thus, an HPC enumeration does not indicate pathogen presence, rather it provides an indication of loss of disinfection or biostability in drinking water distribution systems.

Hammes et al. (2010) measured chemical and biological parameters of an undisinfected Swiss drinking water supply over 18 months to study biological stability. The researchers found direct correlations between changes in total cell count (measured using flow cytometry), dissolved organic carbon, and AOC. However, changes in HPC (using a Swiss standard of incubation at 30°C over three days) did not correspond to the other markers of biostability. On average, the HPC method measured 0.0009 percent to 0.001 percent of microbial cells in comparison to those measured by total cell count. As such, this parameter is not suitable for public health target setting or as sole justification for issuing a boil-water advisory.

Elevated HPC levels occur in areas with less-than-optimal water quality, especially in stagnant parts of piped distribution systems. Those high levels may indicate inadequate treatment and maintenance of disinfectant residual that could cause problems with water-based opportunistic pathogens.

Chowdhury's (2012) review of research on drinking water HPCs notes that, with respect to enumeration, the spread plate method typically yields the highest results followed by membrane filtration and pour-plate methods. Selection of the appropriate method can depend on cost (pour-plate is less expensive) and analytical needs (Chowdhury 2012).

While the most common HPC method in the United States relies on standard plate count (SPC) agar, Reasoner's 2A (R2A) agar (with incubation at 20°C for seven days) provides more sensitivity (Chowdhury 2012; Allen, Edberg, and Reasoner 2004; Reasoner 2004). Figure 3-4 shows a comparison of HPC measurements from an undisinfected groundwater system using SPC agar versus R2A agar. The R2A results were generally two orders of magnitude greater than the SPC results. If only the SPC method had been used, this water utility might not have realized that they had a significant biofilm problem within their distribution system. The more sensitive R2A method revealed significant differences between monitoring locations and provided a better baseline for assessing the effectiveness of subsequent mitigation strategies.

Health Canada (2013b) guidance describes multiple HPC methods but does not specify one that must be used.

From an operational perspective, HPC results are useful for assessing trends in water quality and distribution system management, for example, biofilm mobilization or detachment, hydraulic regimes, mobilization of sediments, and inadequate disinfectant residual levels (Payment, Sartory, and Reasoner 2003; Reasoner 1990; Allen, Edberg, and Reasoner 2004). From a microbiological perspective, it may be important to identify the dominant bacterial groups present in the corresponding distribution and storage systems in response to sustained high levels of heterotrophs or significant regrowth events.

When assessing the significance of the HPC assay, it is important to consider that the commonly cited level of concern, that is, HPC >500 cfu/mL, has no link to a health-based limit. Such high levels may pose potential interference with analytical recovery by membrane-based methods that use lactose fermentation to assay coliform bacteria (Payment, Sartory, and Reasoner 2003; Allen Edberg, and Reasoner 2004). There is no indication that defined-substrate methods suffer from coliform suppression by high HPC counts (Allen Edberg, and Reasoner 2004).

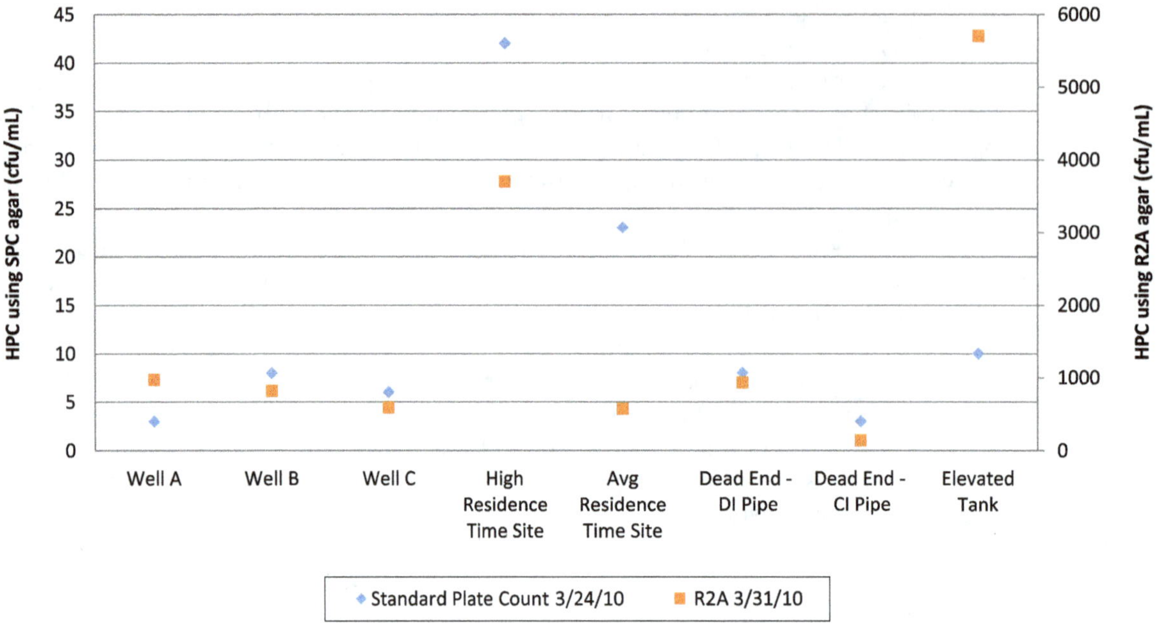

Source: Confluence Engineering Group, LLC project files. Reprinted with permission from Confluence Engineering Group, LLC.

Figure 3-4 Importance of the heterotrophic plate count method: standard plate count agar versus Reasoner's 2A agar

TECHNIQUES FOR CHARACTERIZING MICROBIAL COMMUNITIES

New methods are available that allow for more specific characterization of microorganisms, and the drinking water industry is developing procedures to take advantage of this new knowledge in order to respond to unusual circumstances. For example, in the future, systems with nitrification may be able to characterize the specific organism responsible and use new methods to identify a nitrification event before it becomes established. When faced with a microbial challenge in its distribution system, a utility may consider using a nonstandard analysis method because the parameter is not regulated, a more sensitive method is required, or the investigation requires a different type of result than offered by standard methods. In these cases, water systems should consider the following criteria:

- Amount of time required to obtain results,
- Reliability of results,
- Type of result (such as presence/absence, enumeration, indicative of presence of specified species versus overall microbial activity),
- Whether the analysis needs to be conducted by a specialized laboratory, and
- Cost.

Figure 3-5 summarizes the following detection, enumeration, community composition, and microbial activity methods as described by Douterelo et al. (2014):

- Detection–Enumeration: The majority of regulatory-based analyses conducted by drinking water systems fall into the category of detection–enumeration and are culture-based methods. Douterelo et al. point out that while culture-dependent methods are useful for detecting the presence of indicator organisms at a reasonable cost, they provide little information on the microbial community and do not capture the microbial cells that can be categorized as viable but not culturable, as measured by flow cytometry and other methods.

- Microbial community composition and characterization: These methods are used to detect specific microorganisms such as pathogens or those that cause corrosion. These techniques can be used to compare microbial communities spatially in the distribution system or to evaluate temporal changes that result from changes in system operations.
- Microbial activity and functional genes: These techniques can be used to assess the viability or sustainability of microbial communities as systems make changes to treatment processes or distribution system operations. One example of this type of technique, which is commonly used by drinking water systems, is measurement of ATP. The ATP bioluminescence assay is described in *Standard Methods for the Examination of Water and Wastewater* as a bioluminescence, rapid microbial detection test. Results obtained from ATP assays cannot be considered direct indicators of water quality (Liu et al. 2013). Instead, ATP should be used as a tool for drinking water treatment process control (e.g., ATP determination has been successfully used in the development, control, and maintenance of filters and biofilters) or for the rapid assessment of the general microbial stability of drinking water (Magic-Knezevand van der Kooij 2004; Velten et al. 2007; van der Wielen and van der Kooij 2010; Vital et al. 2012).

Appendix A provides basic information on characterization of microbial communities. This includes an overview of microbial detection methods, information on detection of coliform and heterotrophic bacteria, selected culture-independent methods, as well as information on investigating other microbial communities.

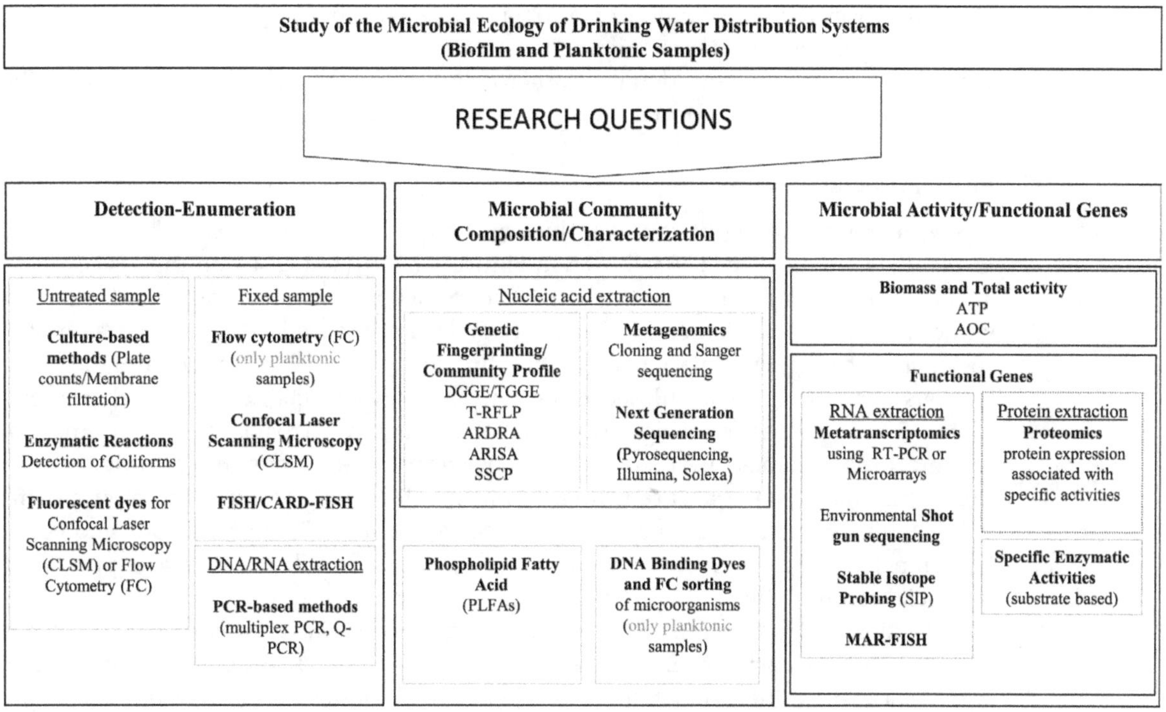

Source: Douterelo et al. 2014.

Figure 3-5 Scheme showing the different techniques available to characterize microbial communities in drinking water distribution systems

DSWQ Challenge:

Monitor Typical Indicators

- Total coliform
- *E. coli*
- HPC
- ATP
- Nitrification indicators

Determine Cause Based on Indicator

- Process issue
- Distribution system integrity
- Distribution system O&M
- Nitrification
- Pathogen/fecal contamination presence

Further Characterize Microbial Communities, if necessary

CONTRIBUTING FACTORS, MITIGATION, AND CORRECTIVE ACTIONS FOR MICROBIAL OCCURRENCE PROBLEMS

This section provides a brief overview of potential distribution system conditions and O&M practices that contribute to microbial issues within distribution systems. If applicable, specific mitigation strategies and corrective actions are discussed. Many of these topics are covered in more detail in other chapters, and the reader is directed to those chapters for additional information.

Pipe Materials

Piping material can significantly influence the quantity and types of bacteria present in the distribution system. As shown in Figure 3-6, unlined cast-iron pipe is considered "scale forming" and can develop very thick deposits of iron oxides, other metal and inorganic substrates, and biomass. Refer to Chapter 7 for a more in-depth discussion of corrosion and scaling. Friedman et al. (2010) estimated a median solids accumulation rate of 6.3 tons per mile of 8-in. unlined cast-iron pipe. Reiber (2008) estimated that approximately 5 to 10 percent by weight of a typical iron-based corrosion scale is comprised of organic carbon/biomass. Cement-lined and plastic pipes are "nonscale-forming," meaning that accumulated materials are primarily precipitated or adsorbed at the piping surface. While the mass of accumulated solids may be significantly lower for plastic and cement-lined pipe, a huge amount of surface area and nutrients are still available to support robust microbial communities.

Douterelo, Husband, and Boxall (2014) studies the bacterial composition of material flushed from cast-iron and polyethylene pipes that carried the same water and that were in near proximity. They found that the bacterial communities flushed from the pipes were significantly different. More material was mobilized from the cast-iron pipes, and the bacterial community included an abundance of bacteria that can use iron and manganese compounds, such as *Lysinibacillus* species, *Geobacillus* species, and *Magneto bacterium* species. The bacterial community collected from the polyethylene pipes was more

Source: Confluence Engineering Group, LLC project files. Reprinted with permission from Confluence Engineering Group, LLC.

Figure 3-6 Examples of scales formed on (a) unlined cast-iron, (b) cement-lined ductile-iron, and (c) plastic piping materials

diverse than that collected from the cast-iron pipes and included, as the most represented bacteria, *Spirochaeta* species, *Methylobacterium* species, *Clostridium* species, and *Desulfobacterium* species.

Other researchers have studied the effects of pipe materials on biofilm formation potential and microbial communities in biofilms using bench-scale techniques (e.g., coupons, annular reactors; Yu, Kim, and Lee 2010; Jang, Choi, and Ka 2011) or model distribution systems (Rożej et al. 2015). Yu and colleagues (2010) used pipe coupons to study biofilm formation potential and microbial communities established on copper, chlorinated PVC, polybutylene, polyethylene, stainless steel, and steel coated with zinc. The investigators found that pipe material affected biofilm formation potential (steel pipes have the highest potential, and copper pipes have the lowest potential) and microbial diversity. The researchers recommend use of plastic piping such as chlorinated PVC because of its low biofilm formation potential and small amount of microbial diversity in the biofilm.

Jang and colleagues (2011) also used coupons to investigate microbial growth associated with different piping materials. Over 15 months, the investigators measured growth on steel, copper, stainless steel, and PVC piping. They found that the HPC count of steel piping was about 100 times greater than on the other materials throughout the study and that the lowest growth was found on stainless steel piping. Jang and colleagues (2011) recommend stainless steel as the best material with respect to microbial growth, which is in conflict with the findings of Yu, Kim, and Lee 2010.

Rożej et al. (2015) used three model distribution systems to study microbial communities present on the piping materials after two years of operation. Each model was composed of one of three materials: unplasticized PVC, silane cross-linked polyethylene, or high-density polyethylene. The investigators found that biofilm growth was most extensive on high-density polyethylene piping and least extensive on chlorinated PVC pipes. The greatest number of cells was quantified in the silane cross-linked polyethylene system. The researchers noted that the microbial community present in the high-density polyethylene system included *Pseudomonas aeruginosa* and *Acinetobacter* species, which can be opportunistic pathogens to humans. However, total biomass is one issue for control, another is the type of microbiota that may support more opportunistic pathogens and form under different distribution conditions. The type of secondary disinfectant affects formation of nontuberculous mycobacteria and antibiotic resistant bacteria (Gomez-Alvarez et al. 2016). Premise plumbing material, such as copper pipe or PVC may affect occurrence and growth of *Legionella pneumophila*. (Buse et al. 2014).

With respect to biofilm removal from piping materials, strategies such as increasing disinfectant dose or changing disinfectant type are less effective on established biofilms. Systems may need to rely on methods that dissolve, shear, or otherwise dislodge biofilm from the pipe surface. For smooth piping and pipe diameters of 12 in. or less, unidirectional flushing at high velocities (>6 fps) can be effective. However, often more aggressive techniques such as ice pigging or swabbing are needed for more effective biofilm removal (Friedman et al. 2016). A summary of useful applications for flushing water distribution system mains is provided in Appendix B. Additional water quality considerations and mitigation strategies associated with aging infrastructure are discussed in Chapter 4.

Storage Facilities

There are external and internal mechanisms for microbial contamination of storage facilities. Animals, most commonly midge flies, birds, squirrels, raccoons, and possums, can enter storage facilities directly through open hatches, damaged or missing screens on vents, or overflow piping, bringing with them microbial contaminants. It is critical to regularly inspect tanks to make sure that all screens are intact and that the hatch and anti-climb device are securely locked. Sediment should be removed periodically.

Microbes can continue to multiply within storage facilities if excess water age causes loss of disinfectant residual and temperature increases. Microbes can also develop biofilms on storage facility walls and within accumulated sediments.

With respect to water quality problems associated with microorganisms, improved operational practices such as increased turnover and mixing in storage facilities can help reduce water age, increase disinfectant residual levels, and reduce sediment accumulation. In cases where improved turnover and mixing cannot be achieved or if the water quality is degraded before the water enters the tank, booster disinfection should be considered. This type of disinfection can be temporary or permanent. Impacts on DBP formation and other water quality conditions must also be considered. Chapter 2 provides recommendations for water age management in storage facilities.

Nutrient Loading

Although sometimes overlooked as a cause of total coliform presence, nutrient loading to the distribution system can contribute significantly to persistent biofilm growth, loss of disinfectant residual, and coliform occurrences. Limiting nutrient levels in the treated water is critical to providing biologically stable drinking water. As described earlier, organic nutrients are measured using AOC and BDOC. Select microorganisms can also use inorganic nutrients such as ammonia, iron, manganese, sulfur, and other chemicals. Nutrient levels affect the biostability of finished water, and the different processes used to treat source waters can greatly impact the water's biostability. Microbial growth associated with the presence of nutrients in drinking water should be addressed at the treatment plant, before the water enters the distribution system.

Secondary Disinfectant Residual Type

Disinfectant type can affect the quantity and type of microbes present in the distribution system. LeChevallier and colleagues (1996) observed significantly greater densities of coliform in positive samples collected from chlorinated systems than those collected from chloraminated systems. Secondary disinfectant type can also affect which microorganisms are present in the distribution system (Hwang et al. 2012).

Studies of switching from use of free chlorine to monochloramine as a secondary disinfectant have revealed the ecology in the distribution system and at the tap. A switch from free chlorine to chloramines led to a significantly reduced presence of *Legionella* in the distribution system (Pryor et al. 2004; Moore et al. 2006) and at the point of use

(Pryor et al. 2004; Weintraub et al. 2008). However, a water system in Florida experienced an increase in *Mycobacteria* after switching to chloramines (Pryor et al. 2004), which is consistent with pilot studies by Revetta et al. (2013) and Wang et al. (2014).

Finally, the type of disinfectant can affect inactivation of pathogens in the distribution system. Yang et al. (2011), using quantitative microbial risk analysis, evaluated the risks associated with intrusion of norovirus into a distribution system and compared the effectiveness of selected mitigation practices. The level of monochloramines present, 0.5 mg/L monochloramine, did not appear to reduce the risk of pathogen exposure, while free chlorine ≥0.2 mg/L reduced exposure. Management of disinfectant residuals is discussed in more detail in Chapter 8.

Inadequate Disinfectant Residual

As noted earlier, low to no disinfectant residual in the distribution system can contribute to bacterial growth. LeChevallier, Welch, and Smith (1996) found that low to no residual (<0.2 mg/L for free chlorine and <0.5 mg/L for monochloramine) in distribution system dead ends corresponded to more positive coliform samples. The following recommendations have been made with respect to secondary disinfectant residual maintenance in the distribution system:

- Friedman et al. (2010) recommend minimum levels of 0.2 mg/L for free chlorine and 0.5 mg/L for total chlorine in 95 percent of samples collected from locations including TCR and other recommended sites. TCR sites are typically selected to capture "representative" water quality conditions, not necessarily problematic sites such as downstream of storage facilities, blending zones, or areas with known water age problems. These types of sites should be included in a secondary disinfection evaluation.
- LeChevallier, Welch, and Smith (1996) recommend maintaining at least 0.2 mg/L free chlorine or 0.5 mg/L chloramines at far ends of the distribution system. The authors cite chlorine threshold criteria for increased coliform occurrence as <0.5 mg/L free chlorine and <1.0 mg/L chloramines.

Many factors influence the concentration of disinfectant residual in the distribution system, including:

- Bulk water chlorine demand and decay reactions,
- Water age and hydraulic conditions,
- Accumulated sediments, including in storage tanks,
- Blending of sources,
- Nitrification, and
- Reactions with pipe walls.

Water systems that experience difficulty with disinfectant residual levels in the distribution system can make the following evaluations:

- Consider increased dosage for the overall system.
- Assess the need for boosting within specific areas.
- Consider water age management techniques such as
    - Reduce demand,
    - Implement or optimize flushing program, and
    - Enhance storage turnover and mixing.

- Control blending of sources.
- Assess and control nitrification.

Additional water quality issues associated with water age are discussed in more detail in Chapter 2. Management of disinfectant residuals is discussed in more detail in Chapter 8. and Chapter 6 addresses nitrification.

Pressure Management

Temporary or sustained loss of pressure can allow contaminants to enter the distribution system through breaches in infrastructure or other pathways (Besner, Prevost, and Regli 2011; Teunis et al. 2010; Friedman et al. 2004). Contaminants can be drawn into the distribution system through leaks, submerged air-vacuum valves, or other infrastructure breaches if the pressure within the main drops significantly. Contamination under these circumstances is referred to as "intrusion." Additionally, pressure losses can be accompanied by or caused by sudden changes in velocity, which can scour biofilm and sediments, potentially releasing coliform bacteria into the bulk water stream.

Below-grade/submerged air-vacuum valves can serve as pathways for contamination during low-pressure events. Contaminated water can be drawn into the distribution system when the valve opens in response to the low-pressure condition. Air-vacuum vaults in the same pressure zone or within the vicinity of coliform occurrence should be inspected for standing water, and the valves should be designed so that they are vented and screened above grade.

As recommended by Friedman et al. (2004), water systems can evaluate potential impacts of pressure loss on coliform occurrence by doing the following:

- Reviewing operational best practices including control valve and hydrant operation speeds, pump shut-down speeds, and pump shut-down sequencing; avoid changing the demand rapidly, which can result in low or negative pressures. See Chapter 2 for more information.
- Implementing engineering strategies to minimize or reduce the impacts of pressure events when they occur. Strategies include installation of control devices such as surge tanks or valves, looping dead ends, providing overhead storage, modeling the system to identify areas vulnerable to pressure events, and optimizing steady-state pressures.
- Using maintenance practices to minimize the likelihood of contamination when pressure events occur. These include improved leak detection programs, main repair strategies, increased inspections of vaults, and enhanced cross-connection inspections in areas determined to be susceptible to pressure events.

Additional water quality considerations associated with pressure management and control strategies are discussed in Chapter 9.

Cross-Connection Control

Implementation of a cross-connection control and backflow prevention (CCCBFP) program, including installation of backflow-prevention assemblies and devices, can prevent the flow of nonpotable substances into the distribution system. Direct contamination of the distribution system can occur from back-siphonage or backflow from pressurized service connections. Contamination through unprotected cross-connections should always be investigated if coliform samples are positive for *E. coli* and/or if an acute MCL violation occurs. Satisfactory test results from all known high-hazard sites should be verified, and additional surveys and inspections may be appropriate in the region of the contamination event. Components and recommended elements of a CCCBFP program are discussed in more detail in Chapter 10.

Main Repair and Installation

Main breaks and repairs can contribute to contamination of the distribution system due to the loss of pressure that accompanies a main break, as well as unsanitary conditions that can cause contamination if good repair practices are not followed. The risk of contamination caused by main breaks can increase as the degree and duration of depressurization increases.

Water quality issues associated with aging infrastructure are discussed further in Chapter 4.

AWWA Standard C651 (2014) should be followed throughout main repairs. In the case of sewage contamination, special precautions must be followed, including super-chlorination of the repair trench, super-chlorination of the main, and potentially issuing a boil-water advisory.

Responding to the DSWQ Challenge

- Flush, rehabilitate, and/or replace unlined cast-iron mains if chlorine cannot be maintained and biofilm problems are evident. Avoid flushing practices that stir up sediment and biofilm without actually purging from the distribution system.
- Make the following operational adjustments at storage facilities:
    - Increase turnover and mixing to help reduce water age.
    - Increase disinfectant residual levels. Consider temporary or permanent booster disinfection of storage facilities. Monitor any DBP formation as a result.
    - Remove sediment.
- Maintain minimum disinfectant residual levels of 0.2 mg/L for free chlorine and 0.5 mg/L for total chlorine at TCR sites and far ends of the distribution system. This is done for the following reasons:
    - For microbial control,
    - To maintain pipe scale stability, and
    - To reduce the potential for metals/biofilm release.
- Remediate potential impacts of pressure loss (see Chapter 9) by:
    - Controlling valve/hydrant operation speeds, pump shut-down speeds, and pump shut-down sequencing;
    - Using leak detection equipment to find and repair leaking mains; and
    - Conducting inspections of valve vaults and potential cross-connections.

CASE STUDY: REGROWTH

This case study describes a water system that experienced and resolved a nonacute coliform violation. It also summarizes the strategies that useful for understanding the factors that contributed to the event and the effectiveness of mitigation approaches.

System Description

This large, western water system serves more than 950,000 retail and wholesale customers in a geographic area of approximately 225 square miles. The majority of the water supply is unfiltered surface water, with additional supply from groundwater during emergencies and to supplement during summer peak season demand. The water system relies on free chlorine to provide primary disinfection and adds ammonia to produce chloramines for secondary disinfection.

Event Description

In mid-September 2013, the system experienced a nonacute TCR violation, with TC-positive samples exceeding 5 percent of the total monthly TCR samples. One initial positive routine TCR sample cascaded into more positive coliform samples as the system collected repeat samples nearby. No samples were positive for *E. coli* during this event. All but one of the positive samples were collected at or near water quality sampling station (WQSS) A, which had been in use for less than one year. Forty of 42 positive coliform samples (95 percent) were collected from areas served by unlined cast-iron pipe. The state regulator issued a Tier 2 violation within two weeks of the initial positive sample.

In late October, monitoring results indicated that corrective actions were effective and that the system could resume TCR compliance monitoring in this area. During the event, the system was able to continue TCR compliance monitoring throughout the unaffected areas of the distribution system.

Evaluation of Causes

The water system worked with a consultant during the following year to evaluate the system's response efforts and, if possible, to identify potential causes of the coliform occurrence. The evaluation of potential causes of the TCR event included consideration of the contaminant pathways (described previously in the Microbial Occurrence Pathways section of this chapter), and the following determinations were made:

- Given the bulk water and pipe wall conditions that were related to the TCR event, the event was likely primarily due to regrowth and release of coliform within pipe biofilms.
- An upstream nitrification event may have caused the initial loss of chlorine residual in the affected area, which took months to reestablish and allowed biofilms to flourish when temperatures were warmer. Dissolved oxygen and pH also decreased during nitrification. The combined loss of residual and dissolved oxygen and pH decreases likely caused a significant change in oxidation–reduction potential, which could have impacted the stability of the accumulated cast-iron scale–biofilm complex. Destabilized scales are more susceptible to hydraulic/physical release during changes in flow direction and velocity.

Effectiveness of the Response

The evaluation of the system's response focused primarily on the following areas:

- Tank operations: After observing nitrification in a finished water reservoir, the system responded by emptying the tank. This and other mitigation activities resulted in rapidly improved water quality in the tank. However, a slug, or large volume, of nitrified water may have been pumped from the tank into downstream portions of the distribution system, possibly distributing this water along with lower-quality water throughout the system.

- Spot flushing as a response to biofilm/regrowth TCR event: Data suggest that spot flushing (without valve isolation) conducted during the TCR event provided very short-term improvement at many locations and barely any improvement at others. It is possible that this activity caused the following:
  - Biofilm/sediment that contained coliform bacteria to be stirred up and released into the distribution system, creating more widespread positives.
  - Poor-quality water from the tank that was experiencing nitrification was further spread around the immediate vicinity.
- Unidirectional flushing: The system implemented a targeted unidirectional flushing (UDF) program in response to the TC-positive samples after spot flushing appeared to be an unsuccessful mitigation strategy. UDF was an appropriate response to the TCR event. Data collected during implementation demonstrated that UDF rapidly removed significant quantities of turbidity from the system.
- System-wide chloramine boosting: In early October, the water system increased the chloramine residual target in the water leaving the water treatment plant from 1.8 mg/L to 3.0 mg/L (as total chlorine). While increases in residual were observed in the distribution system, significantly improved residuals were not measured within the affected area for nearly two weeks after chloramine boosting. Given how difficult it was to increase chlorine residual in the affected pressure zone, the system cannot expect dose increases at the water treatment plant to be a quick reactive response to coliform problems; however, total chlorine target increases should be used as a preventive nitrification and biofilm control strategy. Now, the total chlorine target is increased in late summer so that increased residuals are present system-wide prior to the onset of this system's nitrification season.

2014 Status

System managers indicate that nitrification and regrowth are ongoing issues for the distribution system, but they are trying to be very proactive with both monitoring and mitigations (distribution system optimization) to keep nitrification levels low. The system's current disinfection goal is 2.2 mg/L as total chlorine leaving the water treatment plant, with an increased goal during summer months. During 2014, the system conducted targeted UDF in select pressure zones where water quality had been problematic in 2013. Staff also conduct spot flushing on an as-needed basis.

The system experienced improved water quality during the 2014 nitrification season. While the system did observe positive coliform samples within the distribution system during 2014, none occurred in the area affected by the 2013 event.

BEST PRACTICES TO MITIGATE AND MANAGE MICROBIAL GROWTH

Table 3-4 presents the important elements of a distribution system water quality program, focusing on controlling microbial growth categorized by level of importance. Basic practices should be included in such a program because they are foundational best practices for microbial control. Advanced elements should be considered by water systems that focus on specific microbial problems or are being proactive in this area. Because advanced elements are used to focus on specific microbial problems and/or chronic issues, they require greater resources and more significant effort.

Table 3-4 Best practices to control microbial growth

Best Practice	Element Importance
Main cleaning	
• Understand the distribution of pipe materials and pipe conditions	Basic
• Proactively clean water mains using UDF	Basic
• Implement more aggressive cleaning techniques (e.g., ice pigging, swabbing) if UDF unsuccessful for microbial control on plastic or cement-lined pipes or implement swabbing	Advanced
• Replace or line unlined cast-iron mains if residual maintenance cannot be achieved through other means	Advanced
• Avoid spreading poor-quality water and stirring up sediments when conducting conventional flushing	Basic
Storage facilities	
• Develop standard operating procedures to increase turnover and mixing, increase disinfectant residual levels, and reduce sediment accumulation	Basic
• Properly maintain screens on all tank openings	Basic
• Optimize mixing in storage tanks by using predictive models and/or adding physical mixing equipment	Advanced
• Consider booster disinfection of storage facilities, evaluating impacts on disinfection by-product formation; other water quality conditions must also be considered	Advanced (see Chapter 8)
Secondary disinfectant	
• Assess need for increased dosage	Basic
• Assess need and identify suitable locations for residual boosting within specific areas of the distribution system	Basic/Advanced
• Evaluate efficacy of different disinfectants to achieve case-by-case microbial management needs	Advanced (see Chapter 8)
Maintain distribution residuals	
• Understand and control water age	Basic (see Chapter 2)
• Maintain minimum levels of 0.2 mg/L for free chlorine and 0.5 mg/L for total chlorine in 95 percent of samples collected from Total Coliform Rule sites	Basic
• Maintain at least 0.2 mg/L free chlorine and 0.5 mg/L for total chlorine at far ends of the distribution system	Advanced
• Understand and control blending of sources	Basic
• Understand and control nitrification	Basic (see Chapter 6)
• Implement or optimize main cleaning program	Basic/Advanced (see Chapter 2)
Pressure management (see Chapter 9)	
• Perform operational practices to control flow direction and velocity	Basic/Advanced
• Develop engineering strategies to mitigate the propagation and magnitude of pressure events	Basic/Advanced
• Perform maintenance practices to reduce contamination potential during a pressure event	Basic/Advanced

Abbreviation: UDF, unidirectional flushing.

REFERENCES

Allen, M.J.; S.C. Edberg; & D.J. Reasoner. 2004. Heterotrophic Plate Count Bacteria—What Is Their Significance in Drinking Water? *International Journal of Food Microbiology*, 92: 265–274.

Ashbolt, N.J. 2015a. Microbial Contamination of Drinking Water and Human Health from Community Water Systems. *Current Environmental Health Reports*, 2:1:95–106.

———. 2015b. Environmental (Saprozoic) Pathogens of Engineered Water Systems: Understanding Their Ecology for Risk Assessment and Management. *Pathogens*, 4:2:390–405.

Ashbolt, N.J.; W.O.K. Grabow; & M. Snozzi. 2001. Indicators of Microbial Water Quality. In *Water Quality: Guidelines, Standards and Health*. London: IWA Publishing.

AWWA. 2013. M56. *Nitrification and Control in Drinking Water.* Denver, CO: AWWA.

Bartram, J.; J. Cotruvo; M. Exner; C. Fricker; & A. Glasmacher. 2003. *Heterotrophic Plate Counts and Drinking-Water Safety: The Significance of HPCs for Water Quality and Human Health*. London: IWA Publishing.

Bartrand, T.A.; J.J. Causey; & J.L. Clancy. 2014. *Naegleria fowleri*: An Emerging Drinking Water Pathogen. *Journal of the American Water Works Association*, 106:10:E418–E432.

Bautista-de Los Santos, Q.M.; J.L.Schroeder; O. Blakemore; J. Moses; M. Haffey; W. Sloan; & A.J. Pinto. 2016. The Impact of Sampling, PCR, and Sequencing Replication on Discerning Changes in Drinking Water Bacterial Community Over Diurnal Time-Scales. *Water Research*, 90:216–224.

Beer, K.D.; J.W. Gargano; V.A. Roberts; V.R. Hill; L.E. Garrison; P.K. Kutty; E.D. Hilborn; T.J. Wade; K.E. Fullerton; & J.S. Yoder. 2015. Surveillance for Waterborne Disease Outbreaks Associated with Drinking Water—United States, 201–2012. *Morbidity and Mortality Weekly Report*, 64:31:842–848.

Besner, M.C.; M. Prévost; & S. Regli. 2011. Assessing the Public Health Risk of Microbial Intrusion Events in Distribution Systems: Conceptual Model, Available Data, and Challenges. *Water Research*, 45:3:961–979.

Boe-Hansen, R.; H.J. Albrechtsen; E. Arvin; & C. Jorgensen. 2002. Bulk Water Phase and Biofilm Growth in Drinking Water at Low Nutrient Conditions. *Water Research*, 36:18:4477–4486.

Boehm, A.B.; J.A. Soller; & O.C. Shanks. 2015. Human-Associated Fecal Quantitative Polymerase Chain Reaction Measurements and Simulated Risk of Gastrointestinal Illness in Recreational Waters Contaminated With Raw Sewage. *Environmental Science & Technology Letters*, 2:10:270–275.

Borchardt, M.A.; S.K. Spencer; B.A. Kieke; E. Lambertini; & F.J. Loge. 2012. Viruses in Nondisinfected Drinking Water From Municipal Wells and Community Incidence of Acute Gastrointestinal Illness. *Environmental Health Perspectives*, 120:9:1272–1279.

Boulay, N. & M. Edwards. 2011. Role of Temperature, Chlorine, and Organic Matter in Copper Corrosion By-product Release in Soft Water. *Water Research,* 35:3:683–690.

Buse, H.Y.; J. Lu; I.T. Struewing; & N.J. Ashbolt. 2014. Preferential Colonization and Release of *Legionella pneumophila* From Mature Drinking Water Biofilms Grown on Copper Versus Unplasticized Polyvinylchloride Coupons. *International Journal of Hygiene and Environmental Health,* 217:1:219–225.

Cabral, J. 2010. Water Microbiology. Bacterial Pathogens and Water. *International Journal of Environmental Research and Public Health,* 7:3657–3703.

CDC (Centers for Disease Control and Prevention). 2013. Surveillance for Waterborne Disease Outbreaks Associated With Drinking Water and Other Nonrecreational Water—United States, 2009–2010 (including editorial note). *Morbidity and Mortality Weekly Report,* 62:35:714–720.

———. 2015a. Recommendations for Users of *Naegleria fowleri*-Contaminated Drinking Water Systems. (accessed May 31, 2016). www.cdc.gov/parasites/naegleria/public-water-systems.html

———. 2015b. Surveillance Reports for Drinking Water-Associated Disease Outbreaks: 2011–2012; supplemental figures: deficiencies assigned to drinking water outbreaks, 2011–2012. Excel spreadsheet. http://www.cdc.gov/healthywater/surveillance/drinking/2011-2012-figures.html (accessed Jan. 26, 2015).

Characklis, W.G. & K.C. Marshall. 1990. Biofilms: A Basis for an Interdisciplinary Approach. In *Biofilms*. New York: John Wiley & Sons; p. 3–15.

Chowdhury, S. 2012. Heterotrophic Bacteria in Drinking Water Distribution: A Review. *Environmental Monitoring and Assessment,* 184:10:6087–6137.

Coetser, S.E. & T.E. Cloete. 2005. Biofouling and Biocorrosion in Industrial Water Systems. *Critical Reviews in Microbiology,* 31:4:213–232.

Collier, S.A., L.J. Stockman, L.A. Hicks, L.E. Garrison, F.J. Zhou; & M.J. Beach. 2012. Direct Healthcare Costs of Selected Diseases Primarily or Partially Transmitted by Water. *Epidemiology and Infection,* 140:11:2003–2013.

Costerton, J.W.: K.J. Cheng: G.G. Geesey: T.I. Ladd: J.C. Nickel: M. Dasgupta; & T.J. Marrie. 1987. Bacterial Biofilms in Nature and Disease. *Annual Review of Microbiology,* 41:435–464.

Craun, G.F.: J.M. Brunkard: J.S. Yoder: V.A. Roberts: J. Carpenter; T. Wade; R.L. Calderon; J.M. Roberts; M.J. Beach; & S.L. Roy. 2010. Causes of Outbreaks Associated With Drinking Water in the United States From 1971 to 2006. *Clinical Microbiology Reviews,* 23:3:507–528.

Culotti, A. & A. Packman. 2014. *Pseudomonas aeruginosa* Promotes *Escherichia coli* Biofilm Formation in Nutrient-Limited Medium. PLOS ONE, 9:9. http://journals.plos.org/plosone/article?id=10.1371/journal.pone.0107186 (accessed May 20, 2015).

Dirk van Elsas, Jan; Semenov, Alexander V; Costa, Rodrigo; Trevors, Jack T. 2010. Survival of Escherichia Coli in the Environment: Fundamental and Public Health Aspects. *The ISME Journal.* Published online 2010 Jun 24. doi: 10.1038/ismej.2010.80. Corrected ISME J. 2011 Feb; 5(2): 173–183. https://www.ncbi.nlm.nih.gov/pmc/articles/PMC3105702/ (Accessed May 23, 2017).

Donohue, M.J.; J.H. Mistry; J.M. Donohue; K. O'Connell; D. King,; J. Byran; T. Covert; & S. Pfaller. 2015. Increased Frequency of Nontuberculous Mycobacteria Detection at Potable Water Taps within the United States. *Environmental Science & Technology,* 49:10:6127–6133.

Dornas FP, Silva LC, de Almeida GM, Campos RK, Boratto PV, Franco-Luiz AP, et al. *Acanthamoeba polyphaga* mimivirus stability in environmental and clinical substrates: implications for virus detection and isolation. PLoS One. 2014;9(2):e87811, as cited in Ashbolt, N.J. 2015a.

Douterelo, I.; J.B. Boxall; P. Deines; R. Sekar; K.E. Fish; & C.A. Biggs. 2014a. Methodological Approaches for Studying the Microbial Ecology of Drinking Water Distribution Systems. *Water Research,* 65:2015:134–156.

Douterelo, I.; S. Husband; & J.B. Boxall. 2014b. The Bacteriological Composition of Biomass Recovered by Flushing an Operational Drinking Water Distribution System. *Water Research,* 54:100–114.

Escobar, I.C.; A.A. Randall; & J.S. Taylor. 2001. Bacterial Growth in Distribution Systems: Effect of Assimilable Organic Carbon and Biodegradable Dissolved Organic Carbon. *Environmental Science & Technology,* 35:17:3442–3447.

Falkinham III, J.O. 2011. Nontuberculous Mycobacteria From Household Plumbing of Patients With Nontuberculous Mycobacteria Disease. *Emerging Infectious Diseases,* 17:3:419–424.

Flemming, H.-C. 2011. Microbial Biofouling–Unsolved Problems, Insufficient Approaches and Possible Solutions. In *Biofilm Perspectives.* Heidelberg, NY: Springer International.

Flemming, H. C., S.L. Percival; & J.T Walker. 2002. Contamination Potential of Biofilms in Water Distribution Systems. *Water Supply,* 2:1:271–280.

Flemming, H.-C.; & J. Wingender. 2010. The Biofilm Matrix. *Nature Reviews Microbiology,* 8:623–633.

Friedman, M.; A. Hill; S. Booth; M. Hallett; L. McNeill; J. McLean; D. Stevens; D. Sorensen; T. Hammer; W. Kent; M. De Haan; K. MacArthur; & K. Mitchell. 2016. *Metals Accumulation and Release within the Distribution System: Evaluation and Mitigation.* Denver, CO: Water Research Foundation.

Friedman, M.; M. LeChevallier; J. Rosen; G. Gagnon; M.C. Besner; L. Truelstrup-Hansen; T. Hargy; A. Hanson; K. Dewis; & G. Kirmeyer. 2009. *Strategies for Managing Total Coliform and* E. coli *in Distribution Systems.* Denver, CO: AwwaRF and AWWA.

Friedman, M.; M. LeChevallier; S. Seidl; J. Routt; G. Kirmeyer; & J. LeMieux. 2010. *Criteria for Optimized Distribution Systems.* Denver, CO: Water Research Foundation.

Friedman, M.; L. Radder; S. Harrison; D. Howie; M. Britton; G. Boyd; H. Want; R. Gullick; M. LeChevallier; D. Wood; & J. Funk. 2004. *Verification and Control of Pressure Transients and Intrusion in Distribution Systems.* Denver, CO: AwwaRF.

Gomez-Alvarez, V.; B.W. Humrighouse; R.P. Revetta; & J.W. Santo Domingo. 2015. Bacterial Composition in a Metropolitan Drinking Water Distribution System Utilizing Different Source Waters. *Journal of Water and Health,* 13:1:140–151.

Gomez-Alvarez, V.; S. Pfaller; J.G. Pressman; D.G. Wahman; & R.P. Revetta. 2016. Resilience of Microbial Communities in a Simulated Drinking Water Distribution System Subjected to Disturbances: Role of Conditionally Rare Taxa and Potential Implications for Antibiotic-Resistant Bacteria. *Environmental Science: Water Research & Technology,* 2:4:645–657.

Gourabathini P, Brandl MT, Redding KS, Gunderson JH, Berk SG. Interactions between food-borne pathogens and protozoa isolated from lettuce and spinach. Appl Environ Microbiol. 2008;74(8):2518–25. doi: 10.1128/AEM.02709-07, as cited in Ashbolt, NJ 2015a.

Hammes, F.; C. Berger; O. Köster; & T. Egli. 2010. Assessing Biological Stability of Drinking Water Without Disinfectant Residuals in a Full-Scale Water Supply System. *Journal of Water Supply: Research and Technology—AQUA,* 59:1:31–40.

Hammes, F.; T. Broger; H.U. Weilenmann; M. Vital; J. Helbing; U. Bosshart; P. Huber; R.P. Odermatt; & B. Sonnleithner. 2012. Development and Laboratory-Scale Testing of a Fully Automated Online Flow Cytometer for Drinking Water Analysis. *Cytometry Part A,* 81:6:508–16.

Health Canada. 2013a. *Guidance on the Use of the Microbiological Drinking Water Quality Guidelines.* Ottawa, Ontario: Water and Air Quality Bureau, Healthy Environments and Consumer Safety Branch, Health Canada, Ottawa, Ontario (catalogue no. H144-12/2013E-PDF).

———. 2013b. *Guidance on the Use of Heterotrophic Plate Counts in Canadian Drinking Water Supplies.* (accessed June 1, 2016).

Helmi, K.; F. Barthod; G. Meheut; A. Henry; F. Poty; F. Laurent; & N. Charni-Ben-Tabassi. 2015. Methods for Microbiological Quality Assessment in Drinking Water: A Comparative Study. *Journal of Water and Health,* 13:1:34–41.

Hwang, C.; F. Ling; G. L. Andersen,; M. W. LeChevallier; & W. T. Liu. 2012. Microbial Community Dynamics of an Urban Drinking Water Distribution System Subjected to Phases of Chloramination and Chlorination Treatments. *Applied and Environmental Microbiology,* 78:22:7856–7865.

Jang, H.J.; Y.J. Choi; & J.O. Ka. 2011. Effects of Diverse Water Pipe Materials on Bacterial Communities and Water Quality in the Annular Reactor. *Journal of Microbiology and Biotechnology,* 21:2:115–123.

Juhna, T.; D. Birzniece; S. Larsson; D. Zulenkovs; A. Sharipo; N.F. Azevedo; F. Menard-Szczebara; S. Castagnet; C. Feliers; & C.W. Keevil. 2007. Detection of *Escherichia coli* in

Biofilms From Pipe Samples and Coupons in Drinking Water Distribution Networks. *Applied and Environmental Microbiology,* 73:22:7456–7464.

LeChevallier, M.W.; W. Schulz; & R.G. Lee. 1991. Bacterial Nutrients in Drinking Water. *Applied and Environmental Microbiology,* 57:3:857–862.

LeChevallier, M.W., N.J. Welch; & D.B. Smith. 1996. Full-Scale Studies of Factors Related to Coliform Regrowth in Drinking Water. *Applied and Environmental Microbiology,* 62:7):2201–2211.

Lehtola, M.J.; I.T. Miettinen; & P.J. Martikainen. 2002. Biofilm Formation in Drinking Water Affected by Low Concentrations of Phosphorus. *Canadian Journal of Microbiology,* 48:494–499.

Li, W.; F. Wang; J. Zhang; Y. Qiao; C. Xu; Y. Liu; L. Qian; W. Li; & B. Dong. 2016. Community Shift of Biofilms Developed in a Full-Scale Drinking Water Distribution System Switching From Different Water Sources. *Science of the Total Environment,* 544:499–506.

Ling, F.; Hwang, C.; LeChevallier, M.W.; Andersen, G.L.; & Liu, W.T. 2015. Core-Satellite Populations and Seasonality of Water Meter Biofilms in a Metropolitan Drinking Water Distribution System. *ISME Journal,* 10(3): 582-595.

Liu, G.; G.L. Bakker; S. Li; J.H.G. Vreeburg; J.Q.J.C. Verberk; G.J. Medema; & J.C. Van Dijk. 2014. Pyrosequencing Reveals Bacterial Communities in Unchlorinated Drinking Water Distribution System: An Integral Study of Bulk Water, Suspended Solids, Loose Deposits, and Pipe Wall Biofilm. *Environmental Science & Technology,* 48:5467–5476.

Liu, G.; E.J. Van der Mark; J.Q.J.C. Verberk; & J.C. Van Dijk. 2013. Flow Cytometry Total Cell Counts: A Field Study Assessing Microbiological Water Quality and Growth in Unchlorinated Drinking Water Distribution Systems. BioMed Research International. http://dx.doi.org/10.1155/2013/595872 (accessed Sept. 21, 2016).

Liu, W.; H. Wu; Z. Wang; S.L. Ong; J.Y. Hu; & W.J. Ng. 2002. Investigation of Assimilable Organic Carbon (AOC) and Bacterial Regrowth in Drinking Water Distribution System. *Water Research,* 36:891–898.

Lu, J.; I. Struewing; S. Yelton; & N.J. Ashbolt. 2015. Molecular Survey of Occurrence and Quantity of *Legionella* spp., *Mycobacterium* spp., *Pseudomonas aeruginosa* and Amoeba Hosts in Municipal Drinking Water Storage Tank Sediments. *Journal of Applied Microbiology,* 119:1: 278–288.

Magic-Knezev, A. & D. van der Kooij. 2004. Optimisation and Significance of ATP Analysis for Measuring Active Biomass in Granular Activated Carbon Filters Used in Water Treatment. *Water Research,* 38:18:3971–3979.

Miller, H.C.; J. Wylie; G. Dejean; A.H. Kaksonen; D. Sutton; K. Braun; & G.J. Puzon. 2015. Reduced Efficiency of Chlorine Disinfection of *Naegleria fowleri* in a Drinking Water Distribution Biofilm. *Environmental Science & Technology,* 49:18:1125–1131.

Miller, R.; M. Friedman; D. Koci; & R. Moore. 2013. Biofilm Control Develop an Effective Strategy to Preserve Quality, Ensure Compliance. *Opflow,* 39:9:10–13.

Moore, M.R.; M. Pryor; B. Fields; C. Lucas; M. Phelan; & R.E. Besser. 2006. Introduction of Monochloramine into a Municipal Water System: Impact on Colonization of Buildings by *Legionella* spp. *Applied and Environmental Microbiology,* 72:1:378–383.

Murphy, H.M.; M.K. Thomas; D.T. Medeiros; S. McFayden; & K.D. Pintar. 2016b. Estimating the Number of Cases of Acute Gastrointestinal Illness (AGI) Associated With Canadian Municipal Drinking Water Systems. *Epidemiology and Infection,* 144:7:1371–1385.

Murphy, H.M.; M.K. Thomas; P.J. Schmidt; D.T. Medeiros; S. McFadyen; & K.D. Pintar. 2016a. Estimating the Burden of Acute Gastrointestinal Illness Due to *Giardia, Cryptosporidium, Campylobacter, E. coli* O157 and Norovirus Associated With Private Wells and Small Water Systems in Canada. *Epidemiology and Infection,* 144:7:1355–1370.

National Collaborating Centres for Public Health. 2011. Water-borne Disease Outbreaks in Canadian Small Drinking Water Systems. http://www.ncceh.ca/documents/evidence-review/water-borne-disease-outbreaks-canadian-small-drinking-water-systems (accessed May 31, 2016).

Neumeister, B.; M. Faigle; M. Sommer; U. Zahringer; F. Stelter; R. Menzel; C. Schutt; & H. Northoff. 1998. Low Endotoxic Potential of *Legionella pneumophila* Lipopolysaccharide Due to Failure of Interaction With the Monocyte Lipopolysaccharide Receptor CD14. *Infection and Immunity,* 66:9:4151–4157.

Niquette, P.; P. Servais; & R. Savoir. 2001. Bacterial Dynamics in the Drinking Water Distribution System of Brussels. *Water Research,* 35:3:675–682.

Norton, C.D. & M.W. LeChevallier. 2000. A Pilot Study of Bacteriological Population Changes Through Potable Water Treatment and Distribution. *Applied and Environmental Microbiology,* 66:1:268–276.

Ormeci, B. & K.G. Linden. 2002. Comparison of UV and Chlorine Inactivation of Particle and Non-Particle Associated Coliform. *Water Science & Technology: Water Supply,* 2:5–6:403–410.

Payment, P.; D.P. Sartory; & D.J. Reasoner. 2003. History and Use of HPC in Drinking Water Quality Management. *HPC and Drinking Water Safety: The Significance of HPCs for Water Quality and the Human Health.* London: IWA Publishing on behalf of the World Health Organization.

Pinto, A.J.; D.N. Marcus; U.Z. Ijaz; Q.M. Bautista-de Lose Santos; G.J. Dick; and L. Raskin. 2016. Metagenomic Evidence for the Presence of Comammox Nitrospira-Like Bacteria in a Drinking Water System. *mSphere,* 1:1.

Prest, E.I.; J. El-Chakhtoura; P. Saikaly; M. van Loosdrecht; F. Hammes; & H. Vrouwenvelder. 2014. Quantitative Monitoring of Biological Stability in Drinking Water Distribution Systems. Proceedings AWWA Water Quality Technology Conference. Denver, CO: AWWA.

Prest, E.I.; F. Hammes; M.C. van Loosdrecht; and J.S. Vrouwenvelder. 2016. Biological Stability of Drinking Water: Controlling Factors, Methods, and Challenges. *Frontiers in Microbiology,* 7:45.

Pryor, M.; S. Springthorpe; S. Riffard; T. Brooks; Y. Huo; G. Davis; & S.A. Satter. 2004. Investigation of Opportunistic Pathogens in Municipal Drinking Water Under Different Supply and Treatment Regimes. *Water Science & Technology,* 50:1:83-90.

Public Health Agency of Canada. 2013. Chief Public Health Officer's Report on the State of Public Health in Canada, 2013: Infectious Disease–The Never-ending Threat. http://www.phac-aspc.gc.ca/cphorsphc-respcacsp/2013/index-eng.php . Accessed: 5/31/16.

Reasoner, D.J. 1990. Monitoring heterotrophic bacteria in potable water. *Drinking Water Microbiology: Progress and Developments.* New York: Springer–Verlag.

———. 2004. Heterotrophic plate count methodology in the United States. *International Journal of Food Microbiology,* 92:3:307–315.

Reiber, S. 2008. Water Distribution Issues–Contaminant Accumulation and Re-equilibrium: What We Think We Know. Proceedings AWWA Water Quality Technology Conference. Denver, CO: AWWA.

Revetta, R.P.; V. Gomez-Alvarez; T.L. Gerke; J.W. Santo Domingo; & N.J. Ashbolt. 2013. Changes in Microbial Community Structure Associated With Monochloramine-Treated Drinking Water Biofilms. *FEMS Microbiology Ecology,* 86:3:404–414.

Rice, E.W., Baird, R.B., Eaton, A.D., & Clesceri, L.S. (editors). 2012. *Standard Methods for the Examination of Water and Wastewater.* 22nd ed. ISBN: 9780875530130. American Public Health Association, American Water Works Association, & Water Environment Federation: Washington, D.C.

———. 2016. Changes in Bacterial Composition of Biofilm in a Metropolitan Drinking Water Distribution System. *Journal of Applied Microbiology,* 121:1:294–305.

Rożej, A.; A. Cydzik-Kwiatkowska; B. Kowalska; & D. Kowalski. 2015. Structure and Microbial Diversity of Biofilms on Different Pipe Materials of a Model Drinking Water Distribution Systems. *World Journal of Microbiology and Biotechnology,* 31:1: 37–47.

Safe Drinking Water Act. 2002. Ontario Regulation 170/03 Drinking Water Systems. https://www.ontario.ca/laws/regulation/030170 (accessed June 1, 2016).

Schwartz, T.; S. Hoffmann; & U. Obst. 2003. Formation of Natural Biofilms During Chlorine Dioxide and UV Disinfection in a Public Drinking Water Distribution System. *Journal of Applied Microbiology,* 95:3:591–601.

Sales-Ortells H, Medema G. Screening-level microbial risk assessment of urban water locations: a tool for prioritization. Environ Sci Technol. 2014;48(16):9780–9. doi: 10.1021/es5020407, as cited in Ashbolt, N.J. 2015a.

Simoes, L.C.; M. Simoes; & M.J. Vieira. 2010. Influence of the Diversity of Bacterial Isolates From Drinking Water on Resistance of Biofilms to Disinfection. *Applied and Environmental Microbiology,* 76:19:6673–6679.

Singer, E.; B. Bushnell; D. Coleman-Derr; B. Bowman; R.M. Bowers; A. Levy; E.A. Gies; J.F. Cheng; A. Copeland; H.P. Klenk; S.J. Hallam; P. Hugenholtz; S.G. Tringe; & T. Woyke. 2016. High-Resolution Phylogenetic Microbial Community Profiling. *ISME Journal,* 10:8:2020–2032.

Smeets, P.W.M.H., G.J. Medema; & J.C. van Dijk. 2009. The Dutch Secret: How to Provide Safe Drinking Water Without Chlorine in the Netherlands. *Drinking Water Engineering and Science*, 2:1:1–14.

Sommerstein, R.; C. Ruegg; P. Kohler; G. Bloemberg; S.P. Kuster; & H. Sax, H. 2016. Transmission of *Mycobacterium chimaera* From Heater-Cooler Units During Cardiac Surgery Despite an Ultraclean Air Ventilation System. *Emerging Infectious Diseases*, 22:61008–1013.

Sorlini, S.; D. Palazzini; A. Mbawala; M.B. Ngassoum; & M.C. Collivignarelli. 2013. Is Drinking Water From 'Improved Sources' Really Safe? A Case Study in the Logone Valley (Chad-Cameroon). *Journal of Water and Health*, 11:4:748–761.

Iron Corrosion Products Under the Influence of Bacterial Communities in a Simulated Water Distribution System. *Environmental Science: Processes & Impacts*, 16:3:576–585.

State of Louisiana. 2013. DHH Issues Emergency Rule Requiring Drinking Water Systems in Louisiana to Raise the Level of Disinfectant in their Water, Increase Monitoring by 25 Percent. http://www.dhh.louisiana.gov/index.cfm/newsroom/detail/2906. (accessed Dec. 21, 2016.

Teunis, P.F.; M. Xu; K.K. Fleming; J. Yang; C.L. Moe; & M.W. LeChevallier. 2010. Enteric Virus Infection Risk From Intrusion of Sewage Into a Drinking Water Distribution Network. *Environmental Science & Technology*, 44:22:8561–8566.

Thomas V, McDonnel G, Denyer SP, Maillard J-Y. Freeliving amoebae and their intracellular pathogenic microorganisms: risk for water quality. *FEMS Microbiol Rev.* 2010;34:231–59. doi: 10.1111/j.1574-6976.2009.00190.x, as cited in Ashbolt, N.J. 2015a.

Thompson, R.M.; R. Carter; C. Tolson; C. Coulter; F. Huygens; & M. Hargreaves. 2013. Factors Associated With the Isolation of Nontuberculous Mycobacteria (NTM) From a Large Municipal Water System in Brisbane Australia. *BMC Microbiology*, 13:89.

USEPA. 2002. Health Risks from Microbial Growth and Biofilms in Drinking Water Distribution Systems. http://www.epa.gov/sites/production/files/2015-09/documents/2007_05_18_disinfection_tcr_whitepaper_tcr_biofilms.pdf. (accessed Jan. 13, 2016).

USEPA. 2006. Distribution System Indicators of Drinking Water Quality. Washington, DC: EPA; 2006 Dec.

USEPA. 2013. National Primary Drinking Water Regulations: Revisions to the Total Coliform Rule; Final Rule. 40 CFR parts 141 and 142. *Federal Register*, 78:30: February 13, 2013.

USEPA. 2016. Contaminant Candidate List 4 – CCL 4. https://www.epa.gov/ccl/contaminant-candidate-list-4-ccl-4-0 (accessed Dec. 22, 2016).

USEPA. 2016a. Six-Year Review 3 of Drinking Water Standards. https://www.epa.gov/dwsixyearreview/six-year-review-3-drinking-water-standards (accessed March 10, 2017).

van der Kooij, D. 1999. Potential for Biofilm Development in Drinking Water Distribution Systems. *Journal of Applied Microbiology Symposium Supplement 1999*, 85:39S–44S.

———. 2003. Managing Regrowth in Drinking-Water Distribution Systems. In: *Heterotrophic Plate Counts and Drinking-Water Safety: The Significance of HPCs for Water Quality and Human Health*. London: IWA Publishing.

van der Kooij, D. & H. Veenendaal. 1993. Assessment of the Biofilm Formation Potential of Synthetic Materials in Contact With Drinking Water During Distribution. Proceedings AWWA Water Quality Technology Conference. Denver, CO: AWWA.

van der Wielen, P. & D. van der Kooij. 2010. Effect of Water Composition, Distance and Season on the Adenosine Triphosphate Concentration in Unchlorinated Drinking Water in the Netherlands. *Water Research*, 44:17:4860–4867.

Velten, S.; F. Hammes; M. Boller; & T. Egli. 2007. Rapid and Direct Estimation of Active Biomass on Granular Activated Carbon Through Adenosine Tri-Phosphate (ATP) Determination. *Water Research*, 41:9:1973–1983.

Vital, M.; M. Dignum; A. Magic-Knezev; P. Ross; L. Rietveld; & F. Hammes. 2012. Flow Cytometry and Adenosine Tri-Phosphate Analysis: Alternative Possibilities to Evaluate Major Bacteriological Changes in Drinking Water Treatment and Distribution Systems. *Water Research*, 46:15:4665–4676.

Wang, H.; S. Masters; M.A. Edwards; J.O. Falkinham III; & A. Pruden. 2014. Effect of Disinfectant, Water Age, and Pipe Materials on Bacterial and Eukaryotic Community Structure in Drinking Water Biofilm. *Environmental Science & Technology*, 48:3 1426–1435.

Wang, Z.; O. Choi; & Y. Seo. 2013. Relative Contribution of Biomolecules in Bacterial Extracellular Polymeric Substances to Disinfection Byproduct Formation. *Environmental Science & Technology*, 47:17:9764–9773.

Weintraub, J.M.; B. Flannery; D.J. Vugia; L.B. Gelling; J.J. Salerno; M.J. Conroy; V.A. Stevens; C.E. Rose; R.E. Besser; B.S. Fields; & M.R. Moore. 2008. *Legionella* Reduction After Conversion to Monochloramine for Residual Disinfection. *Journal of the American Water Works Association*, 100:4:129–139.

Whiley, H.; A. Keegan; S. Giglio; & R. Bentham. 2012. *Mycobacterium avium* Complex–The Role of Potable Water in Disease Transmission. *Journal of Applied Microbiology*, 113: 223–232.

WHO (World Health Organization). 2011 (4th edition). *Guidelines for Drinking-Water Quality*. Geneva, Switzerland: World Health Organization.

Wingender, J. & H.-C. Flemming. 2011. Biofilms in Drinking Water and Their Role as a Reservoir for Pathogens. *International Journal of Hygiene and Environmental Health*, 214:6:417–423.

Wolfe, R.L.; N.I. Lieu; G. Izaguirre; & E.G. Means. 1990. Ammonia-Oxidizing Bacteria in a Chloraminated Distribution System: Seasonal Occurrence, Distribution, and Disinfection Resistance. *Applied and Environmental Microbiology*, 56:2:451–462.

Yang, F.; B. Shi; Y. Bai; H. Sun; D.A. Lytle; & D. Wang. 2014. Effect of Sulfate on the Transformation of Corrosion Scale Composition and Bacterial Community in Cast Iron Water Distribution Pipes. *Water Research,* 59:46–57.

Yang, J.; M.W. LeChevallier; P.F.M. Teunis; & M. Xu. 2011. Managing Risks From Virus Intrusion Into Water Distribution Systems due to Pressure Transients. *Journal of Water and Health,* 9:2:291–305.

Yu, J.; D. Kim; & T. Lee. 2010. Microbial Diversity in Biofilms on Water Distribution Pipes of Different Materials. *Water Science and Technology,* 61;1:163–171.

Zhang, W. & F.A. DiGiano. 2002. Comparison of Bacterial Regrowth in Distribution Systems Using Free Chlorine and Chloramine: A Statistical Study of Causative Factors. *Water Research,* 36:6:1469–1482.

Chapter 4

Infrastructure Integrity and Water Quality

Sonya M. Semanuik, Ontario Clean Water Agency, Mississauga, Ontario, Canada
Peter Fiske, PAX Water Technologies, Richmond, Calif.
Kira S. Smith, City of Houston Drinking Water Operations, Houston, Texas
Jennifer Wade, Wachs Water Services, Buffalo Grove, Ill.

Water quality affects the longevity and integrity of a water utility's infrastructure. Conversely, the quality of the infrastructure has significant effects on the quality of the water at the customer's tap (Water RF 2014). As distribution system infrastructure begins to fail, water quality risks may be incurred as a result of contaminants entering the system through failures (breaks) or through internal chemical or biological changes, resulting in undesirable water conditions.

This chapter looks at the factors that affect infrastructure integrity, describes how these factors affect water quality, and provides best practices for mitigating impacts.

DSWQ Challenge:

As distribution system infrastructure deteriorates, water quality risks may increase through internal chemical or biological changes or as a result of contaminants that enter the system through failures (breaks).

FACTORS THAT AFFECT PHYSICAL INTEGRITY

The distribution system's ability to act as a physical barrier and prevent external contamination from affecting the quality of the internal drinking water supply is referred to as physical integrity (CPWSDS 2006). Physical integrity depends on several factors including pipeline age, installation methods, soil conditions and external corrosion, water chemistry, and operations and maintenance conditions.

Pipeline Age

Average life expectancies of water mains are variable for several reasons including the factors listed above that affect physical integrity (e.g., age, installation methods, soil conditions, and water chemistry). Pipeline conditions and materials are variable, and age alone is not necessarily a good predictor of condition. Different maintenance and management strategies (i.e., the decision to repair or replace) can also result in different life spans (Ellison et al. 2014).

Using multiple methods (e.g., statistical modeling, deterministic modeling, and benchmarking) to assess pipeline life expectancy allows for greater confidence in predictions. Table 4-1 lists typical benchmark life expectancies for distribution pipelines according to pipe material. As shown, some pipelines can function for at least 100 years. Benchmark estimates are based on statistical analysis or expert opinion and work best if a system is relatively young and experiencing few failures (Ellison et al. 2014). The AWWA Distribution Partnership for Safe Water considers an optimized water system to be one that experiences fewer than 15 reported breaks and leaks per 100 miles of pipe per year (AWWA 2011).

Installation Methods

Regardless of material, pipeline performance can be reduced if installed or operated under adverse conditions. A number of resources are available that provide guidance for proper installation and operation practices, including materials that can be obtained directly from pipe manufacturers who supply installation guides for their specific product.

AWWA has published a number of manuals regarding the proper design and installation for distribution pipelines, based on material including:

- AWWA M9, Concrete Pressure Pipe,
- AWWA M11, Steel Pipe: A Guide for Design and Installation,
- AWWA M23, PVC Pipe—Design and Installation,

Table 4-1 Pipeline life expectancy benchmarks

Pipe Material	Life Expectancy, years
Cast iron	115
Cast iron–cement lined	75–100
Ductile iron	60–110
Polyvinyl chloride	70
Asbestos–cement	75–105
Concrete and prestressed concrete cylinder	75
Steel	95

Source: Ellison et al. 2014; Table 2.1.

- AWWA M41, Ductile-Iron Pipe and Fittings,
- AWWA M45, Fiberglass Pipe Design, and
- AWWA M55, PE Pipe—Design and Installation.

There are also AWWA manuals that describe appropriate selection and installation of other distribution system components such as hydrants, valves, and services.

Soil Conditions and External Corrosion

External corrosion can lead to structural failure of pipes and joints and can be due to stray currents, dissimilar metals, and/or soil redox and pH conditions (CPWSDS 2006). The pipe material used, thickness of the pipe wall, use of outer wraps or coatings, cathodic protection, and backfill materials and techniques also play a role in the degree of external corrosion conditions (CPWSDS 2006).

Iron pipes are also prone to graphitic corrosion that occurs in soil conditions favorable to anaerobic bacterial growth. Graphitic corrosion is not always visible in exhumed pipe sections as eroding pipe material is replaced by a graphite matrix plug. Although this graphite plug may prevent leaking, it has less structural strength than the host pipe and can contribute to pipe failure (USEPA 2009). AWWA Manual M27 (2013) provides information on external corrosion and controls.

Water Chemistry

Similar to external corrosion, internal corrosion can lead to structural failures (CPWSDS 2006). Internal corrosion can be influenced by pH, alkalinity, disinfectant type and dose, biofilm makeup, velocity, water use patterns, use of inhibitors, and many other factors (CPWSDS 2006).

Metal lines are more prone to corrosion than plastic materials such as polyvinyl chloride (PVC) and high-density polyethylene. Depending on the pH and the presence of ions, tuberculation can result. Tuberculation is a corrosion by-product that results in a reduced cross-sectional area that, in turn, reduces flows and may also cause water quality issues. Although tuberculation can be removed with aggressive cleaning, it is generally not recommended unless the pipe is lined afterward because the pipe wall will be pitted beneath the tuberculation and, once cleaned, that exposed wall will be unprotected (USEPA 2009).

Operations and Maintenance Conditions

Distribution pipeline operations and maintenance (O&M) generally include activities such as flushing, swabbing, and valve exercising. During these activities, flows are temporarily altered by the action of opening and/or closing valves. The resulting operating conditions can contribute to the deterioration of distribution pipelines, particularly if the system pressure cannot be maintained within the design range. O&M activities that create transient pressure surges, particularly in deteriorated pipe, increase the likelihood of failure (USEPA 2009).

Valves are designed to control the flow of water and to protect other assets in the water distribution system. To work as designed, valves need to be physically locatable, accessible, and operable and they must be operated as designed to prevent the creation of a transient pressure surge. An additional benefit of locatable, accessible, and operable valves is the efficiency of isolating a section of pipe in the event of a main break. Delays in isolating a break increase the time that the water system is exposed to contaminants that may enter through the broken pipe.

If there is a failure in the water main, either in the form of a leak or break, and repairs are necessary, the repair activities can lead to system contamination and reduced water quality. To address this possibility, AWWA published Standard C651 (2014) that includes disinfection procedures to be used when cutting into or repairing existing pipe. Appropriate procedures largely depend on whether or not pressure can be maintained in the system, with more vigorous requirements applied when it cannot.

FACTORS THAT AFFECT WATER QUALITY

Loss of physical integrity in distribution systems, such as the introduction of microbial and chemical contaminants, debris, and particulate matter, can impact water quality. These are sometimes accompanied by changes in water color, turbidity, taste, and odor (CPWSDS 2006).

According to a Water Research Foundation report (Ellison et al. 2014), potential water quality impacts associated with deteriorated infrastructure include:

- Entry of pathogens during main break repairs;
- Depletion of disinfectant due to interaction with materials, sediments, and biofilms;
- Accumulation of sediments within corroded and scaled pipes and associated problems with water discoloration and biofilm growth;
- Leaching of dissolved metals from pipe walls and appurtenances; and
- Episodic releases of contaminants accumulated within scales due to changes in water chemistry or operations.

Pathogen Entry

If the pipe deterioration is external, there may be little effect on water quality until a failure event occurs and water pressure can no longer be maintained. In addition, boundary valves between pressure zones that are left open can result in low pressure. When water pressure cannot be maintained, contaminants can enter the system through a breach in the barrier between the water and the pipe wall, through a defective backflow-prevention device, or through an unknown and undetected cross-connection. Cross-connection control is discussed in detail in Chapter 10.

According to the National Academy of Sciences (NRC 2006), contamination from cross-connections and back siphonage causes the most waterborne disease outbreaks, followed by contamination of water mains due to breaks and contamination of storage facilities.

Storage facilities are exposed to contaminants via tank vents and rooftop openings. Tanks require venting to allow air to displace water as a tank drains. Vent screens may be damaged by environmental factors such as ambient changes (e.g., in colder areas, winter frost buildup can result in occlusion of the vent opening and stretching and breaking of the screen). If the vent screen is torn, the roof vent structure may become an attractive location for nesting birds and insects, which introduce their own foreign matter into the tank.

Disinfectant Loss

In a pipeline, increased iron due to internal corrosion can lead to an increased chlorine demand, reducing the amount of expected disinfectant in the line. This leads to premature water aging and an associated loss of water quality. Corrosion and its effects on water quality are described in depth in Chapter 7.

Like the pipeline component of the distribution system, valves deteriorate over time. Deterioration can occur anywhere, whether on the valve body or an individual component such as an O-ring, disk, or seat, and can impact water quality. Additionally, it is important that valves always be set in their correct position. Inline main valves that are inadvertently left in a closed position create dead ends that have a negative effect on chlorine residuals.

Accumulation of Scale and Sediment

Scale, when still attached to pipe walls, can reduce hydraulic capacity and also create surfaces that are ideal for growing bacteria. Although biofilms are generally expected to occur in any distribution system, water quality issues can be created when unmanageable quantities of bacteria begin to form and mature. Biofilms tend to exist to a greater extent in aged, tuberculated distribution pipes, providing a medium for the bacteria to more readily grow and be protected from the disinfectant (NRC 2006). Chapter 3 provides more information on microbial aspects of distribution systems.

Sediment may also accumulate in water tanks due to prolonged periods of very low velocity, allowing fine particles that have been carried through the treatment and/or distribution systems to settle. Sediment may impact water quality by increasing chlorine demand, providing a large surface area for bacteria to proliferate, and providing buried areas within the sediment where bacteria are protected from residual disinfectant.

Internal Corrosion

Corrosive conditions can lead to leaching, or metal release, which occurs when materials from the pipeline or appurtenances dissolve into the water stream. In some cases, this can lead to aesthetic effects. Cement pipe, for example, may cause dissolved aluminum to enter the drinking water supply, which may result in discoloration. Some leached materials do have associated health concerns, the most obvious example being lead and copper. PVC lines may leach vinyl chloride, and it is also possible for protective liners to leach chemicals into the water under certain conditions (FCM 2003).

In storage tanks, internal corrosion can create new surface areas for bacteria to grow. Delamination of the tank coating also creates new surface area between the coating and the tank wall where bacteria can proliferate and exert demand on residual disinfectant.

Chapter 7 and AWWA Manual M58 (2017) contain more discussion on internal corrosion and its potential impacts on water quality.

Scale and Sediment Release

When detached from the pipe wall, scale can increase turbidity and metal release as well as cause color issues, especially for metal lines. Risk factors that contribute to release events are system dependent and vary between pressure zones that are fed by different sources or that have different pipe materials (Friedman et al. 2016). The progression of accumulation and release of materials is shown in Figure 4-1 (Friedman et al. 2016).

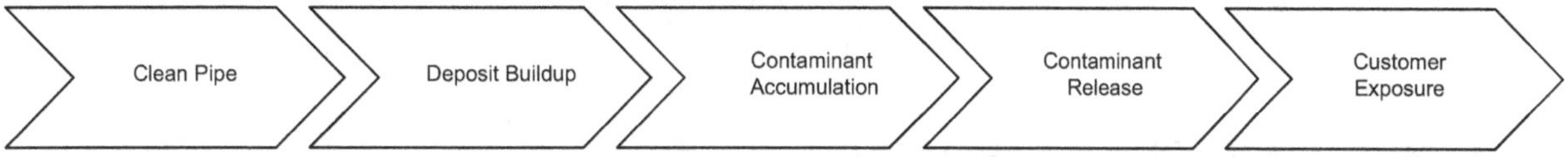

Source: Friedman et al. 2016.

Figure 4-1 Progression of accumulation and release of pipe materials

The exact concentrations of contaminants present will vary. Friedman et al. (2016) found iron, manganese, zinc, copper, lead, and arsenic in removed scale/biofilm as well as iron, arsenic, calcium, and silica in loose, weakly attached particles.

WATER QUALITY INDICATORS

Basic indicators of water quality impacts due to loss of physical integrity include disinfectant residual levels, microbial growth, customer complaints, and frequency of main breaks. These factors are discussed in the following paragraphs.

Disinfectant Residual Levels

As water travels through the distribution system, chlorine disinfectants react with material in both the water and on the pipe wall, reducing the amount available to ensure continued disinfection. An increased disinfectant demand, particularly in metal pipe, may be an early indication of pipeline deterioration.

Microbial Growth

Corrosion and excessive growth of biofilm can be signs of infrastructure deterioration. Consequently, monitoring total coliforms and microbial communities can be beneficial. Chapter 3 discusses the relationship between these parameters and monitoring approaches in more detail.

Customer Complaints and Main Breaks

In addition to water quality monitoring, levels of service parameters can be used to gauge the performance of the distribution system. Internal corrosion may noticeably affect water quality, resulting in customer complaints of water taste and/or odor.

Customer complaints regarding aesthetics or low pressure, the number of service interruptions, and break history can also be tracked to determine areas where infrastructure is deteriorating and needs attention. Corrective and preventative maintenance records are also useful for this purpose.

Characterizing the DSWQ Challenge:

- Monitor integrity loss indicators for changes in disinfectant residual and microbial growth
- Review customer complaints for patterns
- Maintain and review records of breaks and leaks

ADDRESSING WATER QUALITY CHALLENGES

The definitive solution to addressing issues associated with compromised infrastructure is replacement; however, immediate replacement may be costly. Other responses including main flushing and cleaning techniques. Also, targeted condition assessment and renewal techniques can increase the expected life span of the distribution system and reduce overall life-cycle costs (Ontario Ministry of Infrastructure 2012).

Main Flushing and Cleaning

To reduce tuberculation and buildup of sediment in a pipeline, regular flushing or other cleaning programs can be implemented, particularly in dead ends and regions with slow turnover. A number of options are available based on the desired outcome (Friedman et al. 2016). Note that it is not always advantageous to pursue the most aggressive cleaning procedure as this may incur a number of unwanted side effects including:

- Increase in complaints of discolored water,
- Plugged service connections,
- Increased pressure transients (or water hammer), and
- Damage to older, weaker pipe regions.

For this reason, it is preferable to perform frequent, regular inspections with more moderate cleaning techniques. The techniques typically used are unidirectional flushing (UDF), swabbing or pigging, and ice pigging. Appendix B provides a summary of flushing techniques and their application to cleaning. AWWA Manual M28 (2014) also provides information on flushing and distribution system cleaning practices.

Condition Assessment Programs

To protect infrastructure from deterioration and associated water quality impacts, it is necessary to understand the infrastructure's present condition. There are many approaches to condition assessment that can be implemented to accommodate the system requirements and available budget.

Condition assessment programs present two main challenges to some water systems:

- Where do we begin? For water systems that have been reactively managing their distribution program using a run-to-failure approach, starting a condition assessment program to predict where failure is imminent or to diagnose corroded areas that are affecting water quality can be daunting. Although condition assessment with targeted rehabilitation is normally less expensive than replacement, it can be perceived as more expensive than doing nothing. Therefore, water system operators must determine how condition assessment programs can be implemented in order to show value early in the process and justify further implementation.
- Which areas should be rehabilitated first? Once the system's condition is known, there may be a large list of improvements needed. Funding for capital and major maintenance projects can be limited, particularly for smaller systems, and it may be tempting to address the "low-hanging fruit," such as relatively inexpensive spot repairs first.

A proper risk management approach can be valuable to a water system manager who needs to determine which pipe sections are at the highest risk and to prioritize them for a renewal program. Most risk management approaches determine risk based on a score for each distribution system asset; this is calculated as the product of the likelihood of failure (LOF) and the consequence of failure (COF). The LOF score quantifies the conditions that led to failure, addressing the first challenge listed above. The COF score quantifies the damage that will occur as a result of failure, addressing the second challenge.

LOF criteria include variables that influence asset performance, such as age, break history, soil conditions, pipe material, wall thickness, and operating and transient pressures. The availability of LOF variables is dependent on good record-keeping practices and greatly assist the effectiveness of rehabilitation planning. Pipe material, diameter, age, soil conditions, operating pressures, and break histories should be available and ideally tracked within a geospatial format. COF criteria include volume of water lost, response time, number of users affected, proximity to vital facilities such as hospitals, ease of access, and similar items. Once the LOF and COF parameters are defined for each asset, they can be used to create a risk assessment rating. Higher-rated risks can then be prioritized for condition assessment, further monitoring, or renewal.

Repair Damaged Infrastructure

There are a number of approaches for repairing distribution pipelines. The most appropriate choice depends on a number of factors such as pipe material, extent of deterioration, density of service connections, depth, accessibility, and cost (FCM 2003a). Damaged infrastructure can be repaired locally using a variety of spot-repair methods such as clamps, short liners or seals, and spray-on epoxies. For pipelines that require renewal over a longer distance, a variety of coatings and liners can be used in lieu of full replacement.

Replacing Compromised Infrastructure

When an asset nears the end of its life or has failed catastrophically, total replacement is likely necessary. Pipelines can be replaced using open-cut or trenchless methods depending largely on the surrounding terrain, adjacent utilities (i.e., gas or sewer lines), and budget. More information on repair and replacement approaches is provided later in this chapter as best practices.

DSWQ Response Actions:

- Flush and/or clean mains
- Implement condition assessment programs to prioritize renewal efforts
- Perform repairs and rehabilitation
- Replace compromised infrastructure

BEST PRACTICES

Water quality impacts that are associated with deteriorated infrastructure can be reduced through an infrastructure management plan that includes:

- Monitoring and modeling of changes in quality as water moves through the system to help in development of operational and capital plans;
- Implementing water main cleaning programs that regularly remove sediment from the system, particularly from unlined cast-iron mains;
- Adopting pipeline rehabilitation programs that are geared toward cleaning and lining all unlined cast-iron pipes that are not scheduled for replacement;
- Adopting infrastructure renewal programs that keep the frequency of breaks and the number of leaks at a moderate, manageable level; and
- Creating a plan for lead and copper reduction, as needed to achieve regulatory compliance and meet public health goals, which may include use of corrosion inhibitors, replacement of lead service lines, and other strategies (Ellison et al. 2014).

A disciplined asset management approach that incorporates regular maintenance with condition assessment to determine appropriate rehabilitation activities may extend the service life and reduce water quality issues caused by corroding or failing infrastructure.

Condition Assessment Programs

As discussed earlier in this chapter, condition assessment programs can be used to prioritize renewal projects and reduce life-cycle infrastructure costs. There are a number of approaches to distribution condition assessment, including leak detection, wall integrity measurement, and long-term monitoring.

Leak detection programs. Leak detection programs can be useful in response to main breaks and leaks in order to locate the source of the break. Different techniques are available, some more permanent and costly than others. Methods are based on use of the following:

- Gas tracer or injection,
- Acoustics (aboveground and inline), and
- An accelerometer.

These techniques are summarized in Table 4-2.

As with any technology, performance is based on a number of factors. In this case, leak detection success can vary based on pipe material, diameter, operating pressure, leak size, and surrounding soil conditions. Table 4-3 lists leak detection method pipe based on material and diameter. The table does not address the effectiveness of each technology or the ease of implementation. Therefore, the decision regarding the "best" technology is left with the water system to make based on their individual circumstances.

A systematic approach to leak detection based on prioritized assets can be a cost-effective method for addressing leak sources before they become less manageable. It also has the added benefit of identifying water losses that can reduce nonrevenue water and conservation efforts for water systems. In critical areas, a long-term monitoring program can be implemented using technologies that are permanently or semipermanently installed to continuously measure parameters such as pressure, flow, and/or chlorine residuals or to determine the occurrence of a new leak in real time. Techniques are summarized in Table 4-3 and discussed extensively in AWWA Manual M36 (2016).

Wall integrity measurement. Wall integrity, as it pertains to water quality, is more important for metal pipes because corrosion by-products can directly affect suspended solids, water age, and pipeline hydraulics. By reducing the rate at which the pipeline corrodes, it may be possible to not only increase the longevity of the pipeline asset but also lessen the adverse effects of the aging process on water quality.

The first step in assessing wall integrity for metal pipe is to ascertain its current condition. In some cases, poor water quality may be the first indication that there is an issue. To detect which areas are most affected or to identify corroded areas before they affect water quality, a wall thickness assessment may be performed. Numerous methods for this assessment are currently available, as shown in Table 4-4.

Table 4-2 Leak detection methods

Method	Description
Gas injection	Gases such as helium and hydrogen are injected into the region of pipeline under study. The gas then escapes the pipe via the same routes as leaking water and takes the path of least resistance to the surface where it is detected by a gas detector.
External listening devices (manual and amplified)	Listening devices can be manual, such as a listening stick, or more complicated, such as ground microphones that amplify the leak sounds. In each case, an operator places the device above the pipeline in the region of the suspected leak to pinpoint the leak's location.
Correlators	Correlators position two sensors on the pipeline at appurtenances such as valves or fittings. Each sensor detects the leak sound or vibration at a moment in time. Using the known velocity of sound and the measured time of flight, the distance from each sensor can be calculated.
Loggers and monitors	Loggers are listening devices that are permanently or semipermanently fixed to a system or pressure zone to detect leaks, normally at night when other noise is minimal. Operators collect logged data for later analysis. Monitors are usually more sophisticated systems that are placed on high-risk pipelines. These systems automatically perform the analysis and transmit data and alerts to water system operators in real time.
Inline techniques	Unlike correlators and loggers where the leak noise travels to the sensor, inline technologies increase detection abilities by bringing the sensor to the leak where the noise is strongest. Inline techniques may be tethered or free swimming and require an access point such as a 2-in. or 4-in. tap.
Ground-penetrating radar	Ground-penetrating radar transmits radar frequency waves into the ground. These waves are reflected back to the tool and vary based on the reflected surface (e.g., rock, soil, water). The operator analyzes the reflected signal to determine if and where leaks exist in the study area.
Thermal imaging	Thermal imaging can be used for larger areas (e.g., transmission pipelines) and provides a color map that displays radiant energy gradients (i.e., temperature changes). As water is typically a different temperature than the surrounding soil, it is possible to determine where water is exiting a main.

Courtesy of Sonya Semanuik, Ontario Clean Water Agency.

Table 4-3 Recommended leak detection method based on type of pipe

Pipe Diameter, in.	3–8	10	12–14	16	18–20	24–28	32+
Metal	A, B, C, D, F, G	A, B, C, D, F, G	A, C, D, F, G	A, C, D, E	C, D, E	D, E	E
Concrete	A, C, D	A, D	A, D, E	A, D, E	E	E	E
Asbestos cement	A, C, D	A, D	A, D, E	A, D, E	E	E	E
Polyvinyl chloride/ polyethylene	A, D	A, D	A, D, E	A, D, E	E	E	E

Adapted from Hamilton and Charalambous (2013).

Method: A, gas injection; B, manual listening stick; C, correlator (accelerometer); D, correlator (hydrophone); E, inline technique; F, noise logger; G, amplified ground microphone.

Table 4-4 Examples of wall thickness measurement methods

Method	Description
Wall thickness averaging	This is the least invasive approach and uses principles that are similar to those for leak detection (velocity of sound between two sensors) to obtain a "stiffness" measurement. The stiffness measurement is converted to an average wall reading and compared to other average wall readings. This process provides a "relative" condition and is used to prioritize similar sections of distribution piping.
Mathematical models	Small portions of pipeline are excavated, and detailed measurements are taken to determine local wall thicknesses. The results are then combined with other information such as soil condition in a mathematical model to predict the condition of the rest of the pipeline.
Trenchless ultrasound	Ultrasound is a wall measurement technology that has been recently introduced as a trenchless technology in North America. It uses a tethered entry through an existing hydrant to inspect up to 330 ft in each direction.
Trenchless electromagnetics	This technique was originally developed for use in the oil and gas industry and is very accurate. It is used for metal pipeline (i.e., cast, ductile, steel) in either tethered or free-swimming applications, increasing the amount of pipeline that can be inspected in a single deployment.

Courtesy of Sonya Semanuik, Ontario Clean Water Agency.

The list in Table 4-4 is not exhaustive. Wall thickness averaging is being tried by inline leak detection companies, with some positive early results. Variations that exploit the electromagnetic properties of metal pipe can also be found in tools that use broadband electromagnetics or magnetic flux leakage.

As with leak detection options, wall thickness measurement methods do not offer a perfect solution. Each solution performs best in different applications, and there are always trade-offs between accuracy and ease of implementation. A condition assessment approach that uses a mix of the available options will likely provide the most effective results.

Monitoring and modeling. Unless multiple inspections are conducted over time, leak detection methods and condition assessment technologies provide information regarding the pipeline at a single point in time and do not provide a timeline for future deterioration. If a pipeline has been identified as high risk, the utility may wish to monitor the pipe in real time. Many technologies can provide this type of service. One of the most important parameters to monitor is transient pressure change. This is done using high-speed electronic pressure data loggers (NRC 2006). Large-diameter, prestressed concrete mains can be monitored using acoustic fiber optics; wire break activity is identified and reported to the utility. In addition, leaks can be detected over a network by installing acoustic leak loggers or similarly networked leak detection devices. More sophisticated leak monitoring systems can also measure chlorine residual, transient pressures, pH, flows, and similar characteristics.

Water Main Cleaning Programs

In most cases, cleaning programs can assist with water quality improvement, particularly in nonmetallic pipelines. Cleaning programs range from simple flushing of dead ends to remove stagnant water and in turn improve water quality to UDF, which is better able to clean the pipe in addition to removing older water. More information on flushing can be found in Appendix B. Where flushing is not enough, swabbing programs can be implemented. As mentioned previously, care should be taken when cleaning tuberculated cast- and ductile-iron pipe so that pitted pipe is not exposed and put at risk of further corrosion. If this is a risk, cleaning should be followed by lining, if possible.

Scheduled Maintenance Programs

Valves. Valves should be inventoried and operated regularly to ensure they are both operable and in the correct position. Valves that are difficult to operate may indicate tuberculation or sediment buildup in the line. Not only does poor valve maintenance affect system hydraulics, it also increases the time necessary to isolate a line during a break event. More information can be found in AWWA Manual M44 (2016)and section 4 of AWWA Standard G200 (2015).

Hydrants. In many systems, hydrants are audited for overall condition and flow rates as part of a fire protection plan. It is important that hydrants and hydrant isolation valves be tested regularly to prevent tuberculated laterals (isolation valves) that can lead to low flow and poor water quality. Water systems can use hydrant maintenance activities to collect information that can be used to identify areas with water quality issues. By incorporating the collection of information on chlorine residuals and turbidity as part of the inspection process and integrating the information into a geographic information system, areas that may need more frequent flushing, UDF, or more aggressive swabbing techniques can be identified. Flow rate values can also be mapped to indicate the likelihood of inappropriately closed valves or severely tuberculated water mains. AWWA Manual M17 (2016) provides more extensive guidance on this topic.

Storage tanks. Properly maintained, well-mixed water storage tanks provide optimal conditions for maintaining water quality. As discussed below, it is critical to maintain tank coatings, vents, and mechanical mixing systems.

Tank coatings. Tank coatings are the primary sanitary barrier in a tank. Coatings provide a physical barrier between the water and the underlying material of tank construction, which is essential for corrosion protection for steel tanks. In addition, a well-applied tank coating provides a smooth surface with minimal roughness, which reduces the surface area available for bacteriological activity and biofilm growth.

As tank coatings age, stress fractures can expose the underlying materials of construction to corrosion, as mentioned earlier in this chapter. Patching small areas of failure in tank coatings provides a temporary repair and can delay continued propagation of cracks and coating failure. However, deteriorated tank coatings should eventually be removed and reapplied in order to restore the tank to its original as-designed state.

Tank vents. Because neither the roof vent nor other rooftop penetrations are visible from the ground, water systems should conduct regular tank inspections of the roof and the vent screens. Recently, unmanned aerial vehicles with remote video have been successfully used to inspect hard-to-reach areas at the top and along the sides of water tanks.

Tank mixing systems. Where present, tank mixing systems should be operable in order to provide water quality benefits. Tank mixing systems are a relatively recent introduction to distribution system assets. Components of these systems are potentially subject to degradation due to corrosion. Tank mixing equipment should be added to the water system's asset management plan when installed.

Nonfunctioning, weak, or undersized tank mixing systems will fail to provide adequate velocity to overcome thermal and chemical stratification and to keep fine particles in suspension. Because tank mixing systems are difficult to inspect when the tank is in operation, operators are advised to ensure that the mixing systems are regularly inspected per the manufacturer's specifications. Supervisory control and data acquisition integration of the mixing system can alert operators when such systems are off line; physical inspection of electronic data collection equipment (during regular tank interior inspections) is recommended.

Infrastructure Rehabilitation and Renewal Programs

Renewal programs include options for repair, rehabilitation, or replacement, depending on the current and expected future condition of the system infrastructure. According to AWWA Manual M28 (2014), pipeline renewal methods are generally selected based on:

- The reason for renewal,
- Cost,
- Site conditions, and
- Expected life-cycle performance.

Renewal methods range from minimally intrusive cleaning regimes to full pipeline replacement. In some cases, it may be possible to slow the rate of deterioration by implementing external corrosion control methods until suitable renewal can be implemented. Physical corrosion control methods include coating application, polyethylene wraps, and cathodic protection (Rajani and Kleiner 2003)

Repair processes generally restore the asset to an operable condition with no overall improvement or expected life-cycle extension. Examples include clamps, patches, and joint seals.

Rehabilitation methods generally extend an asset's current life expectancy. Some structural liners can add as much as 50 to 60 years of life if installed appropriately. Appropriate materials for drinking water systems (see applicable AWWA standards and NSF/ANSI Standard 61 [2015]) should be used to minimize leaching of chemicals. Table 4-5 lists some common water main rehabilitation methods.

Lead and Copper Reduction

Lead and copper plumbing has been found to have adverse health effects under some conditions (refer to Chapter 7 for more specific information on lead line identification and replacement). In 1991, the US Environmental Protection Agency implemented the Lead and Copper Rule to control lead and copper in drinking water systems. This regulation identifies sampling, monitoring, and reporting requirements as well as requirements for corrosion control treatment, source water treatment, and public education.

As the majority of copper and lead is found in the customer-owned portion of the system, system owners provide guidance and tools such as lead testing kits and searchable databases to customers to increase awareness and action.

Summary of Best Practices

Table 4-6 summarizes best practices discussed in this section.

Table 4-5 Common rehabilitation methods

Method	Advantages	Limitations
Cement–mortar lining, spray-applied in situ (ANSI/AWWA Standard C602, 2011)	• Low cost • Proven protection against internal corrosion • Service reconnection not required	• Nonstructural • Not recommended for soft water
Polymer lining, 1 mm thick (epoxy, polyurethane, or polyurea), spray-applied in situ (ANSI/AWWA Standard C620, 2007)	• Low cost • Proven protection against internal corrosion • Service reconnection not required • Rapid setup of some linings may allow same-day return to service (avoiding bypass costs)	• Nonstructural
Polymer lining, 3–8 mm thick (epoxy, polyurethane or polyurea), spray-applied in situ	• Moderate cost • Semistructural • Service reconnection not required • Rapid setup of some linings may allow same-day return to service (avoiding bypass costs)	• Not likely to survive fracturing of the pipe • Ability to serve as fully structural system not yet confirmed
Cured-in-place pipe lining, reinforced with fiberglass, polyester, or carbon fibers	• Fully or semistructural • Appears capable of surviving pipe fracture • Robotic service restoration is possible in many cases	• More costly than spray-applied linings • Service reconnections are required
Tight-fit high-density polyethylene slip lining, using roll-down, swage, or deformed methods	• Fully or semistructural • Capable of surviving pipe fracture • Design criteria and properties are well established	• More costly than spray-applied linings • Service reconnections are required • Limited wall thickness available
Pipe bursting replacement	• Fully structural • Some upsizing possible • Design criteria and properties are well established • Compared to tight-fit lining, pipe materials should be more easily procured (less critical sizing requirements and different materials can be used)	• More costly than most other methods, although competitive market exists (not proprietary) • Service reconnections are required

Source: Ellison et al. 2014.

Table 4-6 Best practices to mitigate aging infrastructure

Best Practice	Level of Service
Condition assessment programs	
• Determine asset risk using the likelihood of failure and the consequence of failure criteria associated with distribution system components	Advanced
• Implement a leak detection program for proactive management of leaks and breaks	Advanced
• Conduct a wall thickness assessment	Advanced
• Determine pipeline rehabilitation and replacement (renewal) schedules based on the reason for renewal, cost, site conditions, and expected life-cycle performance	Advanced
Mains cleaning	
• Conduct localized flushing	Basic
• Implement unidirectional flushing	Advanced
• Conduct swabbing, pigging, or ice pigging to clean mains	Advanced
Scheduled maintenance	
• Inventory and operate valves	Basic
• Collect information on disinfectant residual, turbidity, and flow rates as part of regular hydrant maintenance	Basic
• Incorporate information into a geographic information system	Advanced
• Maintain storage tank coatings and vents	Basic
• Add mixing systems to storage tanks	Advanced
Rehabilitation and renewal	
• Implement external corrosion control	Advanced
• Perform repairs such as clamps, patches, and joint seals	Basic
• Install structural liners	Advanced
Lead service line replacement	Basic

CASE STUDY: CITY OF CALGARY, CANADA'S CONDITION ASSESSMENT AND CAPITAL MANAGEMENT PROGRAM

The City of Calgary, Alberta, Canada, has been developing a condition assessment and capital management program since the 1980s when their break rates were unacceptably high. Through extensive participation in new and evolving assessment technologies, Calgary has progressed to become a global leader in the management of aging distribution infrastructure, ensuring the delivery of quality drinking water to its customers.

City of Calgary Water System

The City of Calgary, at nearly 1.2 million people, is one of Canada's largest cities. It has more than 310 miles of linear water assets that vary from 0.5 in. to 118 in. in diameter. Their distribution piping, which is generally less than 20 in. in diameter, is the focus of their linear infrastructure capital and inspection programs and can be simplified down to the three material types:

- Unprotected metal mains (cast iron, ductile iron, and steel), accounting for approximately 28 percent of the total length of the network and 92 percent of total main breaks (2013);
- Protected metal mains (yellow-jacketed ductile iron), accounting for approximately 13 percent of the total length of the network and 4 percent of total main breaks;
- Nonmetallic mains (PVC and a mix of other materials), accounting for approximately 59 percent of the total length the network and 3 percent of total main breaks.

The focus of Calgary's condition assessment and capital replacement programs has been on the unprotected metal mains where, in most cases, cast iron is gradually being replaced by PVC. Considerable investment has been made in installing and maintaining cathodic protection to their yellow-jacketed ductile-iron network, and the success of this has resulted in a significant deferral of replacement costs associated with this material type. Calgary was an early adopter of PVC and has had good success with this material, enforcing good-quality installation practices that result in a very low main-break rate compared to other materials in the network.

Condition Assessment Programs

Calgary was also an early adopter of condition assessment for water mains and has served as a test bed for the development of many of the technologies used today.

- In 2002 and 2003, Calgary collected more than 100,000 soil-resistivity readings along their metal mains using a soil-resistivity meter. Lower resistivity readings are equivalent to high conductivity, indicating moderate to severe corrosivity.
- Regular electromagnetic inspections of metal mains are conducted to obtain detailed corrosion information in the form of remaining wall thickness. With more than 280 runs completed, the results provided Calgary with the actual condition of the main inspected and an idea of what is happening to similar mains in the same region.
- When necessary, Calgary performs leak detection inspections using a free-swimming internal hydrophone that identifies and locates leaks and air pockets.

Distribution System Capital Programs

In the early to mid-1980s when its distribution network size was approximately 1,500 miles, Calgary was experiencing nearly 1,800 breaks per year. As a result, the water main replacement program was ramped up, peaking at nearly 22 miles per year by the mid-1990s. Anode retrofit started in the late 1990s as an alternative to replacement. With the same budget, Calgary could protect 10 times the length of main compared to replacement. Combined with their early adoption of PVC pipe in the late 1970s, main breaks have been reduced by almost 85 percent to fewer than 250 breaks per year in a system that has doubled in length.

Water Main Replacement Program

Calgary's inspection data have allowed them to adopt an advanced risk-based methodology to strategically target replacements based on a number of criteria. Calgary typically replaces a water main if it has had three or more breaks in the past five years. This criterion was developed using extensive statistical analysis of historical main breaks and their likelihood to reoccur on the same section of pipe. They also look at the following parameters prior to selecting replacements for the annual program:

- Break number and spacing,
- Corrosion assessments,
- Material and age,
- Resistivity values,
- Services,
- Complaints,
- Damage,
- Remote Field Testing,
- Inspection results, and
- Soil data.

Calgary Water Main Anode Retrofit Program

Calgary adds sacrificial anodes to metal mains to redirect the corrosion away from the main until the anode is worn out, generally 20 years. This approach works best for ductile-iron materials but can be used on all metal mains. Calgary generally uses the criteria that are used for the water main replacement program to select appropriate lines for the anode retrofit program. However, the city also looks for pipes with approximately five years left before failure so they can survive the retrofit process and have enough metal remaining to extend life to the same degree as the anode.

Calgary Water Main Anode Replacement Program

With this program, Calgary replaces anode banks on mains, such as yellow-jacketed ductile iron, where banks were installed at the same time as the pipe. Calgary uses readings from test points to determine when the anode banks are depleted, then replaces them.

Calgary Water Service Anode Retrofit Program

With this retrofit program, Calgary installs anodes on copper service lines. Although it does not strictly reduce water main breaks, it does reduce water outages in the network due to downtime associated with service repairs.

The City of Calgary is using several options from today's commercially available inspection technologies and rehabilitation solutions to ensure that their aging infrastructure continues to provide high-quality drinking water to its residents.

REFERENCES

AWWA. 2011. *Partnership for Safe Water Self-Assessment Guide for Distribution System Optimization*. Denver, CO: AWWA.

———. 2013a. M27, *External Corrosion Control for Infrastructure Sustainability*. Denver, CO: AWWA.

———. 2014a. M28, *Rehabilitation of Water Mains.* Denver, CO: AWWA.

———. 2008. M9, *Concrete Pressure Pipe.* Denver, CO: AWWA.

———. 2004. M11, *Steel Pipe: A Guide for Design and Installation*. Denver, CO: AWWA.

———. 2002. M23, *PVC Pipe—Design and Installation*. Denver, CO: AWWA.

———. 2009. M41, *Ductile-Iron Pipe and Fittings*. Denver, CO: AWWA.

———. 2013b. M45, *Fiberglass Pipe Design*. Denver, CO: AWWA.

———. 2006. M55, *PE Pipe—Design and Installation*. Denver, CO: AWWA.

———. 2014b. Standard C651-14, *Disinfecting Water Mains*. Denver, CO: AWWA.

———. 2017. M58, *Internal Corrosion Control in Water Distribution Systems*. Denver, CO: AWWA.

———. 2016. M44, *Distribution Valves: Selection, Installation, Field Testing, and Maintenance*. Denver, CO: AWWA.

———. 2015. Standard G200-15, *Distribution Systems Operation and Management*. Denver, CO: AWWA.

———. 2016. M17, *Fire Hydrants: Installation, Field Testing, and Maintenance*. Denver, CO: AWWA.

The American National Standards Institute (ANSI) and AWWA. 2011. Standard C602, *Cement–Mortar Lining of Water Pipelines in Place—4 In. (100 mm) and Larger*. Denver, CO: AWWA.

ANSI and AWWA. 2007. Standard C620, *Spray-Applied In-Place Epoxy Lining of Water Pipelines, 3 In. and Larger*. Denver, CO: AWWA.

Ellison, E.; Bell, G.; Reiber, S.; & Spencer, D. 2014. *Answers to Challenging Infrastructure Management Questions*. Denver, CO: Water Research Foundation.

FCM (Federation of Canadian Municipalities and National Research Council). 2003. *Deterioration and Inspection of Water Distribution Systems*. Federation of Canadian Municipalities and National Research Council.

———. 2003a. *Selection of Technologies for the Rehabilitation or Replacement of Sections of a Water Distribution System*. Federation of Canadian Municipalities and National Research Council

———. 2003b. *Water Quality in Distribution Systems*. Issue 1.0. Federation of Canadian Municipalities and National Research Council.

Friedman, M.; Hill, A.; Booth, S.; & Hallett, M. 2016. *Metals Accumulation and Release Within the Distribution System: Evaluation and Mitigation*. Denver, CO: Water Research Foundation.

Hamilton, S. & Charalambous, B. 2013. *Leak Detection: Technology and Implementation*. London: IWA Publishing.

National Research Council. 2006. *Drinking Water Distribution Systems: Assessing and Reducing Risks*. Washington, DC: The National Academies Press.

NSF International. 2015. NSF/ANSI Standard 61. *Drinking Water System Components—Health Effects*. Ann Arbor, Mich.: NSF International.

Ontario Ministry of Infrastructure. 2012. *Building Together–Guide for Municipal Asset Management Plans.*

Rajani, B. & Kleiner, Y. 2003. Protection of Ductile Iron Water Main Against External Corrosion: Review of Methods and Case Studies. Institute for Research in Construction, National Research Council Canada, Ottawa, Ontario Canada.

US Environmental Protection Agency. 2009. *Condition Assessment of Ferrous Water Transmission and Distribution Systems.* State of Technology Review Report. United States Environmental Protection Agency. Cincinnati, OH: USEPA.

Water Research Foundation. Methods for Renewing, Restoring & Replacing Pipeline. Asset Management Renewal Fact Sheet. Denver, CO: Water Research Foundation.

This page intentionally blank.

Chapter 5

Taste, Odor, and Appearance

Justin Sutherland, Carollo Engineers, Inc., Austin, Texas

Ricard Devesa, Aigues de Barcelona (Agbar), Barcelona, Spain

Andrea Dietrich, Virginia Tech, Blacksburg, Vir.

Francesc Ventura, IDAEA-CSIC, Barcelona, Spain

Pure water is clear and odorless. However, as water moves through the natural environment and human-made infrastructure, it picks up particulate and dissolved matter that may impart undesirable aesthetic qualities in terms of taste, odor, and appearance. In some cases, this may produce good-tasting water and in other cases objectionable water. In general, water system customers judge the cleanliness of their tap water based on these characteristics, regardless of whether it meets all other drinking water standards. In other words, taste, odor, and appearance typically have the lowest health risk but the highest impact on customer satisfaction. Furthermore, aesthetic degradation may be the first sign of issues that can result in health risks.

Objectionable taste, odor, and appearance of tap water are considered to be minimized if general parameters, for example, odor, color, turbidity, and specific constituents (e.g., iron, manganese, and total dissolved solids [TDS]), are at acceptable levels. In the United States, the US Environmental Protection Agency (USEPA) sets unenforceable guidelines rather than enforceable maximum contaminant levels (MCLs) for aesthetic constituents, termed secondary MCLs (SMCLs), because these constituents do not have adverse health effects. Similarly, Canada sets guidelines rather than requirements.

In general, a US water system is considered to be in compliance with the SMCLs at the federal level (some US States enforce them) based on measurements at the entry point to the distribution system. However, the sources and/or causes for objectionable taste, odor, and appearance in tap water may be more complex and require more than removal of a secondary contaminant within the water treatment system. Physical, chemical, and biological phenomena in a distribution system and premise plumbing may have a significant impact on the taste, odor, and appearance of tap water. Another way of looking at this is that objectionable taste, odor, and appearance may be indicators of issues within the distribution system that require corrective action. Therefore, properly monitoring for and responding to aesthetic water quality issues is an important first step to identifying other water quality challenges presented in this manual.

DSWQ Challenge:

Water system customers judge the cleanliness of their tap water based on taste, odor, and appearance. Aesthetic degradation can be the first sign of underlying health risk issues.

In this chapter, several aesthetic water quality challenges related to drinking water distribution systems are presented. Several common aesthetic issues are described in this chapter; for each issue, the chapter provides a summary of the following:

- Aesthetic challenges;
- Recommended monitoring;
- Considerations for addressing the challenge with respect to design, operations, and maintenance activities;
- Best management practices for minimizing the challenge; and
- Emerging issues (if any).

The challenges described in this chapter are:

- Taste and odor issues:
  - Disinfectant derived: chlorinous, nous, and medicinal odors
  - Polymer derived and organic material derived: chemical and hydrocarbon odors
  - Metal pipe corrosion derived: metallic flavor and astringent tastes (e.g., mouthfeel [copper, iron, zinc])
  - Mineral derived: salty, concrete tastes (e.g., aluminum, calcium, magnesium)

- Appearance issues:
  - Turbidity (cloudiness), particulate matter, and sediments
  - Color (brown, black, red, green, blue)

Some taste and odor issues are not caused by distribution system problems but may still be present in tap water. Common issues include earthy, musty, moldy, and fishy odors or complaints that the water smells or tastes like "lake water." In general, these taste and odor issues are specific to algae and bacteria found in raw surface water sources. Guidance for taste and odors derived from raw water sources is found in AWWA Manual M7 (2004), Suffet, Mallevialle, and Kawczynski (1995), and Rashash et al. (1996).

AESTHETIC WATER QUALITY GOALS

Drinking water customers judge the quality of their water by its taste, odor, color, and appearance. They expect drinking water to meet health-based regulatory standards and also want the water to have acceptable flavor and clarity. Drinking water that is perceived as safe and that has good flavor will maintain trust in drinking water quality and water providers (Taste and Odor Committee 2002; Kirmeyer et al. 2002; Dietrich 2006; Doria 2010; Hu, Morton, and Mahler 2011; Dietrich and Burlingame 2015). The purpose of aesthetic water quality goals, such as SMCLs in the United States, is to provide guidance for contaminants that can cause a low-quality drinking water and also to maintain customer confidence in the drinking water and its provider.

While broadly referred to as "aesthetic" guidance, the SMCLs and other similar contaminant guidelines (Table 5-1) address issues related to aesthetics (taste, odor, color, foaming), technical issues (scaling, sedimentation, and staining, often related to corrosion), and cosmetic issues (bodily discoloration) (USEPA 1979, 2012; Dietrich and Burlingame 2015). Aesthetic guidance can take the form of a specific contaminant at a specific concentration, such as limiting chloride to 250 mg/L Cl^- in order to avoid salty-tasting water (Table 5-1). The guidance can also be broader, such as minimizing odors in drinking water by setting overall odor guidance such as categorizing the odor quality for the customer as "acceptable" or "inoffensive" (Table 5-1).

Odor can also be guided by an overall test to assess odor, such as the threshold odor number (TON <3). Unfortunately, TON does not correlate with customer complaints and is hard to implement and interpret. A replacement method(s) should be considered, one that is based on the number of taste and odor–related consumer complaints or specific guidelines for select odorants and tastants (Taste and Odor Committee 2002; Burlingame and Dietrich 2015).

GENERAL IDENTIFICATION AND MONITORING OF WATER QUALITY AESTHETICS IN DISTRIBUTION SYSTEMS

There are several techniques to characterize or monitor aesthetic water quality. These techniques typically involve sensory analyses, laboratory or physicochemical analyses, and/or customer feedback. The sections below describe each of these in more detail.

Sensory Analysis: Taste, Odor, and Color

There are many sensory analysis methods, and water system personnel should choose the right tool for the job. Methods available for assessing water samples through sensory analyses include:

- Threshold analyses that determine when a taste or odor can be detected,
- Descriptive analyses that qualitatively and quantitatively describe tastes and odors in a water sample,
- Effective analyses that determine the acceptability of water's flavor, and
- Discrimination analyses that determine if samples are different.

The specific sensory methods that are used in the drinking water industry as indicators of water quality and customer satisfaction are listed below. For more information on the methods associated with these tests, see Appendix C.

Table 5-1 Regulations for drinking water aesthetics for the World Health Organization, European Union, Canada, and the United States

Water Quality Parameter	Aesthetic Quality	World Health Organization*	European Union†	Canada‡	United States¶
Color	Visual tint	15 CU	Acceptable‖	15 "true" CU	15 CU
Odor, threshold odor number	Odor	—	Acceptable‖	Inoffensive	<3
Low pH High pH (unit: $-\log [H^+]$)	Low: bitter, metallic taste High: soda taste, slippery mouthfeel	6.5–9.5	6.5–9.5	6.5–8.5	6.5–8.5
Taste	—	—	Acceptable‖	Inoffensive	—
Total dissolved solids, mg/L	Salty taste, scaling, staining	1,000	~1,600 mg/L (2,500 μS/cm at 20°C**)	≤500	500
Aluminum, mg/L	Color, scaling, sedimentation	—	—	<0.1 or <0.2	0.05–0.2
Chloride, mg/L	Salty taste	250	250	≤ 250	250
Copper, mg/L	Metallic flavor, bitter taste, astringent, green/blue color	— (2.0, health-based)	2.0	≤ 1.0	1.0 (1.3, health-based)
Corrosivity	Corrosion, staining	—	—	—	Noncorrosive
Fluoride, mg/L	Tooth discoloration or pitting	—	—	— (1.5, health-based)	2.0 (4.0, health-based)
Foaming agents, mg/L	Odor, bitter taste, color, frothing	—	—	—	0.5
Hardness, mg/L as $CaCO_3$	Scaling	200	—	None required	—
Iron, mg/L	Metallic taste or flavor; red, brown, or yellow tint	0.3	0.2	≤0.3	0.3
Manganese, mg/L	Black/brown color	0.1 (0.4, health-based)	0.05	≤0.02	0.05
Silver, mg/L	Skin discoloration, graying of white of eye	—	—	None required	0.1
Sodium, mg/L	Salty taste	200	200	≤200	—
Sulfate, mg/L	Salty taste, temporary laxative effects	250	250	≤500	250
Zinc, mg/L	Metallic taste	4.0	—	≤ 5.0	3.0

Abbreviation: CU, color unit.

* Values cited are non-health–based indicators of likely level at which human detection occurs. World Health Organization guidelines have not been established for constituents in water without direct adverse health impacts (WHO 2011).

† EC DWD (1998).

‡ Health Canada (2012).

¶ USEPA (2012).

‖ Acceptable and no abnormal changes.

** The concentration (mg/L) of total dissolved solids is calculated based on conductivity (μS/cm at 20°C).

- Odor
  - Threshold odor number (TON)
  - Total intensity of odor (TIO)
  - Attribute rating test
- Taste
  - Flavor threshold test
  - Flavor profile analysis (FPA)
  - Flavor rating analysis
- Color
  - Visual comparison
- Various attributes
  - Discrimination test

Physicochemical Analysis

The methods associated with analyzing and measuring the physicochemical properties of water that are related to aesthetic challenges, for example, turbidity and organics, are covered in detail in Appendix C.

Customer Feedback

The odorous (licorice like) chemical spill that occurred in West Virginia reminds us that customers are at the forefront of being water quality monitors (Foreman et al. 2015; Gallagher et al. 2015; Manuel 2014; McGuire, Suffet, and Rosen 2014; Whelton et al. 2015). Within hours of the spill, crude 4-methylcyclohexane methanol, with its sweet licorice-like odor, was distributed to more than 300,000 residents of Charleston, W.V. (Gallagher et al. 2015). The result was an astounding fear of drinking or using treated and distributed water, loss of consumer confidence, and a "do not drink the water" order that lasted several days (Manuel 2014; Schade et al. 2015). Because customers are present throughout the water distribution network 24 hours a day and 7 days a week, they are able to assess aesthetic water quality and provide valuable water quality feedback. Research with gas chromatography/mass spectrometry (GC/MS), nuclear magnetic resonance (NMR), and gas chromatography–olfactory (GC-O) identified *trans*-4-methylcyclohexyl, methanol (Gallagher et al. 2015), and *trans*-methyl-4-methylcyclohexanecarboxylate as the major licorice and sweet odorants, which had air odor threshold concentrations in the parts-per-trillion range.

Consumer complaint surveillance (CCS) in the water industry is a protocol for collecting, recording, and interpreting customer feedback with the intent to harness a valuable sensory data stream that augments chemical and microbiological water quality monitoring (Lauer 2014; Dietrich 2006; USEPA 2008; Speight, Grayman, and Khanal 2010; PWD and CH2M HILL 2013; Gallagher and Dietrich 2014). Determining patterns or increases in complaints provides valuable water quality information. CCS can simultaneously serve several purposes, including aiding a contaminant warning system (CWS) designed to detect serious threats to drinking water (USEPA 2008; PWD 2013) and detect routine or unusual aesthetic issues.

Using a standardized form or check sheet, a water system should capture the following data regarding customer complaints (PWD and CH2M HILL 2013):

- Date and time of complaint;
- Customer name and address;

- Geographic information system (GIS) information for the exact location;
- Nature of the complaint, including placing into subcategories for tastes, odors, colors; and
- Resolution of complaint.

Customer complaints can be assessed in the following ways (PWD and CH2M HILL 2013; Gallagher and Dietrich 2014; Dietrich, Phetxumphou, and Gallagher 2014):

- Collecting electronic data;
- Performing repeated, regular review of all complaints on daily, weekly, and monthly bases;
- Visualizing data in straightforward graphs that show the number of complaints versus time;
- Assessing trends and patterns using visualization techniques such as spider (radar) plots, tag clouds, and GIS;
- Developing exceedance plots that demonstrate the probability of exceeding a complaint level; and
- Establishing triggers or detection systems to determine when the complaint pattern is high or unusual.

Customer feedback pertaining to water quality includes taste, odor, color, and turbidity. For a CCS to be effective, customer feedback needs to be properly categorized electronically so that the numbers and types of complaints can be tracked. Maintaining long-term records of complaints and systematically examining them for trends and changes is essential to an effective CCS (PWD and CH2M HILL 2013; Gallagher and Dietrich 2014).

Complaints should be organized into specific and widely used categories such as those shown in Figures 5-1 and 5-2. These color wheels are based on a conceptual range of four tastes (sweet, salty, sour, and bitter), eight odors (earthy–musty, chlorinous, grassy–woody, swampy–sulfurous, fragrant, fishy, medicinal, and chemical), and mouthfeel (APHA 2012, method 2170). A water system could develop a system-specific wheel after analyzing the particular issues that needed investigation. Color, turbidity, and particulate matter can be categorized using decision trees and images such as provided by Booth and Brazos (2004). Major colors in water include black, brown, red, orange, and yellow.

Source: APHA (2012), method 2170 B.

Figure 5-1 Drinking water taste-and-odor wheel

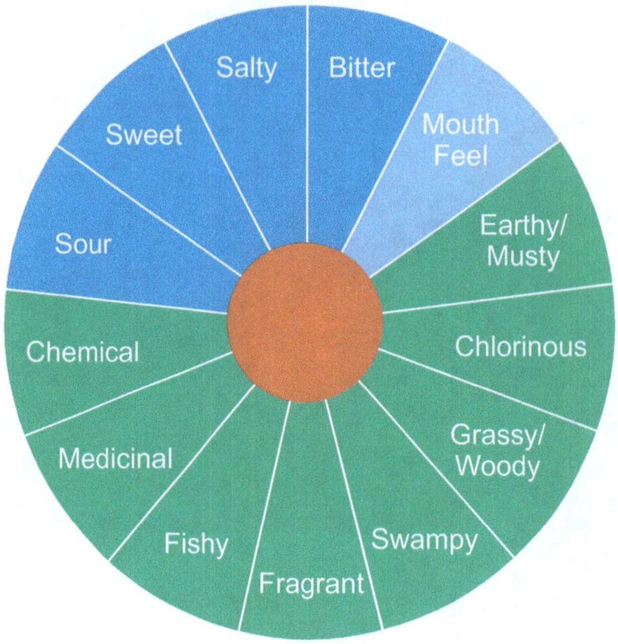

Source: APHA (2012), method 2170 B.

Figure 5-2 Simplified taste-and-odor wheel for drinking water

---

Characterizing the DSWQ Challenge

Monitoring Techniques

- Sensory analyses
- Physicochemical or laboratory analyses
- Customer complaints

Determine Source Based on Monitoring

- Taste and odor
  - Sulfur derived: marshy, swampy, septic, rotten eggs
  - Disinfectant derived: chlorinous, ozonous, medicinal
  - Polymers and organic material derived: chemical, hydrocarbons
  - Metal pipe corrosion derived: copper, iron, zinc
  - Mineral derived: salty, concrete (aluminum, calcium, magnesium)
- Appearance
  - Turbid or cloudy
  - Discolored

While paper copies can be a short-term collection mechanism, paper cannot serve as part of long-term data analysis. A computerized check sheet with drop-down menus is most desirable for ease and consistency of data collection (Whelton et al. 2007). When gathering written data, multiple descriptors for the same aesthetic issue can be used and misspellings can occur, both of which make tracking the data a challenge (Dietrich Phetxumphou, and Gallagher 2014).

Once collected, the frequency for total number of complaints or number of complaints in specific categories (e.g., sulfur odors, chlorine flavor, rusty particles, earthy–musty odors) is plotted against time and assessed for patterns. Both the pattern of complaints and the typical/atypical nature for the pattern can be examined using exceedance plots (Figure 5-3) Gallagher and Dietrich 2014). These plots show the probability of exceeding a certain number of complaints (*y*-axis) within a defined time period (*x*-axis). Varying time periods such as per day, per weekend, per week, or per month can be plotted. Exceedance plots are useful for establishing preliminary alarms or triggers.

Triggers can be established in order to determine when the actual number of complaints exceeds an expected number of complaints. One method is to manually decide the number of complaints (USEPA 2011). Preferably, changes above background complaint volumes should be monitored using event detection algorithms and event detection systems that integrate data, algorithms, and management (USEPA 2010; PWD and CH2M HILL 2013).

Event detection algorithms and event detection systems for aesthetic issues in drinking water are not yet available, but some are under development. When developed, detection algorithms and systems will be used to detect aesthetic events by incorporating the number of customers in the distribution system, and complaint temporal–spatial locations (Proulx et al. 2007), complaint category (Dietrich, Phetxumphou, and Gallagher 2014), plus statistical evaluation of historical data for complaints (Gallagher and Dietrich 2014).

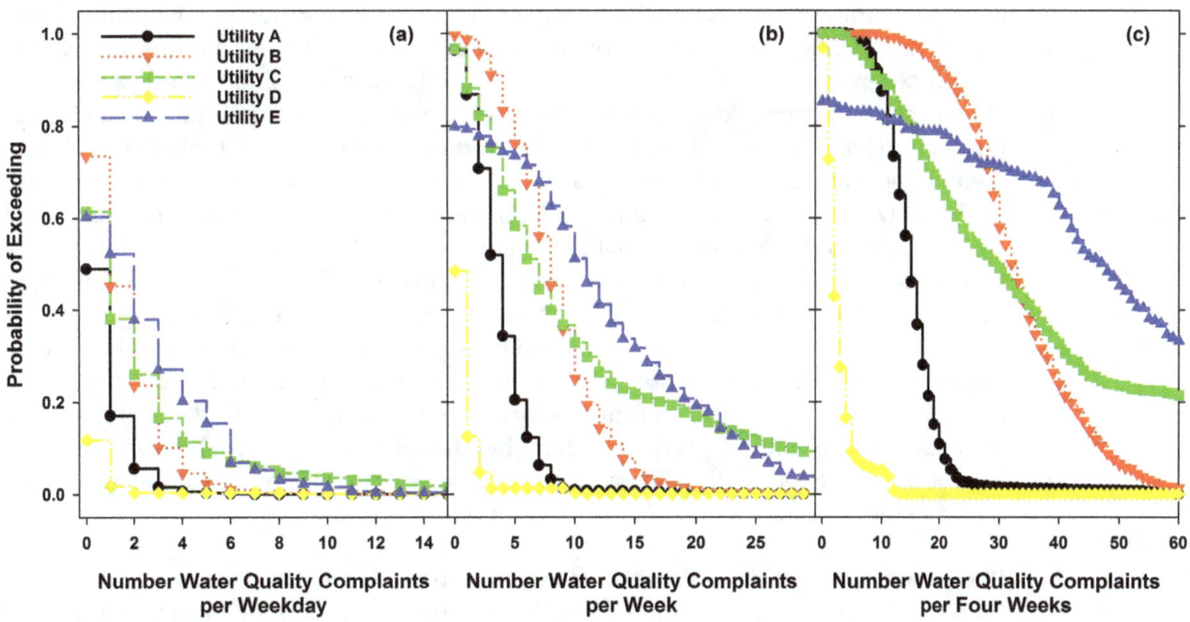

Source: Gallagher and Dietrich (2014).

Figure 5-3 Exceedance plots for the total number of water quality complaints that occurred per day, week, and month (28 days)

Collecting, categorizing, and interpreting customer complaints requires a systematic and dedicated process, similar to collecting mandatory regulatory data. One way a water system can aid this process is to provide a simple taste and odor wheel (Figure 5-2) to its customers on-line, in the consumer confidence report, or in a special mailing. Having access to a list of common descriptors should improve the accuracy of the dialogue between consumers and their water system for collection of customer complaints.

TASTE AND ODOR

While the four main tastes (sweet, salty, sour, and bitter) and colors are familiar and usually can be described, water system personnel should be aware that it may be difficult for customers to accurately describe odors (Dietrich, Phetxumphou, and Gallagher 2014). The sense of smell is more complex than vision or taste, and there are trillions of odors and olfactory stimuli (Bushdid et al. 2014). The consumer does not readily expect their tap water to taste like licorice, cucumber, or plastic; to be musty; or to have any of the myriad odors that may occur. The unexpected nature of a new odor in tap water and the lack of memory or vocabulary associated with this experience impacts an individual's ability to accurately report odors. The sections below discuss typical taste and odor descriptors and what each might indicate, along with design considerations and best practices.

Sulfur Derived: Marshy, Swampy, Septic, Rotten Eggs

If marshy, swampy, or septic odors are detected, volatile organic sulfur compounds (VOSCs) are most likely the cause. VOSCs are a combination of methylated oligosulfide compounds derived from hydrogen sulfide, with sulfide through trisulfide species being the most common. VOSCs are produced by aerobic and anaerobic organisms in distribution system biofilms (Wajon et al. 1985; Heitz et al., 2000; Franzmann et al. 2001; Heitz et al. 2002). Common sources of sulfur compounds are the conversion of inorganic sulfate to sulfide and degradation of sulfur-containing proteins including methionine and cysteine.

A severe episode of VOSCs occurred from the decomposition of a cyanobacterial bloom that was triggered by combined domestic and industrial pollution, which added nutrients. The decomposed cyanobacteria produced odorous dimethylpolysulfides and methyl thiols that degraded water quality in lake and tap waters serving 2 million people in China (Zhang et al. 2010; Ma et al. 2012). Research from a well-documented case in Perth, Australia (Heitz et al., 2000; Franzmann et al. 2001; Heitz et al. 2002), found that a sulfurous odor was attributed primarily to dimethyl trisulfide. Dimethyl disulfide and dimethyl sulfide were also found. In research performed in the laboratory with reactors that used biofilms isolated from Perth's distribution system, dimethyl disulfide and dimethyl trisulfide were produced from select organic precursors including the amino acid methionine. When the biofilm reactors were fed inorganic sulfide or oligosulfides, biomethylation occurred, producing dimethyl disulfide and dimethyl trisulfide. Abiotic production of dimethyl disulfide also occurred when chemical methylating agents were added to the biofilm.

Recommendations for monitoring. Monitoring for sulfurous odors is best achieved with sensory analyses as these compounds have extremely low odor detection thresholds (Table 5-2), which make analytical measurement a challenge. Monitoring for hydrogen sulfide is not sufficient as this inorganic species is usually not the main cause of the odor. Headspace solid-phase microextraction–gas chromatography–mass spectrometry (SPME-GC/MS) can be used for odorous volatile organic sulfur compounds in water (Kristiana et al. 2010), although the oligosulfide compounds can change their distribution during sampling, sample processing, and measurement. Closed-loop stripping analysis (CLSA) and/or purge and trap coupled with GC/MS can also be suitable methods for determination at trace levels (APHA 2012 methods 6040B, 6040C). An established air method (ASTM [2012]) will measure VOSCs at picograms of sulfur in air samples.

Design, operation, and/or maintenance considerations. Experiments to determine the effectiveness of oxidation of dimethyl disulfide and dimethyl trisulfide (5 μg/L) by free chlorine (0.2–0.6 mg/L as chlorine [Cl_2]) in distributed water showed that both compounds are rapidly and completely oxidized in water that contains a chlorine residual >0.4 mg/L (Heitz 2002). Water systems may wish to consider temporarily increasing or switching to free chlorine as a secondary disinfectant as a response action.

In source waters, potassium permanganate and activated carbon have been applied to remove VOSCs. For the incident in China related to degradation of cyanobacteria, an emergency water treatment process applied potassium permanganate (3–5 mg/L) at the intake of the source water influent pipeline to oxidize the VOSCs during transit to the drinking water treatment plant. This oxidation was followed by addition of powdered activated carbon (30–50 mg/L) to adsorb other pollutants and decompose the residual potassium permanganate (Zhang et al. 2007). Dimethyl trisulfide, one of the VOSCs present in the Chinese sulfurous odor incident, was shown in the laboratory to be efficiently oxidized by potassium permanganate and followed second-order kinetics (Ma et al. 2012). A design response may be to consider implementing these treatment technologies for recurring events.

Best management practices. Best management practices include controlling biofilms (see Chapter 3), minimizing water age (see Chapter 2), and source water management to minimize algal blooms and manage nutrients in the watershed and water (AWWA Manual M7 [2004]; AWWA Standard G300 and associated Operational Guide [2010]).

Emerging issue: premise plumbing. An emerging issue is that extended water age in premise plumbing due to low water usage can promote biofilm growth. If sulfate and sulfate-reducing bacteria are present, VOSCs and sulfurous odors can form.

Table 5-2 Odor threshold and descriptors for volatile inorganic and organic sulfur compounds

Compound	Formula	Odor Description	Odor Threshold in Air, μg/m^3	Odor Threshold in Water, ng/L	Natural Sources*
Hydrogen Sulfide	H_2S	Cooked eggs, rotten eggs	0.7–14†	—	Rotten eggs, swamp
Methyl sulfide	$(CH_3)SH$	Rotten cabbage	0.04–82†	20‡	Cheese, nuts, algae, tomato juice
Dimethyl sulfide	$(CH_3)_2S$	Decayed cabbage, canned corn	2.5–508†	42¶	Corn, cabbage, beet root, seafoods, "smell of the sea"
Dimethyl disulfide	$(CH_3)_2S_2$	Rotten cabbage	0.1–346†	640¶	Limburger cheese, broccoli, cauliflower, cabbage
Dimethyl trisulfide	$(CH_3)_2S_3$	Garlic, onion, rotten onion	6.2†	1.7¶	Onion, leek, broccoli, cabbage, stale beer

* www.thegoodscentscompany.com; http: Wikipedia; Parliament, Kolor, and Rizzo 1982; Buttery et al. 1976.
† Ruth 1986.
‡ Odor threshold from Guadagni, Miers, and Venstrom 1969.
¶ Water odor thresholds courtesy of Tsair-Fuh Lin, National Cheng Kung University, Tainan, Taiwan.

Responding to the DSWQ Challenge: Sulfur-Derived Taste and Odor

- Temporarily switch to free chlorine
- Implement potassium permanganate and/or powdered activated carbon treatment

Disinfectant Derived: Chlorinous, Ozonous, Medicinal

Chlorinous, earthy–musty (62 percent), and medicinal (25 percent) are among the most common odor and taste complaints reported by customers in the United States (Suffet et al. 1995) and Paris, France (Bruchet 1999). More recently (AWWA 2012), the most common odor and taste events detected in tap water in the United States also were chlorinous followed by earth–musty. Customer complaints concentrated on earthy–musty, chlorinous, and metallic. The main causes of taste and odor episodes in source and tap water were attributed to algal blooms (66 percent), disinfectant agent (57 percent), domestic plumbing (44 percent), the distribution system (34 percent), and reservoir or lake turnover (32 percent).

Disinfectant residuals. Major complaints about the taste and odor of drinking water are directly related to chlorine (bleach-like smell) and chloramines (swimming pool smell). The chlorinous odor is pH dependent; odor intensity decreases as pH increases from 5.5 to 8.5.

Several studies carried out in different countries have shown differences in chlorine perception by consumers attributed more to habituation than to intrinsic sensitivity differences (Piriou et al. 2014). This is shown in Table 5-3, which lists the odor thresholds and descriptors for several relevant compounds that cause chlorinous, ozonous, and medicinal tastes and odors in water. The odor threshold of free chlorine in water ranges from 0.20 to 1.1 mg/L as Cl_2 in the United States (Krasner and Barrett 1984; Mackey et al. 2004; Piriou et al. 2004), whereas an odor threshold of 0.15 mg/L as Cl_2 has been reported for dichloroamine (Krasner and Barrett 1984). When chloramination is practiced correctly, dichloramine should not be present at any significant level (see Chapter 6 for more information on proper chloramination).

Chlorine dioxide can also form detectable tastes and odors when residual concentrations reach 0.20–0.25 mg/L as Cl2 (Krasner and Barrett 1984; Piriou et al. 2004). Offensive odors (e.g., chlorinous, kerosene-like, and cat urine) from dioxichlorinated waters have been reported (Dietrich et al. 1992; Hoehn, Rosenblatt, and Gates 1996). This problem was attributed to chlorine dioxide regeneration at the consumer's tap by a reaction between chlorite and residual chlorine. Strong chlorinous–bleachy odors were produced by chlorine dioxide itself, while the others were caused by a reaction in the gas phase between the disinfectant and volatile organic compounds in household air.

Ozone has been reported to impart odor to waters (ozonous, oxidant), even when an ozone residual is not present (Bruchet et al. 1991).

Disinfection by-products. Another source of taste and odor arises from disinfection by-products (DBPs) formed when disinfectants are added to water. The best known DBPs of chlorine oxidation that are related to taste and odor problems are aldehydes, N-chloroaldimines, phenols, chlorophenols, and iodinated trihalomethanes (I-THMs).

Aldehydes and N-chloroaldimines are formed by chlorination of amino acids (Freuze et al. 2005). Chloramines can have the same reactions with amino acids, which are present in source waters (Dotson and Westerhoff 2009). Amino acids are often associated with extensive algal growth or wastewater discharge (How et al. 2014; Ishizawa et al. 2010), and can be major constituents of dissolved nitrogenous organic matter (Hagedorn et al. 2000).

Table 5-3 Compounds that cause chlorinous, ozonous, and medicinal tastes and odors in water

Compound	Origin	Odor	Odor Threshold in Water*
Chlorinous/ozonous			
Free chlorine	Disinfection of water	Chlorinous	0.25–1.1 mg/L Cl_2[†,‡,¶] France: 0.14–0.20 mg/L Cl_2[‡,‖,**] Spain: 0.56 mg/L Cl_2[‖]
Monochloramine	Disinfection of water	Chlorinous	1.8,[‡] 0.65[¶] mg/L Cl_2 France: 0.18 mg/L Cl_2[‡]
Dichloramine	Disinfection of water	Swimming pool	0.15mg/L Cl_2[¶]
Chlorine dioxide	Disinfection of water	Chlorinous, kerosene-like, cat-urine like	0.20–0.25 mg/L ClO_2[††] France: 0.20 mg/L ClO_2[‡]
Aldehydes (higher molecular weight >C_7)	Ozonation	Fruity/fragrant	Octanal: 0.32 μg/L[‡‡] Nonanal: 0.32 μg/L[‡‡] Decanal: 0.08 μg/L[‡‡] Undecanal: 0.03 μg/L[‡‡] Dodecanal: 0.13 μg/L[‡‡]
Aldehydes (low molecular weight)	Chlorination of amino acids, chloramination	Swimming pool	2-methyl propanal: 0.9–2.3 μg/L[¶¶] 2-methyl butanal: 0.15–2.0 μg/L[¶¶] 2-methyl butanal: 12.5 μg/L[¶¶] Phenylacetaldehyde: 4 μg/L[¶¶]
N-chloroaldimines	Chlorination of amino acids	Swimming pool	N-chlorophenylacetaldimine: 3 μg/L[‖ ‖] N-chloroisobutyraldamine: 0.20 μg/L[‖ ‖] N-chloroisovaleraldimine: 0.25 μg/L[‖ ‖]
Medicinal			
Iodinated trihalomethanes	Chloramination, chlorination	Medicinal, sweet, solvent	Iodoform: 0.03 μg/L[***] Bromodiiodomethane: 0.10 μg/L[***] Chlorodiiodomethane: 0.20 μg/L[***]
Phenols	Chlorination of phenols	Medicinal, plaster-like	2-iodophenol: 0.2 μg/L[†††] 2-chlorophenol: 0.047 μg/L[‡‡‡] 2-bromophenol: 0.043,[‡‡‡] 0.03[¶¶¶] μg/L 2,4-dibromophenol: 4.5 μg/L[‡‡‡] 2-iodo-4-cresol: 0.003 μg/L[†††]
		Medicinal	2,6-dibromophenol: 0.0005 μg/L[¶¶¶] 2,4,6-tribromophenol: 0.6 μg/L[¶¶¶]

NOTES: All levels are US values, unless otherwise denoted.

* Odor thresholds not associated with a country's regulations are reported based on experimental data.

† Mackey et al. 2004.

‡ Piriou et al. 2004.

¶ Krasner and Barrett 1984.

‖ Piriou et al. 2014.

** Puget et al. 2010.

†† Hoehn et al. 1990.

‡‡ Fabrellas, Matia, and Ventura 2004.

¶¶ Hrudey, Gac, and Daignault 1988; Hrudey 2009.

‖ ‖ Freuze et al. 2005.

*** Cancho et al. 2001.

††† Strube, Guth, and Buettner.

‡‡‡ Strube, Buettner, and Czerny 2012.

¶¶¶ Whitfield 1988.

Odor characteristics of aldehydes are usually associated with several descriptors, even for a single compound (Suffet et al. 1995). Low-molecular-weight aldehydes (see Table 5-3) with very low odor thresholds (low microgram per liter levels) are described as swimming pool and sweaty (Hrudey et al. 1988).

In the treatment plant, N-chloroaldimines may react with amino acids that have high Cl_2:N molar ratios (Brosillon et al. 2009). Their odor thresholds are close to 1 μg/L or lower and have strong swimming pool odors with a floral background (Freuze et al. 2005). Phenylacetonitrile, a specific odorant from the chlorination of phenylalanine, was recently identified in drinking waters in the United States and China (Ma et al. 2016).

Phenols leached from materials used in distribution systems or that come from urban runoff, industrial wastes, or natural organic compounds such as vanillic acid and *p*-hydroxybenzoic acid can react with chlorine and chloramines to form chlorophenols in the treatment plant (Mallevialle and Suffet 1987). Also, chlorophenols can be present as biocides or as degradation products of phenoxy herbicides. The formation of odorous chloro-and bromophenols is pH dependent (optimum formation is at pH 8–9) (Bruchet 1999), presenting a wide range of odor thresholds, with medicinal as the main taste and odor descriptor. These phenols can also create medicinal, plaster-like, or ink-like odors, and they exhibit low odor threshold concentrations (Strube, Buettner, and Czerny 2012). Other related compounds such as the 2-halogenated 4-methylphenols are odorants with the lowest odor threshold concentrations (Strube, Buettner, and Czerny 2012). The most relevant odorous halogenated phenolic compounds are summarized in Table 5-3, with values well below or close to the nanogram per liter level.

Another group of DBPs that impart medicinal odors to water are I-THMs (mainly iodoform and chloro- and bromodiiodinated species) that come from the reaction of chlorine (chloramines, ozone) with organic matter in the presence of bromides and iodides (see Table 5-4). Cancho et al. (2001) determined that the odor thresholds of I-THMs decrease with increasing iodine substitution, in other words:

iodoform > bromodiiodomethane >chlorodiodomethane >> monoiodinated

The chlorination process creates less medicinal odor than chloramination because the predominant THM is bromoform (odor threshold, 0.3 mg/L). At levels most commonly encountered in water systems, the regulated THMs do not have any discernible taste, odor, or color.

Aldehydes, ketones, and carboxylic acids are typical ozone DBPs. Several aldehydes (chains longer than seven carbon atoms [$>C_7$]) have been observed to impart tastes and odors in treated water (Bruchet et al. 1991), with fruity, fragrant, and orange-like odors as the main descriptors and odor thresholds as low as 1 μg/L (Fabrellas, Mantia, and Ventura 2004). Tastes and odors have not been identified with other ozone by-products (Suffet et al. 1995).

Recommendations for monitoring. See Chapter 8 and Appendix C for discussions on the physicochemical analyses for disinfectants and DBPs. DBPs that cause taste and odor events in distribution systems can be analyzed efficiently at trace levels (nanograms per liter) with a wide range of analytical techniques.

Design, operation, and/or maintenance considerations. Consumers are accustomed to the type of disinfectant that is in their tap water, or lack thereof. They may notice a change in this agent and complain (Muylwyk, MacDonald, and Klawunn 2005; García and Devesa 2015). Maintaining a consistent type of disinfectant and consistent concentration at consumers' taps can prevent potential complaints. In order to improve the water's flavor, the disinfectant residual should be maintained at the minimum needed to ensure that the water is sanitary. See Chapters 3 and 8 for more information on disinfectant residuals.

Table 5-4 Odorous chemicals that leach from polymer pipes and have or may have regulatory limits

Chemical Contaminant	Polymer Material Source	Regulatory Concern
Benzene	HDPE, epoxy, PEX*	PMCL† = 0.005 mg/L
Ethylbenzene	HDPE, epoxy	PMCL = 0.7 mg/L
Styrene	HDPE	PMCL = 0.1 mg/L
Trichloroethene	HDPE	PMCL = 0.005 mg/L
Toluene	PEX, epoxy	PMCL = 1 mg/L
Xylenes (total)	HDPE, epoxy, PEX class B, PEX	PMCL = 10 mg/L
n-Propylbenzene	HDPE	Proposed on CCL3‡
Methyl-t-butyl ether	PEX class A, PEX class C, PEX	Proposed on CCL3
Ethyl-t-butyl ether	PEX class B	Odorant
Bisphenol compounds	HDPE, epoxy	Potential endocrine disruptor

Sources: Satchwill 2002; USEPA 2002a; Skjevrak et al. 2003; Durand and Dietrich 2007; Heim and Dietrich,2007; Deb et al. 2010; Whelton and Nguyen 2013.

Abbreviations: CCL, Contaminant Candidate List; HDPE, high-density polyethylene; PEX, polyethylene; PMCL, primary maximum contaminant level.

* PEX type was not specified.

† USEPA 2012.

‡ USEPA 2015.

A regular flushing program can be useful to maintaining good water quality and minimizing customer complaints. See Appendix B for more information on flushing techniques.

Maximum disinfectant levels are regulated by the USEPA in the United States and include 4.0 mg/L as Cl_2 for chlorine, 4.0 mg/L as Cl_2 for chloramines (based on the running annual average of all distribution samples), and 0.8 mg/L as Cl_2 for chlorine dioxide (based on a routine or triggered monthly three-sample set from the distribution system). In Canada, chloramines are limited to 3.0 mg/L measured as total chlorine. The European Commission Drinking Water Directive does not regulate disinfectant levels, but several EU members states have implemented maximum admissible concentrations (EC DWD 1998). Most international regulations mention the concept of a minimum disinfectant concentration in the distribution system.

In general, blending waters treated with different secondary disinfectants, especially free chlorine and chloramines, should be avoided (Muylwyk 2005). Blending chlorinated and chlorine dioxide–treated waters may also be a concern (Hoehn, Rosenblatt, and Gates 1996; Gates 1998). It has been shown that the addition of chloramines as a secondary disinfectant with chlorine dioxide used as a primary disinfectant can be effective (Hoehn et al. 1990, 1996; Gates 1998). Switching from chlorination to chloramination is, in principle, recommended strictly as it relates to the sensory aspect of water, as consumers are less sensitive to chloramines than to free chlorine.

Formation of low-molecular-weight odorous aldehydes peaks at chlorine-to-amino acid molar ratios of 1.5 (at pH 7 and 2-hour reaction times), and lower concentrations are formed at higher molar ratios (Hrudey et al. 1989). The reaction of chloramines with amino acids also yields aldehydes, but at longer reaction times (24 hours). Aldehyde generation by chlorine, chloramines, ozone, and chlorine dioxide was compared, and it was shown that only free chlorine and monochloramine produced substantial yields of the odorous aldehydes under conditions relevant to drinking water treatment (Froese, Wolanski, and Hrudey 1999). In general, chloramination is a better option than chlorination for aldehyde minimization. If the disinfectant is changed (such as during breakpoint free chlorination), consumers should be notified proactively in order to lessen potential complaints.

Ozonation leads to the complete elimination of fishy odors and a significant decrease of earthy and muddy descriptors. However, ozonation also causes the formation of aldehydes, including long-chain aldehydes (>C_7) that produce fruity and fragrant odors (Anselme, Suffet, and Mallevialle 1988). In treatment plants that use ozonation, biologically active granular activated carbon filters can effectively remove aldehydes (Weinberg et al. 1993).

N-chloroaldimines with half-lives of about 50 hours at 20°C are stable enough to be present in tap water (Freuze et al. 2005). Brosillon et al. (2009) studied the influence of contact time on the formation of by-products from chlorination of model amino acids and showed that their concentrations decreased after 10 hours. Therefore, N-chloroaldimines are likely present near the outlet of the treatment plant but are less likely to exist far from the treatment plant.

To avoid medicinal odors (ie, from I-THMs) from the chloramination process, it has been proven that chlorine must be added before ammonia in order to form bromoform instead of iodoform, which is usually produced if ammonia is added first (Hansson et al. 1987). Since most THM precursors are consumed during the chlorination process, chlorination is more efficient than chloramination in removing medicinal odors (Duguet et al. 1995).

The efficient removal of phenols and chlorophenols (>80 percent) from water using ozone has been documented (Manojlovic et al. 2007). Efficient removal has also been documented for granular activated carbon (Zogorski, Faust, and Haas 1976; Hamdaoui and Naffrechoux 2007). Zogorski and colleagues (1976) studied the kinetics of adsorption of phenols on granular activated carbon and observed that 60 percent to 80 percent of the adsorption occurs within the first hour of contact.

Responding to the DSWQ Challenge: Disinfectant-Derived Taste and Odor

- Maintain consistent disinfectant type and concentration at customer taps
- Implement flushing programs
- Avoid blending waters with different secondary disinfectants

Best management practices for minimizing aesthetic challenges. New or repaired pipelines and tank disinfection, imbalance in the chlorine-to-ammonia ratio in the chloramination process, residual boosting, blending of waters (i.e., different qualities or chlorinated/chloraminated), continuing reactions, and DBP formation are possible sources of chlorinous taste and odor problems (Khiari et al. 2002; Lauer 2014). Corrective actions suggested for each potential source include:

- Flushing water mains with dechlorinated water or discharging superchlorinated water, if necessary;
- Optimizing the weight ratio of chlorine to ammonia nitrogen for chloraminated water;
- Chloraminating the distribution system; and
- Optimizing treatment to diminish odorous DBPs.

Regarding medicinal/phenolic taste and odor events, it has been proven that chlorinated water in a reservoir and tank linings may leach halogenated phenols and yield medicinal flavors or further transform into halogenated anisoles to produce earthy–musty events. Proper curing times on new coatings and flushing newly coated tanks and pipelines after filling and storage are possible corrective measures (Khiari et al. 2002).

Polymers and Organic Material Derived: Chemical and Hydrocarbons

Odorous compounds that leach from polymer pipes and distribution materials (e.g., gaskets and lubricants) into drinking water are well-known aesthetic issues (Burlingame and Anselme 1995; Rigal and Danjou 1999; USEPA 2002a; Khiari et al. 2002; Skjevrak et al. 2003; Marchesan and Morran 2004; Tomboulian et al. 2004; Dietrich 2006; Deb et al. 2010; Whelton and Nguyen 2013). To date, more than 150 contaminants that migrate from polymeric pipes have been identified and compiled (Whelton and Nguyen 2013). Reports of odors from polymer materials are increasing as these materials are more frequently used in distributions systems and premise plumbing for both new and replacement materials in order to address the aging drinking water infrastructure (USEPA 2002b; see Chapter 4 for more information on aging infrastructure).

Odors from polymer materials can arise from several sources, including:

- Leaching or migration of odorous chemicals directly from the material,
- Chemical conversion of nonodorous compounds that leach or migrate from a material to an odorous chemical through chemical reaction in the distribution system, and
- Biochemical conversion of nonodorous compounds that leach or migrate from a material to an odorous chemical through a microbial reaction in the distribution system.

An overall aesthetic issue is the leaching of plastic–chemical–solvent odors directly from polymer materials. In some instances of direct leaching of odorous chemicals, the identity of the odorant is known. As shown in Table 5-4, specific odorous aromatic chemicals including benzene, toluene, xylenes, propyl benzene, styrene, and bisphenols are known to leach from polymers into drinking water. Polyethylene (PEX) pipes, or cross-linked polyethylenes, are common in drinking water distribution systems. PEX is designated A, B, or C depending on the process used to cross link the polyethylene. Leaching of ethyl-t-butyl ether, which has a solvent-like odor, from cross-linked PEX class B pipe (Skjevrak et al. 2003; Durand and Dietrich 2007) and leaching of methyl-t-butyl ether, which has a sweet solvent-like odor, from PEX class A, B, and C pipes (Skjevrak et al. 2003; Whelton and Nguyen 2013) are other sources of plastic–chemical–solvent odors. Hindered phenols, which are used as antioxidants in polyethylene pipes, are also leached from pipes into water (Viebke and Gedde 1997; Heim and Dietrich 2007; Kelley et al. 2014).

Once leached into the distribution system, phenolic compounds can react with chlorine and bromine to form more odorous halogenated phenols that have medicinal odors at parts-per-billion levels. The halophenols can subsequently be biomethylated to form haloanisoles that have musty odors at parts-per-trillion levels (Díaz et al. 2005). Similarly, chlorine and chloramines react with oleate-based pipe lubricants (vegetable soaps) to form rancid odors (Burlingame et al. 1995). Thus, the odors that result from distribution system materials can be from odorants that are directly leached or from odorous reaction products formed during distribution.

Epoxy liners are used in distribution systems, and there are no odor guidelines associated with their installation or use (AWWA 2015). Satchwill (2002) found that odorous benzene, toluene, ethylbenzene, and xylene leached into water from five National Sanitation Foundation- and AWWA-approved new epoxy resins. While new epoxy liners cause plastic- and solvent-like odors that may last for weeks, they also release bisphenol A (Deb et al. 2010). Bisphenol A produces medicinal odors and also has potential health effects; it is a potential endocrine disruptor (Tomboulian et al. 2004; Heim and Dietrich 2007). New PVC pipes have minimal odors, although pipe-joining compounds do release odors (Wiesenthal et al. 2007), and leaching of vinyl chloride should be considered for health reasons. One study on polyurethane liners demonstrated that this polymer results in minimal odors (Deb et al. 2010).

Another concern with polymer pipes is degradation caused by reaction with chlorine and chloramine (Hassinen et al. 2004; Heim and Dietrich 2007; Whelton and Dietrich 2009; Castillo-Montes et al. 2012; Yu et al. 2012; Mitroka et al. 2013) or chlorine dioxide (Colin et al. 2009; Yu et al. 2011). Chloramines attacked and severely degraded elastomer materials in the distribution system and premise plumbing; chloramines also attacked some thermoplastic materials, but the degradation was not as severe (Reiber 1993). This degradation can lead to odors as well as pipe failures. Likewise, polymer pipes and liners can sorb and then desorb odorants; this creates the potential for odors to linger in the distribution system (Whelton and Nguyen 2013; Ahart et al. 2016)

Recommendations for monitoring. Prior migration/leaching tests with sensory analysis of materials in contact with drinking water with and without disinfectants is the key to avoiding taste and odor issues associated with materials. The Utility Quick Test was developed for this application (Schweitzer et al. 2004) and has been successfully performed (Durand and Dietrich 2007; Heim and Dietrich 2007). Other static migration/leaching protocols such as British Standard 6920-1:2014, European Standard 1420, American National Standards Institute/National Sanitation Foundation Standards 60 and 61, and Australia/New Zealand StandardTM 4020:2005 can also be applied if sensory analysis is included.

Once water is contaminated with tastes and odor from materials, a sensory test can be performed (see the section Sensory Analysis: Taste, Odor, and Color earlier in this chapter). FPA (APHA 2012, method 2170) can be performed in order to describe any odor and assess its intensity. Other sensory methods can provide data for overall odor by rating specific odors or tastes, determining differences, and determining acceptability. Routine sensory monitoring of finished drinking water benefits both water systems and customers by identifying and avoiding aesthetic issues. Two field guides (Lauer 2014; Burlingame et al. 2011) are available for assessing aesthetic issues in source water, during treatment, and at customer's homes.

Design, operation, and/or maintenance considerations. Contaminant migration from polymer pipes is highest during the initial exposure of water to the new materials. For new materials—pipes, plumbing, gaskets, lubricants, mains, asphaltic coatings, epoxy/solvent-based linings—the recommendation is to fill, hold, and flush newly lined mains. Contamination from permeation of buried plastic pipes has been reported, and an appropriate corrective action is to replace or remove the contaminated soils that surround the pipelines (Khiari et al. 2002). Other problems arise from changes in drinking water quality, that is, blending waters from different sources or switching drinking water treatment practices that could increase contaminant migration from polymer pipes. Many of these water quality issues have historically been handled on a case-by-case basis by the water systems, and an overview of the problem is lacking.

Responding to the DSWQ Challenge: Polymers and Organic Material Derived

- Fill, hold, and flush newly lined mains
- Remove contaminated soils that surround polymer-based pipelines
- Replace deteriorated sections of polymer-based pipes

Best management practices for minimizing aesthetic challenge. Water age, or the length of time that drinking water is in contact with material, will affect the amount of odors and contaminants that are leached. Therefore, minimizing water age can limit the production of odors. Control of biofilms is needed because it has been proven that biofilms that line the inner walls of pipe can affect contaminant migration. See Chapter 2 for a discussion of water age and capacity issues.

Emerging issues and technologies. Development of materials for distributions systems is an ongoing process. Production of odors during and after polymer formation should be considered prior to use of these materials in water distribution systems. The polymer industry should develop materials that have low odor potential and that are resistant to attack from disinfectants used in drinking water. An emerging issue requiring further research is the role of polymer pipe contamination in biofilm formation and how this affects contaminant migration from the pipe into the water.

Also, for regulatory planning purposes, water systems must have access to information regarding which contaminants migrate and at what quantities, taking into account that manufacturers throughout the world use different ingredients and manufacturing processes. There is insufficient information available to predict how polymer pipes will affect long-term water quality, impact long-term water system regulatory compliance (Whelton and Nguyen 2013), and how their long-term performance is impacted by use of current and future disinfectants (Hassinen et al. 2004; Colin et al. 2009).

Metal Pipe Corrosion Derived: Copper, Iron, and Zinc

Aesthetic issues due to corrosion of metal pipes are widespread (Ömür-Özbek 2012; Dietrich and Burlingame 2015). Corrosion of metal from public water distribution system pipes is the primary source of iron in tap water (Kirmeyer et al. 2002; Imran et al. 2005; Burlingame et al. 2006; Benson, Dietrich, and Gallagher 2012) and can also be a source of zinc (DeZuane 1997). Corrosion of premise plumbing is the primary source of copper (Edwards, Schock, and Meyer 1996; Dietrich et al. 2004c), although corrosion of brass (e.g., in fittings, propellers) can contribute both copper and zinc (Lauer 2014; Sarver, Zhang, and Edwards 2011). As shown in Table 5-1, worldwide, aesthetic guidance for copper ranges from 1 to 2 mg/L, from 0.2 to 0.3 mg/L for iron, and from 3 to 5 mg/L for zinc.

The aesthetic guidance for copper and iron are designed to control both metallic taste and color (see the section on Discolored Appearance below) and have existed for decades. The USEPA SMCLs (1979) were established in 1962 (USPHS 1962). In the 21st century, research revealed that cupric, cuprous, and ferrous iron cause both taste and odor—or "flavor"—in drinking water at levels below their SMCLs (Glindemann et al. 2006; Dietrich 2009). Aesthetic guidance for copper should be lowered to between 0.4 and 0.8 mg/L to avoid the metallic/bitter flavor, astringent mouthfeel, and blue–green staining of household plumbing fixtures, sinks, and toilets (Cuppett, Duncan, and Dietrich 2006; Hong et al. 2006, 2010a, 2010b; Dietrich, Cuppett, and Duncan 2008; Ömür-Özbek and Dietrich 2011; Dietrich and Burlingame 2015).

Dissolved copper has substantial flavor, while particulate copper has very little flavor. Ferrous iron typically occurs as a colorless dissolved ion in drinking water and causes metallic flavor (Ömür-Özbek and Dietrich 2011; Mirlohi, Dietrich, and Duncan 2011; Dietrich and Burlingame 2015). Customers can detect the ferrous flavor at 0.003 to 0.17 mg/L, which is below aesthetic guidance. Systems should therefore consider a lower value as a goal if ferrous flavors are a system-specific issue.

Ferric iron is largely insoluble and has nearly no taste in drinking water. Ferric iron is primarily responsible for the yellow–orange color of iron that is visually detectable at concentrations below the SMCL (Sain and Dietrich 2015; Dietrich 2015).

Zinc is leached from galvanized pipes in premise plumbing and water mains and also from corrosion of brass (Lauer 2014; Sarver, Zhang, and Edwards 2011). At concentrations of 5 to 30 mg/L of zinc in drinking water, a milky appearance can occur from precipitation (USEPA 1979; Lauer 2014). Using hot water to make beverages such as tea can also form a greasy surface scum when zinc is present (DeZuane 1997). Zinc imparts an astringent taste at >5 mg/L (DeZuane 1997; Keast 2003).

Recommendations for monitoring. Copper, iron, and zinc can be monitored at or below their aesthetic guidelines using atomic absorption spectrophotometry, atomic emission spectrophotometry, or inductively coupled plasma mass spectrometry (APHA 2012; Dietrich 2015). Phenathroline can be used to measure ferrous ion colorimetrically. Copper has health-based action levels of 1.3 to 2 mg/L in drinking water (Table 5-1) that are slightly higher than aesthetic guidelines. For example, the USEPA secondary guideline for copper is 1.0 mg/L.

Copper is mainly a premise plumbing issue; therefore, monitoring for copper must be performed at customers' taps according to the USEPA Lead and Copper Rule (USEPA 1991). Likewise, iron and zinc occur at the tap primarily from corrosion of premise plumbing or distribution system materials that produce ferrous ion, ferric ion, iron particulates, and zinc ion in tap water. This suggests that particulate and dissolved iron and zinc be monitored at customers' taps in order to detect and respond to aesthetic issues. For groundwater sources, ferrous ion can be present in the groundwater, in which case monitoring at the source and tap is recommended.

Design, operation, and/or maintenance considerations. Design, operation, and maintenance considerations for minimizing taste and odor related to metal pipe corrosion is similar to that used to mitigate internal corrosion of these pipelines, as discussed in Chapter 7. Design of metallic distribution system pipelines typically accounts for structural considerations such as flow rate and pressure class requirements; external pipe corrosion is based on local soil conditions. Internal corrosion is dealt with as a water quality issue because adequate methods to predict corrosion phenomena that could adequately direct the design of noncorrodible metallic pipelines do not exist (Water Research Foundation 1996).

Operational and maintenance considerations to minimize taste and odor from corroded metallic pipe include flushing (see Appendix B), pigging, chemical treatment for corrosion control, and rehabilitation or replacement of severely corroded pipe segments. The cause of corrosion in a distribution system pipeline should be carefully investigated before committing to one or more of these operation and maintenance options.

Responding to the DSWQ Challenge: Metal Pipe Corrosion Derived

- Flushing
- Pigging
- Chemical treatment for corrosion control
- Rehabilitation or replacement of corroded pipe segments

Best management practices for minimizing aesthetic challenge. Best management practices for minimizing taste and odor issues related to corroded metals in the distribution system include (AWWA 2010) the following:

- Develop a prevent-and-response program for internal pipe corrosion that includes a water quality monitoring and sampling plan, pipeline condition assessment plan, lead and copper control procedures, and guidelines for controlling zinc and ferric iron.
- Develop an action plan to receive taste and odor complaints from customers by trained personnel who use a call system that can properly identify and track complaints and relay the information to the appropriate response team for timely resolution.
- Develop policies and procedures for reviewing potential water quality degradation issues related to design and construction of new and rehabilitated distribution system infrastructure.

Emerging issues and technologies. Metallic corrosion in water distribution systems is a complex interaction of water with the metal pipeline that transports it. Controlling its effects by the time the water reaches customers' taps is no less complicated as the control strategy for water quality, metal type, and combination thereof, and may not be as effective in another situation. Research into the internal corrosion of metallic pipes and, by extension, the taste and odor issues it causes is ongoing in the water industry. Recent research has focused on improving the water industry's understanding of corrosion mechanisms in distribution systems and premise plumbing, including microbially induced corrosion mechanisms.

Corrosion is described in depth in Chapter 7 and microbial issues are discussed in Chapter 3.

Mineral Derived: Salty, Concrete (Aluminum, Calcium, Magnesium)

Minerals occur naturally in source water and drinking water and have mainly geological origin from weathering and erosion of rocks and soils. Minerals in drinking water are desirable, as consumers prefer water that has some mineral content, typically between 80 and 350 mg/L TDS depending on the customer's preference (Bruvold 1968; Bruvold and Daniels 1990; Burlingame, Dietrich, and Whelton 2007; Teillet et al. 2010; Platikanov et al. 2013; García et al. 2014). An appropriate level of minerals is necessary to avoid corrosion (see Chapter 7). Too much mineral content, which is often measured as TDS, conductivity, or by individual cations and anions, results in a salty or mineral taste that customers find undesirable.

Another aspect to the aesthetic challenge is providing the right amount of total minerals and individual anions and cations. Pure water may be colorless and odorless but it can have an objectionable taste. Table 5-5 presents a summary of what is known about specific minerals and their corresponding tastes based on their concentrations in room temperature water. Providing taste guidance presents challenges, as the taste of minerals varies as a function of temperature. Room temperature water has substantially more taste than chilled or hot water (Pangborn and Bertolero 1972; Gallagher and Dietrich 2010). Temperature also affects the ability to discriminate between water samples with differing TDS; customers can most readily discriminate the taste of TDS in room temperature water (Gallagher and Dietrich 2010).

It is important to note that much guidance in the drinking water industry was developed prior to 1980 and focused on which anions and cations produce a salty taste. The leading contenders were chloride and sulfate, for which there is worldwide guidance on concentrations that are not to be exceeded. In fact, in the 1980s, health and sensory scientists revealed that sodium is primarily responsible for the salty taste, and its intensity is moderated by the anion present (National Academies of Science 2010, citing Bartoshuk 1980; Weiffenbach, Baum, and Burghauser 1982; USEPA 2003). Thus, the drinking water industry needs to update its guidance on salty tastes and sodium.

Anthropogenic sources. A number of anthropogenic sources can present a noticeable contribution to the mineral content of source water and drinking water, such as deicing agents used on roads (sodium and potassium chloride); industrial discharges; agricultural water, particularly from irrigation return; and leaching from cementitious materials in pipes. Customers become accustomed to the mineral level of their drinking water. When one of these anthropogenic sources impacts a drinking water supply, customers may notice the change and complain. The effects of sodium increase on salty taste due to road salt have been discussed elsewhere (Dietrich and Burlingame 2015).

A classic case study of salinity remediation is that for the River Llobregat, in Catalonia, Spain. A 124-km brine collector was constructed to prevent salt spills and impurities from sodium and potassium chloride mines. The salty taste of the water was dramatically reduced (Martín-Alonso 1994; Matía 1995).

Concrete and cement-based pipes. About 50 percent of the total length of water distribution systems in the United States consist of cement-based piping (USEPA 2007), basically concrete and cement-lined pipes. Significant release of calcium and aluminum from cement-based materials in the distribution system has been reported related to aggressive waters and new pipes (Berend and Trouworst 1999; Deb et al. 2010; Zielina et al. 2014). This circumstance is especially important when water systems introduce desalinated water into their distribution systems (Zhang et al. 2012; Blute et al. 2008). To control the aggressiveness of desalinated waters, it is crucial to perform proper remineralization. More detailed information about corrosive water is provided in Chapter 7.

Remineralization is also important in maintaining acceptable taste (Devesa, Garcia, and Matia 2010; Duranceau, Wilder, and Susaye 2012). Studies have shown that (zinc) orthophosphate corrosion inhibitors (used to mitigate lead and copper release) tend to decrease the degradation of cement-based materials, although this effect is pH dependent (Parks et al. 2012). In any case, the degradation of these cement-based materials is primarily a sanitary issue (Berend and Trouwborst 1999; Berend, van de Voet, and Boer 2001) rather than a taste and odor issue. Aluminum comes mainly from the coagulant used to treat water, which has a taste threshold of 4 to 10 mg/L (Young et al. 1996), while the permitted level in international regulations and the US SMCL is 50 to 200µg/L.

Table 5-5 Taste characteristics and regulations of major components in waters

Parameter	Threshold Taste Concentration, mg/L	Perception	Aesthetic Guidance, mg/L			
			United States	Canada	European Union	World Health Organization
Sodium (Na^+)	30–460* 51 ($-NO_3^-$) 59–82 ($-Cl^-$) 65–81 ($-SO_4^{2-}$) 115–290 ($-HCO_3^-$)	Salty	NA	≤200	≤200	≤200
Potassium (K^+)	162 (Cl^-)	Salty/bitter	NA	NA	NA	NA
Calcium (Ca^{2+})	100–300*	Bitter	NA	NA	NA	≤80*
Magnesium (Mg^{2+})	100–500*	Bitter/ astringent	NA	NA	NA	NA
Chloride (Cl^-)	200–300* 91–127 (Na^+)	Salty, cation modulated	≤250	≤250	≤250	≤250
Sulfate (SO_2- 4)	169 (Na^+)–705 (Ca^{2+}) 83 (Ca^{2+})–159 (Na^+)	Salty, gypsum	≤250	≤500	≤250	≤250
pH‡	NA	Low: sour, bitter, metallic High: soda, slippery feel	6.5 to 8.5	6.5 to 8.5	6.5 to 9.5	6.5 to 9.5
Total dissolved solids	NA	Objectionable taste for high values	≤500	≤500	≤1,600¶	≤1,000

Data based on Matia 1995; Burlingame, Dietrich, and Whelton 2007; and Dietrich and Burlingame 2015. A plus sign (+) indicates that the taste threshold depends on the co-ion. The co-ion is indicated within parentheses.

Abbreviation: NA, not applicable.

* Corresponding to a hardness of 200 mg/L $CaCO_3$.

† 10 mg/L as nitrogen.

‡ Units are pH units.

¶ Corresponding to a conductivity of 2,500 μS/cm (20°C).

Recommendations for monitoring. Water systems should continuously monitor TDS levels in their source water and finished water. Conductivity measurement is generally used for this purpose due its simplicity. TDS/conductivity provide quick and extremely useful information for the initial evaluation of taste changes in the distribution network. The next step involves identification and monitoring of the specific species responsible for the change in TDS/conductivity. This requires physicochemical analysis for the determination of parameters such as pH, alkalinity, hardness, and individual cations and anions; sodium, potassium, calcium, magnesium, chloride, and sulfate are among the most important.

Sensory tests, described earlier in this chapter, can be used to monitor the mineral taste of drinking water. FPA (APHA 2012; method 2170) can be used to estimate salt intensity and, less frequently, for acid and bitter tastes. The established flavor threshold test (APHA 2012; method 2160 B) can be used to determine the dilution value at which a taste is not perceptible. Flavor rating analysis (APHA 2011; method 2160) can be used to determine if the water has a taste that customers find acceptable. The attribute rating test (Dietrich et al. 2004) can be used to determine if the drinking water meets or exceeds a target TDS or individual anion/cation concentration at which customers would find the water objectionable. Discrimination tests such as the triangle test and 2-of-5 test (Meilgaard, Carr, and Civille 2006; Dietrich et al. 2004; Devesa, Garcia, and Matia 2010) can be used to determine when samples are different in their TDS or specific mineral content.

Design, operation, and/or maintenance considerations. Water systems need a detailed management program of their available resources. Special attention should be paid to the TDS of the distributed water. Blending may be an option to homogenize the water supply and avoid delivery of salty water to certain areas of the system. However, care should be taken for any unintended consequences.

The installation of membranes can be considered when decreasing TDS is a major objective because it is the only treatment technique capable of removing all of the major and minor dissolved ions. Consideration of desalination is especially important if new sources are to be treated or treatment improvements are to be made. In this case, some previous taste experiments with trained panels or customers have been shown to be extremely useful (McGuire et al. 2007a, 2007b; Biggs et al. 2007; Devesa, Garcia, and Matia 2010).

When relevant, marine intrusion, also referred to as saltwater intrusion, can be mitigated using different methods such as reduction of the abstraction flow, artificial recharge, and physical water barriers. A proper hydrological study must be performed in order to choose the best option. Brine and industrial collectors should be properly maintained, and the circulating waters regularly monitored to minimize breaks and cross-connections.

Responding to the DSWQ Challenge: Mineral Derived

- Perform a hydraulic study to determine the best mitigation method
- Properly maintain brine and industrial collectors
- Monitor circulating waters regularly to minimize breaks and cross-connections

Best management practices for minimizing aesthetic challenge. Customers want tap water with consistent quality and homogeneous taste. For water systems that manage several water sources, it is common to supply different waters to different distribution areas. Although spatial homogeneity of the supply is not possible, maintaining the characteristics of the delivered water over time is emphasized. Water systems should avoid extensive changes in the physicochemical characteristics of the water supplied because it is a common cause of complaints. Blending of different waters, including desalted seawater (Kilmore 2009; García et al. 2014), can be useful for managing available resources and for mitigating any incidence or special circumstance in the distribution system.

In the case of new concrete mains or cement-mortar lining, initial conditioning is necessary. Water systems should monitor changes in water quality due to leaching (e.g., pH, alkalinity, aluminum, free chlorine). This is especially important when aggressive waters are distributed. In general, cement-mortar lining should be avoided in dead ends and in mains with high hydraulic residence time (Deb et al. 2010). Initial and scheduled flushings are also highly recommended (Deb et al. 2002, 2010; Firat Server 2012).

Emerging issues and technologies. The increasing use of reclaimed waters and climate change are both significant reasons for the increase in TDS in drinking water sources worldwide (Stanford et al. 2014). Hydraulic fracturing involves the use of substantial quantities of fresh water and the conversion of that water to brines with >200,000 mg/L TDS. Remineralization of permeates is another fundamental method for controlling the water's corrosivity but also for increasing water's palatability. A variety of technical methods are available for desalination and remineralization (Duranceau, Wilder, and Susaye 2012). Water systems should choose the most appropriate system and operating conditions for each water and supply target.

Regarding the presence of undesirable compounds from products and materials that come in contact with drinking water, organoleptic and chemical migration tests and approval schemes have been developed for different types of materials. However, none of these methods have approved standard methods.

APPEARANCE

Turbidity

A turbid appearance (or turbidity) in different types of waters originates from natural and anthropogenic sources. These sources include clays and silts from the weathering of rocks and soils; inorganic precipitates such as hydrous oxides of iron and manganese, calcium carbonate, and basic salts of aluminum and zinc; natural organic matter from soils (humic), plants, and animal debris; and microorganisms (e.g., algae, cyanobacteria, protozoa, biofilm detachment).

Water turbidity is measured using nephelometry, which is based on the intensity of scattered light produced by the particles suspended in the water (methods are described in Appendix C. High turbidities generally correspond to a large number of particles. However, it is not possible to establish a correlation between these parameters. Scattered light is an aggregate response that depends primarily on the concentration of particles, with influence from their size, shape, color, and refractive index and on some characteristics of the measurement instrument. Therefore, a decrease in turbidity does not necessarily imply a lower particle concentration if a significant change in particle size has taken place, including aggregation, flocculation, and disaggregation processes.

Turbidity is a crucial parameter in water monitoring for both aesthetic and health reasons. Consumers reject water with aesthetic alterations or when they perceive a significant change. Consumers look for consistency in their tap water just as they do for other products (Dietrich 2006). The main terms that consumers use to describe their concerns about tap water quality include unusual appearance, dirty, turbid, cloudy, yellow, black, rusty, muddy, colored, sand, sediments, tar, and foreign particles (Whelton et al. 2007; Lauer 2014). Consumers can visually detect turbidity at home when it is approximately 4 ntu (WHO 2011)

Turbidity is a primary parameter monitored by USEPA and Health Canada and is an indicator (secondary) parameter in Europe. Maximum acceptable levels depend on the particular regulation, type of source water, and type of treatment. The reference value for water in the distribution system is 1.0 ntu worldwide. Due to its nonspecific nature, turbidity has no health effects per se. Most of the particles commonly found in drinking water are not considered to be dangerous. However, turbidity is clearly associated with water's microbiological quality, although at the network uses to be more indicator of the system disturbance rather than presence of pathogens). Suspended matter can harbor microorganisms, protecting them from the action of disinfectants. Turbidity measurements are used in conjunction with disinfectant residuals and microbiological indicators to check the quality of the water in the distribution system. High turbidity values mean a potential risk of pathogens; in most instances, values <1.0 ntu are recommended. For this reason, in some countries such as the United States and Canada, turbidity is included in the microbiological list of parameters. The measurement is less relevant for chemical health risks (Ritter et al. 2002), for example, due to the adsorption of heavy metals and other contaminants by the suspended matter.

There is a strong relationship between turbidity and color. Turbidity spikes in the distribution system are often due to metallic particles and colloids from corrosion processes, especially in the case of unlined cast-iron pipes; this is referred to as red water. Turbidity profiles during unidirectional flushing have been used to evaluate corrosion of iron mains in several water systems in the United Kingdom (Husband and Boxall 2011). Sometimes instances of elevated turbidity are associated with bad taste and odor. Appendix B describes unidirectional flushing in more detail.

Turbidity that has a white or milky appearance at a customer's tap may be caused by air in the water. There is an easy way to confirm this. When a glass of water is filled, the bubbles will rise to the top in 1 to 2 minutes and the cloudiness will gradually disappear. The bubbles are formed when the solubility of air decreases because of a decrease in pressure or an increase in temperature that may occur in the distribution system or household plumbing. This situation is more common in faucets that have aerators, and the phenomenon has no relationship with the quality of the water (Mc Donald 2000; Lauer 2014).

Recommendations for monitoring. Turbidity monitoring is essential in water treatment as an indicator of the effectiveness of treatment processes, particularly filtration. Individual filtration units (including membrane modules) should be systematically monitored. High turbidity values or significant changes and fluctuations are indicative of alterations in source water, problems in treatment (including insufficient dissolution of lime for remineralization), or disturbances in the distribution system. Turbidity monitoring is extremely useful for checking the water quality and/or the situation within the water supply network in conjunction with other parameters such as disinfectant residual, microbiological indicators, color, and taste and odor (Kirmeyer et al. 2002). Turbidity measurement is simple, reliable, rapid, and inexpensive and can be performed in situ. Automatic online turbidimeters are typically the best option for continuous monitoring of the distribution system.

Design, operation, and/or maintenance considerations for addressing challenge. A sudden increase in turbidity may indicate a lack of system integrity or a significant disturbance within the distribution system, such as a treatment malfunction, contaminant intrusion and cross-connections, main breaks, valve failures, system maintenance or repairs, or flushing (Lauer 2014; WHO 2011). Increased turbidity may also be caused by release of corrosion products and biofilm from water mains (Husband and Boxall 2011). In any case, when an unexpected increase in turbidity occurs, the cause must be identified and proper corrective actions implemented.

Flushing helps maintain the "freshness" of water that reaches the customers' taps. Chapter 2 discusses the degradation of water quality with excessive water age, and Appendix B discusses different flushing techniques. Spot flushing of water mains is a common corrective action in response to customer complaints about turbidity and other aesthetic issues (Burlingame et al. 2011). In addition to spot flushing, preventive flushing programs have been shown to be effective in improving the quality of the distributed water, in particular in removing sediments and biofilm, decreasing disinfectant demands, lowering DBPs, and minimizing customer complaints (see Appendix B).

The flushing velocity depends on the type of distribution system maintenance needed. A recent study that took a microbial risk approach (Yang et al. 2015) confirmed the importance of flushing velocity in removing dense soil particles after main break repair.

When flushing appears to be ineffective (e.g., in heavily turberculated pipes) or even detrimental (Friedman et al. 2015), other cleaning methods, rather than replacement of old pipes, should be considered. Ice pigging and swabbing have shown good efficiency and require less water than flushing techniques (Moore 2013).

Responding to the DSWQ Challenge: Turbidity

- Identify the cause, for example, treatment malfunction, cross-connection, main break, repair, and perform proper repair and take corrective action
- Consider flushing and/or main cleaning techniques

Best management practices for minimizing aesthetic challenge. Flushing the distribution system can be preventive or in response to a specific problem. Preventive flushing should concentrate on sensitive areas where historical records indicate common customer complaints and/or distribution incidents. Adequate flushing following water main repairs has been shown to be effective in minimizing contamination (Besner et al. 2008). Local regulations and/or environmental reasons may require neutralization of disinfectant in the discharge water that is flushed from the system. See Appendix B for more information on different flushing techniques.

Turbidity values do not provide information about the size of particles in drinking water. In some cases, it is useful to know the particle's size distribution, for example, at the point of entry to the distribution system or the effluent of a water treatment plant, in order to check filtration efficiency (Huck et al. 2001, 2002). A laser particle counter should be used. No correlation exists between particle counts and turbidity values (Bridgeman, Simms, and Parsons 2002) because principles of measurement for both techniques are different.

Discolored Appearance

The SMCL for color in drinking water is 15 platinum-cobalt (Pt-Co) units. The Pt-Co standard has a yellow tint that, in some cases, presents a challenge for quantitatively and qualitatively measuring color in distribution system water quality samples. In general, the Pt-Co standard is limited to measuring discolored water that exhibits varying intensities of yellow. This seems to be consistent with the original intent of the standard, which was to provide a surrogate for measuring organic matter in drinking water (Black and Christman 1963). In addition, the color observed by the naked eye may be from dissolved constituents, particulate matter (or turbidity), or both. The combination of color derived from both dissolved and particulate (or turbidity) constituents is referred to as the apparent color. The true color is the measure of the color after the particulate matter (or turbidity) has been removed by a 0.45-µm filter. The drinking water standards do not clarify whether the SMCL is based on apparent or true color. Further complicating the analysis of color in distribution systems is the fact that color that originates from within the pipe network may have many sources and resulting colors. The yellow tint of the Pt-Co standard is inadequate to properly describe all of them. Customers may simply describe discolored water, whether dissolved or particulate constituents are present, as "dirty water" or use other common descriptions, as presented in Table 5-6. Also, as the concentration of dissolved or particulate matter of a contaminant increases, its color may change. An example of this is iron; at low concentrations, water appears yellow and changes to red or brown as the iron concentration increases.

Table 5-6 Common descriptions and potential sources of discolored water events

Common Description	Potential Source
"Black water" or particles	Activated carbon fines and granules*
	Anthracite particles*
	Black rubber*
	Bitumastic pipe coating material*
	Manganese accumulation in pipes*,†, ‡
	Lead-rich scale from distribution system plumbing materials or in pipes*
"Blue–green water" or particles	Copper-based particles¶
	Sand or dirt particles (very low concentrations may be observed as a "greenish" hue in water) ∥
"Brown water" or particles	Iron rust particles from water mains*, ‡
	Manganese accumulation in pipes*,†,††, **
	Biofilm or bacterial clumps attached to particles*
	Sand or dirt particles*
"Red water" or particles	Iron rust particles from water mains**, ††, ‡‡
"White or gray water" or particles	Aluminum-based particles‡, ¶¶
	Air in water*
	Dip tube plastic (plastic pipe that brings cold water in water heater tanks)*
	Mineral deposits (such as calcium carbonate)*
	Zinc-based particles*, ∥ ∥
"Yellow water" or particles	Iron rust particles (low concentrations)‡‡
	Manganese accumulation in pipes**
	Dissolved organic carbon***
	Sand or dirt particles*

* Booth and Brazos 2005.
† Sly, Hodgkinson, and Arunpairojana 1990.
‡ Cerrato et al. 2006
¶ Edwards, Jacobs, and Taylor 2000.
∥ Sutherland 2015.
** Brandhuber et al. 2015.
†† Imran et al. 2005.
‡‡ Yang et al. 2014.
¶¶ Snoeyink et al. 2003.
∥ ∥ Sutherland et al.2009.
*** Black and Christman 1963.

The sources of discolored water are varied and need to be identified in order to mitigate discolored water events. Sources include internal corrosion of metal pipes, deterioration of plastic parts, and accumulation and release of metals, among others (see Table 5-6). In addition to identifying the source of color, the other major challenge is to identify the treatment or operational cause(s) for its appearance. These causes may be internal or external to the distribution system pipe network and also be continuous or temporary. Example causes include:

- Inadequate corrosion protection of metallic pipe;
- Accidental or intentional change in chemical dosages at the treatment plant or in the distribution system;
- Particle or turbidity breakthrough during treatment that accumulates in the distribution system (Vreeburg et al. 2008);
- Resuspension of settled particulate matter in typically low-flow and dead-end areas in the distribution system or storage tanks (Hasit et al. 2006);
- Vibration of existing pipe segments by nearby construction of new pipe or other infrastructure that dislodges material on the pipe wall;
- Water hammer or surges that may dislodge accumulated material on the pipe wall; and
- Hydraulic releases/changes that may result from flow reversals, filling storage tanks, and improper use of spot flushing.

Recommendations for monitoring. As mentioned previously, the standard measurement for color in water, whether apparent or true, relies on the yellow-tinted Pt-Co unit and should be used with caution for color derived from the distribution system. In some cases, apparent and true color have been used as surrogates for iron corrosion (Lytle and Snoeyink 2002; Imran et al. 2005). However, discolored water found in distribution systems and at customers' taps may be yellow, red, black, gray, white, or blue–green. In many of these cases, the Pt-Co standard is not a meaningful quantitative or qualitative measure of color. In fact, the presence of these colors in drinking water may be an indication of a more serious problem in the distribution system; this is discussed Chapters 2, 3, 5, 6, and 9.

In response to a discolored water event, samples are collected from the customer's tap, if practical, and/or corresponding hydrant nearby on the connecting distribution pipe at low flow after the hydrant barrel clears. If particulate matter is present, measuring the turbidity may be a good way to quantify the intensity and to track discolored water events. The particulate matter found in the sample can also be analyzed using microscopy. Booth and Brazos (2005) provided several procedures for using microscopy and other methods to identify particulate matter. If further analysis is needed, X-ray diffraction is suitable for identifying the mineralogy of the sample and X-ray fluorescence or Scanning Electron Microscopy and Energy Dispersive Spectrometry may be used to quantify the chemical composition. Dissolved metals may be analyzed using atomic absorption spectroscopy or ion chromatography. For color suspected to be related to biological growth, heterotrophic plate counts and adenosine triphosphate methods may be used to confirm and measure the level of biological activity.

Design, operation, and/or maintenance considerations. The design, operation, and maintenance considerations for color in drinking water are dependent on the source and cause of its appearance, as discussed earlier. These include inadequate internal corrosion protection of metallic pipes, resuspension of settled particulate matter in typically low-flow areas, and detachment of biofilm from the pipe wall from sudden increases in flow velocity or intermittent disinfectant residual concentration (Sly, Hodgkinson, and Arunpairojana 1990; Lytle and Snoeyink 2002; Imran et al. 2005; Hasit et al. 2006; Lehtola et al. 2006). Identifying the source and cause of the discolored water is generally the first step to determining the appropriate corrective action, whether design or operational. Examples of design considerations include targeting concentrations lower than the SMCL for a contaminant (such as manganese) within the respective treatment system, changing hydraulic requirements for distribution system piping, and providing separate pipe networks for industrial and residential use (Sly, Hodgkinson, and Arunpairojana 1990; Hasit et al. 2006). Examples of operational and maintenance considerations include optimizing chemical treatment, flushing and/or cleaning techniques, changing storage tank cycling practices, and replacing aged infrastructure (Lytle and Snoeyink 2002; Ellison et al. 2003; Imran et al. 2005; Hasit et al. 2006). The cause of discolored water in a distribution system pipeline should be carefully investigated before committing to a corrective action.

Responding to the DSWQ Challenge: Discolored Appearance

- Identify the source or cause to perform appropriate corrective action
- Optimize chemical treatment
- Consider flushing and/or main cleaning techniques
- Optimize storage tank cycling practices
- Replace aging or deteriorated infrastructure

Best management practices for minimizing color. Best management practices for minimizing discolored water in the distribution system include (AWWA 2010):

- Develop an action plan to receive discolored water complaints from customers by trained personnel who use a call system that can properly identify and track complaints and relay the information to the appropriate response team for timely resolution.
- Include, as part of the response to complaints, sample and chemical analysis of discolored water. Review data to identify potential sources of the problem. Keep a log of the data, date, and location, preferably with the complaint log. Review the dataset annually to identify both spatial and temporal trends.
- Develop a distribution system flushing program that includes both spot and routine flushing to address problem areas identified through customer complaints and trends in system data. See Appendix B for information on different flushing techniques.

Emerging issues and technologies. Some emerging issues that may impact discolored water in the distribution system include:

- Evaluation of accumulation and release of inorganic contaminants in distribution systems (Friedman, Kirmeyer, Antoun 2010);
- Reevaluation of health impacts of manganese in drinking water, which may have an influence on the manganese that is released to a distribution system (Bouchard et al. 2011);
- Release and control of corrosion products from iron-based pipes; and
- Control of biofilm development and release in distribution systems.

SUMMARY AND RECOMMENDATIONS

Objectionable taste, odor, and appearance of drinking water may range from an exceedance of aesthetic guidelines for water entering the distribution system to an indicator of an issue with potential public health risk. Properly monitoring for and responding to aesthetic water quality challenges is an important first step for a water system in identifying issues that require corrective action. Table 5-7 provides a summary of monitoring and response actions for each aesthetic challenge based on the following two levels of service:

- Basic: all water systems perform this action as part of standard practice for addressing an aesthetic issue and
- Advanced: water systems with persistent taste, odor, and appearance issues may need to perform this action, which is beyond standard practice.

Table 5-7 Recommended practices to address taste, odor, and appearance issues

Aesthetic Challenge		Best Management Practice	Level of Service
Sulfur-derived T&O: marshy, swampy, septic, rotten eggs	Monitoring	• Sensory analysis of affected T&O sites in distribution system	Basic
	Design, O&M considerations	• Provide and maintain free chlorine residual in distribution system	Advanced
		• Control biofilms	Basic
		• Minimize algal blooms and nutrients in source water(s)	Advanced
Disinfectant-derived T&O: chlorinous, ozonous, medicinal	Monitoring	• Laboratory analysis (e.g., gas chromatography/mass spectrometry) of T&O sites for aldehydes, N-chloroaldemines, iodinated trihalomethanes, or phenols	Basic
	Design, O&M considerations	• Minimize or eliminate potential for blending finished waters treated with different disinfectants	Basic/Advanced
		• Add chlorine before ammonia and optimize chlorine-to-nitrogen ratio to form chloramines	Basic
		• Use biologically active filters after ozonation	Advanced
		• Use proper curing times on new or repaired pipelines and tanks	Basic
		• Flush newly coated pipelines and tanks before putting into service	Basic
Polymers and their organic material–derived T&O: chemical, hydrocarbon	Monitoring	• Sensory analysis of affected T&O sites in distribution system	Basic
	Design, O&M considerations	• Fill, hold, and flush newly installed plastic pipelines or liners with water	Basic
		• Remove contaminated soil surrounding pipelines for areas impacted by permeation of contaminant(s) through the pipe wall	Advanced
		• Minimize water age to limit odor production	Basic
		• Control biofilms to minimize contaminant migration from the pipe wall	Basic

(continued)

Table 5-7 Recommended practices to address taste, odor, and appearance issues *(continued)*

Aesthetic Challenge		Best Management Practice	Level of Service
Metal pipe corrosion–derived T&O: copper, iron, zinc	Monitoring	• Laboratory analysis (e.g., inductively coupled plasma mass spectrometry) of T&O sites for copper, iron, and/or zinc	Basic
	Design, O&M considerations	• Flush and/or ice pig corroded metal pipelines	Basic/ Advanced
		• Add corrosion control chemicals	Basic
		• Rehabilitate or replace severely corroded pipe segments	Advanced
Mineral-derived T&O: salty, concrete (aluminum, calcium, magnesium)	Monitoring	• Continuous monitoring of TDS or conductivity	Basic
		• Laboratory analysis (e.g., ion chromatography) of T&O sites for alkalinity, hardness, sodium, potassium, calcium, magnesium, chloride, and sulfate	Basic
		• Sensory analysis of T&O sites for intensity of salty taste	Advanced
	Design, O&M considerations	• Blend salty water with water with lower TDS before distribution	Basic/ Advanced
		• Desalinate partial or full flow of source water	Advanced
		• Mitigate marine intrusion into source water	Advanced
		• Properly maintain brine and industrial water collection systems to minimize breaks and cross-connections	Basic
		• Monitor water in contact with concrete pipelines or linings for changes in pH, alkalinity, aluminum, and free chlorine	Basic
		• Avoid cement-mortar linings in dead ends of the distribution system	Basic
		• Flush newly coated pipelines and tanks before putting into service	Basic
		• Minimize water age to limit taste issues	Basic
Turbid appearance	Monitoring	• Laboratory or field analysis using a nephelometer (turbidimeter)	Basic
	Design, O&M considerations	• Use spot and unidirectional flushing as appropriate	Basic
		• Monitor the point of entry to the distribution system with a laser particle counter to check filtration efficiency	Advanced

(continued)

Table 5-7 Recommended practices to address taste, odor, and appearance issues *(continued)*

Aesthetic Challenge	Best Management Practice		Level of Service
Discolored appearance	Monitoring	• Laboratory or field analysis using a nephelometer (turbidimeter) to track intensity of the event	Basic
		• Laboratory analysis using microscopy to identify relatively large particles	Advanced
		• Filter water sample and send solids sample to laboratory for X-ray diffraction/X-ray fluorescence or SEM-EDS analysis	Advanced
		• For suspected biological growth, use laboratory analysis for heterotrophic plate counts and adenosine triphosphate to measure the level of biological activity	Advanced
	Design, O&M considerations	• Target treatment goals lower than the secondary maximum contaminant level for contaminant of concern	Advanced
		• Minimize water age in affected areas	Basic
		• Optimize chemical treatment	Basic
		• Implement unidirectional flushing, ice pigging, swabbing of affected areas	Basic
		• Change storage tank cycling practices	Basic
		• Replace aged infrastructure	Advanced

Abbreviations: O&M, operations and maintenance; TDS, total dissolved solids; T&O, taste and odor.

REFERENCES

Ahart, M.; D.L. Gallagher; P. Scardina; & A.M. Dietrich. 2016. Industrial Spills and Water Distribution: MCHM Sorption and Desorption in Polymer Pipes and Linings. *Journal of Environmental Engineering.* , 2016.

Allard, S.; J.W.A. Charrois; C.A. Joll; & A. Heitz. 2012. Simultaneous Analysis of 10 Trihalomethanes at Nanogram per Liter Levels in Water Using Solid-Phase Microextraction and Gas Chromatography Mass-Spectrometry. *Journal of Chromatography A,* 1238:15–21.

Anselme, C.; I.H. Suffet; & J. Mallevialle. 1988. Effects of Ozonation on Tastes and Odors. *Journal of the American Water Works Association,* 80:10:45–51.

APHA & American Water Works Association. 1946 (9th ed.). *Standard Methods for the Examination of Water and Sewage.* Washington, DC: American Public Health Association.

APHA (American Public Health Association); American Water Works Association; & Water Pollution Control Federation. 2011 (21st ed.). *Standard Methods for the Examination of Water and Wastewater. Standard Methods.* 6040 Constituent Concentration by Gas Extraction. Washington, DC: American Public Health Association.

———. 2012 (22nd ed.) *Standard Methods for the Examination of Water and Wastewater.* Washington, DC: American Public Health Association.

American Standard Testing Materials (ASTM). 2012. D5504, *Standard Test Method for Determination of Sulfur Compounds in Natural Gas and Gaseous Fuels by Gas Chromatography and Chemiluminescence.* West Conshohocken, PA: ASTM. http://www.astm.org/ (accessed Dec. 19, 2014).

Australia/New Zealand StandardTM 4020:2005. *Testing of products for use in contact with drinking water.* Jointly published by Standards Australia, GPO Box 476, Sydney, NSW 2001 and Standards New Zealand, Private Bag 2439, Wellington 6020. ISBN 0 7337 6993 4.

AWWA (American Water Works Association). 2004. M7, *Problem Organisms in Water: Identification and Treatment.* Denver, CO: AWWA.

———. 2010. AWWA Standard G200. *Distribution System Operation and Management.* Denver, CO: AWWA.

———. AWWA. 2012. *Global Taste and Odor Survey of Water Utilities.* Prepared by P. Ömur-Özbek. 2012, Denver, CO: AWWA.

———. 2014. Standard G300, *Source Water Protection.* Denver, CO: AWWA.

———. 2015. Standard 213-15. *Fusion-Bonded Epoxy Coatings and Linings for Steel Water Pipe and Fittings.*Denver, CO:AWWA.

AwwaRF and TZW (Awwa Research Foundation and DVGW-Technologiezentrum Wasser). 1996. Internal Corrosion of Water Distribution Systems, 2nd ed. Denver, Colo.: AwwaRF.

Bae, B. U.; Y. I. Kim; D.W. Dugas; G.A. Burlingame; & A.M. Dietrich. 2002. Demonstration of New Sensory Methods for Drinking Water Taste-and-Odor Control. *Water Science and Technology: Water Supply,* 2:5–6:241–247.

Bartoshuk, L.M. 1980. Biological and Behavioral Aspects of Salt Intake: Chapter 7—Sensory Analysis of the Taste of NaCl (M.R. Kare; M.J. Fregly; & R.A. Bernard, editors). New York: Academic Press; pp. 83–98.

Benson, A.S.; A.M. Dietrich; & D.L. Gallagher. 2012. Evaluation of Iron Corrosion Release Models for Water Distribution Systems. *Critical Reviews in Environmental Science and Technology,* 42:1:44–97.

Berend, K. & T. Trouwborst. 1999. Cement-Mortar Pipes as a Source of Aluminum. *Journal of the American Water Works Association,* 9:7:91–100.

Berend, K., G. van de Voet; & W.H. Boer. 2001. Acute Aluminum Encephalopathy in a Dyalysis Center Caused by a Cement Mortar Distribution Pipe. *Kidney International,* 59:2:746–753.

Besner, M.C.; J. Lavoie; C.H. Morissette; P. Payment; & M. Prevost. 2008. Effect of Water Main Repairs on Water Quality. *Journal of the American Water Works Association*, 100:7: 95–109.

Biggs, J.; T. Thomure; K. Arnol; & C. Cotton. 2007. Public Outreach Tool Used in Determining Customer Preferences of Water Aesthetics. Proceedings AWWA Water Quality Technology Conference, Charlotte, N.C.

Black, A. & R. Christman. 1963. Characteristics of Colored Surface Waters. *Journal of the American Water Works Association*, 55:6:753–770.

Blute, N.K.; M.J. McGuire; N. West; N. Vouthkov; P. Maclaggan; & K. Reich. 2008, Integration of Desalinated Water Into a Distribution System: A Corrosion Pilot Study. *Journal of the American Water Works Association*, 100:9:117–131.

Booth, S. & B. Brazos. 2004. *Qualitative Procedures for Identifying Particles in Drinking Water.* Denver, CO: Water Research Foundation.

Bouchard, M.; S. Sauve; B. Barbeau; M. Legrand; M.-E. Brodeur; T. Bouffard; E. Limoges; D. Bellinger; and D. Mergler. 2011. Intellectual Impairment in School-Aged Children Exposed to Manganese From Drinking Water. *Environmental Health Perspectives*, 119:1:138–143.

Brandhuber, P.; S. Craig; M. Friedman; A. Hill; S. Booth; & A. Hanson. 2015. *Legacy of Manganese Accumulation in Water Systems.* Denver, CO: Water Research Foundation.

Brindgeman, J.; J.A. Simms; & S.A. Parsons. 2002. Practical Theoretical Analysis of Relationships Between Particle Count Data and Turbidity. *Journal of Water Supply: Research and Technology Aqua*, 51:5:263–271.

British Standards Institution. 2014. BS 6920-1, *Suitability of non-metallic materials and products for use in contact with water intended for human consumption with regard to their effect on the quality of the water.* British Standards Institution. London, United Kingdom.

Brosillon, S.; M. Lemasle; E. Renault; D. Tozza; V. Heim; & A. Laplanche. 2009. Analysis and Occurrence of Odorous Disinfection By-products From Chlorination of Amino Acids in Three Different Drinking Water Treatment Plants and Corresponding Distribution Networks. *Chemosphere*, 77:1035–1042.

Bruchet, A. 1999. Solved and Unsolved Cases of Taste and Odor Episodes in the Files of Inspector Cluzeau. *Water Science and Technology*, 40:6:15–21.

Bruchet, A.; K. N'Guyen; M.F. Legrand; & J. Mallevialle. 1991. Variations in Organic and Organoleptic Water Quality During Treatment of Artificial Recharged Groundwater. *Journal of the American Water Works Association*, 83:3:63–70.

Bruvold, W.H. 1968. Scales for Rating the Taste of Water. *Journal of Applied Psychology*, 52:3:245–253.

Bruvold, W.H. & J.I. Daniels. 1990. Standards for Mineral Content in Drinking Water. *Journal of the American Water Works Association*, 82:2:59–65.

Burlingame, G.A. 1999a. Solving Customers' Taste and Odor Complaints–Part 1: The Importance of the First Response. *Opflow*, 25:10:10–11.

———. 1999b. Solving Customers' Taste and Odor Complaints–Part 2: Tracking Odors to Their Source. *Opflow*, 25:11:6–7.

Burlingame, G.A. & C. Anselme. 1995. Distribution System Tastes and Odors. In: *Advances in Taste- and-Odor Treatment and Control*. Denver, CO: AwwaRF.

Burlingame, G.A.; D.J. Booth; A. Bruchet; A.M. Dietrich; D.L. Gallagher; D. Khiari; I.H. Suffet; & S.B. Watson. 2011. Diagnosing Taste and Odor Problems: Source Water and Treatment Field Guide. Denver, CO: AWWA.

Burlingame, G.A.; A.M. Dietrich; & A.J. Whelton. 2007. Understanding the Basics of Tap Water Taste. *Journal of the American Water Works Association*, 99:5:100–110.

Burlingame G.A., D.A. Lytle; & V.L. Snoeyink. 2006. Why Red Water? Understanding Iron Release in Distribution Systems. *Opflow,* 32:12:12–16.

Burlingame, G.A.; J.J. Muldowney; & R.E. Maddrey. 1992. Cucumber Flavor in Philadelphia's Water. *Journal of the American Water Works Association*, 84:8:92–97.

Bushdid, C.; M.O. Magnasco; L.B. Vosshall; & A. Keller. 2014. Humans Can Discriminate More than 1 Trillion Olfactory Stimuli. *Science*, 343:6177: 1370–1372.

Buttery, R.G.; D.G. Guadagni; L.C. Ling; R.M. Seifert; & W. Lipton. 1976. *Journal of Agriculture and Food Chemistry*, 24:4:829–832.

Cancho, B.; C. Fabrellas; A. Díaz; & F. Ventura. 2001. Determination of the Odor Threshold Concentrations of Iodinated Trihalomethanes in Drinking Water. *Journal of Agriculture and Food Chemistry*, 49:1881–1884.

Cancho, B. & F. Ventura. 2005. Optimization of Methods for the Determination of DBPs. *Global NEST Journal*, 7:1:72–94.

Cancho, B.; F. Ventura; M.T. Galceran; A. Diaz; & S. Ricart. 2000. Determination, Synthesis and Survey of Iodinated Trihalomethanes in Water Treatment Processes. *Water Research*, 34:13:3380–3390.

Castillo Montes, J.; D. Cadoux; J. Creus; S. Touzain; E. Gaudichet-Maurin; & O. Correc. 2012. Ageing of Polyethylene at raised temperature in contact with chlorinated sanitary hot water. Part I – Chemical aspects. *Polymer Degradation and Stability*, 97:149–157

Cerrato, J.; L. Reyes; C. Alvarado; & A. Dietrich. 2006. Effect of PVC and Iron Materials on Mn(II) Deposition in Drinking Water Distribution Systems. *Water Research*, 40:2720–2726.

Colin, X., Audouin, L., Verdu, J., Rozental-Evesque, M., Rabaud, B., Martin, F., & F. Bourgine. 2009. Aging of polyethylene pipes transporting drinking water disinfected by chlorine dioxide. I. Chemical aspects. *Polymer Engineering & Science*, 49(7):1429–1437

Cuppett, J.D.; S.E. Duncan; & A.M. Dietrich. 2006. Evaluation of Copper Speciation and Water Quality Factors That Affect Aqueous Copper Tasting Sensitivity, *Chemical Senses*, 31:7:689–697.

Dale, M.S; R.D. Desrochers; S.M. Labernik; & M.S. Moylan. 2011. Threshold Odor Number Versus Flavor Profile Analysis: A Case Study in the Endeavor to Measure Consumer Acceptance. 9th IWA Symposium on Off-Flavours in the Aquatic Environment, Aberdeen, Scotland.

Deb, A.; F.M. Grablutz; Y.J. Harit; J.K. Snyder: G.V. Loganathan; & N. Agbenowsi. 2002. *Prioritizing Water Main Replacement and Rehabilitation*. Denver, CO: AWWA and AWWARF.

Deb, A.; S.B. Mc Cammon; J. Snyder; & A.M. Dietrich. 2010. *Impacts of Lining Materials on Water Quality*. Denver, CO: Water Research Foundation.

Desrochers, R. 2008. Sensory Analysis in the Water Industry. *Journal of the American Water Works Association*, 100:10:50–54.

Devesa, R.; V. García; & L. Matía. 2010. Water Flavour Improvement by Membrane (RO and EDR) Treatment. *Desalination*, 250:1113–117.

DeZuane, J. 1997 (2nd ed.). *Handbook of Drinking Water Quality*. New York: John Wiley & Sons; p. 575.

Díaz, A.; C. Fabrellas; F. Ventura; & M.T. Galceran MT. 2005. Determination of the Odor Threshold Concentrations of Chlorobrominated Anisoles in Water. *Journal of Agricultural and Food Chemistry*, 53:2:383–387.

Dietrich, A.M. 2006. Aesthetic Issues for Drinking Water. *Journal of Water and Health*, 4 (suppl. 1):11–16.

———. 2009. The Sense of Smell: Contributions of Orthonasal and Retronasal Perception Applied to Metallic Flavor of Drinking Water. *Journal of Water Supply: Research and Technology–AQUA*, 58:8:562–570.

———. 2015. *USEPA Secondary Maximum Contaminant Limits: A Strategy for Drinking Water Quality and Consumer Acceptability*. Denver, CO: Water Research Foundation.

Dietrich, A.M. & G.A. Burlingame. 2015. Critical Review and Rethinking of USEPA Secondary Standards for Maintaining Consumer Acceptability of Organoleptic Quality of Drinking Water. *Environmental Sciences and Technology*, 49:2:708–720,

Dietrich, A.M.; G.A. Burlingame; & R.C. Hoehn. 2003. Strategies for Taste-and-Odor Testing Methods. *Opflow*, 29:10:10–14.

Dietrich, A.M.; J.D. Cuppett; & S.E. Duncan. 2008. How Much Copper is Too Much? *Opflow* 34:9:8–30.

Dietrich A. & R. Devesa. Characterization and removal of minerals that cause taste. To be published by IWA in brief.

Dietrich, A.M. & C.D. Gallagher. 2013. Consumer Ability to Detect the Taste of Total Dissolved Solids. *Journal of the American Water Works Association*, 105:5:E255–263.

Dietrich, A.M.; D. Glindemann; F. Pizarro; V. Gidi; M. Olivares; M. Araya; A. Camper; S. Duncan; S. Dwyer; A.J. Whelton; T. Younos; S. Subramanian; G.A. Burlingame; D. Khiari; & M. Edwards. 2004. Health and Aesthetic Impacts of Copper Corrosion on Drinking Water. *Water Science and Technology*, 49:2:55–62.

Dietrich, A.M.; R.C. Hoehn; G.A. Burlingame; & T. Gittelman. 2004a. *Practical Taste-and-Odor Methods for Routine Operations: DECISION TREE*. Denver, CO: AWWA Research Foundation.

Dietrich, A.M.; M.P. Orr; D.L. Gallagher; & R.C. Hoehn. 1992. Tastes and Odors Associated With Chlorine Dioxide. *Journal of the American Water Works Association*, 84(6): 82–88.

Dietrich, A.M.; K. Phetxumphou; & D.L. Gallagher. 2014. Systematic Tracking, Visualizing, and Interpreting of Consumer Feedback for Drinking Water Quality. *Water Research*, 66:63–72.

Dietrich, A.M.; A. Thomas; Y. Zhao; E. Smiley; N. Shanaiah; M. Ahart; Charbonnet, K.A.; DeYonker, N.J.; Alexander, W.A.; & Gallagher, D.L.. 2015. Partitioning, Aqueous Solubility, and Dipole Moment Data for *cis*- and *trans*-(4-Methylcyclohexyl) Methanol, Principal Contaminants of the West Virginia Chemical Spill. *Environmental Science & Technology Letters*, 2:123–127.

Dietrich, A.M.; A. Whelton,; R.C. Hoehn; R. Anderson; & M. Wille. 2004b. Attribute Rating Test for Sensory Analysis. *Water Science and Technology*, 49:9:61–67.

Doria, M. 2010. Factors Influencing Public Perception of Drinking Water Quality. *Water Policy*, 12:1–19.

Dotson, A. & P. Westerhoff. 2009. Occurrence and Removal of Amino Acids During Drinking Water Treatment. *Journal of the American Water Works Association*, 101:9:101–115.

Duguet, J.P.; J. Mallevialle; J. Ho; & I.H. Suffet. 1995. *Advances in Taste-and-Odor Treatment and Control: Chapter 3 – Oxidation Processes*. (I.H. Suffet, J. Mallevialle, & E. Kawczynski, editors). Denver, CO: AWWA Research Foundation and Lyonnaise des Eaux; pp. 75–156.

Duranceau, S.; R.J. Wilder; & D. Susaye. 2012. A Survey of Desalinated Permeate Post-Treatment Practices. *Desalination and Water Treatment*, 37:185–199.

Durand, M. & A. Dietrich. 2007. Contributions of silane cross-linked PEX pipe to chemical/solvent odours in drinking water. Water Science & Technology, 55 (5):153-160.

EC DWD (European Commission Drinking Water Directive). 1998. Council Directive 98/83/EC of 3 November, on the Quality of Water Intended for Human Consumption. Brussels, Belgium; Luxembourg, Luxembourg.

Edwards, M.; M.R. Schock; & T.E. Meyer. 1996. Alkalinity, pH and Copper Corrosion By-product Release. *Journal of the American Water Works Association*, 88;3:81–94.

Edwards, M.; S. Jacobs; & R. J. Taylor. 2000. The Blue Water Phenomenon. *Journal of the American Water Works Association*, 92:7:72–82.

Ellison, D.; S. Duranceau,; S. Ancel; &G. Deagle. 2003. Investigation of Pipe Cleaning Methods. Denver, CO: AWWA Research Foundation.

European Commission. 2013. Review of Sensors to Monitor Water Quality. ERNCIP Thematic Area, Chemical & Biological Risks in the Water Sector. European Commission, J. Raich.

Fabrellas, C.; L. Matía; & F. Ventura. 2004. Determination of Odor Threshold Concentrations and Dose-Response Relations in Water of Several Minor Disinfection By-products: Aldehydes and Alkyl nitriles. *Water Science and Technology*, 49:267–272.

Faludi, T.; A. Vasanits-Zsigrai; G. Záray; & I. Molnár-Perl. 2015. Identification, Quantification and Distribution of Substituted Phenols in the Dissolved and Suspended Phases of Water Samples by Gas Chromatography Tandem Mass Spectrometry: Derivatization, Mass Fragmentation and Acquisition Studies. *Microchemical Journal*, 118:45–54.

Firat Server, V. 2012. Trenchless Water Main Rehabilitation Design Considerations. *Waterworld*, 1:28.

Franzmann, P.D.; A. Heitz; L.R. Zappia; J.E. Wajon; & K. Xanthis. 2001. The Formation of Malodorous Dimethyl Oligosulphides in Treated Groundwater: The Role of Biofilms and Potential Precursors. *Water Research*, 35:7:1730–1738.

Freuze, I.; S. Brosillon; A. Laplanche; D. Tozza; & J. Cavard. 2005. Effect of Chlorination on the Formation of Odorous Disinfection By-products. *Water Research*, 39:2636–2642.

Froese, K.L.; A. Wolanski; & S.E. Hrudey. 1999. Factors Governing Odorous Aldehyde Formation as Disinfection By-products in Drinking Water. *Water Research*, 33:6:1355–1364.

Foreman, W.T.; D.L. Rose; D.B. Chambers; A.S. Crain; L.K. Murtagh; H. Thakellapalli; & K.K. Wang. 2015. Determination of (4-methylcyclohexyl)Methanol Isomers by Heated Purge-and-Trap GC/MS in Water Samples From the 2014 Elk River, West Virginia, Chemical Spill. *Chemosphere*, [Epub ahead of print].

Friedman, M. 2015. Metals Accumulation and Release: Significance to the Industry. Proceedings AWWA Water Quality Technology Conference, Salt Lake City, Utah.

Friedman, M.; A. Hill; S. Reiber; R. Valentine; G. Larsen; A. Young; G. Korshin; & C.Y. Peng. 2010. *Assessment of Inorganics Accumulation in Drinking Water System Scales and Sediments*. Denver, CO: Water Research Foundation.

Friedman, M.; G.J. Kirmeyer; & E. Antoun. 2005. Developing and Implementing a Distribution System Flushing Program. In: *Water Quality in the Distribution System* (W.C. Lauer, editor). Denver, CO: AWWA.

Fritz, J & D.T. Gjerde. 2009 (4th ed.). *Ion Chromatography*. Wiley-VCH Verlag.

Fuh Lin, Tsair; Watson, Sue; Suffer, Mel. *Taste and Odour in Source and Drinking Water Causes, Controls, and Consequences*. IWA publication announced for January 15, 2018.

Gallagher, C.D. & A.M. Dietrich. 2010. TDS and Temperature Affect Consumer Taste Preferences. *Opflow,* 36:11:20–22.

Gallagher, D.L. 2004. Statistical Comparison of the Triangle Test and the Two-of-Five Test for Taste and Odor Evaluation. *Water Science and Technology,* 49:9:107–114.

Gallagher D.L. & A.M. Dietrich. 2014. Statistical Approaches for Analyzing Customer Complaint Data to Assess Aesthetic Episodes in Drinking Water. *Journal of Water Supply: Research and Technology–AQUA,* 63:5:358–367.

Gallagher, D.L.; K. Phetxumphou; E. Smiley; & A.M. Dietrich. 2015. Tale of Two Isomers: Complexities of Human Odor Perception for cis- and trans-4-Methylcyclohexane Methanol From the Chemical Spill in West Virginia. *Environmental Science and Technology;* DOI: 10.1021/es5049418.

Gates, D. 1998. *The Chlorine Dioxide Handbook.* Denver, CO: American Water Works Association.

García, V. & R. Devesa. 2015. Are Consumers Able to Differentiate Chlorine Dioxide From Chlorine? AWWA Water Quality Technology Conference, Salt Lake City, Utah.

García, V.; A. Fernández; M.E. Medina; O. Ferrer; J.L. Cortina; F. Valero; & R. Devesa. 2014. Flavour Assessment of Blends Between Desalinated and Conventionally Treated Sources. *Desalination and Water Treatment,* DOI: 10.1080/19443994.2013.875943.

Glindemann, D.; A.M. Dietrich; H.-J. Staerk; & P. Kuschk. 2006. The Two Smells of Touched or Pickled Iron (Skin) Carbonyl-Hydrocarbons and Organophosphines, *Angewandte Chemie (European Journal Applied Chemistry),* 45:7006–7009.

Good Scents Company, www.thegoodscentscompany.com; (accessed May 23, 2017).

Guadagni D.G.; J.C. Miers; & D.W. Venstrom. 1969. Concentration Effect on Odor Addition or Synergism in Mixtures of Methyl Sulfide and Tomato Juice. *Journal of Food Science,* 34:6:630–632.

Hagedorn, F.; P. Schleppi; P. Waldner; & H. Flühler. 2000. Export of Dissolved Organic Carbon and Nitrogen From Gleysol Dominated Catchments—The Significance of Water Flow Paths. *Biogeochemistry,* 50:2:137–161.

Hamdaoui, O. & E. Naffrechoux. 2007. Modeling of Adsorption Isotherms of Phenol and Chlorophenols Onto Granular Activated Carbon Part I. Two-Parameter Models and Equations Allowing Determination of Thermodynamic Parameters. *Journal of Hazardous Materials,* 147:381–394.

Hansson, R.C.; M.J. Henderson; P. Jack; & R.D. Taylor. 1987. Iodoform Complaints in Chloramination. *Water Research,* 21:10:1265–1271.

Hasit, Y.; J. Anderson; A. Parolari; T. Rockaway; & M. French. 2006. *Distribution Water Quality Issues Related to New Development or Low Usage.* Denver, CO: Water Research Foundation.

Hassinen, J.; M. Lundbäck; M. Ifwarson; & U.W. Gedde. 2004. Deterioration of polyethylene pipes exposed to chlorinated water. *Polymer Degradation and Stability*, 84(2):261–267

Heim, T. & A. Dietrich. 2007. Sensory aspects and water quality impacts of chlorinated and chloraminated drinking water in contact with HDPE and cPVC pipe. *Water Research*, 41(4):757–764

Heitz, A. 2002. Malodorous Dimethylpolysulfides in Perth Drinking Water. Doctoral thesis, Curtin University of Technology, Perth, Western Australia. http://espace.library.curtin.edu.au:80/R?func=dbin-jump-full&local base=gen01-era02&object id=12576. (accessed May 23, 2017)

Heitz, A.; R. Kagi; & R. Alexander. 2000. Polysulfide Sulfur in Pipewall Biofilms: Its Role in the Formation of Swampy Odour in Distribution Systems. *Water Science and Technology*, 41:4–5:271–278.

Hill, S.J. (editor). 2006 (2nd ed.). *Inductively Coupled Plasma Spectrometry and Its Applications*. Chichester, United Kingdom: John Wiley and Sons.

Hoehn, R.C.; A.M. Dietrich; W.S. Farmer; M.P. Orr; R.G. Lee; E.M. Aieta; D.W. Wood; & G. Gordon. 1990. Household Odors Associated With the Use of Chlorine Dioxide. *Journal of the American Water Works Association*, 82:4:166–172.

Hoehn, R.C.; A.A. Rosenblatt; and & D.J. Gates. 1996. Considerations for Chlorine Dioxide Treatment of Drinking Water. Proceedings AWWA Water Quality Technology Conference, Boston, Mass.

Hong, J-H.; S.E. Duncan; A.M. Dietrich; & S. O'Keefe. 2006. Effect of Copper on Volatile Flavour Compounds in a Model Mouth System. *Journal for Agricultural and Food Chemistry*, 54:9168–9175.

Hong J.H.; S.E. Duncan; & A.M. Dietrich. 2010b Effect of Copper Speciation at Different pH on Temporal Sensory Attributes of Copper. *Food Quality and Preference*, 21:1:132–139.

Hong J-H.; S.E. Duncan; A.M. Dietrich; & S. O'Keefe. 2010. Evaluation of the Operationally Defined Soluble, Insoluble, and Complexing Copper Consumed Through Drinking Water in Human Saliva. *European Food Research and Technology*, 231:977–984.

How, Z.T.; F. Busetti; K.L. Linge; I. Kristiana; C.A. Joll; & W.A. Jeffrey. 2014. Analysis of Free Amino Acids in Natural Waters by Liquid Chromatography–Tandem Mass Spectrometry. *Journal of Chromatography A*, 1370:135:146.

Hrudey, S.E. 2009. Chlorination Disinfection By-products, Public Health Risk Tradeoffs and Me. *Water Research*, 43: 2057–2092.

Hrudey, S.E.; S.A. Daignault; A.J. Gac; D. Poole; G. Walker; & D. Birkholz. 1989. Swampy or Swimming Pool Odours Caused by Chlorination and Chloramination. AWWA Annual Conference, Los Angeles, Calif.

Hrudey, S.E.; A. Gac; & S.A. Daignault. 1988. Potent Odor-Causing Chemicals Arising From Drinking Water Disinfection. *Water Science and Technology*, 20:8/9:55–61.

Hu, Z.; L. Morton; & R. Mahler. 2011. Bottled Water: United States Consumers and Their Perceptions of Water Quality. *International Journal of Environmental Research and Public Health*, 8:2:565–578.

Huck, P.M.; B.M. Coffey; M.B. Emelko; D.D. Maurizio; R.M. Slawson; W.B. Anderson; J.V. Den Oever; I.P. Douglas; & C.R. O'Melia. *Effects of filter operation on Cryptosporidium removal.* 2002. Denver, CO: AWWA Research Foundation and AWWA.

Huck, P.M.; M.B. Emelko; B.M. Coffey; D.D. Maurizio; & C.R. O'Melia. 2001. *Filter Operation Effects on Pathogen Passage.* Denver, CO: AWWA Research Foundation and AWWA.

Husband, P.S. & J.B. Boxall. 2011. Asset Deterioration and Discolouration in Water Distribution Systems. *Water Research*, 45:113–124.

Imran, S.; J. Dietz; G. Mutoti; J. Taylor; A. Randall; & C.D. Cooper. 2005. Red Water Release in Drinking Water Distribution Systems. *Journal of the American Water Works Association*, 97:9:93–100.

Ishizawa, S.; Y.Z. Yamamoto; T. Denboh; & H. Ueda. 2010. Release of Dissolved Free Amino Acids From Biofilms. *Fisheries Science*, 76:4:669–676.

Kawaguchi, M.; Y. Ishii; N. Sakui; N. Okanouchi; R. Ito; K. Saito; & H. Nakazawa. 2005. Stir Bar Sorptive Extraction With In Situ Derivatization and Thermal Desorption–Gas Chromatography–Mass Spectrometry for Determination of Chlorophenols in Water and Body Fluid Samples. *Analytica Chimica Acta*, 533:57–65.

Keast, R. 2003. The Effect of Zinc on Human Taste Perception. *Journal Food Science*, 68:1871–1877.

Kelley, K.; A. Stenson; R. Dey; & A. Whelton. 2014. Release of drinking water contaminants and odor impacts caused by green building cross-linked polyethylene (PEX) plumbing systems. *Water Research*, 67:19–32

Khiari, D.; S. Barrett; R. Chinn; A. Bruchet; P. Piriou; L. Matia; F. Ventura; I.H. Suffet; T. Gittelman; & P. Leutweiler. 2002. In *Distribution Generated Taste-and-Odor Phenomena*. Denver, CO: AWWA Research Foundation and UK Water Industry Research Foundation.

Kilmore, G. 2009. Challenges Associated With Mixing Desalinated Sea Water and Conventionally Treated Surface Waters to Minimise Customer Complaints. Proceedings AWWA Water Quality Technology Conference, Seattle, Wash.

Kirmeyer G.J.; M. Friedman; K. Martel; G. Thompson; J. Clement; & M. Frey. 2002. *Guidance Manual for Maintaining Distribution System Water Quality.* Denver, CO: AWWA Research Foundation and AWWA.

Krasner, S.W. 1988. Flavor-Profile Analysis—An Objective Sensory Technique for the Identification and Treatment of Off-Flavors in Drinking Water. *Water Science and Technology*, 20:8–9:31–36.

Krasner, S.W. & S.E. Barrett. 1984. Aroma and Flavor Characteristics of Free Chlorine and Chloramines. *In Proc. AWWA Water Quality Technology Conference*. Denver, CO: AWWA. Proceedings AWWA Water Quality Technology Conference.

Kristiana, I.; A. Heitz; C. Joll; & A. Sathasivan. 2010. Analysis of Polysulfides in Drinking Water Distribution Systems Using Headspace Solid-Phase Microextraction and Gas Chromatography-Mass Spectrometry. *Journal of Chromatography A*, 1217:38:5995–6001.

Lan, J.; M. Hu; C. Gao; A. Alshawabkeh; & A.Z. Gu. 2015. Toxicity Assessment of 4-Methyl-1-Cyclohexanemethanol and Its Metabolites in Response to a Recent Chemical Spill in West Virginia, USA. *Environmental Science & Technology*, 49:6284–6293.

Lauer, W.C. 2014 *Water Quality Complaint Investigator's Field Guide, Second Edition*. Denver, CO: AWWA.

Lee, J.; E. Ewa Kleczyk; D.J. Bosch; A.M. Dietrich; V.K. Lohani; & G.V. Loganathan. 2013. Homeowners' decision- making in a premise plumbing failure–prone area. *Journal of the American Water Works Association*, 105:5:E236–E241.

Lehtola, M.; M. Laxander; I. Miettinen; A. Hirvonen; T. Vartiainen; & P. Martikainen. 2006. The Effects of Changing Water Flow Velocity on the Formation of Biofilms and Water Quality in Pilot Distribution System Consisting of Copper and Polyethylene Pipes. *Water Research*, 40:2151–2160.

Llompart, M.; M. Lourido; P. Landın; C. Garcıa-Jares; & R. Cela. 2002. Optimization of a Derivatization–Solid-Phase Microextraction Method for the Analysis of Thirty Phenolic Pollutants in Water Samples. *Journal of Chromatography A*, 963:137–148.

Lytle, D. & V. Snoeyink. 2002. Effect of Ortho- and Poly-Phosphates on the Properties of Iron Particles and Suspensions. *Journal of the American Water Works Association*, 94:10:87–99.

Ma, X.; J. Deng; J. Feng; N. Shanaiah; E. Smiley; & A.M. Dietrich. 2016. Identification and Characterization of Phenylacetonitrile as a Nitrogenous Disinfection Byproducts Derived From Chlorination of Phenylalanine in Drinking Water. *Water Research*, 102:202-10.

Ma, X.; S. Hu; H. Wang; J. Li; J. Huang; Y. Zhang; W. Lu; & Q. Li. 2012. Kinetics of Oxidation of Dimethyl Trisulfide by Potassium Permanganate in Drinking Water. *Frontiers of Environmental Science & Engineering*, 6:2:171–176.

MacKeague, D. 2007. Automation Works—Automatic Hydrant Flushers Save Hundreds of Man-Hours. *Opflow*, 33:6:10–14.

Mackey, E.D.; H. Baribeau; G.F. Crozes; I.H. Suffet; & P. Piriou. 2004. Public Thresholds for Chlorinous Flavors in U.S. Tap Waters. *Water Science Technology*, 49:9:335–340.

Malleret, L. & A. Bruchet. 2002. A Taste and Odor Episode Caused by 2,4,6 Tribromoanisole. *Journal of the American Water Works Association*, 94:7:84–95.

Malleviale, J. & I.H. Suffet. (editors). 1987. *Identification and Treatment of Tastes and Odors in Drinking Water*. Denver, CO: AWWA Research Foundation and Lyonnaise des Eaux.

Manojlovic, D.; D.R. Ostojic; B.M. Obradovic; M.M. Kuraica; V.D. Krsmanovic; & C. Puric. 2007. Removal of Phenol and Chlorophenols From Water by New Ozone Generator. *Desalination,* 213:116–122.

Manuel, J. 2014. Crisis and Emergency Risk Communication: Lessons From the Elk River Spill. *Environmental Health Perspectives,* 122:8:A214–219.

Marchesan, M. & J. Morran. 2004. Tastes associated with products in contact with drinking water. Water Science & Technology, 49 (9): 227-231.

Martín-Alonso, J. 1994. Barcelona's Water Supply Improvement: The Brine Collector of the Llobregat River. *Water Science and Technology,* 30:10:221–227.

Martín-Alonso, J.; R. Devesa; A. Bernal; & L. Matía. 2007. Managing an Odour Episode in Barcelona's Water Supply: Strategies Adopted, the Causative Agent (Diacetyl) and Determination of Its Organoleptic Properties. *Water Science and Technology,* 55:5209–216.

Matía, L. 1995. Treatment of Tastes in Drinking Water: Causes and Control. In: *Advances in Taste-and-Odor. Treatment and Control* (I.H. Suffet, J. Mallevialle, & E. Kawczynski, editors). Denver, CO: AWWA Research Foundation and Lyonnaise des Eaux.

McDonald, J. 2000. Question of the Month—Why Does My Water Look Milky? *Opflow,* 26:6:44.

McGuire, M.J.; K.F. Arnold; J. Biggs; & M.S. Pearthree. 2007b. Using a Consumer Panel as a Tool for Making Water Resource Blending Decisions. Proceedings AWWA Water Quality Technology Conference, Charlotte, NC.

McGuire, M.J.; J. Loveland; E.G. Means; & J. Garvey. 2007a. Use of Flavour Profile and Consumer Panels to Determine Differences Between Local Water Supplies and Desalinated Seawater. *Water Science and Technology,* 55:5:275–282.

McGuire, M.J.; J. Rosen; A.J. Whelton; & I.H. Suffet. 2014. An Unwanted Licorice Odor in a West Virginia Water Supply. *Journal of the American Water Works Association,* 106:6:72–82.

McGuire, M.J.; I.H. Suffet; & J. Rosen. 2014. Customer Panel Estimates of Odor Thresholds for Crude 4-Methylcyclohexanemethanol. *Journal of the American Water Works Association,* 106:10:E445–E458.

McPherson, L. 2012. Amperometric vs. Colorimetric Methods for On-line Measurement of Chlorine. *Waterworld,* 28:8.

Meilgaard, M.C.; B.T. Carr; & G.V. Civille. 2006 (4th ed.). *Sensory Evaluation Techniques.* Boca Raton, FL: CRC Press.

Mirlohi, S.; A.M. Dietrich; & S.E. Duncan. 2011. Age-Associated Variation in Sensory Perception of Iron in Drinking Water and the Potential for Overexposure in the Human Population. *Environmental Science and Technology,* 45:15:6575–83.

Mitroka, S.; T. Smiley; J. Tanko; & A. Dietrich. 2013. Reaction mechanism for oxidation and degradation of high density polyethylene in chlorinated water. *Polymer Degradation and Stability*, 98(7):1369–1377.

Moore, R. 2013. Ice Pigging Offers Sustainable Main Cleaning Technology. *Opflow* 39:314–16.

Muylwyk, Q. 2005 (1st ed.). Blending Chlorinated and Chloraminated Water in Multiple Distribution Systems. In: *Water Quality in the Distribution System*. (W.C. Lauer, editor). Denver, CO: American Water Works Association.

Muylwyk, Q.; J. MacDonald; & M. Klawunn. 2005 (1st ed.). Success: Switching From Chloramines to Chlorine in the Distribution System: Results From a One-Year Full-Scale Trial. 2005. In: *Water Quality in the Distribution System*. (W.C. Lauer, editor). Denver, CO: American Water Works Association.

NSF International. 2015. NSF/ANSI Standard 61. *Drinking Water System Components—Health Effects*. Ann Arbor, Mich.: NSF International.

———. 2016. NSF/ANSI Standard 60. *Drinking Water Treatment Chemicals—Health Effects*. Ann Arbor, Mich.: NSF International.

National Academies of Sciences. 2010. Taste and Flavor Roles of Sodium in Foods: A Unique Challenge to Reducing Sodium Intake. In: *Strategies to Reduce Sodium Intake in the United States*. Institute of Medicine (US) Committee on Strategies to Reduce Sodium Intake (J.E. Henney, C.L. Taylor, & C.S. Boon editors). Washington, DC: National Academies Press.

National Research Council. 2006. Water Quality Integrity. In: *Drinking Water Distribution Systems: Assessing and Reducing Risks*. Committee on Public Water Supply Distribution Systems: Assessing and Reducing Risks. Washington, DC: National Academies Press; pp. 221–268.

Ömür-Özbek, P. 2012. *Global Taste and Odor Survey of Water Utilities*. Denver, CO: AWWA.

Ömür-Özbek, P. & A.M. Dietrich. 2008. Developing Hexanal as an Odor Reference Standard for Sensory Analysis of Drinking Water. *Water Research*, 42:2598–2604.

Ömür-Özbek, P. & A.M. Dietrich. 2011. Retronasal Perception and Flavor Thresholds of Iron and Copper in Drinking Water. *Journal of Water and Health*, 9:1:1–9.

Parks, J.L.; A. Kashyap; A. Atassi; O. Schneider; & M. Edwards. 2012. Effect of Zinc and Orthophosphate Corrosion Inhibitors on Cement-based Pipes. *Journal of the American Water Works Association*, 104:1:37–38.

Parliament, T.; M.G. Kolor; & D.J. Rizzo. 1982. Volatile Components of Limburger Cheese. *Journal of Agriculture and Food Chemistry*, 30:6:1006–1008.

Pérez, A. & H. An. 2005 (1st ed.). Automatic Flushing: A Solution for Enhancement of Distribution System Water Quality. In: *Water Quality in the Distribution System*. (W.C. Lauer, editor). Denver, CO: American Water Works Association.

Phetxumphou, K.; A.M. Dietrich; N. Shanaiah; E. Smiley; & D.L. Gallagher. 2016. Subtleties of Human Exposure and Response to Chemical Mixtures From Spills. *Environmental Pollution*, 214:618–626. http://dx.doi.org/10.1016/j.envpol.2016.04.056.

PWD (Philadelphia Water Department) and CH2M HILL. 2013. *Philadelphia Water Department Contamination Warning System Demonstration Pilot Project: Customer Complaints Surveillance Guidance*. White paper submitted to USEPA as part of the Water Security Initiative grant awarded to Philadelphia Water Department. https://www.ch2m.com/sites/default/files/content/article/attachments/CH2M-HILL-Customer-Complaints-Surveillance-Guidance.pdf (accessed May 23, 2017).

Piriou, P.; R. Devesa; S. Puget; T. Thomas-Danguin; & F. Zraick. 2015. Evidences of Regional Differences in Chlorine Perception by Consumers: Sensitivity Differences or Habituation? *Journal of Water Supply: Research and Technology—AQUA*, 64(7): 783-792.

Piriou, P.; E.D. Mackey; I.H. Suffet; & A. Bruchet. 2004. Chlorinous Flavor Perception in Drinking Water. *Water Science Technology*, 49:9:321–328.

Platikanov, S.; V. Garcia; I. Fonseca; E. Rullán; R. Devesa; & R. Tauler. 2013. Influence of Minerals on the Taste of Bottled and Tap Water: A Chemometric Approach. *Water Research*, 47:693–704.

Proulx F.; M.J. Rodriguez; J. Sérodes; & C. Bouchard. 2007. A Methodology for Identifying Vulnerable Locations to Taste and Odour Problems in a Drinking Water System. *Water Science and Technology*, 55:5:177–183.

Puget, S.; N. Beno; C. Chabanet; E. Guichard; & T. Thomas-Danguin. 2010. Tap Water Consumers Differ From Non-Consumers in Chlorine Flavor Acceptability but not Sensitivity. *Water Research*, 44:956–964.

Rashash, D.; R. Hoehn; A.M. Dietrich; T. Grizzard; & B. Parker. 1996. *Identification and Control of Odorous Algal Metabolites*. Denver, CO: Water Research Foundation.

Reiber, S. 1993. Investigating the effects of chloramines on elastomer degradation. *Journal American Water Works Association*, 85(8):101-111.

Rigal, S. &.J. Danjou. 1999. Tastes and odors in drinking water distribution systems related to the use of synthetic materials. *Water Science and Technology*, 40(6):203-208.

Ritter, L.; K. Salomon; P. Sibley; K. Hall; P. Kleen; G. Mattu; & B. Linton. 2002. Sources, Pathways and Relative Risks of Contaminants in Surface Water and Groundwater: A Perspective Prepared for the Walkerton Inquiry. *Journal of Toxicology and Environmental Health A*, 65:1:1–142.

Romero, R.C.; F. Ventura; J. Caixach; J. Rivera; L.X. Godé; & J.M. Niñerola. 1998. Identification of 1,3-Dioxanes and 1,3-Dioxolanes as Malodorous Compounds at Trace Levels in River Water, Groundwater, and Tap Water. *Environmental Science & Technology*, 36:206–216.

Rouessac, F. & A. Rouessac, A. 2007 (2nd ed.). *Chemical Analysis: Modern Instrumentation Methods and Techniques*. Chichester, United Kingdom: John Wiley and Sons.

Ruth, J. 1986. Odor Thresholds and Irritation Levels of Several Chemical Substances: A Review. *American Industrial Hygiene Association,* 47:A142–151.

Sain, A.E. & A.M. Dietrich. 2015. Rethinking Aesthetic Guidelines for Manganese and Iron in Drinking Water. *Journal of Water Supply: Research and Technology—AQUA,* 64(7): 775-782 *Water Supply: Research and Technology–AQUA,* 2015 in press (doi: 10.2166/aqua.2014.091).

Sarver, E.; Y. Zhang; & M. Edwards. 2011. *Copper Pitting and Brass Dezincification: Physical and Chemical Effects.* Denver, CO: Water Research Foundation.

Satchwill, T. 2002. The impact of pipe coatings on drinking water quality. *Alberta Water and Wastewater Operators Association.*

Schade, C.P.; N. Wright; R. Gupta; D.A. Latif; A. Jha; & J. Robinson. 2015. Self-Reported Household Impacts of Large-Scale Chemical Contamination of the Public Water Supply, Charleston, West Virginia, USA. *PLOS ONE,* 10:5:e0126744.

Seattle Public Utilities. 2012. Annual Report Card. http://www.seattle.gov/util/AboutUs/Management/AnnualReportCard/index.htm (accessed May 23, 2017).

Skjevrak, I., Due, A., Gjerstad, K. O., & Herikstad, H. 2003. Volatile organic components migrating from plastic pipes (HDPE, PEX and PVC) into drinking water. *Water research, 37*(8), 1912-1920.

Sly, L.; M. Hodgkinson; & V. Arunpairojana. 1990. Deposition of Manganese in Drinking Water Distribution Systems. *Journal of Applied Environmental Microbiology,* 56:3:628.

Snoeyink, V.; M. Schock; P. Sarin; L. Wang; A. Chen; & S. Harmon, 2003. Aluminium-Containing Scales in Water Distribution Systems: Prevalence and Composition. *Journal of Water Supply: Research and Technology–AQUA,* 52:7:455–474.

Speight, V.; W. Grayman; & N. Khanal. 2010. Syngeries Between Security Research and Distribution System Water Quality Management. *Journal of the American Water Works Association,* 102:2:30–33.

Stanford, B.D.; B. Wright; J.C. Routt; J.F. Debroux; & S.J. Khan. 2014. *Water Quality Impacts of Extreme Weather-related Events.* Denver, CO: Water Research Foundation.

Strube, A.; A. Buettner; & M. Czerny. 2012. Influence of Chemical Structure on absolute Odour Thresholds and Odour Characteristics of *ortho-* and *para*-Halogenated Phenols and Cresols. *Flavor and Fragrance Journal,* 27:304–312.

Strube, A.; H. Guth; & A. Buettner. (2009). Identification of a Medicinal Off-Flavour in Mineral Water. *Water Research,* 43:5216–5224.

Suffet, I.H.; D. Khiari; & A. Bruchet. 1999. The Drinking Water Taste and Odor Wheel for the Millennium: Beyond Geosmin and Methylisoborneol. *Water Science & Technology,* 40:6:1–13.

Suffet, I.H.; J. Mallevialle; & E. Kawczynski (editors). 1995. *Advances in Taste-and-Odor Treatment and Control.*Denver, CO: AwwaRF and Lyonnaise des Eaux.

Suffet, I.H.; L. Schweitzer; & D. Khiari. 2004. Olfactory and Chemical Analysis of Taste and Odor Episodes in Drinking Water Supplies. *Reviews in Environmental Science and Biotechnology*, 3:3–13.

Sutherland, J. 2015. Personal communication.

Sutherland, J.; D. Bryant; K. Morrison; R. Jenkins; L. Rothe; G. Juby; & T. Seacord. 2009. Desalination Pilot Study of a Brackish Groundwater in Texas. AWWA Annual Conference and Exhibition. San Diego, CA.

Schweitzer, L., Tomboulian, P., Atasi, K., Chen, T., & Khiari, D. 2004. Utility quick test for analyzing materials for drinking water distribution systems for effect on taste-and -odor. *Water Science and Technology, 49*(9), 75-80.

Taste and Odor Committee of the American Water Works Association. 2002. Options for a Taste and Odor Standard. *Journal of the American Water Works Association*, 946:80–87.

Teillet, E.; C. Urbano; S. Cordelle; & P. Schlich. 2010. Consumer Perception and Preference of Bottled and Tap Water. *Journal of Sensory Studies*, 25:3:463–480.

Tomboulian, P., Schweitzer, L., Mullin, K., Wilson, J., & Khiari, D. 2004. Materials used in drinking water distribution systems: contribution to taste-and-odor. *Water Science and Technology, 49*(9), 219-226.

USEPA (US Environmental Protection Agency). 1979. National Secondary Drinking Water Regulations. Final Rule. *Federal Register*, 44:140:42195.

———. 1991. Maximum Contaminant Level Goals and National Primary Drinking Water Regulations for Lead and Copper; Final Rule. *Federal Register*, 56 F.R. 26460, 1991-06-07. 40 CFR Part 141, Subpart I56 FR 26460, June 7, 1991.

———. 2003. Drinking Water Advisory: Consumer Acceptability Advice and Health Effects Analysis on Sodium. www.epa.gov/safewater/ccl/pdf/sodium.pdf. EPA 822-R-03-006.(accessed May 23, 2017).

———. 2007. Distribution System Inventory, Integrity and Water Quality. Washington, DC: USEPA.

———. 2008. *Water Security Initiative: Interim Guidance on Developing an Operational Strategy for Contamination Warning Systems*. EPA-817-R-08-002. Washington, DC: USEPA.

———. 2009. Method 334.0: Determination of Residual Chlorine in Drinking Water Using an On-Line Chlorine Analyzer. EPA 815-B-09-013, Office of Research and Development, Washington.

———. 2010. *Water Quality Event Detection Systems for Drinking Water Contamination Warning Systems.* EPA/600/R-10/036, Cincinnati, OH: Office of Research and Development.

———. 2011. Alarm Estimation Tool Microsoft Excel File. https://www.epa.gov/waterqualitysurveillance/customer-complaint-surveillance-resources .(accessed May 23, 2017).

———. 2012. Edition of the Drinking Water Standards and Health Advisories. EPA 822-S-12-001. Washington, DC: USEPA Office of Water; p. 12.

———. 2015. Water Quality Surveillance and Response System (WQSRS) Primer. EPA 817-B-15-002. Washington, DC: USEPA Office of Water.

USPHS (US Public Health Service). 1962. *Public Health Service Drinking Water Standards.* Publication No. 956. Washington, DC: US Government Printing Office.

Viebke, J., & Gedde, U. W. 1997. Antioxidant diffusion in polyethylene hot-water pipes. *Polymer Engineering & Science, 37*(5), 896-911.

Vreeburg, J.H.G.; D. Schippers; J.Q.J.C. Verberk; & J.C. van Dijk. 2008. Impact of Particles on Sediment Accumulation in a Drinking Water Distribution System. *Water Research,* 42:4233–4242.

Wajon J.E.; R. Alexander; R.I. Kagi; & B. Kavanagh. 1985. Dimethyl Trisulfide and Objectionable Odours in Potable Water. *Chemosphere,* 14:85–89.

Weiffenbach, J.M.; B.J. Baum; & R. Burghauser. 1982. Taste Thresholds: Quality Specific Variation With Human Aging. *Journal of Gerontology,* 37:3:372–377.

Weinberg, H.S.; W.H. Glaze; S.W. Krasner; & M.J. Sclimenti. 1993. Treatment and Removal of Aldehydes in Plants That Use Ozonation. *Journal of the American Water Works Association,* 85:5:72–85.

Whelton, A. & A. Dietrich. 2009. Critical considerations for the accelerated ageing of high-density polyethylene potable water materials. *Polymer Degradation and Stability* 94(7): 1163–1175.

Whelton, A.; ,A.M. Dietrich; D.L. Gallagher; & J.A. Roberson. 2007. Using Customer Feedback for Improved Water Quality and Infrastructure Monitoring. *Journal of the American Water Works Association,* 99:11:62–76.

Whelton, A. & T. Nguyen. 2013. Contaminant Migration From Polymeric Pipes Used in Buried Potable Water Distribution Systems: A Review. *Critical Reviews in Environmental Science and Technology,* 43(7):679-751

Whelton, A. J.; L. McMillan; M. Connell; K.M. Kelley; J.P. Gill; K.D. White; R. Gupta; R. Dey; & C. Novy. 2015. Residential Tap Water Contamination Following the Freedom Industries Chemical Spill: Perceptions, Water Quality, and Health Impacts. *Environmental Science and Technology,*. DOI: 10.1021/es5040969.

Whitfield, F.B. 1988. Chemistry of Off-Flavours in Marine Organisms. *Water Science and Technology,* 20:8/9:63–74.

WHO (World Health Organization). 2011 (4th ed.). Guidelines for Drinking-Water Quality. Geneva, Switzerland: World Health Organization. Wiesenthal, K. E., & Suffet, I. H. (2007). The effect of white or grey PVC pipe and its joint solvents (primer and cement) on odour problems in drinking water distribution systems. *Water science and technology, 55*(5), 169–176.

Yang, F., Shi, B., Bai, Y., Sun, H., Lytle, D. A., & Wang, D. (2014). Effect of sulfate on the transformation of corrosion scale composition and bacterial community in cast iron water distribution pipes. *Water research, 59,* 46–57.

Yang, J.; O. Schneider; P.K. Jjemba; & M.W. LeChevalier. 2015. *Microbial Risk Modeling for Main Breaks. Journal of the American Water Works Association,* 107:2:E97–E108.

Young, W.F.; H. Horth; R. Crane; T. Odgen; & M. Arnott. 1996. Taste and Odour Threshold Concentrations of Potential Potable Water Contaminants. *Water Research,* 30:2:331–340.

Yu, W., Azhdar, B., Andersson, D., Reitberger, T., Hassinen, J., Hjertberg, T., & Gedde, U. W. 2011. Deterioration of polyethylene pipes exposed to water containing chlorine dioxide. *Polymer degradation and stability, 96*(5), 790-797.

Yu, W., Reitberger, T., Hjertberg, T., Oderkerk, J., Costa, F. R., & Gedde, U. W. 2012. Antioxidant consumption in squalane and polyethylene exposed to chlorinated aqueous media. *Polymer degradation and stability, 97*(11), 2370-2377.

Zhang, X-J.; C. Chen; J-Q. Ding; A. Hou; Y. Li; Z-B. Niu; X.-Y. Su; Y.-J. Xu; & E.A. Laws. 2010. The 2007 Water Crisis in Wuxi, China: Analysis of the Origin. *Journal of Hazardous Materials,* 182:130–135.

Zhang, X-J.; H. Zhang; S.X. Wang; & R.B. Zhang. 2007. Emergent Drinking Water Treatment for Taste and Odor Control in Wuxi City Water Pollution Incident. *Water and Wastewater Engineering,* 33:7-12.

Zhang, Y.; T.J. Tseng; C. Andrews-Tate; R.C. Cheng; & K.L. Wattier. 2012. Pilot-scale Evaluation of Blending Desalinated Seawater Into a Distribution System. *Journal of the American Water Works Association,* 104:7:E416–E429. doi: 10.5942jawwa.2012.104.0109.

Zielina, M.; W. Dabrowski; & E. Radziszewska-Zielina. 2014. Cement Mortar Lining as a Potential Source of Water Contamination. *International Journal of Environmental, Ecological, Geological and Marine Engineering,* 8:10:658–661.

Zogorski, J.S.; S.D. Faust; & J.H. Haas Jr. 1976. The Kinetics of Adsorption of Phenols by Granular Activated Carbon. *Journal of Colloid and Interface Science,* 55: 329–341.

Chapter 6

Nitrification

Janice Skadsen, Ann Arbor, Mich. (formerly with CDM Smith)

Amlan Ghosh, Corona Environmental Consulting, Austin, Texas

David Wahman, US Environmental Protection Agency, Office of Research and Development, Cincinnati, Ohio

Free chlorine use as a secondary (distribution system) disinfectant remains popular (AWWA WQTD 2008b, 2008a). However, water systems now use monochloramine (in this chapter referred to as chloramine or chloramines) to avoid excessive regulated disinfection by-product (DBP) formation (e.g., trihalomethane and haloacetic acid). Surveys of US water systems showed that 30 percent of respondents currently use chloramine for their distribution system residual, and other surveys suggest that another 8 to 12 percent of water systems are contemplating a switch to chloramination (AWWA WQTD 2008b; Seidel et al. 2005).

Because ammonia is present when chloramine is used, chloramination comes with the risk of distribution system nitrification, the biological process in which ammonia is oxidized to nitrite and then nitrate. Nitrification in drinking water distribution systems is undesirable and may result in water quality degradation, including disinfectant depletion, bacterial growth, and nitrite/nitrate formation. This may cause subsequent noncompliance with regulations, for example, the Surface Water Treatment Rule and Total Coliform Rule in the United States (AWWA 2013).

DISCLAIMER

The US Environmental Protection Agency, through its Office of Research and Development, collaborated in the production of this chapter. It has been subjected to the agency's peer and administrative review and has been approved for external publication. Any opinions expressed in this chapter are those of the authors and do not necessarily reflect the views of the agency; therefore, no official endorsement should be inferred. Mention of trade names or commercial products does not constitute endorsement or recommendation for use.

Based on water utility surveys, 30 to 63 percent of the utilities that use chloramine for their secondary disinfectant experience periodic nitrification episodes that affect localized or widespread system areas (Wilczak et al. 1996; Kirmeyer et al. 2004; Seidel et al. 2005; AWWA WQTD 2008b). These findings likely reflect the perception of water personnel about the patterns and frequency of major nitrification episodes, specifically that observable nitrification is sporadic and typically affects only localized areas that are prone to corrosion/tuberculation, biofilm attachment, solids accumulation, and high water age. Such areas may include pipe surfaces or tank walls, especially at locations farther away from the treatment plant where a combination of factors related to high water age (e.g., oversized mains, dead ends, closed valves) allow for biological growth.

Some water utilities and other types of water systems may not realize the extent of nitrification in their distribution systems, depending on their monitoring program, including sample numbers, frequency, location, and parameters monitored. Overall, nitrification has been reported to occur in water distribution systems that use chloramine throughout the United States, Canada, Europe, and Australia (e.g., Cunliffe 1991; Lipponen, Suutari, and Martikainen 2002; Pintar et al. 2005; Sathasivan, Fisher, and Kastl 2005; Carrico et al. 2008; Rosenfeldt, Baeza, and Knappe 2009).

NITRIFICATION

This section provides an overview of the nitrification process. This includes a general description of the process, as well as a discussion of ammonia-oxidizing bacteria and chloramine inactivation in a laboratory setting.

General Process Description

Nitrification is a two-step aerobic (oxygen-using) biological process. First, ammonia-oxidizing bacteria (AOB) biologically oxidize ammonia to nitrite. Second, nitrite-oxidizing bacteria (NOB) biologically oxidize nitrite to nitrate (Ward, Arp, and Klotz 2011). The second step may also occur from chemical oxidation by free chlorine or monochloramine.

For AOB with ammonium (NH_4^+) as the basis, the reaction is (Grady, Daigger, and Lim 1999):

$$NH_4^+ + 1.382\ O_2 + 1.982\ HCO_3^- \rightarrow 0.0182\ C_5H_7O_2N + 0.982\ NO_2^- + 1.036\ H_2O + 1.891\ H_2CO_3 \qquad \text{(Eq. 6–1)}$$

For NOB with NO_2– as the basis, the reaction is:

$$NO_2^- + 0.00256\ NH_4^+ + 0.0104\ H_2CO_3 + 0.00226\ HCO_3^- + 0.487\ O_2 \rightarrow 0.00244\ C_5H_7O_2N + 0.00767\ H_2O + 1.0\ NO_3^- \qquad \text{(Eq. 6–2)}$$

The overall reaction for complete nitrification is:

$$NH_4^+ + 1.856\ O_2 + 1.979\ HCO_3^- \rightarrow 0.0205\ C_5H_7O_2N + 0.979\ NO_3^- + 1.041\ H_2O + 1.876\ H_2CO_3 \qquad \text{(Eq. 6–3)}$$

The AOB and NOB reactions highlight some important issues related to nitrification's direct impacts on drinking water quality (Table 6–1), including:

- Oxygen consumption,
- Alkalinity consumption (and possible pH decrease), and
- Nitrite and/or nitrate production.

Table 6-1 Water chemistry impact of ammonia and nitrite oxidation by AOB and NOB per milligrams of ammonia-nitrogen per liter

Nitrogen Endpoint	Oxygen Consumed, mg O_2/L	Alkalinity Consumed, mg $CaCO_3$/L	Nitrite Produced, mg N/L	Nitrate Produced, mg N/L
Nitrite	3.2	7.1	1.0	–
Nitrate	4.2	7.1	–	1.0

See Eqs 6-1, 6-2, and 6-3.

The majority of oxygen and alkalinity consumption occurs through the first step in the nitrification process (i.e., biological ammonia oxidation). Therefore, any pH decrease and the majority of oxygen consumption from a nitrification event is a result of AOB activity.

DSWQ Challenge: What is nitrification?

The biological conversion of ammonia to nitrite
The biological conversion of nitrite to nitrate

Because AOB represent the first step in nitrification (i.e., biological ammonia oxidation), they are the key bacteria in nitrification episodes and are discussed with regard to nitrification. In addition to AOB, recent research has shown that another group of ammonia oxidizers—archaea—are present in drinking water systems (van der Wielen, Voost, and van der Kooij 2009; Kasuga et al. 2010); however, their importance in distribution system nitrification is not yet known. In addition to archaea, a nitrifying bacteria that can convert ammonia to nitrate (i.e., do both steps of nitrification by themselves) has recently been discovered, and there is evidence of their presence in drinking water (Pinto et al. 2016; van Kessel et al. 2015; Daims et al. 2015). In this chapter, biological ammonia oxidation is discussed with respect to AOB.

Ammonia-Oxidizing Bacteria

AOB are gram-negative bacteria that use ammonia as their sole energy source and inorganic carbon as their carbon source. Figure 6-1 shows an AOB isolated from a drinking water reservoir (Wolfe et al. 1990). Ammonia oxidation is an aerobic process carried out optimally between pH 7.5 and 8.0 and temperatures in the range of 20 to 30°C. AOB have low growth rates and yields because of their small energy gain from ammonia oxidation and large energy investment needed to reduce inorganic carbon, resulting in doubling times that vary from 8 hours to several days (Ward, Arp, and Klotz 2011; Prosser 1986, 1989).

Because the most studied AOB is the pure culture *Nitrosomonas europaea*, it is used as a model AOB in this chapter. *Nitrosomonas europaea* is a gram-negative bacterium in the β subdivision of the proteobacteria. Ammonia oxidation to nitrite provides *N. europaea* with all of its required reductant for energy and biosynthesis; *N. europaea* uses inorganic carbon for all of its required carbon for growth. Figure 6-2 shows the central metabolic pathway of *N. europaea*.

For ammonia oxidation to nitrite, *N. europaea* uses two enzymes in two reaction steps. The first reaction converts ammonia to hydroxylamine and is catalyzed by the enzyme ammonia monooxygenase, which is bound to the cell membrane. The second step converts hydroxylamine to nitrite and is catalyzed by the hydroxylamine oxidoreductase enzyme (Arp, Sayavedra-Soto, and Hommes 2002), which resides in the periplasm.

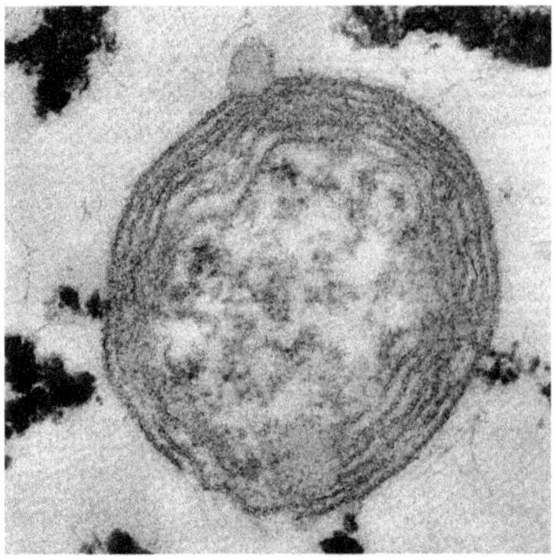

Reprinted with permission of the American Society of Microbiology.

Figure 6-1 *Nitrosomonas* **species isolated from a drinking water reservoir; transmission electron micrograph (bar, 0.1 μm)**

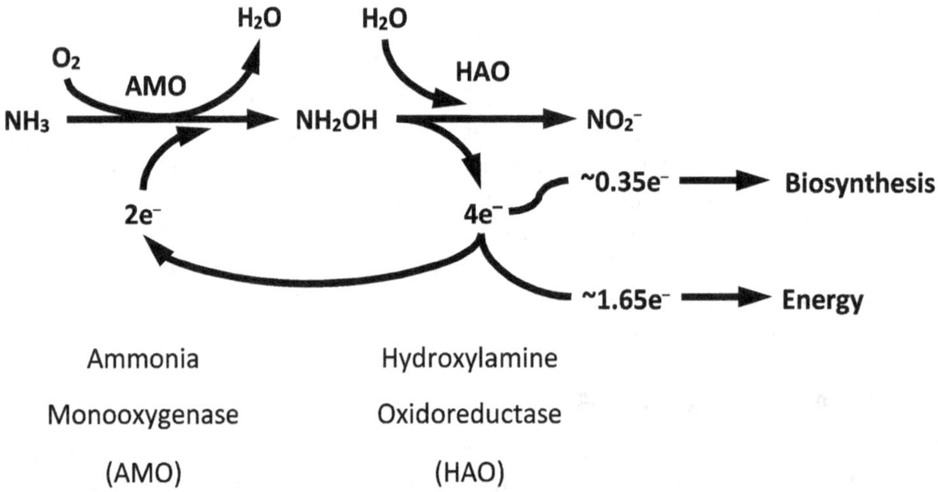

Adapted from Ward, Arp, and Klotz 2011, Arp, Sayavedra-Soto, and Hommes 2002, and Whittaker et al. 2000.

Figure 6-2 **Simplified** ***Nitrosomonas europaea*** **central metabolism**

Chloramine Inactivation in the Laboratory

Several researchers have investigated how fast chloramine disinfection inactivates nitrifier cultures (AWWA 2013; Chauret, Smith, and Baribeau 2008). The change in nitrifier viability after disinfection (Breeuwer and Abee 2000; Lloyd and Hayes 1995; Hoefel et al. 2003) is evaluated using culture-dependent methods and culture-independent methods.

Culture-dependent methods rely on bacterial growth and include plate counts or most-probable number (MPN) techniques. Culture-independent methods include activity measurements such as substrate (food) uptake or oxygen use and other methods that rely on cell membrane integrity as a viability measure.

In general, culture-dependent methods identify faster disinfection kinetics than culture-independent methods. Oldenburg et al. (2002) provided a comparison of estimated disinfection kinetic parameters using both the culture-dependent AOB MPN technique and a culture-independent cell membrane integrity method. The estimated disinfection kinetic parameters based on the AOB MPN method were three orders of magnitude (1,000 times) greater than the culture-independent method. However, the slower disinfection kinetics based on the culture-independent method were more consistent with AOB persistence in chloraminated drinking water distribution systems.

Culture-independent methods that rely on cell membrane integrity as the measure of bacterial inactivation have been used for detailed disinfection experiments with *N. europaea* (Oldenburg et al. 2002; Wahman, Wulfeck-Kleier, and Pressman 2009; Wahman, Schrantz, and Pressman 2010). Oldenburg et al. (2002) found that *N. europaea* inactivation was pH dependent; specifically, pH 7 inactivation was approximately 30 times faster than pH 9 inactivation.

In contrast to these pure culture laboratory studies, AOB are present as mixed cultures in chloraminated drinking water distribution systems. Wahman Wulfeck-Kleier, and Pressman (2009) evaluated a method that could be applied to mixed-culture AOB populations to further define representative monochloramine inactivation kinetics. It remains to be seen how the monochloramine inactivation kinetics of drinking water AOB mixed cultures compare to these pure-culture laboratory studies.

CAUSES OF NITRIFICATION IN DISTRIBUTION SYSTEMS

The following discussion addresses the potential processes that lead to nitrification. This includes general considerations, with a focus on chloramine residual levels, decay, and demand.

General Considerations

Nitrifying bacteria (i.e., AOB and NOB) are found everywhere in nature and can grow under favorable conditions in chloraminated distribution systems. This is because free ammonia (i.e., ammonia not combined in chloramine) is inherently present through chloramine application (a slight excess ammonia is typically used) along with chloramine demand and decay reactions that release free ammonia and diminish disinfectant residuals. In addition to the necessary ammonia substrate, conditions that favor nitrification are determined by source water quality, subsequent treatment, and distribution system conditions and operations.

DSWQ Challenge: What Causes Nitrification?

- Presence of free ammonia
- Low chlorine-to-ammonia nitrogen (Cl_2:N) mass ratio
- High water age
- Insufficient disinfectant residual
- Warm temperature

After chloraminated water leaves the treatment facility and enters the distribution system, water quality and distribution system configuration, condition, and operation affect the chloramine residual stability and potential for nitrification within the system. A distribution system is at risk when it does not receive and maintain water with quality that favors chloramine stability and low free ammonia and when it is not designed or operated for good water turnover and system cleanliness.

The two primary ways water systems attempt to control nitrification involve minimizing free ammonia while maintaining chloramine residuals and minimizing water age. Free ammonia is the essential nutrient for nitrification to establish and continue. Nitrification causes and controls are centered on availability or minimization of free ammonia. This can occur from excess ammonia fed at treatment facilities or released via chloramine demand and/or decay reactions throughout distribution systems. Where free ammonia is routinely present in water systems, nitrifiers are expected to be present.

The extent to which nitrification causes problematic chloramine residual loss depends on a water system's ability to optimize chloramine application while minimizing chloramine demand and decay reactions. Typically, nitrification results from a combination of several factors or conditions; therefore, all possible factors should be evaluated in chloraminated distribution systems and in systems considering chloramination.

Before a water system converts to chloramine, operators should prepare and begin to implement a nitrification detection and control program. In general, distribution system problem areas encountered with free chlorine often continue to generate water quality challenges after chloramine conversion. Hence, work done to identify and address low-flow areas of piping and storage prior to beginning chloramination will prevent future problems.

Optimizing chloraminated distribution system operations typically involves progressive improvement in chloramine formation, water quality monitoring, and operational tools for nitrification control. Sometimes, it may take engineering improvements of the source, treatment, and/or distribution systems to resolve persistent problems. Because of differences between free chlorine and chloramine reactivity and residual stability, systems that previously required booster free chlorination may find that those applications are no longer needed after switching to chloramine. However, they may be useful in boosting chloramine residuals over time, depending on their locations and the extent of nitrification. Therefore, it is advisable to not decommission boosting facilities until the utility is certain that they are no longer needed. Unlike booster-free chlorination, booster chloramination requires additional monitoring and process control. If booster chloramination is conducted, free ammonia should be monitored frequently, and the booster chlorine dose should be adjusted accordingly (i.e., at a desired chlorine-to-ammonia mass ratio).

How chlorine and ammonia are applied influences the amount of free ammonia initially available and the total ammonia (i.e., ammonia bound in chloramine plus free ammonia) available to support nitrification in the downstream distribution system. At the application point, free ammonia reacts with free chlorine to form predominantly monochloramine—the preferred chloramine species—along with very minimal excess residual free ammonia. AOB only use free ammonia as a growth substrate, and ammonia present in chloramines does not directly support nitrification. However, recent research suggests that monochloramine may be biologically transformed by AOB (Maestre, Wahman, and Speitel 2013; Speitel, Maestre, and Wahman 2014).

Chloramine Residual Level

Chloramine impacts nitrification due to its presence as a disinfectant and degradation that releases free ammonia. Chloramine residual is most commonly measured as "total chlorine" by the N,N-diethyl-p-phenylenediamine method, which measures the combined or chloramine concentration.

The USEPA Stage 1 Disinfectant/Disinfection Byproduct Rule (D/DBPR) establishes a maximum residual disinfectant level for total chlorine of 4.0 mg Cl_2/L as a running annual average of distribution system residuals. Systems covered under the USEPA Surface Water Treatment Rule are required to have a "detectable" disinfectant residual in at least 95 percent of samples collected throughout the distribution system on a monthly basis (or heterotrophic plate count [HPC] measurements <500 cfu/mL).

Other countries set somewhat different standards. For example, the province of Ontario, Canada, established a maximum contaminant level (MCL) for total chlorine of 3 mg Cl_2/L but does not have distribution system targets (Ontario Drinking Water Quality Standards 2002). The World Health Organization has a maximum guideline of 3 mg Cl_2/L total chlorine and 5 mg Cl_2/L free chlorine (WHO 2011).

The target chloramine residual is system specific. In general, some water systems have increased their target residual over the years, either seasonally or year-round, to better control nitrification. Higher than "detectable" minimum levels for total chlorine residuals (e.g., 0.5 mg Cl_2/L) have been required or recommended by some states and recommended in water utility research–based guidance (Friedman et al. 2010). A summary of the regulated minimum total chlorine residuals in the distribution system (not entering the distribution system) by state for the United States is presented in Figure 6-3, indicating that several states have implemented regulations that require greater than "detectable."

Maintaining a "minimum detectable" or even a substantially higher chloramine residual is inadequate, in itself, to control nitrification and related chloramine residual loss if free ammonia continues to be present at high enough concentrations to sustain nitrifier growth and if the nitrification product (nitrite) is present, creating a chloramine demand. In a given system, chloramine demand and decay rates, combined with factors that affect water age and microenvironments (e.g., solids and biofilm accumulations, corrosion tubercles), determine the chloramine residual level sufficient to control nitrification and maintain residuals throughout the system.

Chloramine Decay and Demand

An understanding of chloramine chemistry, especially what impacts the chloramine demand and decay reactions, is critical to minimizing free ammonia, minimizing nitrification, and maintaining the chloramine residual in the distribution system. Chloramine formation and subsequent decay occur through a complex series of chemical reactions (Jafvert and Valentine 1992). Both demand- and decay-type reactions decrease the chloramine concentration and release free ammonia. Therefore, the monochloramine residual that is responsible for controlling AOB growth decreases through the distribution system, while the free ammonia residual responsible for supporting AOB growth increases (Harrington et al. 2003).

Once formed, chloramine (i.e., monochloramine) can react (e.g., chloramine dissipates, releasing ammonia) by two general mechanisms: decay reactions and demand reactions. In this chapter, these terms are used to mean the following:

- *Chloramine decay* refers to the inherent chloramine "autodecomposition" (i.e., chloramine reacting with itself) that results in decreasing chloramine concentrations over time. Chloramine decay can be relatively slow and occurs whether or not nitrifiers are present.
- *Chloramine demand* refers to chemical or microbial constituents in the water or on surfaces in contact with the water that react with chloramine (e.g., nitrite, bromide, natural organic matter, and internal pipe surfaces). Chloramine demand reactions can be relatively fast or slow and can also occur whether or not nitrifiers are present, but nitrifier presence can increase and create additional demand reactions.

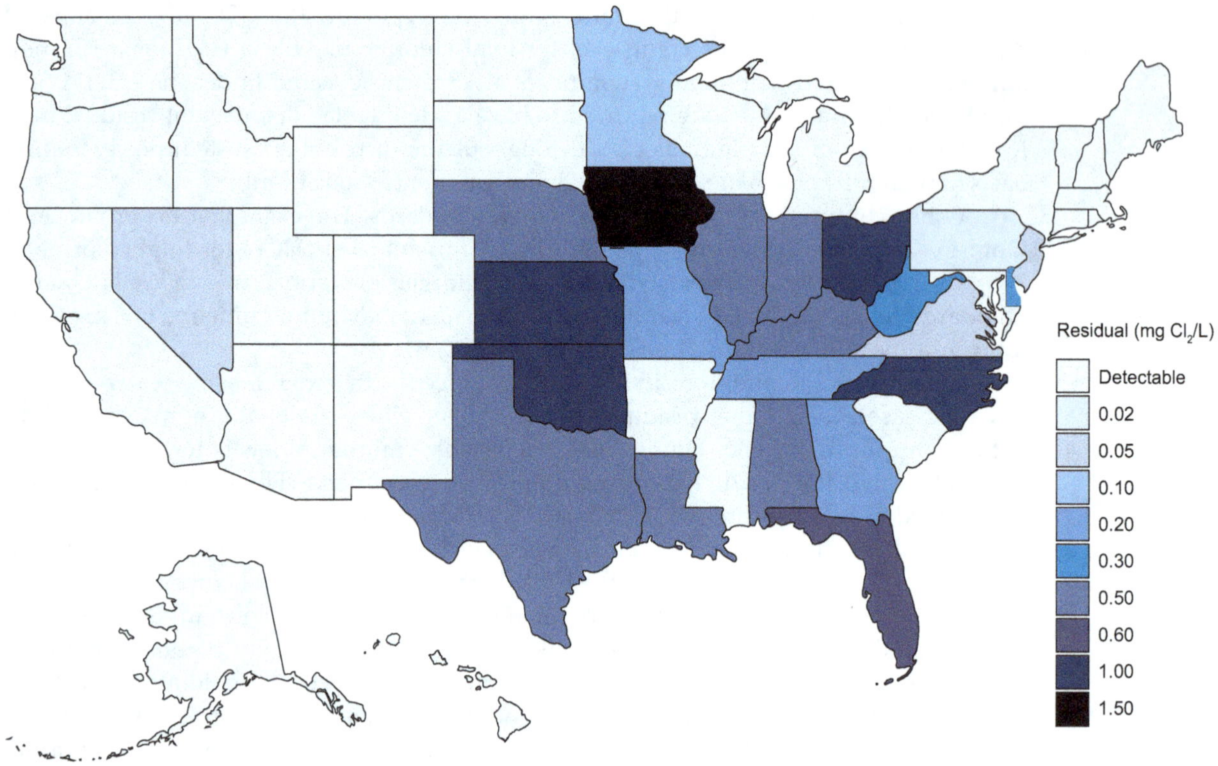

Reprinted from Wahman and Pressman 2015.

Figure 6-3 US minimum total chlorine residuals in distribution systems (not at entry points to the distribution system), by state, for Subpart H Surface Water Treatment Rule systems. Information is based on a review of existing states' rules and regulations as of January 2015.

Notes:

- Colorado has proposed requiring a 0.2 mg Cl_2/L minimum residual throughout the distribution system.
- Iowa provides an exception for dead ends and areas that represent very low use.
- Louisiana requires a "detectable" residual, but they currently have an emergency rule that requires a minimum 0.5 mg Cl_2/L total chlorine residual; therefore, 0.5 mg Cl_2/L is used in the figure.
- Minnesota requires a "detectable" residual, but they also require measurement to the nearest 0.1 mg Cl_2/L in the range below 0.5 mg Cl_2/L; therefore, 0.1 mg Cl_2/L is used in the figure.
- Nebraska allows a minimum 0.25 mg Cl_2/L total chlorine residual if additional conditions are met.
- Nevada requires a "detectable" residual, but they also require measurement to the nearest 0.05 mg Cl_2/L; therefore, 0.05 mg Cl_2/L is used in the figure.
- North Carolina's minimum residual is for Total Coliform Rule compliance samples. North Carolina requires a "detectable" residual at maximum residence times.
- Tennessee currently has no systems that use chloramines, and their rules are specific that systems shall maintain a 0.2 mg Cl_2/L free chlorine residual in all parts of the distribution system; therefore, Tennessee is currently shown as requiring 0.2 mg Cl_2/L total chlorine as well.
- Pennsylvania had proposed requiring a 1.0 mg Cl_2/L total chlorine residual. Since this figure was first published (January 2015), Pennsylvania has proposed a 0.2 mg/L residual (April 2016).
- Vermont requires a "detectable" residual, but states that the residual "should be 0.1 mg Cl_2/L."

Although more stable than free chlorine, chloramines are inherently unstable, and their fate is controlled by a complex set of reactions that oxidize ammonia and decrease active chlorine (total chlorine); this is represented by the model developed by Jafvert and Valentine (1992) and slightly revised by Vikesland, Ozekin, and Valentine (2001). Chloramine decay can be summarized by two reactions (Eq 6-4 and 6-5) that both yield ammonia—AOB's food:

$$3NH_2Cl \rightarrow N_2 + NH_3 + 3H^+ + 3Cl^- \qquad \text{(major pathway)} \qquad \text{(Eq. 6-4)}$$

$$4NH_2Cl + 3H_2O \rightarrow 3NH_3 + NO_3^- + 5H^+ + 4Cl^- \qquad \text{(minor pathway)} \qquad \text{(Eq. 6-5)}$$

Monochloramine's nitrogen is preferentially oxidized to nitrogen gas in addition to ammonia (Eq 6-4), with smaller quantities oxidized to nitrate (Eq 6–5) (Vikesland, Ozekin, and Valentine 1998). If Eq 6-4 dominates, 1 mol of ammonia is produced for every 3 mol of monochloramine lost. Temperature, pH, inorganic carbon, and Cl_2:N mass ratio affect chloramine decay rates. Maximizing chloramine stability minimizes decay-related ammonia release and nitrification potential.

In addition to chloramine decay, several chloramine demand reactions may occur within drinking water distribution systems, including reactions with nitrite (Johnson and Margerum 1991; Margerum et al. 1994; Vikesland, Ozekin, and Valentine 2001; Wahman and Speitel 2012), bromide (Vikesland, Ozekin, and Valentine 2001), natural organic matter (NOM) (Duirk 2003, 2006; Duirk et al. 2002, 2005; Vikesland, Ozekin, and Valentine 1998), microorganisms and their soluble products (e.g., Maestre, Wahman, and Speitel 2013), and pipe surfaces (Vikesland and Valentine 2000, 2002a, 2002b). All of these reactions are generalized by the following reaction:

$$\text{demand constituent (e.g., nitrite, bromide, NOM)} + NH_2Cl \rightarrow NH_3 + Cl^- + \text{products} \qquad (6\text{-}6)$$

For Eq 6-6, 1 mol of free ammonia is released for each mole of monochloramine reacted. Therefore, minimizing monochloramine demand reactions (e.g., minimizing nitrite, bromide, NOM) can minimize ammonia release and nitrification potential to an even greater extent than measures targeted at minimizing decay.

Chlorine (as Cl_2) to ammonia (as N) (Cl_2:N) mass ratio. When making chloramines, the Cl_2:N mass ratio typically varies from 3:1 to 5:1, with current practice recommending ratios between 4.5:1 and 5:1 to limit, yet maintain, some free ammonia (AWWA 2013). The stoichiometric Cl_2:N mass ratio for monochloramine formation is 5.07:1 (1 mol of chlorine to 1 mol of ammonia). This ratio assumes no free chlorine or chloramine "demand" and only monochloramine formation. The Cl_2:N mass ratio should be balanced at the treatment plant to limit excess free ammonia and maximize monochloramine formation. It is important for operations staff to understand that the Cl_2:N ratio is based on the actual free chlorine available to react with free ammonia at a given point in the treatment train or in the system if boosting. If free chlorine demand is not taken into account, setting the feed rate based on the theoretical ratio may result in underestimating the applied Cl_2:N ratio, leaving a greater free ammonia concentration than desired. In addition, any ammonia that occurs naturally in the source water also needs to be factored into Cl_2:N ratio calculations.

As the Cl_2:N mass ratio increases, chloramine decay rates increase (Figure 6-4) and, conversely, as chloramine decays, the Cl_2:N ratio decreases. Therefore, measures may be needed to monitor and recombine liberated free ammonia in the distribution system by booster chlorination to maintain the Cl_2:N mass ratio near 5:1 throughout the system. In general, higher Cl_2:N mass ratios (up to 5:1) are normally needed at the entry to the distribution system because undoubtedly there will be some demand in the source water.

It may be difficult to maintain the ratio very close to 5:1. If free chlorine dosing exceeds the available free ammonia, increased formation of unstable dichloramine and associated residual loss may occur; therefore, some, yet minimal (e.g., < 0.05 mg N/L), free ammonia residual is desired. Routine monitoring of free ammonia at the treatment plant effluent can ensure that a Cl_2:N ratio near 5:1 is maintained.

Total organic carbon/natural organic matter. Total organic carbon (TOC) or NOM concentrations exert a chloramine demand (i.e., react with chloramine, resulting in chlorine residual loss and ammonia release). TOC or NOM can occur in drinking water systems from source water or as a result of microbiological activity (biofilm accumulation) over time.

pH and inorganic carbon (alkalinity). pH is one of the most important factors affecting chloramine formation and decay. Chloramine formation is quickest at pH 8.4; chloramine stability increases with increasing pH (Figure 6-5). Therefore, establishing the proper pH level is essential not only for corrosion control and water stability overall but also for maintaining chloramine residual and limiting nitrification. Chapter 7 describes corrosion control in detail. Chloramine decay is generally acid catalyzed; therefore, water pH and the presence of inorganic carbon (Figure 6-6) impact chloramine stability (Vikesland, Ozekin, and Valentine 2001). Chloramine decay increases with decreasing pH and increasing inorganic carbon concentrations.

Higher pH favors chloramine stability by minimizing decay-related free ammonia release, the key nutrient for nitrification. Some chloraminating water systems often credit maintaining high pH (e.g., 9.0 or higher) as a key factor in successful nitrification control programs (Schrempp et al. 1994; Skadsen 1993; Song et al. 1999). Because nitrification consumes alkalinity and reduces pH while simultaneously producing chloramine-demanding nitrite, it is critical to watch pH closely.

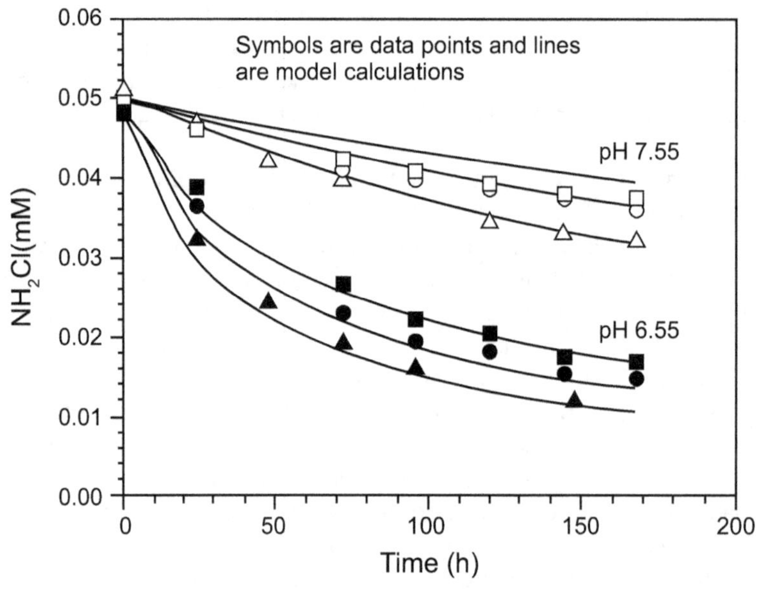

Reprinted from *Water Research*, 37(7):1766–1776, Vikesland et al. Monochloramine decay in model and distribution system waters, 2001 with permission from Elsevier.

Figure 6-4 Monochloramine decay as a function of Cl/N molar ratio.

Cl/N = 0.5 (□, ■) Cl/N = 0.6 (○, ●), and Cl/N = 0.7 (△,). Open symbols are for pH ≈ 7.5 and filled symbols are for pH ≈ 6.5. $[NH_2Cl] = 0.05$ mM, $C_{T, CO3} = 4$ mM, $\mu = 0.1$ M, temperature = 25°C. 0.05 mM NH_2Cl = 3.55 mg Cl_2/L.

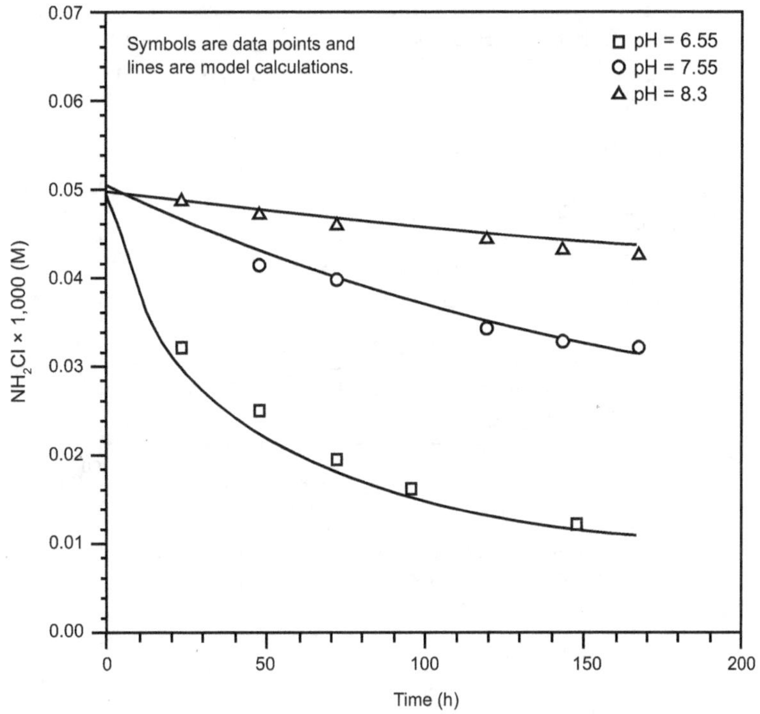

Source: Valentine, Ozekin, and Valentine 1998.

Figure 6-5 Effect of pH on monochloramine decay at 25°C. 0.05 mM NH_2Cl = 3.55 mg Cl_2/L

Temperature. Chloramine decay increases as temperature increases (Figure 6-7), as does the rate at which AOB oxidize free ammonia. Water temperatures in distribution system mains and storage facilities can vary seasonally based on source water and ambient air temperatures, which are out of a water system's control. However, chloraminating water systems need to understand how temperature can impact chloramine decay and nitrification.

For example, water systems often report seasonal nitrification episodes, with nitrification being more likely during the summer or warmest seasons (AWWA WQTD 2008a). Additionally, thermal stratification of stored water can allow excessively high water temperatures and long water detention times, which should be avoided through changes to storage operation and/or configuration (i.e., improving both tank turnover and mixing) to minimize chloramine decay and subsequent nitrification.

As stated earlier, chloramine demand and decay (free ammonia release) reactions are faster at warmer temperatures. This provides more free ammonia to support nitrification, which can also proceed faster at warmer temperatures because of increased biological growth rates. The optimal temperature range for nitrifying bacteria growth is 20 to 30°C; however, nitrification can occur at temperatures ≤5°C in systems with long (many days to weeks) detention times (Pintar, Slawson, and Huck 2000).

Bromide. The presence of bromide does not directly cause nitrification. However, at high enough bromide concentrations, it may have a significant effect on free ammonia release. At bromide concentrations below 0.1 mg/L, the impact is minimal (Vikesland, Ozekin, and Valentine 2001). However, at higher bromide concentrations (Figure 6-8), bromide may increase chloramine demand and free ammonia release due to formation of bromine-substituted haloamines (e.g., bromamines or bromochloramine), which are less stable than monochloramine.

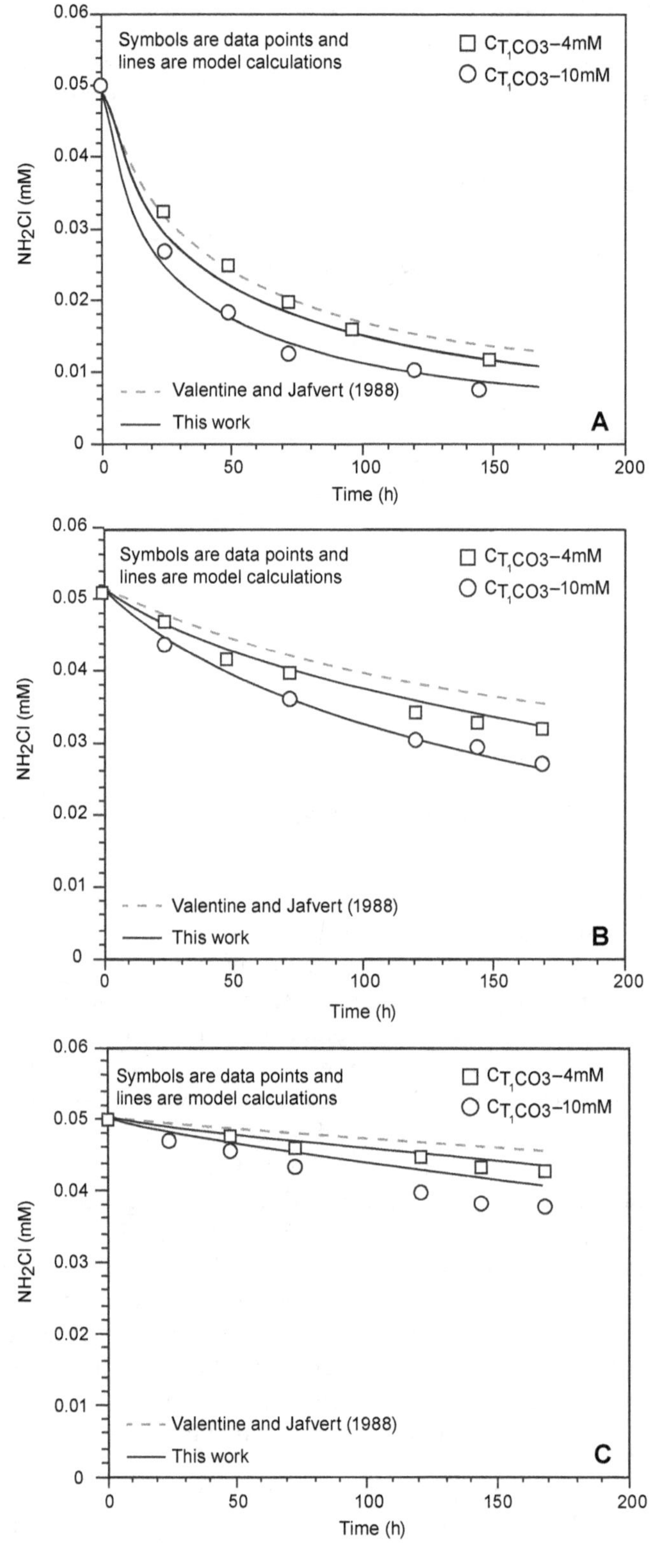

Reprinted from *Water Research*, 37(7):1766–1776, Vikesland et al. Monochloramine decay in model and distribution system waters, 2001 with permission from Elsevier.

Figure 6-6 Effect of total carbonate on monochloramine decay at (A) pH ≈ 6.6, (B) pH ≈ 7.6, and (C) pH ≈ 8.3. Cl/N = 0.7 mol/mol, μ = 0.1 M, temperature = 25°C. 0.05 mM NH_2Cl = 3.55 mg Cl_2/L.

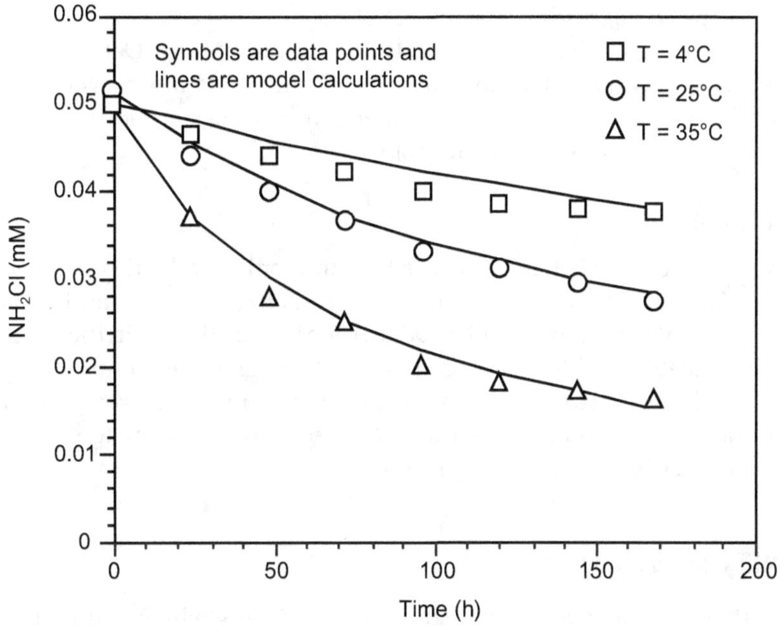

Reprinted from *Water Research,* 37(7):1766-1776, Vikesland et al. Monochloramine decay in model and distribution system waters, 2001 with permission from Elsevier.

Figure 6-7 Effect of temperature on monochloramine decay. Cl/N = 0.7 mol/mol, pH = 7.5, $C_{T,CO3}$ = 10 mM, μ = 0.1 M. 0.05 mM NH_2Cl = 3.55 mg Cl_2/L

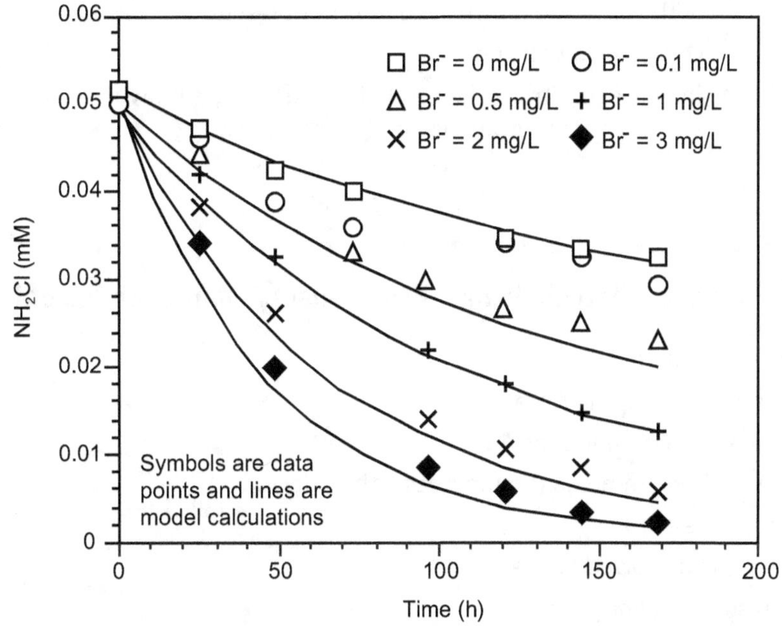

Reprinted from *Water Research,* 37(7):1766-1776, Vikesland et al. Monochloramine decay in model and distribution system waters, 2001 with permission from Elsevier.

Figure 6-8 Effect of 0–3 mg/L bromide ion on monochloramine stability at pH ≈7.5. Cl/N = 0.7 mol/mol, $C_{T,CO3}$ = 4 mM, μ = 0.1 M, temperature = 25°C. 0.05 mM NH_2Cl = 3.55 mg Cl_2/L

Pope (2006) studied the impact of pH and free chlorine contact time prior to ammonia addition on chloramine stability in ultrapure water. Data indicate that pH and chloramine application conditions (e.g., preformed chloramine versus short periods of free chlorination) play an important role in chloramine stability in the presence of high bromide concentrations (e.g., 0.5 mg/L).

Summary

The root causes of nitrification are free ammonia availability as a substrate for nitrifying bacteria (specifically AOB), allowing bacteria to proliferate, and an absence of a disinfectant residual (chloramine) at levels sufficient to inactivate nitrifiers. Free ammonia is inherently available in chloraminated systems through excess ammonia feed and through free ammonia release from chloramine via demand and decay reactions. Understanding chloramine demand and decay for a given water supply is key to understanding approaches used to prevent or control nitrification.

EXAMPLES OF NITRIFICATION

Nitrification is a two-step biological process during which ammonia is converted to nitrite by AOB, then to nitrate by NOB. The first and most commonly detected symptom of nitrification is a decrease in chloramine residual. It is important to note that there are other causes of chloramine residual loss in addition to nitrification, so this single symptom is not sufficient by itself for diagnosing a nitrification event. In addition to a decrease in chloramine residual, the most common symptoms indicating nitrification include the following:

- Decrease from normal free ammonia levels; free ammonia normally increases with water age if nitrification is not present;
- Decrease in pH; and
- Increase in nitrite when AOB is controlling, followed by a decrease in nitrite with an increase in nitrate when NOB controls.

Characterizing the DSWQ Challenge: Water Quality Characteristics of Nitrification

- Increase in nitrite
- Sometimes increase in nitrate
- Decrease in chloramine residual
- Change in free ammonia (first increase, then decrease)
- HPC increase
- Decrease in dissolved oxygen
- Decrease in alkalinity
- Increase in corrosion

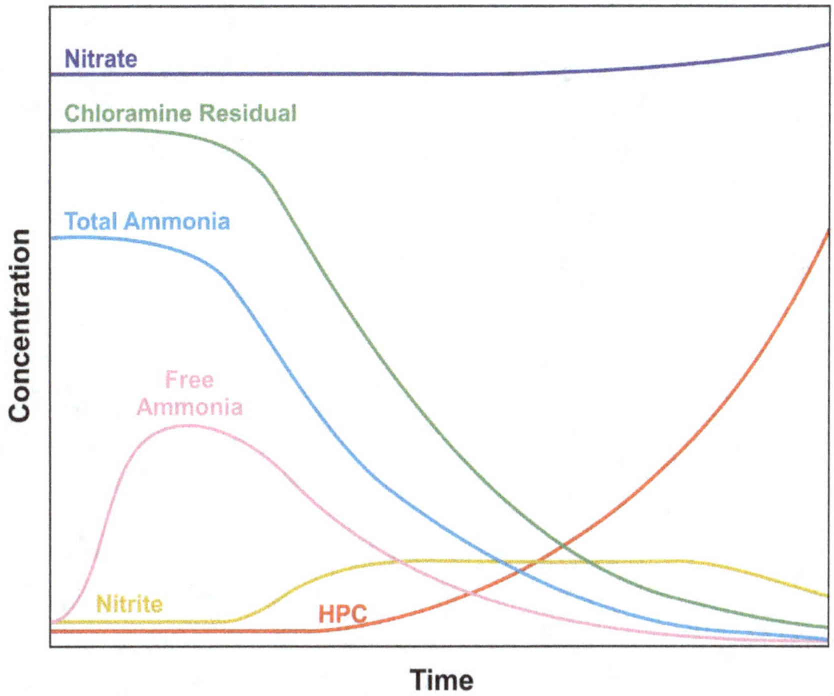

Figure 6-9 Theoretical water quality changes during a nitrification event

Other signs of nitrification include decreases in DO concentration, decreases in alkalinity, and increases in corrosion of distribution materials. Typical nitrification-induced water quality changes are shown in Figure 6-9 (unpublished data, Roberts 2015). The process occurs as follows:

- First, free ammonia increases.
- Next, the chloramine residual decreases and eventually approaches zero while total ammonia decreases as ammonia is consumed by AOB.
- These changes are followed by an increase in nitrite.
- If NOB are present, an increase in nitrate may also occur (note this example assumes there is nitrate in the source water).
- HPC increases as nitrification progresses.

A full-scale water system example of complete nitrification is illustrated in Figure 6-10, which shows samples collected from the chloraminated distribution system at Massachusetts Water Resources Authority (MWRA) (Reilly 2014b). This example shows complete nitrification where the all of the nitrogen is eventually converted to nitrate, and the chloramine residual decreases from more than 2.0 mg Cl_2/L to being nondetectable. At the same time, free ammonia increases and then decreases. Nitrite increases and then decreases as the conversion to nitrate occurs.

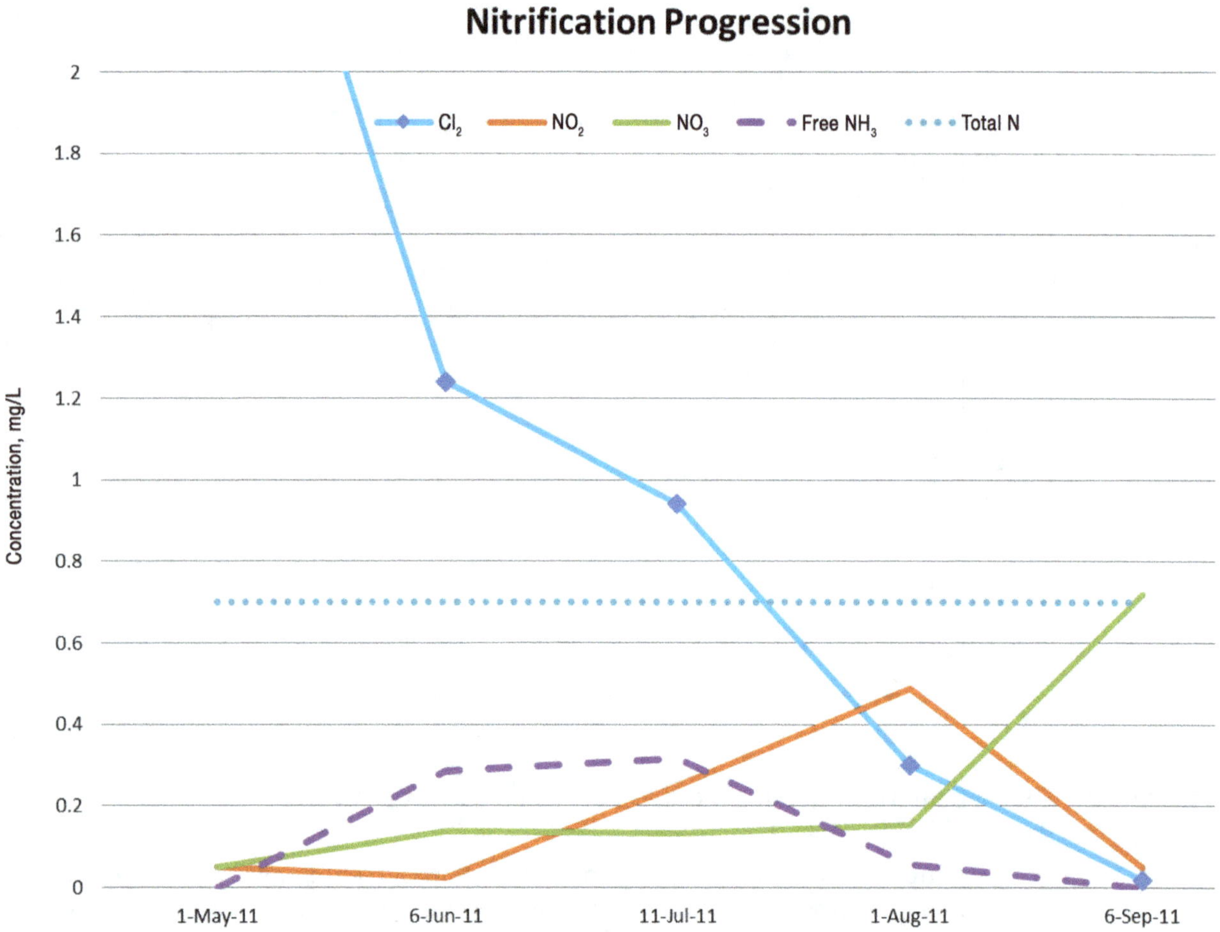

Adapted from Reilly 2014b.

Figure 6-10 Example of complete nitrification in a Massachusetts Water Resources Authority distribution system

RESPONSES TO CONTROL NITRIFICATION

Once started, nitrification is difficult to stop. Thus, it is desirable to both prevent nitrification episodes and detect and correct them early before they gain momentum. When a nitrification episode is well developed, there are limited options that have proven effective at halting the process.

Responding to the DSWQ Challenge: How Do You Respond to a Nitrification Event?

- Localized flushing
- Drain and disinfect impacted storage tanks
- Temporarily convert to free chlorine

Once nitrification is determined to be the cause of low chloramine residuals in the distribution system, responses to halt nitrification should be implemented. While water systems have tried a variety of responses, there are few that have proven effective (AWWA 2013). The most common responses used to halt nitrification after it is established include the following:

- Distribution system flushing: effective for limited, localized nitrification; and
- Storage tank draining and/or disinfection: effective for nitrification in tanks. Temporary conversion to free chlorine (either by ceasing to feed ammonia or by feeding excess chlorine to achieve breakpoint chlorination) is the most effective strategy, although results may be short-lived and the process will need to be repeated.

Flushing

Distribution system flushing is often the first approach used once nitrification has been detected. Flushing is often performed in response to customer water quality complaints, low disinfectant residual, evidence of nitrification, or positive coliform results. When performed as an emergency response to combat the effects of water age in the distribution system (low disinfectant residuals and poor water quality), flushing replaces large volumes of "older" water with higher-quality "newer" water from other parts of the distribution system. There are two main types of flushing (AWWA 2013):

- *Conventional flushing:* typically consists of opening hydrants in a specific area of the distribution system until preselected water quality criteria (e.g., total chlorine residual) are met. In addition to manually opening hydrants, automatic flushing devices may be installed in areas of a distribution system that have chronic water quality problems associated with high water age.
- *Unidirectional flushing (UDF):* consists of defining "flush groups" by isolating a particular pipe section or loop and exercising the hydrants in an organized, sequential manner at high velocity. Within each flush group, the cleanest water with the highest total chlorine residual and low nitrite/nitrate concentrations is flushed out into the system, always moving older water with lower quality out ahead of clean water. Removal of silts and sediments requires velocities equal to or greater than 3 fps. However, to promote scouring and remove biofilm and loose deposits, velocities 5 to 6 fps are desired in UDF. These velocities may be difficult to achieve in some locations.

If a nitrification event is well established, conventional flushing to remove the affected water and reduce the water age in the impacted part of the system may be beneficial. However, the duration of the effect of conventional flushing is variable and the effectiveness is not universal. In areas with chronic problems, automatic flushing may be useful. While UDF may be effective, the level of effort and time to implement this type of flushing makes it useful as a periodic, not routine, strategy.

Even when flushing is used, its effectiveness is limited to the local area. Further nitrification typically returns quickly, making flushing a temporary solution unless the underlying causes of the event are corrected. However, in cold climates where declining temperature may halt the nitrification process, flushing may be effective at buying sufficient time while waiting for the temperature to decrease below the typical nitrification threshold. In areas where water conservation is critical, the ability to flush may be limited and thus a more difficult strategy for those water utilities to implement.

Distribution System Storage Facilities

Often, storage facilities are directly implicated in nitrification events. Stratification, poor mixing, and inadequate drawdown and turnover all contribute to high water age in storage tanks and may lead to nitrification (AWWA 2013). Once nitrification is occurring in a tank, typical management practices such as increased turnover or increased chloramine residuals are usually ineffective at stopping the nitrification. In this case, water systems may also free chlorinate (breakpoint chlorinate) storage tanks. An example procedure is provided in Figure 6-11.

This practice is similar to free chlorination of a portion of the distribution system. The tank is isolated from service and free chlorine is introduced, either by stopping the ammonia feed if the tank is near the plant or by batch-chlorinating the chloraminated tank past the break point. The tank may be either drained to dispose of the free chlorinated water (most common) or the free chlorine water may be introduced into the distribution system, thereby getting free chlorine to the downstream impacted areas (AWWA 2013). This approach is more effective in flow-through tanks than in tanks that fill and draw from the same pipe.

The duration of the breakpoint period can vary depending on the ability of a system to isolate and remove a storage facility from service, given pressure and emergency requirements. Where it is possible to remove a tank from service, a period of 12 to 24 hours is recommended (AWWA 2013).

Batch chlorination of a single tank can be problematic. Injection of chlorine into the tank inlet may be difficult if no facilities such as an injection vault and/or in-line mixer are available. Injection at the inlet generally requires a portable pump with a hose adapted to the pump discharge and injection port and a small generator for power. The pump suction can draw from buckets or carboys. Some water systems have vehicles equipped with portable hypochlorite feed systems. In instances where the inlet configuration does not accommodate chlorine injection, chlorine can be added through access hatches or through recirculation equipment. Chemicals should be transferred to the top of a tank with extreme caution. Recirculation or mixing equipment is necessary if chlorine is not added while the tank is filling and there is insufficient momentum to mix the tank. Because chlorine has a higher specific gravity than water, inadequate dispersion of chlorine throughout the tank is the most common reason for inadequate disinfection and poor outcome. Before a single tank is disinfected, calculations of the amount of chemical present and the amounts needed are absolutely necessary. The basic steps for free chlorination of a storage tank are shown in Figure 6-11.

A more aggressive approach to nitrification in a storage tank is draining, cleaning, and disinfecting (AWWA 2013). First, the storage tank must be isolated from service. Then, all of the water is drained from the tank. It is important to consider disposal of the water and the total water loss, particularly where water conservation is critical. Finally, the tank is cleaned and disinfected to remove corrosion products and biofilm prior to return to service.

This approach typically requires days to complete because of the level of effort involved and the need to perform total coliform testing prior to return to service. Therefore, the system must be able to safely operate without the tank for up to a week. The ability to use these approaches depends on the water system's ability to maintain adequate system pressures and meet emergency flow requirements with the affected tank off line (AWWA 2013).

An example of the impact of tank draining and cleaning at the MWRA is shown in Figure 6-12 (Reilly 2014a). MWRA has distribution system detention times of up to one month with 6,000 miles of piping. MWRA controls pH at >9 and controls the Cl_2:N ratio carefully. In spite of these practices, nitrification occurred in a 20-mil gal storage tank in 2012. MWRA drained, disinfected, and refilled the tank. This successfully stopped nitrification and eliminated its reoccurrence from 2012 until 2015. While draining and disinfection proved effective at this water system, other water systems have reported that this approach is only temporarily effective and requires frequent repetition. It is important that the underlying causes of the nitrification event (i.e., high water age, excess free ammonia) be addressed for this approach to be effective in the long term.

Step 1.
Lower reservoir to as low as possible.

Step 2.
Measure water quality inside tank/reservoir. Calculate required breakpoint chlorination dose.

Step 3.
Begin filling the tank as quickly as possible and add hypochlorite as quickly as possible during the fill cycle.

Step 4.
Isolate the tank or reservoir and let sit for 24 hours (if possible).

Step 5.
Measure chlorine residual to ensure the target free chlorine residual was achieved prior to returning to service.

Step 6.
As long as free chlorine residual is not too high (greater than approximately 1.0 mg/L, depending on local regulations), return tank/reservoir to service. (Note that if the chlorine is too high, the water in the tank will need to be removed or the chlorine allowed to decay to an acceptable level.)

Source: Adapted from AWWA 2013.

Figure 6-11 Example of storage tank breakpoint chlorination procedure

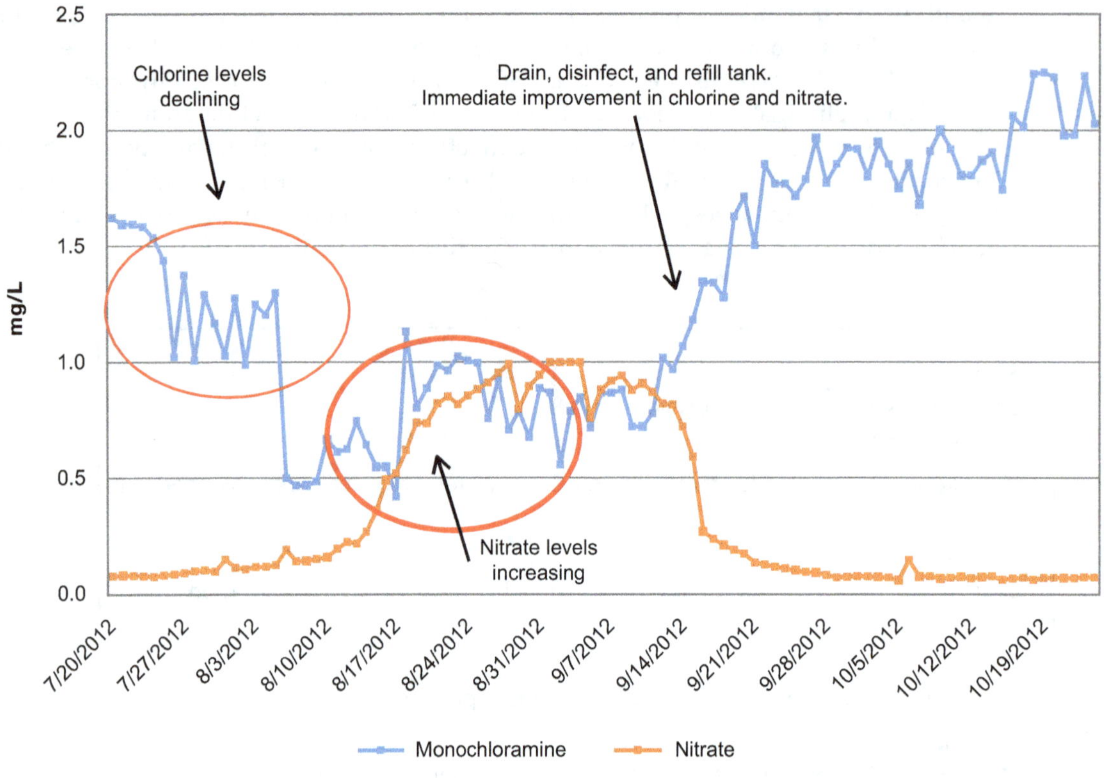

Source: Reilly 2014a.

Figure 6-12 Impact of tank draining and disinfecting on nitrification

Free Chlorination

Temporary conversion to free chlorine occurs when a water system switches from monochloramine to free chlorine for a period of time. Because it removes the food (ammonia) source for AOB, this method is used to prevent the progress of a nitrification episode. While free chlorine conversion may be a very effective short-term nitrification response, it may not prevent nitrification from reoccurring after return to chloramination (Carrico et al. 2008). Free chlorination, or breakpoint chlorination, removes the food and, in turn, stops nitrification. However, if the free chlorine does not penetrate the biofilm, the reintroduction of ammonia to make chloramines starts the process all over again.

If the water system operates the treatment plant feeding the system, free chlorine conversion of the entire distribution system may involve discontinuation of ammonia addition and adjustment of the chlorine dose at the treatment plant (and/or booster stations) to achieve a free chlorine residual throughout the distribution system. For a consecutive, purchased water system, the ammonia feed is not usually under the system's control. Instead, free chlorine conversion involves injecting enough chlorine to overcome the ammonia in the purchased water and achieve a Cl_2:N mass ratio greater than approximately 8:1. A typical breakpoint curve is presented in Figure 6-13.

Theoretically, to achieve a Cl_2:N mass ratio greater than 7.6:1 (often greater in natural waters), free chlorine must be fed at a concentration that will oxidize the ammonia to nitrogen gas. The goal of breakpoint chlorination is to remove the ammonia because if there is no ammonia, there is no food for nitrifiers.

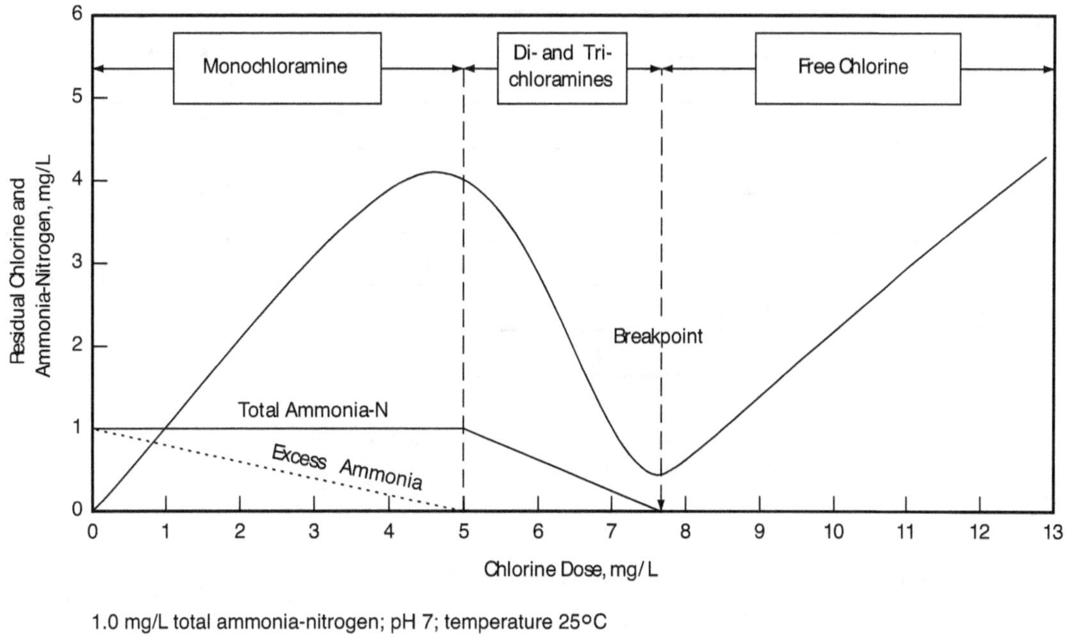

Source: Adapted from Wolfe, Ward, and Olsen 1984.

Figure 6-13 Theoretical breakpoint curve

The desired free chlorine concentration entering the distribution system and at the end of the distribution system varies from system to system. However, a 2- to 3-mg Cl_2/L free chlorine residual entering the system and residual greater than 0.5 mg Cl_2/L at the ends of the system are typical goals (AWWA 2013). The duration of a temporary conversion to free chlorine varies and ranges from days to weeks (Figure 6-14). The procedure by which a water system breakpoint free-chlorinates the entire distribution system is, for the most part, system specific. Figure 6-15 presents an example system-wide breakpoint chlorination protocol.

Breakpoint chlorination may be applied locally to an affected area (e.g., one isolated area of the distribution system or storage tank) or conducted throughout the entire distribution system. When free chlorination is used, flushing should be concurrently practiced to minimize water quality disruptions by quickly displacing chloraminated water out of the distribution system. If free chlorination is practiced at a storage tank, water levels should be decreased to the extent possible to aid in rapid turnover and replacement of chloraminated water with free chlorinated water. The primary reason for this is that the Cl_2:N mass ratio at the chlorinated water and chloraminated water interface often exceeds 5:1 and can result in di- and trichloramine formation (AWWA 2013), which generates taste and odor complaints from customers. Dechlor(am)ination of the flushed water is also necessary prior to disposal and depends on state requirements (AWWA 2013).

When nitrification is localized in an isolated part of the distribution system, targeted breakpoint chlorination can be as effective as system-wide chlorination and can minimize the extent of negative impacts on distribution system water quality. When conducting breakpoint chlorination of a targeted area, it is important to isolate the affected area, to the extent possible, to minimize the mixing of chlorinated and chloraminated water, which can also result in di- and trichloramine formation.

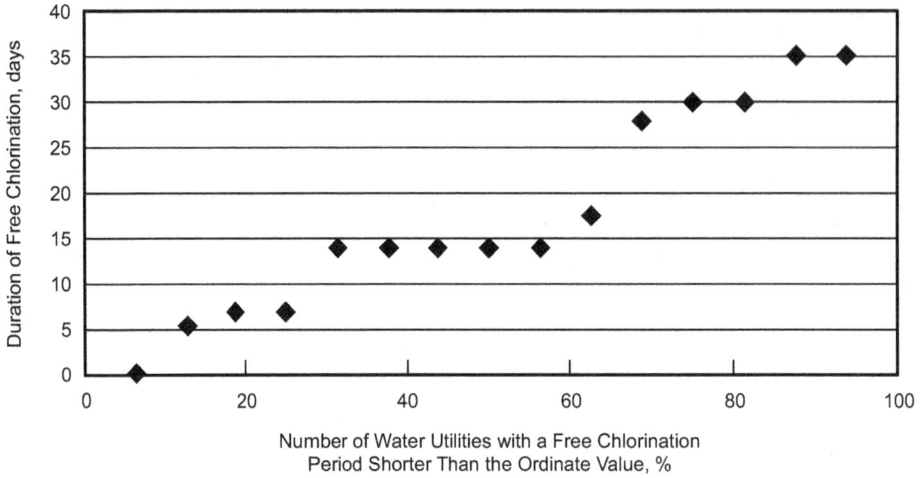

Source: Kirmeyer et al. 2004.

Figure 6-14 Free chlorine period survey results

Step 1. Conduct background monitoring (total chlorine, monochloramine, nitrite, pH, HPC).

Step 2. Determine chlorine dose required to achieve the target free chlorine residual in the problem areas of the distribution system.

Step 3. Determine a planned start date and end date. Conduct any necessary customer or primacy agency notifications.

Step 4. Have flushing crews ready.

Step 5. Reduce tanks levels to lowest level possible or remove tanks from service.

Step 6. Discontinue the ammonia feed at the water treatment plant and any booster disinfection stations and adjust chlorine to the dose determined by step 2 if you operate a WTP. Initiate free chlorine feed if you are a consecutive system that does not operate a WTP.

Step 7. Breakpoint storage tanks if needed and possible. (Effective storage water turnover to all free chlorine should serve this purpose. This may only be readily achievable with fully mixed storage.)

Step 8. Measure total chlorine, monochloramine, ammonia, free chlorine, and HPC in the known nitrifying areas of the distribution system.

Step 9. Continue to operate under free chlorine conditions until desired free chlorine residual and baseline HPC levels are achieved in all areas of the distribution system.

Step 10. Resume ammonia feed (with lowering of storage and system flushing to facilitate turnover back to chloramine and monitoring to verify turnover) if you operate a WTP. Cease free chlorine injection if you are a consecutive water system that does not operate a WTP.

Step 11. Flush the chloraminated water through the system in the same manner that you flushed the chlorinated water through. Measure total chlorine, monochloramine, and ammonia to ensure that the chloramine residual is restored throughout the entire system.

Adapted from AWWA 2013.

Figure 6-15 Example of system-wide breakpoint chlorination protocol

Abbreviations: HPC, heterotrophic plate count; WTP, water treatment plant.

Localized breakpoint chlorination of an isolated area of the distribution system requires a location where the chlorine can be injected, such as a pump station or storage tank. It also requires that the affected area be small enough to be reasonably contained by manipulation of distribution system valves. This response method typically involves addition of chlorine at a dose sufficient to achieve a 1- to 2-mg Cl_2/L free chlorine residual (AWWA 2013). If done properly, temporary conversion to free chlorine is an effective way to respond to nitrification. However, nitrification is likely to return at some point in time. It is necessary to address the underlying causes in an effort to prevent or at least reduce the frequency and severity of episodes in the future.

Switching to free chlorine may create adverse consequences that the water system must address. For example, free chlorination increases DBP formation, potentially to levels above the MCL, may lead to complaints from customers unaccustomed to the taste and odor of free chlorine (or di- and trichloramine, if ratios are off balance), and may alter corrosion chemistry. Customers may also require advanced notification, particularly hospitals that are accustomed to chloraminated water supplies.

NITRIFICATION MONITORING AND CONTROL PLAN

Development of a comprehensive nitrification monitoring and control plan is essential for water system managers and operators to prevent nitrification in their distribution systems. An effective nitrification monitoring plan developed in advance, updated regularly, and executed carefully provides early indications of worsening distribution system water quality. This allows adoption of proactive measures that are significantly less resource intensive and less disruptive to system operations than reactive measures.

This section provides an outline of the major components and considerations that need to be included in a nitrification monitoring and control plan. An example of a nitrification plan and flow chart for illustrative purposes are included in Appendix D. Some primary agencies have provided example plans and guidance, for example, the Texas Commission on Environmental Quality (2015).

Responding to the DSWQ Challenge: What Are the Main Components of a Nitrification Monitoring and Control Plan?

- Description of nitrification
- Objectives
- System details
- Monitoring plan
- Data analysis
- Benchmarks and action plans
- Response strategies
- Prevention strategies

Note that nitrification monitoring plans are very specific to a water system and often to a facility, so the guidelines described here need to be modified appropriately for use by individual water systems. The basic components of a typical nitrification monitoring plan are listed below along with the details that need to be included within each section.

Description of Nitrification

Nitrification is the biological oxidation of ammonia to nitrite followed by the oxidation of nitrite to nitrate. Historical water system–specific observation(s) of nitrification should be described. These can include detection of free ammonia, nitrite, or nitrate in the distribution system; drop in disinfectant residuals; taste and odor issues; customer complaints; and similar items.

Objectives

Well-defined objectives need to be established for the control of nitrification and described within the plan. These objectives can include compliance with regulatory requirements, as well as water system–specific objectives such as minimum disinfectant residual, pH, or free ammonia levels at any location within the distribution system.

System Details

A description of the system background should be provided within the nitrification monitoring plan, such that all parties responsible for executing the plan have a clear understanding of overall system operations. At a minimum, the following details should be included:

- Water treatment plant processes: Briefly describe the treatment processes at the water treatment plant. Any changes to the design or operations of the treatment plant processes that can affect distribution system nitrification (e.g., disinfection processes at treatment plants or booster stations) should be updated within the nitrification monitoring plan.
- Disinfection method: List primary and secondary disinfectants (free chlorine, chloramine, and others if used). Include details such as disinfectant dose, target residual, points of application, free chlorine contact time prior to ammonia addition (for chloraminated facilities), and similar details.
- Distribution system infrastructure: Describe key distribution system infrastructure components that are relevant to nitrification. This includes storage tanks, pump stations, booster disinfection stations, and other components. For storage tanks, it is important to include the inlet/outlet configurations, turnover rates, daily variation in their water levels, and seasonal variation of operations, if any. Also describe average and maximum water ages within the distribution system and location(s) where maximum water age is experienced.
- Historical water quality: Summarize any water quality events that may have occurred in the past. This includes historical nitrification events, areas of the distribution system that result in Total Coliform Rule (TCR) violations (that can occur due to loss of disinfectant residual and subsequent microbial regrowth), or areas of the distribution system that regularly experience disinfectant residual decreases, pH decreases, colored water events, and similar events. Most importantly, describe any corrective actions (e.g., tank cycling, flushing) that may have been taken to prevent nitrification or other water quality events.

- Distribution system sampling locations: Describe sampling locations in the distribution system where regulatory compliance and/or any additional samples are collected. This should include all sampling locations for disinfectant residuals (such as TCR sampling locations) and locations where HPC, free ammonia, nitrite, and nitrate samples are collected. Describe the sampling frequencies for each parameter and seasonal variations of sampling frequency, if any. Geographic information system platforms are useful for creating visual representations of this information.
- Graphical summary of most recent data: If possible, include a graphical summary of the most recent water quality data such as the variation in total chlorine residual at different distribution system sampling locations as a function of time. Figure 6-16 provides an example from a hypothetical water system. Such graphical summaries identify trends and potential "hot spots" within the distribution system. Additionally, it is suggested that the water system perform laboratory bench-scale testing to develop "chlorine demand curves." Chlorine demand is unique to the particular water quality, and an understanding of chlorine demand curves will assist system operators in maintaining adequate disinfectant residuals throughout their distribution system. Figure 6-17 shows the decrease in chlorine residual with increasing water age in a distribution system. Note that source water quality may change throughout the year, and the water's chlorine demand changes accordingly. Consequently, the chlorine demand curves may need to be developed and updated regularly throughout the year.

Monitoring Plan

The monitoring plan should discuss the distribution system sampling locations, water quality parameters, and sampling frequencies, which are site specific and need to be determined on a case-by-case basis. Examples of typical water quality parameters, monitoring locations, and sampling frequencies are illustrated by Kirmeyer et al. (2002). Any changes from the previous year's monitoring plan should be highlighted. Key components of a typical monitoring plan include:

- Identifying and selecting sampling locations. Ideal sampling locations for nitrification prevention should be complementary to a water system's sampling locations for regulatory compliance such as TCR or DBP sampling locations. Some of these locations can also be used for the nitrification monitoring plan. Areas of high water age are key monitoring locations. Monitoring of storage facilities (inflow and out-flow) or low-flow pipes should be included for nitrification control monitoring because these areas are vulnerable to the proliferation of AOB and nitrification generally occurs first at these locations.
- Adopting a consistent flushing approach at the sample tap prior to sampling. This ensures the collection of quality data from the water main. If water is over- or underflushed at a sample location, the sampling results may not accurately characterize water quality at that distribution system location. Alternative procedures may be used if the sample is taken to address premise-plumbing issues.
- Routine monthly monitoring at distribution system sampling locations, including TCR sampling locations, storage tanks, pump stations, and booster chlorination stations. Parameters that should be monitored include free and total chlorine, pH, HPC, free ammonia, nitrite, and nitrate. For storage tanks, samples should be

collected from both the tank influent and tank discharge to characterize what is happening in the tank.

- Additional monitoring that may be necessary seasonally, at "hot spots," or dead ends of the distribution system. Also, in cold weather climates, monitoring frequency may be reduced in the winter.
- Additional monitoring that may be triggered if there is a loss of chlorine residual or detection of high HPC, free ammonia, nitrite, or nitrate.

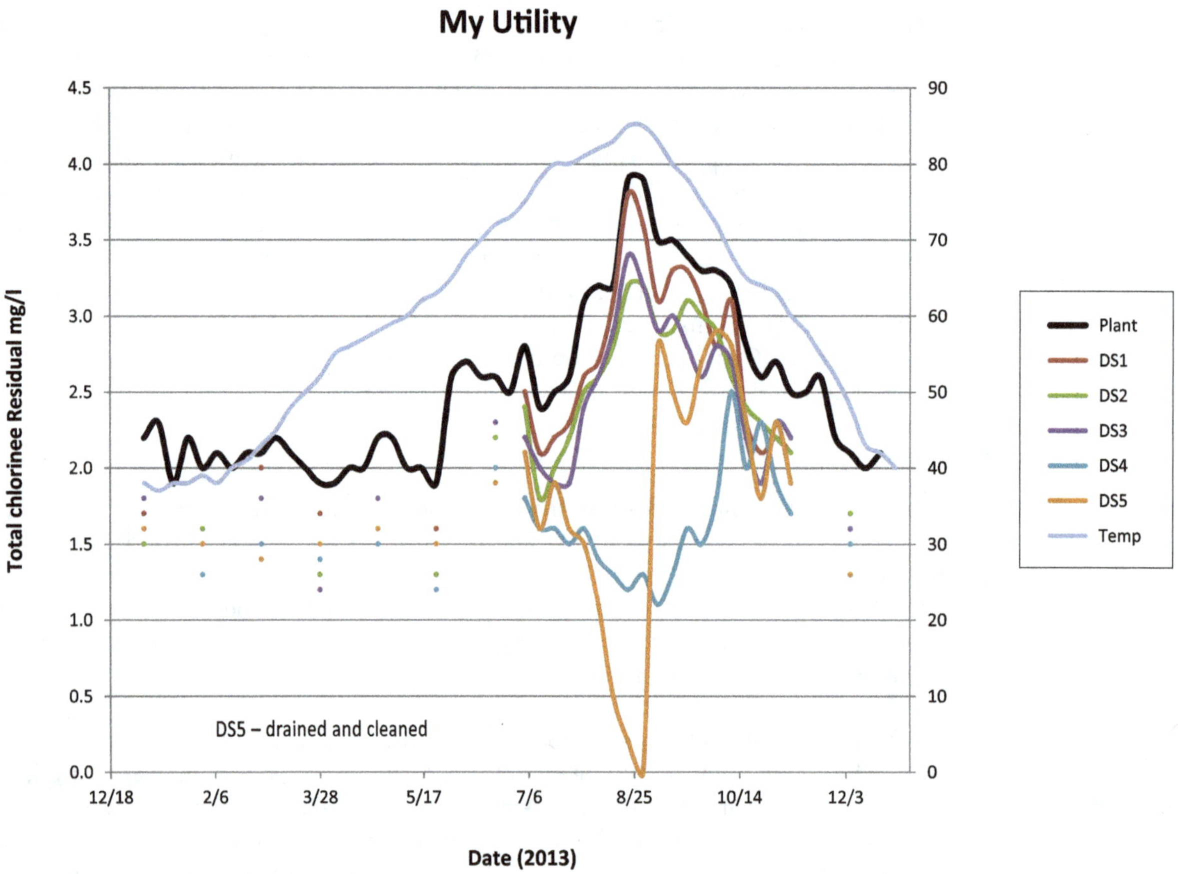

Figure 6-16 Total chlorine residual as a function of time at various distribution system sampling locations

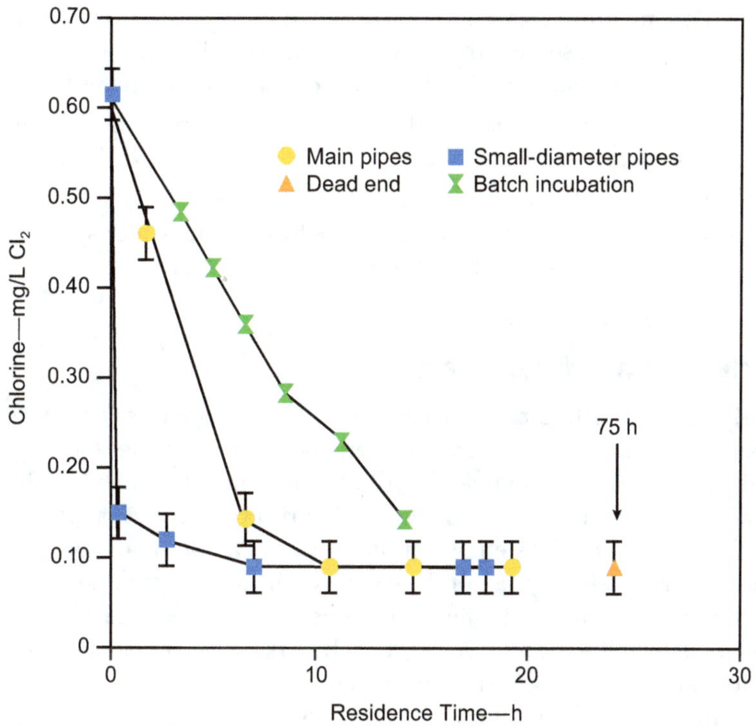

Adapted from Baribeau et al. 2001.

Figure 6-17 Changes in chlorine concentrations with water age in a distribution system

Data Analysis

This section describes the analyses that should be conducted on all collected data. The analyses should include developing graphical summaries of key water quality parameters, such that any outliers or deviations from "normal" conditions can be identified. Monitoring data may also be plotted on a map in order to assess spatial variations in water quality. The following analyses will assist in early identification of a potential nitrification event:

- Monthly analysis from key monitoring locations of the following:
  - Total chlorine (Note: For chloraminated systems, some states or regulatory primacy agencies require monitoring of both free and total chlorine. Recommendations from local regulatory agencies should be implemented during monitoring.)
  - Monochloramine
  - Free ammonia
  - Free chlorine
  - pH
  - Water temperature
  - HPC
  - Nitrite
  - Nitrate (Note: Surface water systems that are subject to agricultural runoff have fluctuating nitrate levels regardless of the nitrification associated with chloramination, so routine baseline data is needed to track distribution trends.)

- During a nitrification event or when early indications of the development of an event are observed, the above analyses should be conducted at a higher frequency, such as weekly or daily, from key distribution system sampling locations.
- Statistical parameters such as maximum, minimum, average, and percentage of data below predetermined threshold values should be calculated to provide additional insight.
- If performed, these analyses are helpful to the water system in their submissions or discussions with the local regulatory agency.

Benchmarks and Action Plans

Clearly stated benchmarks and action plans help a water system prepare in advance and respond efficiently to a nitrification event. Benchmark values (or range of values) for water quality parameters that indicate "normal conditions" should be established. Different alert and action levels that trigger corresponding response or action plans should be developed based on distribution system water quality. AWWA Manual M56 (2013) discusses the usefulness, alert levels, and action levels of water quality parameters such as free and total chlorine, free ammonia, nitrite, and temperature (Table 6-2). The manual should be consulted to determine the appropriate benchmark, alert, and action levels for a specific water quality parameter at a given water system.

Any change to the action plans from previous years should be described in the nitrification monitoring plan. At least one post-event data review and evaluation should be conducted with all involved staff every year at the end of the nitrification season. The collected water quality data, established benchmarks, and action plans should be reviewed, and any needed short-term or long-term improvements (e.g., capital improvements) should be identified during this review session.

Table 6-3 shows an example of alert and action levels for one high water age distribution system location at Loudoun Water (Ashburn, Va.; personal communication, Wright 2015). While each water system needs to develop their own specific goals and alert/action levels for specific locations, this is an example of alert and action levels implemented at a high water age with known nitrification events.

Table 6-2 Usefulness, alert levels, and action levels of total chlorine at varous locations of the distribution system

Location	Plant Effluent/ Distribution Entry Point	Transmission System	Distribution System Pipes	Storage Facilities
Usefulness	Very useful	Very useful	Very useful	Very useful
Typical concentration	1.5–3.5	1.5–3.5	1–3.0	1–2
Alert level*	Based on CT	1.8–2.2	1.8	1.8
Action level	Based on CT	1.6–2.0	1.6	1.6

Adapted from AWWA 2013.

NOTE: CT = Concentration × contact time.

*Values shown assume an entry point value of 2.0 mg/L.

Table 6-3 Example of goals, alert levels, and action levels applied by Loudoun Water at distribution system monitoring locations

Parameter	Goal	Alert Level	Action Level
Total chlorine	>1.5 mg/L	<1.0 mg/L	<0.3 mg/L
Nitrite-N	<10 µg/L	>25 µg/L	>100 µg/L
Free ammonia-N	<0.1 mg/L	>0.1 mg/L	>0.3 mg/L
Temperature	Monitor	>59°F	NA
Optional Parameter	**Goal**	**Alert Level**	**Action Level**
Nitrate-N	NA	Increase of 0.3 mg/L	NA
Heterotrophic plate count (R2A)	<500 cfu/mL	>500 cfu/mL	NA
pH	7.3	Decrease in 0.2 units	NA

Nitrification Response Strategies

This section addresses the actions that the water system will take in the event that nitrification occurs and needs to be halted. It is important to stop nitrification before it degrades water quality to the point where regulatory compliance and public health become threatened.

Approaches such as increasing the monochloramine residual have proven ineffective once nitrification is obvious (Skadsen 1993). Common, potentially effective responses used to control nitrification after it is established include:

- Localized flushing in impacted areas;
- Tank draining, disinfecting, and cleaning; and
- Breakpoint chlorination (either system wide or localized).

The water system will need to decide which of these approaches are appropriate for their situation. In most cases, a combination of approaches may be necessary. These strategies are discussed in more detail above.

Nitrification Prevention Strategies

Ideally, a water utility can prevent nitrification from occurring. Therefore, lessons learned from each year's experience should be reviewed and integrated into the nitrification monitoring plan for the following year, such that nitrification events can be prevented or better managed. The action plans are specific to the water system but can include any or all of the following actions:

- Increase the disinfectant residual leaving the water treatment plant;
- Boost the disinfectant residual at key distribution system locations;
- Increase the frequency of water quality parameter monitoring at key distribution system sampling locations to weekly or daily;
- Optimize treatment processes at the water treatment plant, such as optimizing chlorine to ammonia dose ratios or additional removal of organic matter with coagulation or filtration;
- Implement a flushing program with appropriate frequency at predetermined distribution system locations;

- Increase storage tank cycling and turnover rates;
- Completely drain and refill storage tanks; and
- Initiate a conversion to free chlorine, localized or system-wide.

Examples of completed water system nitrification control plans are provided in Appendix D

NITRIFICATION PREVENTION

The optimal approach to nitrification control is to prevent its occurrence. Prevention consists of those actions that can be implemented to reduce the severity and frequency of nitrification; whereas, a response consists of actions to correct a problem after it has already occurred.

Preventive actions to inhibit nitrification range from best operational practices to innovative approaches (AWWA 2013).

A 2004 report based on a survey of American water utilities (Kirmeyer et al. 2004) indicates the practices that water utilities commonly use in response to nitrification events (Figure 6-18). The results show that water utilities used flushing most frequently, but its success rate is somewhat limited. This is true for the other common approaches used. The success rate is mixed for all but free chlorination, although that strategy is generally only effective for a month or so and must be repeated. Flushing is also a concern in water-scarce regions where water loss may be unacceptable.

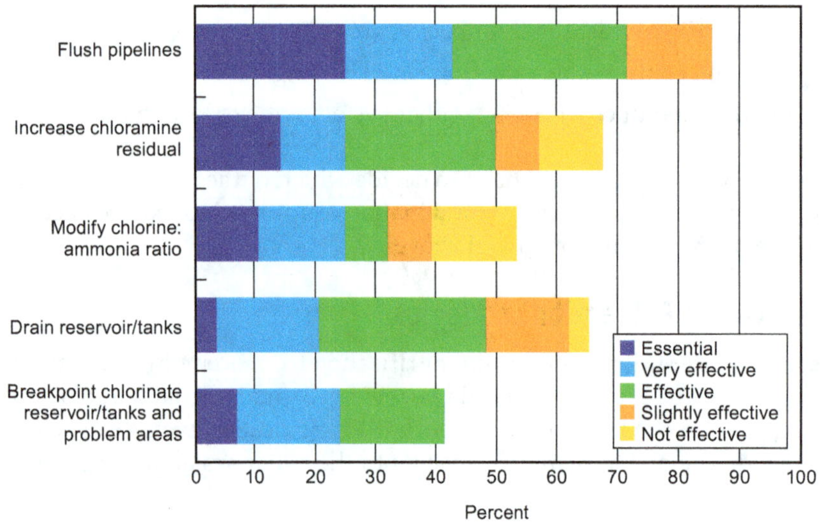

Source: Kirmeyer et al. 2004.

Figure 6-18 Common water system actions to control nitrification from a 2004 survey

Monochloramine Formation

All water utilities and other water systems should use good operational practices. During treatment, it is important to properly control the formation of monochloramine. This optimization consists of correct ammonia and chlorine dosing. The Cl_2:N mass ratio should be between 4.5:1 and 5:1 (AWWA 2013). This prevents excess free ammonia from being introduced into the distribution system. Free ammonia–nitrogen should be limited to less than 0.05 mg N/L (AWWA 2013). If feasible, it is recommended that an even tighter ratio of 4.75:1 and lower free ammonia be achieved. Health Canada (2016) recommends limiting excess free ammonia entering the distribution system to less than 0.1 mg/L and preferably less than 0.05 mg/L, measured as nitrogen. The chemical feed rates must incorporate any chlorine demand exerted prior to ammonia addition. It is important to assess and adjust for any influent ammonia, such as from the source water, and any influent chlorine, such as from primary chlorination, if practiced. If these are not accounted for, the proper Cl_2:N mass ratio will not be achieved. Therefore, tight operational control and monitoring are essential to chloramination practice. To achieve this level of control, frequent monitoring of chlorine and ammonia are essential.

The type of ammonia is important to ensure highly accurate control of the feed rate. Anhydrous or aqua ammonia are recommended (AWWA 2013). The order of chlorine and ammonia addition at the water treatment plant is also important. Most water utilities add chorine first (AWWA WQTD 2008b), as this provides for some disinfection credit and has been observed to form a more stable chloramine residual (Wilczak et al. 2003).

Monochloramine residual

A sufficient monochloramine residual must be provided in the finished water leaving the treatment plant. This is necessary to inhibit nitrifying bacteria. In general, a monochloramine residual of at least 2.0 mg Cl_2/L is recommended (AWWA 2013). The residual is system specific, and higher residuals may be needed to maintain a residual throughout the distribution system. It is recommended that at least 1.0 mg Cl_2/L monochloramine be maintained at all distribution system locations (AWWA 2013; GLUMRB 2007). This concentration is higher than that required by USEPA regulations. Some primacy agencies have adopted specific chloramine residual concentration requirements that are higher than the USEPA rule (see Figure 6-3).

Responding to the DSWQ Challenge: What Strategies Should be Used to Prevent Nitrification?

- Control the chlorine to ammonia nitrogen mass ratio
- Minimize free ammonia leaving the treatment plant
- Maintain sufficient monochloramine residuals
- Provide good corrosion control
- Routinely clean storage facilities and distribution system piping
- Practice spot or automatic flushing of problem areas
- Practice systematic (unidirectional) distribution system flushing
- Minimize distribution system water age
- Consider additional strategies such as use of chlorite, use of ultraviolet light, elevation of pH, alteration of water source and quality, booster chloramination
- Have a written control plan, and budget accordingly

Water Age

Good operational practices for the prevention of nitrification are essential in the distribution system. Because high water age is an important parameter that leads to nitrification (Freidman 2010; Kirmeyer et al. 1995; Skadsen 1993; Negrin, Heyer, and Cheng 1990; Barrios and Stone 1989; Wolfe et al. 1988), all efforts to reduce water age should be routinely pursued. Existing water quality, such as that collected as part of the TCR and the D/DBP rules, can be beneficial in assessing water age. Hydraulic modeling is a useful tool for understanding and identifying potential high water age areas. Because most systems are designed for fire flow protection, there will likely be areas of high water age. Due to a variety of factors, water use is declining across the United States and other parts of the world, leading to oversized systems and hence higher water age. This scenario will promote nitrification.

Practices to reduce water age include proper design of the distribution system, including elimination of dead ends and looping of water mains (see Chapter 2). Additional actions to reduce water age typically include hydrant flushing; this is the most common approach used by water utilities (see Figure 6-18; Kirmeyer et al. 2004). Conventional flushing, where a specific area of the distribution system is flushed, may provide temporary relief of nitrification. However, it only addresses the specific area targeted and potentially misses other impacted locations. Automatic flushing may be used in areas that exhibit chronic problems. UDF (described earlier), a methodical approach to flushing the entire system at sufficient velocity to impact deposits, is recommended to improve the overall condition of the distribution system (Antoun, Dyksen, and Hiltebrand 1999). Any flushing approach is dependent on the ability to successfully operate valves. Therefore, a valve operation program is essential. In addition, unintentionally closed valves may artificially create dead ends, making proper identification of valve position important.

Corrosion

Nitrification has been associated with water main corrosion (Lipponen Suutari, and Martikainen 2002; Kirmeyer et al. 1995). Therefore, corrosion control is important for nitrification prevention. Corrosion control combined with UDF improves overall water quality and helps to prevent nitrification. However, UDF is a time-intensive procedure and therefore needs to be carefully planned, with sufficient resources allocated. Corrosion control is discussed in Chapter 7.

Storage Facilities

A common location for nitrification in the distribution system is within storage facilities (as discussed earlier). Water age can be significant in these locations due to insufficient turnover, insufficient cycling, excess storage capacity, accumulation of sediments and corrosion by-products, and stratification (AWWA 2013). It is critical to assess storage tank operations and water age in order to minimize nitrification. If not controlled at the tank, nitrification will spread to the downstream side in the distribution system. Sample collection and analysis of both influent and effluent storage tank water quality should be routinely performed.

Maintaining a sufficient monochloramine residual in storage tanks is the first indication of satisfactory water quality. Some water utilities have found that a higher residual of 2.0 mg Cl_2/L is beneficial for the prevention of nitrification (Harrington et al. 2002). Storage tanks should be turned over to ensure that old water is routinely removed. In AWWA Manual M56 (2013), at least 30 percent turnover per day is recommended. Analysis of tank stratification, particularly during summer months when temperatures are high, is also recommended (Figure 6-19). Mixing and cycling should be sufficient to prevent

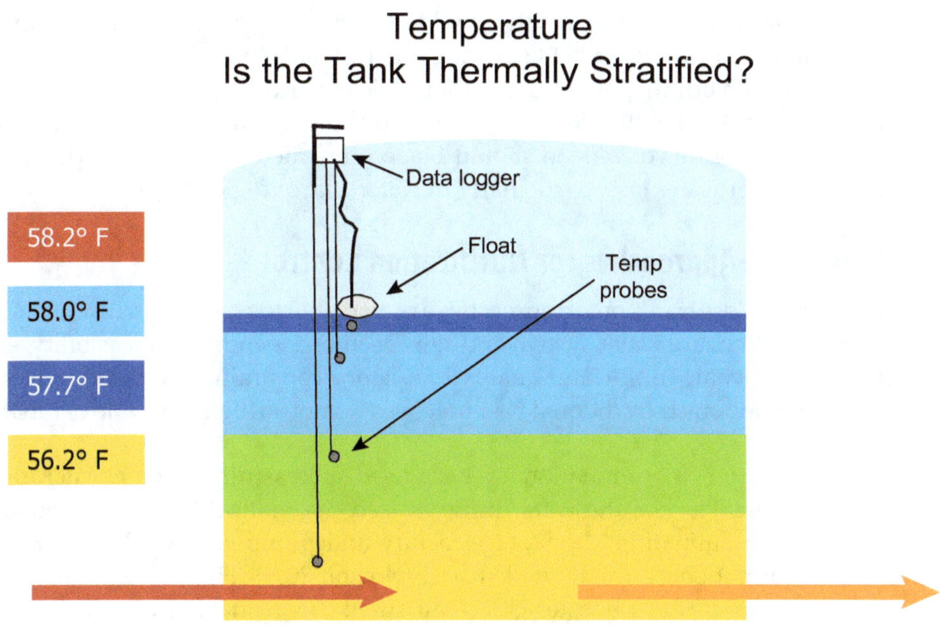

Source: Charlotte Smith and Associates.

Figure 6-19 Example of storage tank stratification occurrence and assessment

short-circuiting (Mahmoud et al. 2005). Effective methods to address these issues include the adding mixers, modifying inlet/outlet structures, and increasing the fill time while maintaining sufficient inlet velocity (Grayman et al. 2004).

Tanks should be routinely inspected and cleaned at least every five years to remove sediment and biofilm that may harbor nitrifying bacteria. Finally, analysis of necessary storage should be conducted. Declining water use patterns may lead to excess existing storage capacity in water systems; this, in turn, leads to high water age and nitrification. Seasonal or permanent decommissioning may be a practical approach to addressing these lower water demands and increased water age (AWWA 2013).

Free Chlorination

Finally, some water utilities annually perform routine free chlorine conversion in the distribution system. In some states, this practice is a regulatory requirement. In a 2008 survey (AWWA WQTD 2008b), it was reported that 23 percent of chloraminated water system practice annual free chlorination. While this is an effective approach to halting nitrification once it has occurred, its effectiveness as a prevention strategy is controversial (AWWA 2013). In many water systems, nitrification occurs in spite of the application of free chlorine during normal (nonnitrifying) water quality periods (Negrin, Heyer, and Cheng 1990; DiGiano et al. 2002; Carrico et al. 2008). One study demonstrated overall reduced bacterial counts with free chlorination but found that hydrant flushing needed to be combined with the free chlorine application in order to reduce the further incidence of nitrification (Rosenfeldt et al. 2009). A variation on this practice has been to free chlorinate only one area of the distribution system or a specific storage tank (Guistino 2004; Song et al. 1999).

In AWWA Manual M56, temporary conversion to free chlorine is recommended as the last strategy to consider for nitrification prevention. If temporary free chlorination is implemented, the water system should evaluate its effectiveness in nitrification occurrence and unintended consequences that could impact water quality. Water quality impacts and concerns include increased DBPs, changes in microbial populations (both inactivation and release from biofilm), possible need for public notification, occurrence of chlorinous taste and odor, and the ability to ensure a distribution system free chlorine residual. In addition, free chlorine conversion should be accompanied by flushing, which may increase concerns over water loss, particularly in water-scarce regions.

Innovative Approaches for Nitrification Control

A variety of less traditional approaches are available to control nitrification (AWWA 2013). These practices are water system and site specific, depending on feasibility, existing capabilities, and competing water quality/regulatory constraints. A water system can use the following options if traditional best practices do not sufficiently control nitrification:

- Booster chloramination has been used successfully at some water utilities. In Key West, Florida, the water utility raised the monochloramine residual by booster chloraminating at a storage facility and pump station, thereby reducing nitrite formation (Coley, Reynolds, and Johnson 2010) (Figure 6-20). There are two basic approaches to booster chloramination: the available free ammonia is combined with additional chlorine fed at the appropriate ratio or both chlorine and ammonia are applied.
- A low dose of chlorite can be applied either directly as chlorite or as a DBP of chlorine dioxide use (McGuire et al. 1999, 2004, 2007, 2009; Zhu, Monks, and Song 2010). This approach should be discussed with regulatory agencies prior to use because chlorite is a regulated substance (1 mg/L MCL). However, in plants that already use or plan to use chlorine dioxide for disinfection, the formation of chlorite may be beneficial for nitrification prevention.
- An increase in pH to above the optimal growth range for AOB (above 9) has been effective in some systems (Skadsen 2002; Harms and Owen 2004). However, such pH adjustment must be evaluated in terms of DBP formation and corrosion/scale-formation potential.
- At water treatment plants with multiple water sources or multiple treatment processes, the waters can be blended to obtain a different water quality. This was demonstrated at the East Bay Municipal Utility District where they altered the TOC concentration (Turner, Lai, and Wilczak 1999; Kirmeyer et al. 2004). This procedure can be detrimental depending on the resultant water quality (Skyes 2003).
- An emerging technique is the use of ultraviolet (UV) light disinfection in storage facilities. Limited experience has shown that the use of UV light lessened nitrification (White and Adams 2010). However, this technique is primarily used to remove waterborne organisms and is less effective at biofilm control. Because there is only one published example of this approach, its potential for more widespread use is not known.

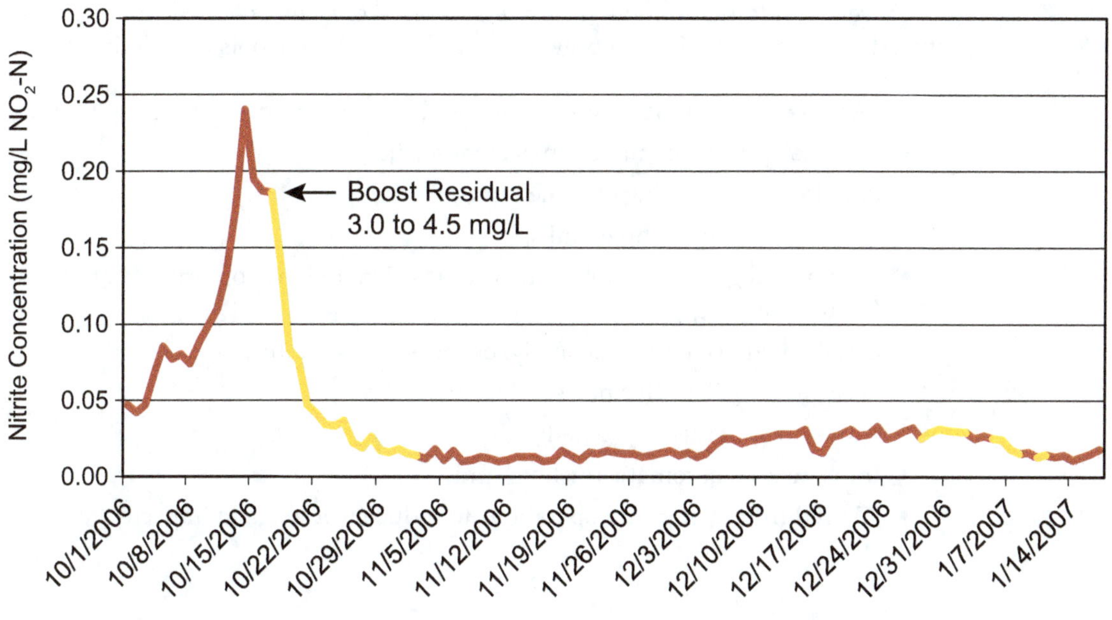

Adapted from Coley, Reynolds, and Johnson 2010.

Figure 6-20 Effectiveness of booster chloramination in reducing nitrite formation at Key West Utility

CASE STUDY

In most cases, a water system must perform a combination of these techniques in order to successfully prevent or at least reduce the occurrence of nitrification. The Portland Water Bureau developed a toolbox of strategies for reducing water age and controlling the cleanliness of their system and thereby improve nitrification prevention (Gupta and Akagi 2014). The Portland water system consists of 180 pressure zones, 70 storage tanks, 39 pumping plants, 3 uncovered finished water reservoirs, and more than 2,000 miles of distribution system pipeline. In 2013, the nitrification prevention strategies that were used included:

- Increased sampling to better understand and capture nitrification events,
- Optimized ammonia and chlorine injection points,
- Minimized free ammonia at the entry point to the distribution system (chlorine to ammonia-N ratio is between 4.76:1 and 4.86:1),
- Lowered tank operating levels,
- Performed conventional flushing,
- Drained and cleaned tanks proactively,
- Removed some storage from service,
- Performed some UDF, and
- Increased chloramine residual leaving the treatment plant.

However, in spite of these practices, nitrification still occurred in the late summer/early fall. Some storage tanks exhibited stratification, which likely contributed to the nitrification.

Therefore, additional tools from the toolbox were implemented in 2014, and these provided improved nitrification prevention and control. The tools included:

- Increased chloramine residual leaving the treatment plant to 2.5 mg Cl_2/L,
- Seasonally removed more tanks from service,
- Practiced deep cycling of some tanks,
- Installed mixers in three tanks,
- Changed regulator settings to increase demand in some parts of the system (reduce water age),
- Installed chlorine residual analyzers at seven key tanks,
- Performed UDF in five pressure zones,
- Adjusted spot flushing schedule,
- Implemented automatic flushers, and
- Altered timing of pump operations to reduce water age in impacted areas.

CONCLUSIONS AND RECOMMENDATIONS

There are many ways to minimize nitrification and improve treatment processes, distribution operations, and overall water quality. Every chloraminating water system should perform nitrification monitoring and prepare a plan for nitrification prevention and response. A toolbox approach that encompasses a list of actions has been proposed and used by some water systems (Gupta and Akagi 2014; Skadsen et al. 2014). These actions are summarized in Table 6-4 and are presented in two levels:

- Basic: All water systems that use chloramines should perform this action and
- Advanced: Water systems that wish to optimize their nitrification control program should perform this action to have a comprehensive program.

If the approaches listed in Table 6-4 are not sufficient to prevent nitrification, the water system may wish to evaluate emerging approaches. The selection of approaches is water system and site specific. These approaches include:

- Blend waters to improve quality,
- Evaluate booster chloramination,
- Evaluate increased pH,
- Evaluate use of chlorite ion, and
- Evaluate use of UV absorbance.

Table 6-4 Recommended practices for nitrification control

Best Practice	Level of Service
Planning	
• Prepare a nitrification control plan that includes both response and prevention actions; review and update plan annually	Basic
• Develop budget cost estimates and prepare capital projects as needed to support the above best practices	Advanced
Monitoring and detection	
• Develop and implement a nitrification monitoring plan	Basic
• Monitor key parameters at Total Coliform Rule/Disinfection By-products Rule sites (e.g., nitrite, nitrate, ammonia, chloramine)	Basic
• Monitor storage tank inlet and outlet water quality	Basic
• Monitor source water quality for nitrification-related parameters	Basic
• Identify critical control points that represent:	
– Fresh water: from water treatment plant or wholesale supplier	Basic
– Average water age	Basic
– High water age: storage tanks, far reaches, low-usage areas	Basic
• Monitor additional indicator parameters to find nitrification (e.g., alkalinity, pH, dissolved oxygen)	Advanced
• Perform a chloramine decay study	Advanced
Operations and maintenance—water treatment plant	
• Control the chlorine to ammonia–nitrogen ratio to be between 4.5 and 5:1 in the finished water leaving the water treatment plant (or request an upstream supplier to do so); if possible, control to be between 4.75 and 5:1	Basic
• Minimize free ammonia leaving the water treatment plant to <0.05 mg N/L or lower if possible	Basic
• Maintain sufficient monochloramine residuals:	
– >2.0 mg Cl_2/L leaving the treatment plant	Basic
– >0.5 mg Cl_2/L in the distribution system, preferably >1.5 mg Cl_2/L	Basic
• Consider seasonal adjustment of chloramine residual (higher residual in nitrification season)	Advanced
• Provide good corrosion control	Basic
Operations and maintenance—distribution system	
• Routinely clean storage facilities and distribution system piping	Basic
• Practice spot or automatic flushing of problem areas	Basic
• Practice systematic (unidirectional) distribution system flushing	Advanced
• Monitor storage tanks for stratification	Advanced
• Perform hydraulic study of distribution system and storage tanks and implement recommendations	Advanced
• Minimize distribution system water age:	
– Eliminate dead ends and low-flow areas	Basic/ Advanced
• Minimize storage facility water age:	
– Maximize storage facility turnover and cycling	Basic
– Maximize hydraulic mixing	Advanced
– Optimize storage volume	Advanced

(continued)

Table 6-4 Recommended practices for nitrification control *(continued)*

Best Practice	Level of Service
Responses to nitrification	
• Perform localized flushing	Basic
• Drain, disinfect, and clean affected storage tanks	Basic
• Develop a free chlorination strategy for temporary conversion to free chlorine and evaluate its effectiveness by monitoring for return of nitrification; free chlorination may be localized or system wide	Basic/ Advanced

REFERENCES

AWWA WQTD (American Water Works Association Water Quality and Technology Division Disinfection Systems Committee). 2008a. Committee Report: Disinfection Survey, Part 1–Recent Changes, Current Practices, and Water Quality. *Journal of the American Water Works Association*, 100:10:76–90.

———. 2008b. Committee Report: Disinfection Survey, Part 2–Alternatives, Experiences, and Future Plans. *Journal of the American Water Works Association*, 100:11:110–124.

———. 2013 (2nd ed.). M56, *Nitrification Prevention and Control in Drinking Water.* Denver, CO: AWWA.

Antoun, E.N.; J.E. Dyksen; & D.J. Hiltebrand. 1999. Unidirectional Flushing: A Powerful Tool. *Journal of the American Water Works Association*, 91:7:62–71.

Arp, D.J.; L.A. Sayavedra-Soto; & N.G. Hommes. 2002. Molecular Biology and Biochemistry of Ammonia Oxidation by *Nitrosomonas europaea*. *Archives of Microbiology*, 1784:250–255.

Baribeau, H.; M. Prevost; R. Desjardins; & P. Lafrance. 2001. Changes in Chlorine and DOX Concentrations in Distribution Systems. *Journal of the American Water Works Association*, 93:12:102–114.

Barrios, M.R. & P. Stone. 1989. Chloramine Depletion in Covered Reservoirs. Proceedings AWWA Water Quality Technology Conference. Austin, Tx.

Breeuwer, P. & T. Abee. 2000. Assessment of Viability of Microorganisms Employing Fluorescence Techniques. *International Journal of Food Microbiology*, 55:1–3:193–200.

Carrico, B.A.; F.A. DiGiano; N.G. Love; P. Vikesland; K. Chandran; M. Fiss; A. Zaklikowski. 2008. Effectiveness of Switching Disinfectants for Nitrification Control. *Journal of the American Water Works Association*, 100:10.

Chauret, C.; C. Smith; & H. Baribeau. 2008. Inactivation of *Nitrosomonas europaea* and Pathogenic *Escherichia Coli* by Chlorine and Monochloramine. *Journal of Water and Health*, 6:3:315–322.

Coley, R.; J. Reynolds; & C. Johnson. 2010. Nitrification Control in the Florida Keys. Proceedings AWWA Water Quality Technology Conference. Savannah, GA.

Cunliffe, D.A. 1991. Bacterial Nitrification in Chloraminated Water Supplies. *Applied and Environmental Microbiology*, 57:11:3399–3402.

Daims, H.; E.V. Lebedeva; P. Pjevac; P. Han; C. Herbold; M. Albertsen; N. Jehmlich; M. Palatinszky; J. Vierheilig; A. Bulaev; R.H. Kirkegaard; M. von Bergen; T. Rattei; B. Bendinger; P.H. Nielsen; & M. Wagner. 2015. Complete Nitrification by Nitrospira Bacteria. *Nature*, 528:7583:504.

DiGiano F.A.; W. Hang; A. Travaglia; D.E. Francisco; & M. Wood. 2002. Occurrence of Bacterial Regrowth and Nitrification in the Raleigh Distribution System and Development of an EPANET Model for Future Assessments. Raleigh, NC: Water Resources Research Institute of the University of North Carolina.

Duirk, S.E. 2003. *Mechanisms and Modeling of Monochloramine Loss, Natural Organic Matter Oxidation, and Disinfectant By-Product Formation in Water.* Doctoral thesis, University of Iowa, Iowa City, IA.

———. 2006. Modeling Monochloramine Loss in the Presence of Natural Organic Matter. *Water Research*, 40:4:851–852.

Duirk, S.E.; B. Gombert; J. Choi; & R.L. Valentine. 2002. Monochloramine Loss in the Presence of Humic Acid. *Journal of Environmental Monitoring*, 4:1:85–89.

Duirk, S.E.; B. Gombert; J.P. Croue; & R.L. Valentine. 2005. Modeling Monochloramine Loss in the Presence of Natural Organic Matter. *Water Research*, 39:14:3418–3431.

Friedman, M.; G. Kirmeyer; J Lemiex; M. LeChevallier; S. Seidl; & J. Routt. 2010. *Criteria for Optimized Distribution Systems*. Denver, CO: Water Research Foundation & Partnership for Safe Water,

GLUMRB (Great Lakes Upper Mississippi River Board of State Public Health and Environmental Managers). 2007. Recommended Standards for Water Works. Albany, N.Y.: Health Education Services. Interim Standard–Use f Chloramine Disinfectant for Public Supplies. Policy Statement Adopted April, 1997; Re-Adopted as Interim Standard April, 2003, Revised October, 2005.
http://10statesstandards.com/waterstandards.html#chloramine

Grady, C.P.L.; G.T. Daigger; & H.C. Lim. 1999 (2nd ed.). *Biological Wastewater Treatment*. New York: Marcel Dekker, Inc.

Grayman, W.M.; R.A. Deininger; C.D. Smith; C.N. Arnold; & J.F. Smith. 2004. Mixing and Aging of Water in Distribution System Storage Facilities. *Journal of the American Water Works Association*, 96:9:70–80.

Guistino, J. 2004. Breakpoint Chlorination of Reservoirs—A Nitrification Band-Aid. AWWA Cal-Nevada Section Spring Conference. Las Vegas, NV.

Gupta, K. & Y. Akagi. 2014. Nitrification Monitoring and Control: A Utility Case Study. Proceedings AWWA Water Quality Technology Conference, New Orleans, La.

Harms, L.L. & C. Owen. 2004. *A Guide for the Implementation and Use of Chloramines*. Denver, CO: Water Research Foundation.

Harrington, G.W.; D.R. Noguera; C.C. Bone; A.I. Kandou; P.S. Oldenburg; J.M. Regan; & D. Van Hoven. 2003. *Ammonia from Chloramine Decay: Effects on Distribution System Nitrification*. Denver, CO: AWWA Research Foundation.

Harrington, G.W.; D.R. Noguera; A.I. Kandou; & D.J. Vanhoven. 2002. Pilot-Scale Evaluation of Nitrification Control Strategies. *Journal of the American Water Works Association*, 94:11:78–89.

Health Canada. 2014. Guidelines for Canadian Drinking Water Quality. https://committees.awwa.org/m68/Lists/Default%20Committee%20Documents/Health% 20Canada_guide-res_recom_2014-10_eng-1.pdf (accessed April 2, 2016).

Hoefel, D.; W.L. Grooby; P.T. Monis; S. Andrews; & C.P. Saint. 2003. Enumeration of Water-Borne Bacteria Using Viability Assays and Flow Cytometry: A Comparison to Culture- Based Techniques. *Journal of Microbiological Methods*, 55:3:585–597.

Jafvert, C.T. & R.L. Valentine. 1992. Reaction Scheme for the Chlorination of Ammoniacal Water. *Environmental Science & Technology*, 26:3:577.

Johnson, D.W. & D.W. Margerum. 1991. Non-Metal Redox Kinetics: A Reexamination of the Mechanisms of the Reaction between Hypochlorite and Nitrite Ions. *Inorganic Chemistry*, 30:4848–4851.

Kasuga, I.; H. Nakagaki; F. Kurisu; & H. Furumai. 2010. Predominance of Ammonia-Oxidizing Archaea on Granular Activated Carbon Used in a Full-Scale Advanced Drinking Water Treatment Plant. *Water Research*, 44:17:5039–5049.

Kirmeyer, G. J.; M. Friedman; K. Martel; G. Thompson; A. Sandvig; J. Clement; & M. Frey. 2002. *Guidance Manual for Monitoring Distribution System Water Quality*. Denver, CO: Water Research Foundation.

Kirmeyer, G.; K. Martel; G. Thompson; L. Radder; W. Klement; M. LeChevallier; H. Baribeau; & A. Flores. 2004 (2nd ed.). *Optimizing Chloramine Treatment*. Denver, CO: Water Research Foundation.

Kirmeyer, G.J.; L.H. Odell; J.G. Jacangelo; A. Wilczak; & R.L. Wolfe. 1995. *Nitrification Occurrence and Control in Chloraminated Water Systems*. Denver, CO: Water Research Foundation and American Water Works Association.

Lipponen, M.T.T.; M.H. Suutari; & P.J. Martikainen. 2002. Occurrence of Nitrifying Bacteria and Nitrification in Finnish Drinking Water Distribution Systems. *Water Research*, 36:4319–4329.

Lloyd, D. & A.J. Hayes. 1995. Vigour, Vitality and Viability of Microorganisms. *FEMS Microbiology Letters*, 133:1–2:1–7.

Maestre, J.P.; D.G. Wahman; & G.E. Speitel, Jr. 2013. Monochloramine Cometabolism by *Nitrosomonas europaea* Under Drinking Water Conditions. *Water Research*, 47:13:4701–4709. doi: http://dx.doi.org/10.1016/j.watres.2013.05.019.

Mahmood, F.; J.G. Pimblett; N.O. Grace; & W.M. Grayman. 2005. Evaluation of Water Mixing Characteristics in Distribution System Storage Tanks. *Journal of the American Water Works Association,* 97:3:74–88.

Margerum, D.W.; L.M. Schurter; J. Hobson; & E.E. Moore. 1994. Water Chlorination Chemistry: Nonmetal Redox Kinetics of Chloramine and Nitrite Ion. *Environmental Science & Technology,* 28:2:331–337.

McGuire, M.J.; P.A. Hayes; G. Qin; N.K. Blute; & X. Wu. 2007. Control of Nitrification Using Chlorite Ion in a Glendale, California Demonstration Area. Proceedings AWWA Annual Conference, city, state

McGuire, M.J.; N.I. Lieu; & M.S. Pearthree. 1999. Using Chlorite Ion to Control Nitrification. *Journal of the American Water Works Association,* 91:10:52–61.

McGuire, M.J.; M.S. Pearthree; N.K. Blute; T. Hoogerwerf; & K.F. Arnold. 2004. Tucson Nitrification Control Pilot Study: A Significant Advance in the State-of-the-Art of Implementing Chloramines in a Desert Environment. Arizona Water & Pollution Control Association Annual Conference.

McGuire, M.J.; X. Wu; N.K. Blute; D. Askenazier; & G. D. 2009. Prevention of Nitrification Using Chlorite Ion: Results of a Demonstration Project in Glendale, Calif. *Journal of the American Water Works Association,* 101:10: 47-59.

Negrin, J.M.; W.M. Heyer; & S. Cheng. 1990. Evaluation of Nitrification Control Strategies at the Los Angeles Department of Water and Power. Proceedings AWWA Annual Conference.

Oldenburg, P.S.; J.M. Regan; G.W. Harrington; & D.R. Noguera. 2002. Kinetics of *Nitrosomonas europaea* Inactivation by Chloramine. *Journal of the American Water Works Association,* 94:10:100–110.

Ontario Drinking Water Quality Standards. 2002. Ontario Regulation 169/03. https://www.ontario.ca/laws/regulation/030169 (accessed Jan. 19, 2015).

Pintar, K.D.M.; W.B. Andersson; R.M. Slawson; E.F. Smith; & P.M. Huck. 2005. Assessment of a Distribution System—Nitrification Critical Threshold Concept. *Journal of the American Water Works Association,* 97:7:116–129.

Pintar, K.D.M.; R.M. Slawson; & P.M. Huck. 2000. Investigation of Conditions and Control Strategies Influencing Nitrification in a Chloraminated Bench-Scale Distribution System. Proceedings AWWA Water Quality Technology Conference, Salt Lake City, UT.

Pinto, A.J.; D.N. Marcus; U.Z. Ijaz; Q.M. Bautista-de lose Santos; G.J. Dick; & L. Raskin. 2016. Metagenomic Evidence for the Presence of Comammox Nitrospira-Like Bacteria in a Drinking Water System. *mSphere,* 1:1.

Pope, P.G. 2006. Haloacetic Acid Formation During Chloramination: Role of Environmental Conditions, Kinetics, and Haloamine Chemistry. Doctoral thesis. Department of Civil, Architectural, and Environmental Engineering, University of Texas at Austin.

Prosser, J.I. 1986. *Nitrification*. Washington, DC: Published for the Society for General Microbiology by IRL Press.

———. 1989. Autotrophic Nitrification in Bacteria. *Advances in Microbial Physiology*, 30:125–181.

Reilly from Skadsen, J. Reilley; B. Gupta; K. Fiske,. 2014. Distribution System Issues, Part 3 of 3; Nitrification Control Strategies. AWWA webinar.

Reilly, B. 2014. Got Chlorine? Controlling Nitrification and Improving Chloramine Residuals in Metro-Boston's Distribution Systems. Proceedings AWWA Water Quality Technology Conference, New Orleans, LA city, state

Roberts, M. 2015. Personal communication, Hazen and Sawyer.

Rosenfeldt, E.J.; C. Baeza: & D.R.U. Knappe. 2009. Effect of Free Chlorine Application on Microbial Quality of Drinking Water in Chloraminated Distribution Systems. *Journal of the American Water Works Association*, 101:10.

Sathasivan, A.; I. Fisher; & G. Kastl. 2005. Simple Method for Quantifying Microbiologically Assisted Chloramine Decay in Drinking Water. *Environmental Science & Technology*, 39:14:5407–5413.

Schrempp, T.; R. Goold; B. Kwan; D. Sarai; & M. Horsley. 1994. Effect of Mechanical Cleaning and Free Residual Chlorine on the Chlorine Demand and Nitrification Process in a Chloraminated System. Proceedings AWWA Water Quality Technology Conference, city, state San Francisco, CA

Seidel, C.J.; M.J. McGuire; R.S. Summers; & S. Via. 2005. Have Utilities Switched to Chloramines? *Journal of the American Water Works Association*, 97:10:87–97.

Skadsen, J.M. 1993. Nitrification in a Distribution System. *Journal of the American Water Works Association*, 85:7:95.

———. 2002. An Assessment of the Effectiveness of High pH for Control of Nitrification. *Journal of the American Water Works Association*, 94:7:73.

Skadsen, J.; B. Reilley; K. Gupta; P. Fiske; & R. Moore. 2015. Nitrification Control Strategies: What is Nitrification and How is it Managed? *Opflow*, 41:1:10–13.

Song, D.J.; A. Sheikholeslami; L.L. Hoover; K.A. Turner; H.H. Lai; & A. Wilczak. 1999. Improvement of Chloramine Stability Through pH Control, TOC Reduction, and Blending at EBMUD, California. Proceedings AWWA Annual Conference. Chicago, IL

Speitel, G.E., Jr.; J.P. Maestre; & D.G. Wahman. 2014. *Monochloramine Cometabolism by Ammonia-Oxidizing Bacteria*. Denver, CO: Water Research Foundation.

Sykes, M. 2003. Issues Related to Mixing Free Chlorine and Monochloramine Residuals in the Distribution System. Proceedings AWWA Water Quality Technology Conference.

Texas Commission on Environmental Quality. 2015. Nitrification Action Plan (NAP) Guidance. https://www.tceq.texas.gov/assets/public/permitting/watersupply/pdw/NAP_Guidance.pdf (accessed April 2, 2016).

Valentine, R.L.; K. Ozekin; & P.J. Vikesland. 1998. *Chloramine Decomposition in Distribution Systems and Model Waters.* Denver, CO: AWWA Research Foundation and American Water Works Association.

van der Wielen, P.W.J.J.; S. Voost; & D. van der Kooij. 2009. Ammonia-Oxidizing Bacteria and Archaea in Groundwater Treatment and Drinking Water Distribution Systems. *Applied and Environmental Microbiology,* 75:14:4687–4695. doi: 10.1128/aem.00387- 09.

van Kessel, M.; D.R. Speth; M. Albertsen; P.H. Nielsen; H.J.M. Op den Camp; B. Kartal; M.S.M. Jetten; & S. Lucker. 2015. Complete Nitrification by a Single Microorganism. *Nature,* 528:7583:555.

Vikesland, P.J.; K. Ozekin; & R.L. Valentine. 2001. Monochloramine Decay in Model and Distribution System Waters. *Water Research,* 35:7:1766–1776.

———. 1998. Effect of Natural Organic Matter on Monochloramine Decomposition: Pathway Elucidation through the Use of Mass and Redox Balances. *Environmental Science & Technology,* 32:10:1409–1416.

Vikesland, P.J. & R.L. Valentine. 2000. Reaction Pathways Involved in the Reduction of Monochloramine by Ferrous Iron. *Environmental Science & Technology,* 34:1:83–90.

———. 2002a. Iron Oxide Surface-Catalyzed Oxidation of Ferrous Iron by Monochloramine: Implications of Oxide Type and Carbonate on Reactivity. *Environmental Science & Technology,* 36:3:512–519.

———. 2002b. Modeling the Kinetics of Ferrous Iron Oxidation by Monochloramine. *Environmental Science & Technology,* 36:4:662–668.

Wahman, D.G. & J.G. Pressman. 2015. Distribution System Residuals—Is "Detectable" Still Acceptable for Chloramines? *Journal of the American Water Works Association,* 107:8:53–63.

Wahman, D.G.; K.A. Schrantz; & J.G. Pressman. 2010. Determination of the Effects of Medium Composition on the Monochloramine Disinfection Kinetics of *Nitrosomonas europaea* by the Propidium Monoazide Quantitative PCR and Live/Dead Baclight Methods. *Applied and Environmental Microbiology,* 76:24:8277–8280. doi: 10.1128/aem.01631-10.

Wahman, D.G. & G.E. Speitel, Jr. 2012. The Relative Importance of Nitrite Oxidation by Hypochlorous Acid Under Chloramination Conditions. *Environmental Science & Technology,* 46:11:6056–6064. doi: 10.1021/es300934x.

Wahman, D.G.; K.A. Wulfeck-Kleier; & J.G. Pressman. 2009. Monochloramine Disinfection Kinetics of *Nitrosomonas europaea* by Propidium Monoazide Quantitative Pcr and Live/Dead Baclight Methods. *Applied and Environmental Microbiology,* 75:17:5555- 5562. doi: 10.1128/aem.00407-09.

Ward, B.B.; D.J. Arp; & M.G. Klotz. 2011. *Nitrification*. Washington, DC: ASM Press.

White, B. & M. Adams. 2010. Battling Nitrification with Blacklights. *Opflow*, 36:2:10-13.

Whittaker, M.; D. Bergmann; D. Arciero; & A.B. Hooper. 2000. Electron Transfer During the Oxidation of Ammonia by the Chemolithotrophic Bacterium *Nitrosomonas europaea*. *Biochimica et Biophysica Acta, Bioenergetics*, 1459:2–3:346–355.

Wilczak, A.; J.G. Jacangelo; J.P. Marcinko; L.H. Odell; G.J. Kirmeyer; & R.L. Wolfe. 1996. Occurrence of Nitrification in Chloraminated Distribution Systems. *Journal of the American Water Works Association*, 88:7:74–85.

Wilczak, A.; C.D. Smith; K.Y. Cohen; & P. Martin. 2003. Strategies for Combining Free Ammonia and Boosting Chloramines—Survey of Utility Practice. Proceedings AWWA Water Quality Technology Conference. Philadelphia, PA

Wolfe, R.L.; N.I. Lieu; G. Izaguirre; & E.G. Means. 1990. Ammonia-Oxidizing Bacteria in a Chloraminated Distribution System: Seasonal Occurrence, Distribution, and Disinfection Resistance. *Applied and Environmental Microbiology*, 56:2:451–462.

Wolfe, R.L.; E.G. Means III,;M.K. Davis; & S.E. Barrett. 1988. Biological Nitrification in Covered Reservoirs Containing Chloraminated Water. *Journal of the American Water Works Association*, 80:9:109–114.

Wolfe, R.L.; N.R. Ward; & B.H. Olsen. 1984. Inorganic Chloramines as Drinking Water Disinfectants: A Review. *Journal of the American Water Works Association*, 76:5:75.

WHO (World Health Organization). 2011 (4th ed.). *Guidelines for Drinking Water Quality*. Geneva, Switzerland: WHO.

Wright, B. 2015. Personal communication. Loudoun Water.

Zhu, E.; V. Monks; & R. Song. 2010. Effectiveness of Early Application of Sodium Chlorite in Nitrification Prevention—A Demonstration-Scale Study. Proceedings AWWA Water Quality Technology Conference. Savannah, GA

Chapter 7

Corrosion Control*

Dave MacNevin, Tetra Tech, Tampa, Fla.
Rich Giani, CH2M, Kansas City, Mo.
Alex Gorzalski, US Army Corps of Engineers, Washington, D.C.
Rebecca Slabaugh, Arcadis, Indianapolis, Ind.

From a water quality perspective, there are many misunderstandings about corrosion in water distribution systems. Just use of the word *corrosion* can be misleading. The main impact on water quality at the tap is elevated metal concentrations that are a by-product of corrosion in the distribution system and/or premise plumbing. Metal release is significant to customers because it can cause undesirable aesthetic effects (e.g. discoloration, particles, tastes) in the water and, in some cases, potential health effects.

DSWQ Challenge:

The main concern associated with corrosion and its impact on water quality at the tap is elevated concentrations of metals. These metals are a by-product ofcorrosion in the distribution system and/or premise plumbing.

This chapter begins with an introduction to corrosion control in water distribution systems, considering water quality impacts, regulatory requirements, health impacts, and secondary aesthetic impacts. The chapter includes an overview of the following three major topics:

* Adapted from AWWA Manual M58, Internal Corrosion Control in Water Distribution Systems.

- Factors affecting corrosion-related water quality;
- Corrosion control methods to reduce metal release in distribution systems; and
- Water quality monitoring tools to diagnose the likely cause, to select an appropriate corrosion control method, and to provide ongoing monitoring.

The primary focus is on corrosion of metal pipe surfaces, solders, and plumbing fixtures, especially those composed of lead, copper, and iron. Other components of the distribution system or premise plumbing that are potentially subject to degradation and release of metals or minerals include water storage tanks, pumps, valves, cement-lined pipe, asbestos–cement (A–C) pipe, concrete pipe, plastic pipe, water meters, galvanized steel service lines, and lead service lines (LSLs).

Corrosion, especially uniform corrosion that is typically the focus of corrosion control studies, is one of several mechanisms that can elevate metal concentrations in distribution systems. Table 7-1 provides a summary of the main corrosion-related mechanisms that can impact distribution system water quality; these mechanisms are described in this chapter. Because the focus here is on metal release, other forms of corrosion such as localized/pitting/pinhole corrosion, degradation of infrastructure, and any other type of corrosion are not discussed at length. Refer to AWWA Manual M58 (2017) for a more thorough discussion of the material in this chapter.

WATER QUALITY IMPACTS

Internal corrosion of distribution system piping and home plumbing can cause several water quality problems including potential health concerns, discoloration, particles, and taste and odor issues. These problems can be the result of corrosion of metal pipe surfaces, pipe solder, A–C pipe, cement linings, and plumbing fixtures or dissolution of existing pipe scales. AWWA Manual M58 (2017) contains more detailed information on types of corrosion on different material surfaces.

Regulatory Requirements and Health Impacts

Corrosion of distribution system components and premise plumbing materials can release contaminants into drinking water. Some of these constituents are regulated for their impacts on human health and others have aesthetic guidelines. A brief summary of international corrosion-related standards is provided in Table 7-2. The regulatory framework in the United States is discussed in limited detail within this chapter.

US Lead and Copper Rule. The Lead and Copper Rule (LCR) was promulgated in 1991 with the purpose of reducing exposures to lead and copper from drinking water (Federal Register 1991). The purpose of the LCR is to protect public health by minimizing lead and copper levels in drinking water, primarily by reducing water corrosivity, avoiding lead-based solders, and replacing LSLs (USEPA 2004). Lead has been demonstrated to cause delays in physical and mental development in infants and children and has been linked to deficits in attention span and learning abilities (USEPA 2006). Short-term copper exposure may cause gastrointestinal distress; long-term exposure may cause liver or kidney damage (NIH 2014).

Table 7-1 Primary corrosion-related mechanisms that influence distribution system water quality

Uniform Corrosion
• Uniform corrosion, which is typically the focus of corrosion control studies, is one mechanism that can cause the release of lead, copper, iron, and other metals or impurities into water. • Naturally occurring constituents such as dissolved inorganic carbon, hydroxides, and oxygen can form solid compounds with dissolved metals. These compounds create protective barriers against uniform corrosion to varying degrees. The adequacy of pipe films as protective barriers depends on the physical and chemical properties of the compounds formed (e.g., solubility, permeability, shear strength). • Calcium carbonate is inconsistent as a protective barrier against uniform corrosion. • Chemicals such as orthophosphate can be added to produce protective barriers against uniform corrosion on pipe walls.
Microbiologically Influenced Corrosion
• Bacterial biofilms drive oxidation-reduction reactions and excrete substances that can drive localized corrosion. Nutrient control, disinfection control, and flushing can help to mitigate the effects of microbiologically influenced corrosion.
Accumulation and Transport
• Metals and other inorganic material can accumulate on pipe walls by both sedimentation and adsorption. This effect is most prominent in unlined iron pipes. • Accumulated materials such as arsenic, lead, and copper can be released by both chemical and physical means back into the water at concentrations higher than those seen at entry points to the distribution system.
Other Metal Release Mechanisms
• The following mechanisms of metal release that should not be overlooked: galvanic corrosion, concentration cells, erosion corrosion, pipe manufacturing quality, pipe installation, and system maintenance. • Localized corrosion, including pitting corrosion, and copper pinhole leaks can be impacted by water quality but generally do not contribute significantly to metal release.
Summary
• Understanding and addressing the various mechanisms of metal release into water in addition to uniform corrosion are the keys to going beyond what is commonly called corrosion control to achieving minimal lead, copper, and iron release.

Table 7-2 International water quality standards for corrosion-related compounds

Compound	US Environmental Protection Agency	Health Canada	European Union	Australia	World Health Organization
Asbestos	7 million fibers per liter ^{H}	None *	None	None *	None *
Copper	1.3 mg/L^{T,†} 1.0 mg/L ^{A}	1 mg/L ^{A}	2 mg/L ^{H}	2 mg/L ^{H} 1 mg/L ^{A}	2 mg/L ^{H}
Iron	0.3 mg/L ^{A}	0.3 mg/L ^{A}	0.2 mg/L^{I}	0.3 mg/L ^{A}	0.3 mg/L ^{A}
Lead	15 μg/L ^{T,†}	10 μg/L ^{H}	10 μg/L ^{H}	10 μg/L ^{H}	10 μg/L ^{T}
Manganese	50 μg/L ^{A}	50 μg/L ^{A}	50 μg/L (I)	500 μg/L ^{H} 100 μg/L ^{A}	400 μg/L ^{H} 100 μg/L ^{A}
Zinc	5 mg/L ^{A}	5 mg/L ^{A}	None	3 mg/L ^{A}	None^{*,H} 3–5 mg/L ^{A}

For more in-depth reviews of international regulatory standards, see 40 CFR § 141 (USEPA), Health Canada (2014), Council Directive (1998), NHMRC (2011), and WHO (2011).

^{A} Aesthetic standard.

^{H} Health-based standard.

^{I} Indicator parameter, requires EU member state to take remedial action if it considers noncompliance to pose a risk to human health.

^{T} Treatment technique, provisional standard, or based on analytical achievability.

* Guideline value not determined to be necessary or insufficient data.

† The USEPA standards for lead and copper are based on action levels (90th percentile).

The LCR established a maximum contaminant level goal (MCLG; unenforceable) of zero for lead and an MCLG of 1.3 mg/L for copper in first-flush samples after a minimum six-hour stagnation time. Any detectable concentration of lead in drinking water is considered harmful to health. The rule also established a treatment technique requirement, including corrosion control treatment, source water treatment, LSL replacement, and public education. The rule set action levels (ALs) of 0.015 mg/L for lead and 1.3 mg/L for copper based on the 90th percentile concentration of all compliance samples. If the 90th percentile concentration for lead or copper is above the AL, public water systems may be required to initiate water quality parameter monitoring for key corrosion control parameters, install corrosion control treatment, begin source water monitoring or treatment, replace LSLs, and/or undertake a public education program.

Revisions to the Lead and Copper Rule. The USEPA revised the LCR in 2007 (Federal Register 2007) with the intent of enhancing implementation in the areas of monitoring, treatment, customer awareness, and LSL replacement. The LCR revisions require that all compliance monitoring results, including those above the required number of samples, be included in the 90th percentile determination and that all samples be taken within the same calendar year. The remainder of the revisions address customer notification for those residents in a water system's sampling program and provisions to deal with LSLs that were previously "tested out" due to levels below the AL. In 2016, USEPA announced plans to release long-term revisions to the LCR in 2017. In March 2016, USEPA published an updated guidance document, Optimal Corrosion Control Treatment Evaluation Technical Recommendations. This guidance is intended to provide technical recommendations that both systems and primacy agencies can use to comply with LCR corrosion control treatment requirements and effective evaluation and designation of optimal corrosion control treatment (USEPA 2016).

US Reduction of Lead in Drinking Water Act. Internal corrosion of LSLs, brass meters and plumbing fixtures, and copper plumbing contributes to the concentrations of lead as well as copper in drinking water. Pure lead pipe (LSLs) were banned by the 1986 Safe Drinking Water Act Amendments, along with wetted components (e.g., pipes, fittings, well pumps) containing greater than 8 percent lead by weight and solder containing greater than 0.2 percent lead by weight. On Jan. 4, 2014, the 2011 Reduction of Lead in Drinking Water Act went into effect. This act reduced the maximum allowable lead concentration in components with wetted surfaces from 8 percent to 0.25 percent. However, fire hydrants and components used exclusively for nonpotable consumption were exempted from the new definition of "lead-free." Legacy piping components installed prior to regulation of leaded materials may also continue to release lead as long as they remain in service.

Health Canada guidelines for lead. The safety of drinking water supplies in Canada is a responsibility that is shared among the provincial/territorial, federal, and municipal governments. The day-to-day responsibility of providing safe drinking water to the public generally rests with the provinces and territories, while the municipalities usually oversee the day-to-day operations of the treatment facilities. Health Canada develops guidelines for Canadian drinking water quality in partnership with the provinces and territories. These guidelines are used by every jurisdiction in Canada and are the basis for establishing provincial drinking water quality requirements for all Canadians (Health Canada 2015).

In Canada, the guideline for lead is a maximum acceptable concentration of 0.01 mg/L (10 μg/L) based on a provisional tolerable weekly intake of lead established by the World Health Organization (Health Canada 1992). However, Health Canada is reviewing the lead guideline to include the most recent health information as well as sampling and monitoring strategies that address exposure from drinking water (Health Canada 2013). Treatment-based guidance for corrosion control was published in 2009 (Health Canada 2009). The recommended sampling protocol found in this guidance is similar to the one outlined in the LCR. The objective of the sampling protocol is to use stagnation times to capture the highest levels of lead; this provides the basis for a system-wide assessment of the corrosion control efficacy.

Asbestos. Asbestos in drinking water is most commonly present as a result of degradation of A–C piping used in drinking water distribution systems. Short-term exposure to asbestos in drinking water is not known to cause any health issues; long-term exposure may result in increased risk of developing benign intestinal polyps. USEPA has established 7 million fibers per liter as both the MCL and MCLG for asbestos. Care should be taken when removing A–C pipe so as to avoid inhalation of asbestos fibers. A–C pipe has not been installed in US distribution systems since the 1980s; however, legacy installations remain.

Biological regrowth and chlorine demand. The Surface Water Treatment Rule was finalized in 1989 and requires maintenance of a "detectable" disinfectant residual in the distribution system of all surface water supplies, among other requirements (Federal Register 1989a). The Total Coliform Rule (TCR) initially imposed monitoring requirements for total coliform as well as for fecal coliform and *Escherichia coli* in 1991 (Federal Register 1989b). The 2013 Revised Total Coliform Rule (RTCR) modified the existing acute MCL into an MCL for *E. coli* and requires assessment and corrective action for systems vulnerable to fecal contamination. The RTCR became effective for water systems on April 1, 2016.

Both biological regrowth and chlorine demand have a complex interaction with internal pipe corrosion. Chlorine and other oxidants impact the oxidation-reduction potential (ORP), which is often a factor in maintaining effective corrosion control. Conversely, corrosion by-products can exert an oxidant demand, diminishing distribution system disinfectant residuals and resulting in increased potential for microbiological regrowth. Microbiological regrowth not only may threaten compliance with the RTCR but also may result in increased potential for microbiologically influenced corrosion (MIC) to occur. Microbial issues are discussed in Chapter 3 and disinfection is discussed in Chapter 8.

Customer health concerns. The majority of health concerns associated with internal corrosion are related to the release of trace concentrations of metals (e.g., lead, copper, cadmium) from corroding metal surfaces. Many of these concerns are aesthetic in nature and are discussed in the next section.

Aesthetic Impacts

Discolored water. Many of the color issues typically encountered in drinking water are attributable to internal corrosion (Kirmeyer, Friedman, and Clement 2000). Corrosion of cast-iron pipe or galvanized steel pipe or dissolution of existing pipe scales may result in rust-colored water or red water due to the presence of ferric iron (Fe^{3+}). Red water (Figure 7-1) can stain laundry and plumbing fixtures. Iron corrosion may also result in yellow or black water, which is a result of the presence of ferrous iron (Fe^{2+}). Black water may also be the result of the presence of manganese or sulfide from source waters. Copper corrosion may result in red or blue water, which can also stain bathroom fixtures and hair. Green, gray, or black water may also be attributable to hot water heaters. Corrosion of zinc in galvanized piping and cement linings may result in a milky-white appearance.

Courtesy of the District of Columbia Water and Sewer Authority.

Figure 7-1 Red water sample from a US water distribution system

Taste and odor. Dissolved and particulate iron can cause taste and odor at low concentrations, including metallic and musty tastes (Khiari et al. 2002; Kirmeyer, Friedman, and Clement 2000). Copper has also been reported to result in a metallic or bitter taste and may be detected at concentrations as low as 0.4 mg/L, depending on an individual's sensitivities (Cuppett, Duncan, and Dietrich 2006). Aluminum and zinc can contribute to an astringent mouthfeel, and zinc can also result in a sour taste at concentrations greater than 4 mg/L (Cohen et al. 1960). More information on taste and odor issues in the distribution system is provided in Chapter 4.

Iron. The USEPA set a secondary maximum contaminant level (SMCL) of 0.3 mg/L for iron entering the distribution system; this represents a "reasonable goal for drinking water quality" (40 CFR § 143.3). SMCLs control contaminants that primarily affect drinking water's aesthetic quality and are not federally enforceable. However, primacy agencies have the authority to include SMCLs in state drinking water regulations. When present at sufficiently high levels, iron may cause rusty colored water, deposit of sediment, metallic taste, and reddish or orange staining.

Zinc. Zinc in drinking water is primarily an aesthetic concern as it can cause a metallic taste. As a result, USEPA established an SMCL of 5 mg/L for zinc. Zinc may be present in drinking water due to erosion of natural deposits or as a result of dezincification of brass plumbing fixtures. In several studies of low-alkalinity waters, it was found that iron and zinc were the principal corrosion by-products of galvanized plumbing (Dangel 1975; JMM 1982).

The zinc coating on galvanized pipe may contain lead, copper, cadmium, chromium, aluminum, barium, and other impurities. There are a number of standards regarding galvanized pipe, most notably by the American Society of Testing and Materials (ASTM). As a result of these impurities, corrosion of galvanized pipe may result in the release of trace metals.

FACTORS AFFECTING CORROSION-RELATED WATER QUALITY

Corrosion-related water quality is influenced by two primary factors: distribution and plumbing system materials and the predominant type of corrosion mechanism. The following sections discuss common distribution system materials, their corrosion resistance, and the primary contaminants associated with each material type. Corrosion mechanisms can be identified as chemically-influenced, biologically influenced, or physically influenced, according to the various symptoms that are observed. A key factor of all corrosion mechanisms is the type and concentration of oxidants present. As materials corrode, the extent and timing of water quality impacts are determined by the rates of accumulation and release of metal corrosion byproducts. More information on factors affecting corrosion-related water quality is available in AWWA Manual M58 (2017).

Distribution and Plumbing System Materials

Table 7-3 provides an overview of the corrosion properties of materials that are frequently used in water distribution systems, including a brief summary of the corrosion resistance and primary contaminants released from different types of pipe. This information can be used to develop a sampling program of pipe contaminants in order to identify the plumbing materials likely to be a source of metal release. More information on materials likely to be sources of metals release is found in AWWA Manual M58 (2017) and the USEPA's 2016 guidance document (2016).

Mechanisms of Corrosion

To help assess corrosion-related water quality problems, a convenient way to classify corrosion for the purpose of diagnosis is as chemically influenced, microbially influenced, or physically influenced. Identifying the cause of a corrosion-related water quality problem requires an evaluation of how these three categories of corrosion contribute to the problem; however, an adequate water quality monitoring program is crucial for determining the cause of corrosion-related water quality problems (see Water Quality Monitoring in this chapter for additional detail). Table 7-4 summarizes common corrosion-related water quality problems and potential reasons for further investigation. For more detailed information regarding the underlying causes of corrosion and metals release issues, refer to AWWA Manual M58 (2017).

Oxidants. Secondary residuals of free chlorine or monochloramine are added for purposes of disinfection (as covered in Chapter 8) but also play an important role in corrosion control. Along with dissolved oxygen (DO), they influence the type of scales formed and ultimately how well those scales remain on the pipe. Distribution scales can be disrupted by changes in disinfectant residual type or DO concentrations, which may accompany source water changes.

Dissolved oxygen. Whether corrosion of lead, copper, and iron is increased or decreased by DO concentrations, it is largely dependent on oxygen gradients. DO gradients can occur on the micro scale, resulting in concentration cells (AwwaRF and TZW 1996).

For oxygenated waters, iron release rates have been shown to be higher during stagnation periods (Sarin et al. 2004). Increasing oxidant concentrations and minimizing stagnation are suggested strategies for minimizing iron release since the Fe^{2+} is more soluble, whereas Fe^{3+} is relatively insoluble.

Table 7-3 Corrosion properties of materials frequently used in water distribution systems

Material	Corrosion Resistance	Primary Contaminant(s) From Pipe
Asbestos–cement, concrete, and cement linings	Good corrosion resistance; immune to electrolysis; aggressive waters can leach calcium from the cement; polyphosphate sequestering agents can deplete the calcium and substantially soften the pipe	Asbestos fibers; increase in pH, aluminum, and calcium; trace metals (cadmium, chromium, barium, and aluminum) due to presence in cement
Brass	Good overall resistance; different types of brass respond differently to water chemistry; subject to dezincification by waters with pH >8.3 with high ratio of chloride-to-carbonate hardness	Lead, copper, zinc, selenium, bismuth, and phosphorus
Copper	Good overall corrosion resistance; subject to corrosive attack from high flow velocities, soft water, chlorine, dissolved oxygen, low pH, and high inorganic carbon levels (alkalinity); subject to microbially influenced corrosion; may be prone to pitting failures	Copper and possibly iron, zinc, tin, antimony, arsenic, cadmium, and lead from associated pipes and solder
Galvanized iron or steel	Subject to galvanic corrosion of zinc by aggressive waters, especially of low hardness; corrosion is accelerated by contact with copper materials; corrosion is accelerated at higher temperatures such as in hot water systems; corrosion is affected by manufacturing process of the pipe and galvanized coating	Zinc and iron; cadmium, chromium, barium, aluminum, and lead are possible due to impurities in the galvanization process
Iron, unlined cast or ductile	Can be subject to surface erosion by aggressive waters and tuberculation in poorly buffered waters	Iron, resulting in red water, complaints, and turbidity
Lead	Corrodes in soft water with pH <8 and in hard waters with high inorganic carbon levels and pH <7.5 or >8.5	Lead
Plastic	Resistant to corrosion	Some pipes contain metals in plasticizers, notably lead; plasticizers are an emerging contaminant issue
Steel, mild	Subject to uniform corrosion; affected primarily by high dissolved oxygen and chlorine levels and poorly buffered water	Iron, resulting in red water, complaints, and turbidity

Source: AwwaRF and TZW 1996.

Table 7-4 Assessment of common corrosion-related water quality impacts

Symptom	Indicators that the symptom may be . . . Chemically Influenced	Microbially Influenced	Physically Influenced
Red water	• Shift in dissolved/ particulate iron speciation but no change in total iron concentration • Loss or reduction in disinfectant residual • Change in ORP • Shift in pH • Change in distributed water alkalinity • Treatment process upsets • Change in finished water chloride-to-sulfate ratio	• Loss or reduction in disinfectant residual • Increase in HPC • Localized shifts in pH but not system wide • Increases in distribution nitrite/nitrate concentrations* • Higher or lower than normal free ammonia concentrations* • Higher or lower DOC concentrations	• Changes in flow • Stagnation • Changes in system pressure • System maintenance • Release of sediments from distribution storage tanks • Release of lower-quality water from poorly mixed storage tanks • Distribution storage tank cleaning
Increased tap lead levels	• Recent change in distribution system disinfectant • Change in ORP • Drop in finished water pH • Shift in pH to outside optimal range for corrosion control chemical • Change in distributed water alkalinity • Change in finished water chloride-to-sulfate ratio • Upset of phosphate addition system, resulting in lower phosphate concentrations • Poor buffer intensity	• Loss or reduction in disinfectant residual • Increase in HPC or ATP • Localized shifts in pH but not system wide • Increases in distribution nitrite/nitrate concentrations* • Higher or lower than normal free ammonia concentrations* • Higher or lower DOC concentrations	• Increase in particulate lead concentration but no corresponding increase in dissolved lead concentrations • Recent lead service line replacement • Major disturbances that shake service lines or plumbing • Hydraulic transients • Installation of new brass plumbing fixture • Long periods of stagnation
Increased tap copper levels	• Recent change in secondary disinfectant • Change in ORP • Drop in finished water pH • Shift in pH to outside optimal range for corrosion control chemical • Change in distributed water alkalinity • Change in finished water chloride-to-sulfate ratio • Upset of phosphate addition system, resulting in lower phosphate concentrations	• Loss or reduction in disinfectant residual • Increase in HPC–R2A or ATP • Localized shifts in pH but not system wide • Increases in distribution nitrite/nitrate concentrations* • Higher or lower than normal free ammonia concentrations* • Higher or lower DOC concentrations • Presence of sulfide	• Increase in particulate copper concentration, but no corresponding increase in dissolved copper concentrations • Long periods of stagnation

(continued)

Table 7-4 Assessment of common corrosion-related water quality impacts *(continued)*

	Indicators that the symptom may be . . .		
Symptom	Chemically Influenced	Microbially Influenced	Physically Influenced
Pinhole leaks in copper plumbing	• pH below 7.2 • High chloride relative to sulfate and bicarbonate • Treatment upsets, resulting in increased residual aluminum concentration in finished water	• Loss or reduction in disinfectant residual • Increase in HPC–R2A • Localized shifts in pH but not system wide • Increases in distribution nitrite/nitrate concentrations* • Higher or lower than normal free ammonia concentrations* • Presence of sulfide • Higher or lower DOC concentrations	• Stagnation • Sediment or debris in plumbing • High velocity (erosion corrosion) • Improper fluxing (leads to "tracks" on inside of pipe)
Blue water (Note: can occur in newly installed or older plumbing and service lines)	• Drop in pH below 7.2 in soft waters • Rise in pH above 8.0 in soft waters • Drop in finished water alkalinity	• Loss or reduction in disinfectant residual • Increase in HPC–R2A • Localized shifts in pH but not system wide • Increases in distribution nitrite/nitrate concentrations* • Higher or lower than normal free ammonia concentrations* • Higher or lower DOC concentrations	• Stagnation
Pink water	• pH shift • Treatment upsets, resulting in increased permanganate concentrations entering distribution system	• Pink water is not likely to be the result of MIC	• Changes in flow • Stagnation • Changes in system pressure • System maintenance • Release of sediments from distribution storage tanks • Release of lower-quality water from poorly mixed storage tanks • Distribution storage tank cleaning
Black or yellow water	• Drop in finished water ORP • Treatment upsets resulting in reduction of manganese dioxide	• Black or yellow water is not likely to be the result of MIC	• Changes in flow • Stagnation • Changes in system pressure • System maintenance • Release of sediments from distribution storage tanks • Release of lower-quality water from poorly mixed storage tanks • Distribution storage tank cleaning

Abbreviations: ATP, adenosine triphosphate; DOC, dissolved organic carbon; HPC, heterotrophic plate count; MIC, microbially influenced corrosion; ORP, oxidation reduction potential; R2A, Reasoner's 2A agar.

*Systems that add chloramine.

Oxidant type and residual. The residual disinfectant in drinking water distribution systems is usually either free chlorine or chloramine, although some systems use chlorine dioxide and others maintain no disinfectant residual. Free chlorine is a strong, reactive oxidizing agent, the implications of which are discussed for lead and iron in the ORP section that follows. However, there are some benefits to disinfectants with lower ORP. Monochloramine has been shown to be orders of magnitude more effective in inactivating biofilm bacteria than free chlorine or chlorine dioxide (LeChevallier, Cawthon, and Lee 1988). Oxidants have often been shown to increase corrosion. Free chlorine concentrations higher than 0.8 mg/L were found to increase corrosion rates of both fresh and aged copper (Stone et al. 1987). Pisigan and Singley (1987) also noted that the corrosion of mild steel and copper was increased by free chlorine.

Oxidation-reduction potential. ORP (also called redox potential) is the measure of a water's capability to oxidize; it is measured as an electrical potential (volts or millivolts). The ORP of various disinfectants is shown in Figure 7-2. The relative oxidizing strength of disinfectants at pH 8 is:

Chlorine dioxide > Free chlorine > Potassium permanganate > Monochloramine > Dissolved oxygen

The ORP of each disinfectant was found to increase with decreasing pH (Copeland and Lytle 2014).

The ORP of residual disinfectants is relevant to the formation of distribution system scales. Many water systems have switched, or are considering switching, from free chlorine to chloramine residuals to control disinfection by-products. Bench-scale experiments (Edwards and Dudi 2004) seemed to indicate that a switch from free chlorine to chloramine can accelerate the dissolution of lead under conditions similar to those in Washington, D.C., when the city experienced an increase in lead levels after switching from chlorine to chloramine in 2000. Further research showed that lead scales were oxidized from divalent lead [lead (II)], which forms scales that are less effective at limiting corrosion, to lead (IV), which forms more effective scales (e.g., plattnerite – PbO_2), by free chlorine and that this process was reversible if free chlorine dissipated (Lytle and Schock 2005).

Pourbaix diagrams, also known as potential/pH or E_H–pH diagrams, show equilibrium phases of a chemical species at various pH levels and oxidizing conditions. Figures 7-3, 7-4, and 7-5 show Pourbaix diagrams for lead, copper, and iron, respectively. The figures for lead and iron show that at a typical water pH (i.e., 6 to 10), lead and iron will be in solid forms under oxidizing conditions and thus less prone to leaching. Maintaining oxidizing conditions within lead scales can be challenging due to loss of residual under stagnant conditions or consumption of oxidant within a scale prior to diffusion to the metal surface.

Copper scales show the opposite pattern of lead, where the less-oxidized copper (I) (Cu^{1+}) compounds are less prone to leaching than the more oxidized copper (II) (Cu^{2+}) compounds. Unlike lead, copper is actually more soluble in its oxidized Cu^{2+} form than in the Cu^{1+} form (Schock, Lytle, Clement 1995). Corrosion inhibitors can help to address this by forming scales that are less dependent on ORP. Phosphate inhibitors can help limit copper release by slowing oxidation of metallic and cuprous Cu^{1+} copper (Schock, Lytle, and Clement 1995). Limitations of Pourbaix diagrams include that nonequilibrium effects can be as important as the equilibrium conditions shown in the diagrams and that the sharp distinctions in phases can be misleading (AwwaRF and TZW 1996).

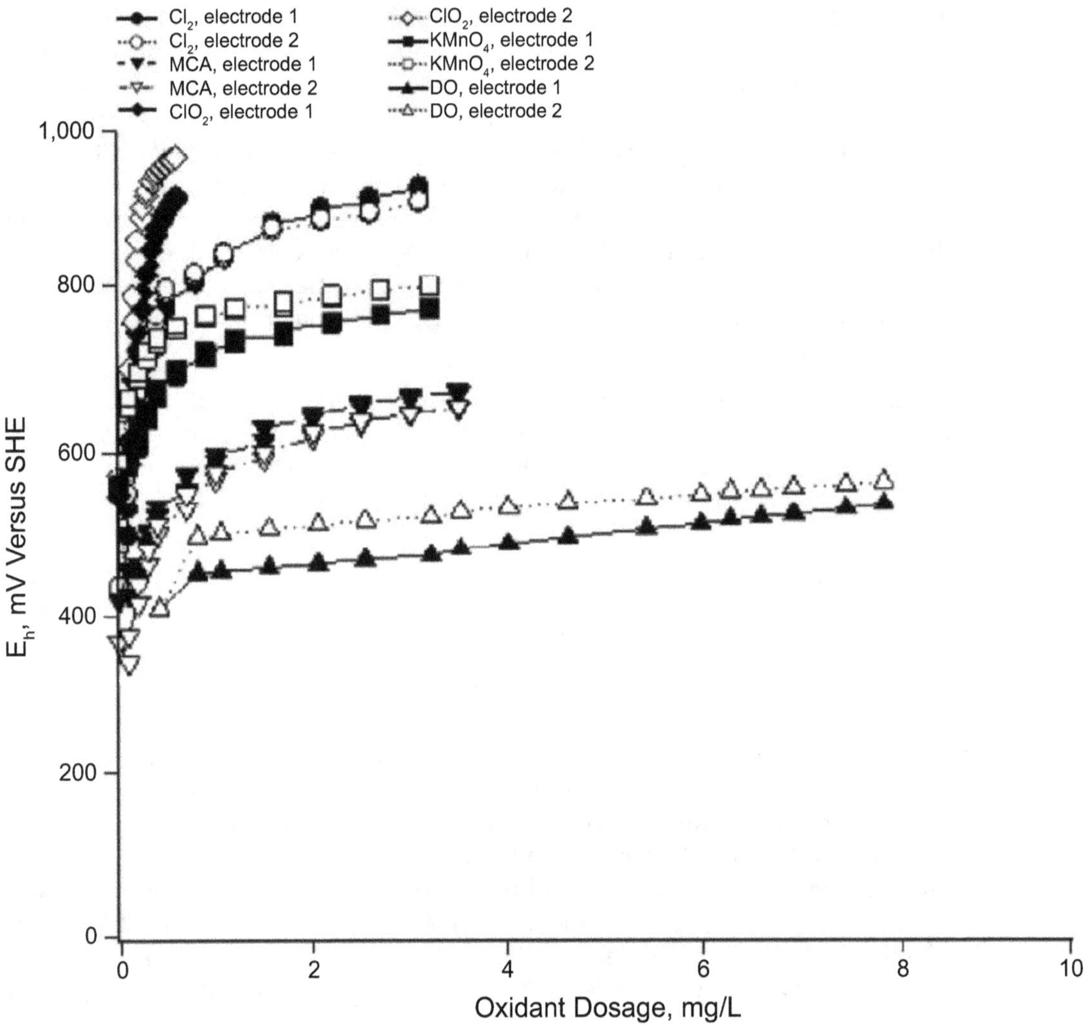

Cl_2—chlorine, ClO_2—chlorine dioxide, DO—dissolved oxygen, Eh—half-reaction electrode potential (measured in millivolts), $KMnO_4$—potassium permanganate, MCA—monochloramine, SHE—standard hydrogen electrode

Dissolved inorganic carbon = 5 mg/L C; pH = 8; temperature = 23°C

Source: Copeland and Lytle (2014).

Figure 7-2 Oxidation-reduction potential of common oxidants at various dosages

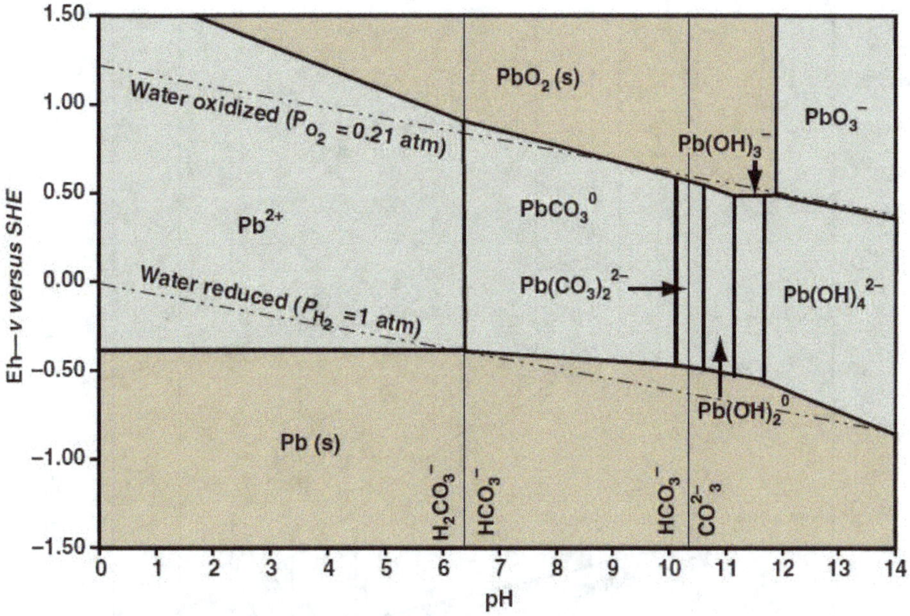

H_2CO_3—carbonic acid, HCO_3^-—bicarbonate ion, P—pressure, Pb—lead, PbO_2 (s)—lead(IV) oxide, (s)—solid, SHE—standard hydrogen electrode

Lead species = 0.015 mg/L, dissolved organic carbon = 10 mg/L C, ionic strength correction = 0

Source: Lytle and Schock 2005.

Figure 7-3 Pourbaix diagram for lead

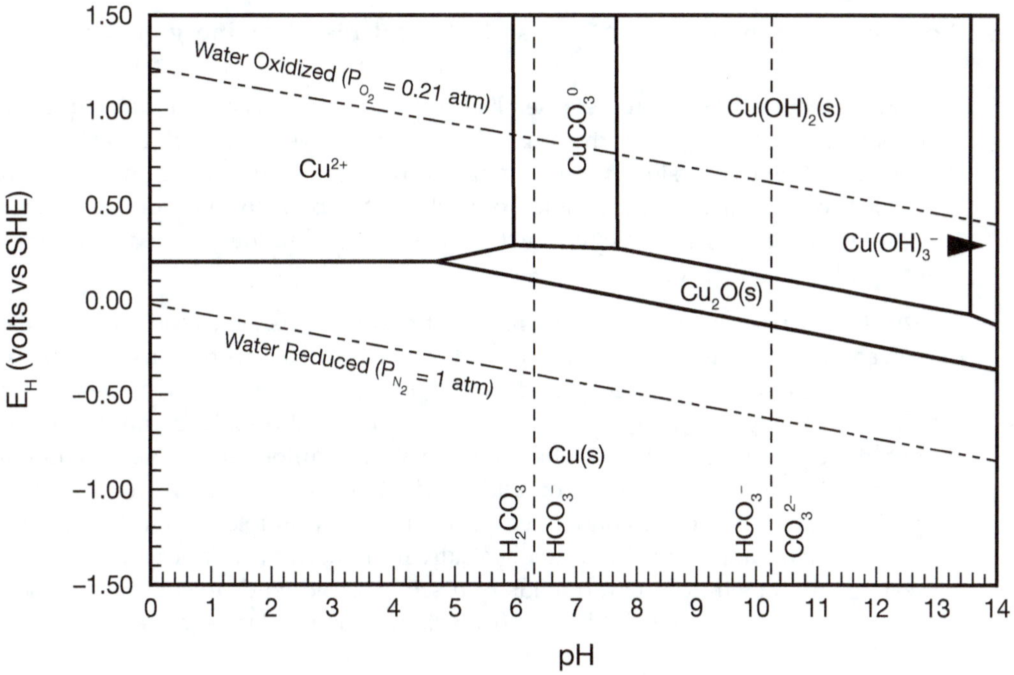

Source: Schock, Lytle, and Clement 1995.

Figure 7-4 Pourbaix diagram for copper

Dissolved inorganic carbon = 96 mg C/L, copper species concentration 1.3 mg/L. Ionic strength correction = 0. Assumes formation of cupric and cuprous hydroxide.

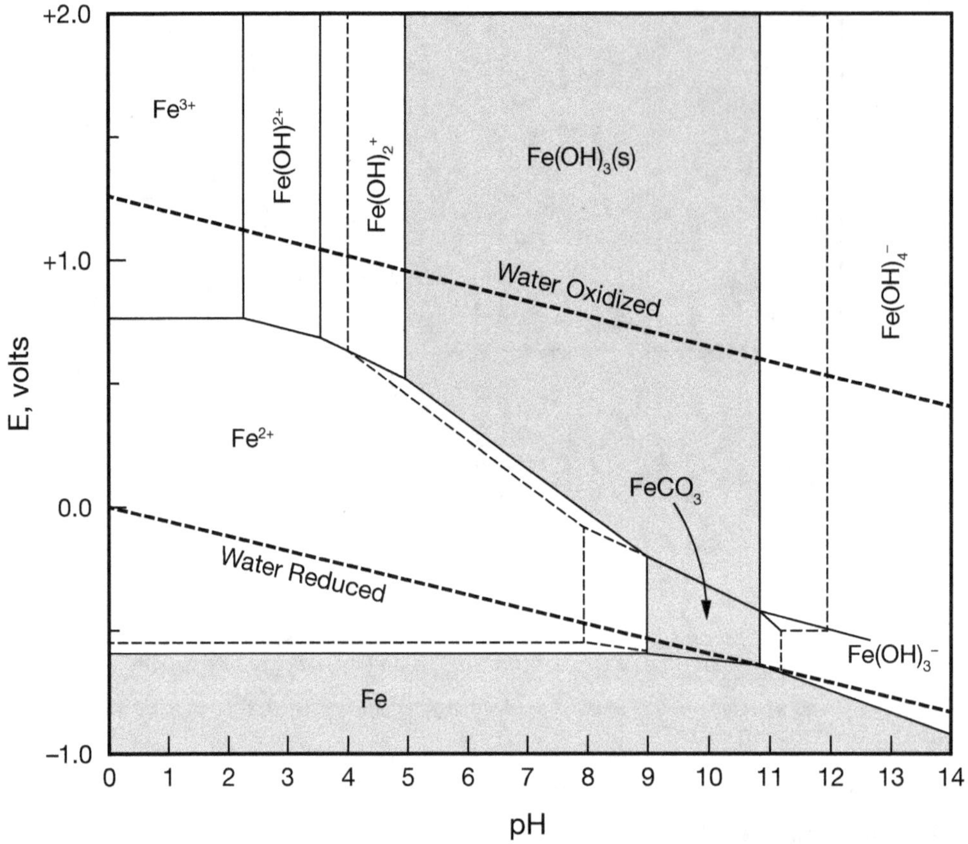

Source: Schock 1999.

Figure 7-5 Pourbaix diagram for iron at 25°C and 4.8 mg/L dissolved inorganic carbon

Metals accumulation and release. The impacts of corrosion on water quality are frequently observed long after the original distribution system and plumbing materials have corroded. Metal corrosion byproducts can accumulate within water lines through buildup of loose corrosion scales, deposition of particulates, or adsorption of dissolved ions. This buildup of metal corrosion byproducts can be released suddenly into the water if a chemical or physical disturbance occurs.

Total and dissolved metals. It is important to distinguish between total metals and dissolved metals, for example, total lead and dissolved lead. The reasons associated with the occurrence of dissolved metals, as opposed to particulate metals, may be vastly different. Similarly, it is possible to see changes in the fraction of dissolved and particulate metals but to see no change in total metals concentration. Cantor (2006) noted the importance of distinguishing between dissolved and particulate metals when trying to determine the potential source of corrosion-related problems. For example, a high fraction of particulate lead in a sample may indicate physically influenced metal release such as scouring or water hammer, while a high fraction of dissolved lead tends to suggest chemically influenced metal release, as would result from stagnation in a lead service line.

Lead. Boyd et al. (2004) simulated partial lead service line replacements with old lead pipes collected from service in New Orleans, Louisiana. They observed that particulate lead made up 38 to 54 percent of total lead and dissolved lead made up 46 to 60 percent. Cantor (2006) found that dissolved lead concentrations in Madison, Wisconsin, were low after full lead service line replacement. It was also found that total lead concentrations were erratic and often high years after replacement, presumably due to the persistence of lead-laden particulates that are flushed out over time. It was later suggested that manganese and iron served as high-capacity sinks for lead in Madison and that future mobilization of these deposits was responsible for elevated particulate lead levels after service line replacement (Schock et al. 2014).

Aluminum. Many primary coagulants, including alum and polyaluminum chloride, are aluminum based. The solubility of aluminum is very low under typical drinking water conditions, and thus very little aluminum passes through filtration. Residual aluminum will react with orthophosphate, such that orthophosphate has been considered a control strategy for residual aluminum (Frommell et al. 2004). Precipitates formed by the reaction of aluminum with phosphate can result in cloudy water in the distribution system (AWWA 2005), especially in dead-end and low-flow areas where precipitates can accumulate.

Public health and regulated trace metals. The role of corrosion scales in concentrating regulated metals that can be periodically released in high concentrations is the subject of increasing research. Since monitoring for many trace metals occurs at the entry point to the distribution system, accumulation and release mechanisms are not accounted for in the current regulatory framework. Figure 7-6 summarizes the R^2 value for trace contaminant to matrix element pairs collected during main cleaning trials at a water system in Utah. For each trace inorganic compound, the substrate element that provides the strongest (positively sloped) correlation is shaded. Manganese correlated well with most metals, indicating that it is a strong adsorber of other trace metals and is a controlling factor in metals release. These results suggest that manganese solids exert a controlling effect on lead and thallium and, to a lesser extent, zinc. Total unidentified elements (TUE) comprised most of the solids and included common elements such as oxygen, inorganic carbon, phosphorus, and sulfur.

AST SITE

Trace Contaminants	Common Matrix Elements: Fe	Mn	Ca	TUE
As	0.91	0.23	0.15	0.83*
Pb	0.30	0.90	0.00*	0.24*
Tl	0.01	0.86	0.09*	0.24*
Cu	0.83	0.46	0.13	0.77*
Zn	0.51	0.75	0.07	0.50*

**Indicates negative slope*

UPA SITE

Trace Contaminants	Common Matrix Elements: Fe	Mn	Ca	TUE
As	0.28	0.05	0.01	0.00
Pb	0.13	0.72	0.03*	0.14*
Tl	0.03	0.84	0.07*	0.23*
Cu	0.29	0.90	0.01*	0.32*
Zn	0.04	0.59	0.20*	0.26*

**Indicates negative slope*

Source: Friedman et al. 2016.

Figure 7-6 Correlations (R^2 values) between trace inorganic compound release in two water systems' distribution systems

Characterizing the DSWQ Challenge:

- Distribution and plumbing system materials
- Mechanisms of corrosion
  - Chemical, microbial, and/or physical
  - Oxidants
  - Metals accumulation and release

WATER QUALITY MONITORING

Water quality monitoring is an essential tool for identifying problems in the distribution system and for distinguishing between the different causes of internal corrosion and metal release. The following sections discuss the drivers of water quality monitoring, traditional monitoring parameters, corrosion indices, and the use of microbial parameters. More information on additional water quality parameters is available in AWWA Manual M58 (2017) and the 2016 USEPA guidance on corrosion control (2016).

Monitoring Drivers

Drivers of water quality monitoring include regulatory requirements and secondary standards, process control, and improvement in the level of service to improve aesthetics and protect public health. Aesthetic concerns are described in Chapter 5. Regulatory requirements provided here are for informational purposes only. Systems should contact their local regulatory agencies for specific regulatory guidance.

Regulatory requirements, public health, and secondary standards. The USEPA LCR is a treatment technique rule enforced by biannual monitoring of lead and copper at customers' taps. As noted previously, lead and copper action levels are 0.015 and 1.3 mg/L, respectively, for the 90th percentile concentration (40 CFR § 141.80). Under different circumstances, systems are also required to measure a specific subset of water quality parameters (WQPs) (40 CFR § 141.87). Table 7-5 provides a summary of WQP sampling requirements. Water systems must collect samples for designated WQPs at customers' taps every six months. Systems are eligible to reduce monitoring frequency at taps if they demonstrate compliance with optimized corrosion control treatment over a sustained period. WQP monitoring at the entry point(s) to the distribution system must be performed at least every two weeks. For more information on the regulatory requirements and water quality parameter monitoring, see USEPA's recent guidance on optimal corrosion control (USEPA 2016).

<H3>Process control. Many systems monitor more frequently for process control than required for regulatory purposes. Online monitoring of pH and corrosion inhibitors can aid plant operations staff in maintaining consistent quality of the water entering the distribution system and in identifying process upsets. In addition, controlling disinfectant residual parameters is also important to corrosion control. As discussed later in this section, residual disinfectant type and concentration have important impacts on corrosion control. Monitoring and minimizing free ammonia in chloraminated systems is also a crucial component of limiting distribution nitrification and associated MIC for copper corrosion and metal and particulates release. Nitrification is discussed in detail in Chapter 6.

Table 7-5 Summary of USEPA Lead and Copper Rule sampling requirements for water quality parameters

Location	Parameter	Frequency
Customers' taps	pH, alkalinity, orthophosphate,* silica,† calcium‡	Every 6 months
Locations where treated finished water enters the distribution system	pH, alkalinity dose and concentration,¶ orthophosphate dose and concentration,* silica dose and concentration†	No less frequently than every two weeks

Source: Adapted from 40 CFR § 141.87.

* When an inhibitor containing a phosphate compound is used.

† When an inhibitor containing a silicate compound is used.

‡ When calcium carbonate stabilization is used as part of corrosion control.

¶ When alkalinity is adjusted as part of corrosion control.

Monitoring Parameters

Traditional parameters. Maintaining a consistent pH is critical to corrosion control in the distribution system. Typical drinking water pH ranges between 6 and 10, although primacy agencies may require a narrower range of between 6.5 and 8.5. Most metals (e.g., iron, copper, and lead) tend to dissolve more at lower pH, so corrosion control treatment often involves increasing the pH of treated waters.

Changes to pH also affect the speciation of dissolved inorganic carbon (DIC). DIC is the sum of all inorganic carbon species, including carbonate, bicarbonate, and dissolved carbon dioxide (aq), that reacts with water to form carbonic acid. In many natural water systems, bicarbonate is the primary species that contributes to a water's alkalinity, which is defined as the amount of strong acid needed to reduce a solution to a certain pH and is often expressed in units of mg/L as $CaCO_3$ (calcium carbonate). Additionally, the USEPA's 2016 guidance document recommends using DIC to estimate saturation pH and scaling potential; this, in turn, can be used for selecting treatment process(es) for lead and/or copper corrosion control (USEPA 2016).

Measurements of alkalinity may be used to evaluate causes of corrosion, or a minimum may be required for corrosion control, given the importance of carbonate species in buffer intensity. Increasing alkalinity is one of the most effective ways to mitigate red water problems, even in the presence of significant chlorides and sulfate (Imran et al. 2005a, 2005b; Taylor et al. 2005). A more detailed discussion of the importance of pH, alkalinity, carbonate species, and buffer intensity to corrosion control can be found in Snoeyink and Wagner (1996), USEPA (2016), and AWWA Manual M58 (2017).

Additional water quality parameters. There are other water quality parameters that affect corrosion and metals release, including chloride, sulfide, sulfate, aluminum, and temperature. These are discussed below. Water quality parameters are discussed in detail in the USEPA's 2016 guidance (2016) and AWWA Manual M58 (2017).

Chloride. Increased chloride concentrations have been shown to increase iron release and red water occurrence (Larson and Skold 1957; Lytle, Sarin, and Snoeyink 2003; Imran et al. 2005a). However, it has also been suggested that copper passivates in the presence of chloride at pH levels greater than or equal to 7 (Edwards, Meyer, and Rehring 1994). Cuprous copper has been shown to be sensitive to chloride and particularly sensitive to ammonia, which both increase copper's solubility through complexation (Schock et al. 1995).

Sulfide. Sulfide has also been shown to significantly increase copper corrosion both in laboratory tests and water system case studies (Jacobs et al. 1998). Sulfides may be present in source waters or microbially mediated reactions by sulfate-reducing bacteria (MIC).

Sulfate. Although increases in sulfate concentrations have been linked to red water (Imran et al. 2005b; Taylor et al. 2005) and copper corrosion (Edwards, Meyer, and Rehring 1994), it is commonly discussed for lead in terms of chloride-to-sulfate mass ratio ($Cl^-:SO_4^{2-}$). An increase in the $Cl^-:SO_4^2$ mass ratio has been shown to increase lead release from lead plumbing (Edwards, Jacobs, and Dodrill 1999) as well as brass and solder joints (Triantafyllidou and Edwards 2006). A survey found that all water systems with $Cl^-:SO_4^{2-}$ ratios less than 0.58 met the lead action level while only 36 percent of those with ratios greater than 0.58 met the action level (Edwards, Jacobs, and Dodrill 1999).

Aluminum. Switching coagulants, such as from aluminum sulfate (alum) to ferric chloride or polyaluminum chloride, or blending new alternative water supplies into the distribution system can significantly alter the $Cl^-:SO_4^2$ ratio. Carlson et al. (2000) gave an example of a treatment plant that switched from alum to ferric chloride in order to reduce disinfection by-products. The coagulant change decreased sulfate concentrations from 74 to 34 mg/L and increased chloride concentrations from 24 to 38 mg/L. Subsequently, the number of tap samples that exceeded the lead action level increased from 0 percent when alum was used to 17 percent with ferric chloride.

Temperature. As noted by McNeill and Edwards (2001), temperature is an often overlooked component of corrosion control that influences numerous other water quality parameters. For example, DO solubility, solution viscosity (e.g., shearing), diffusion rate, chemical reaction rate, and biological activity are all temperature dependent. Typically, chemical reaction rates double for each 10°C rise in temperature. Increasing temperature has been shown to increases the corrosion rate of both aged copper and zinc (Stone et al. 1987).

Corrosion Indices

The monitoring of individual water quality parameters can provide significant insight into corrosion control. In some cases, ratios of individual water quality parameters have been used as an indicator of corrosivity. In addition to the $Cl^-:SO_4^2$ mass ratio discussed earlier, numerous other indices have been developed, including the Langelier saturation index (LSI) and the $CaCO_3$ precipitation potential (CCPP). These corrosion indices can be calculated with spreadsheet software using AWWA's Tetra Tech (RTW) model (2011).* Both LSI and CCPP have significant limitations as discussed below. No single index has universal applicability; therefore, system-specific field data should be consulted prior to use of a corrosion index to make operational decisions (Singley 1981).

Calcium carbonate indices. Many corrosion indices are based on the prediction of $CaCO_3$ scaling. The LSI is the most common corrosion index (Langelier 1939) and is calculated as follows:

$$LSI = pH_M - pH_S \qquad \text{(Eq. 7-1)}$$

Where:

> pH_M is measured pH and pH_S is saturation pH for $CaCO_3$, which is dependent on calcium and bicarbonate concentrations and activities. In concept, positive LSIs should result in the deposition of $CaCO_3$, while negative LSIs would result in the dissolution of $CaCO_3$, assuming that deposited $CaCO_3$ forms a protective scale that prevents corrosion.

* American Water Works Association (2011) Tetra Tech (RTW) Model for Water Process & Corrosion Chemistry.

LSI may be useful in predicting $CaCO_3$ precipitation, which may result in pressure loss due to friction or pipe clogging (see Figure 7-7). However, despite its wide use, LSI has failed to predict the corrosivity of water in contact with iron pipes, and studies have shown that no reliable correlation exists between the index and corrosion rates (AwwaRF and TZW 1996; AwwaRF 2002).

CCPP is often considered the most effective index in predicting $CaCO_3$ precipitation. The Tetra Tech model notes that CCPP values greater than 10 mg/L as $CaCO_3$ are undesirable. Other $CaCO_3$-based indices have been proposed to estimate corrosivity in addition to the LSI, including the aggressiveness index and the Ryznar index. However, these indices are simply reformulations of the same basic $CaCO_3$ saturation conditions that are explained by the LSI and CCPP.

Ratios of corrosive agents to inhibiting agents. Another set of indices, mostly empirical, are calculated as a ratio of corrosive ions to inhibiting ions. These indices apply to iron pipe and are useful for estimating the relative corrosivity of two similar waters or the changing corrosivity over time of a blend of water sources. Two examples of these are the Larson ratio (LR) and the $Cl^-:SO_4^{2}$ mass ratio (discussed in detail in the Traditional Monitoring Parameters section).

Courtesy of CH2M.

Figure 7-7 Calcium carbonate precipitation in a distribution main

Larson and Skold (1957, 1958) observed that the ratio of corrosive agents (i.e., chloride and sulfate) to corrosion-inhibiting agents (i.e., bicarbonate) was correlated with iron corrosion. The LR may be applicable to waters that contain dissolved solids ranging from 250 to 1,000 mg/L (the range of dissolved solids in Larson's experimental solutions). They proposed two indices; the following equation is used most often and is included in the Tetra Tech model:

$$LR = \frac{2[SO_4^{2-}] + [Cl^-]}{[HCO_3^-]} \quad \text{(Eq 7-2)}$$

where higher ratios are associated with greater iron corrosion. Empirical evidence has shown that a higher LR is correlated with red water complaints (AwwaRF 2002).

Imran et al. (2005b) introduced the following modified Larson–Skold ratio (LRM, meaning modified Larson ratio) that accounts for sodium, temperature, and stagnation:

$$LRM = \frac{\sqrt{Cl^- + SO_4^{2-} + Na^+}}{Alkalinity}\left(\frac{T}{25}\right) HRT \quad \text{(Eq 7-3)}$$

where *T* is temperature (°C) and *HRT* is hydraulic residence time (days).

One limitation of conceptual ratios of corrosive agents to corrosion-inhibiting agents for iron is that there can be unintended consequences to other metals. As discussed previously, sulfate appears to have corrosion-inhibiting, rather than corrosion-enhancing, effects for lead. For copper, Edwards et al. (1996) observed that higher bicarbonate concentrations increased copper corrosion rates in new pipe (less than a few years old) and that LR or LSI strategies would counterintuitively increase copper corrosion.

Microbial Parameters

Corrosion that can be partially attributed to biological activity is often referred to as microbiologically influenced corrosion. There are numerous examples of MIC in the literature. Lee, O'Conner, and Banerji (1980) observed that microbial growth increased cast-iron corrosion compared to stable water. MIC of copper tends to occur more often in waters that are soft, weakly buffered, have neutral or neutral-alkaline pH, and have low chlorine concentrations, often manifesting as pitting (Bremer, Webster, and Wells 2001). Lytle and colleagues (2005) found that sulfate-reducing bacteria can produce sulfide in reducing conditions within biofilms and that reduced sulfur was a significant component of iron scales in a groundwater system. Sulfide has also been linked to increased corrosion of copper (Jacobs, Reiber, and Edwards 1998). Because of the role of microbial activity in corrosion, it can be beneficial to monitor indicators of MIC.

Heterotrophic plate count. HPC analyses are used to estimate the concentration of bacteria in drinking water that are heterotrophs, meaning that they rely on nitrogen and organic carbon present in the water for food rather than producing their own. Reasoner's 2A agar (R2A agar) (Reasoner and Geldreich 1985) is the most sensitive growth medium to measure HPCs.

Areas with low disinfectant residual such as stagnant areas of distribution and or home plumbing are more likely to have elevated HPCs. High HPCs in certain areas may indicate the need for more active management of portions of the distribution system.

Conditions that encourage microbial regrowth include high concentrations of assimilable organic carbon (AOC) and other nutrients.

Assimilable organic carbon and biologically degradable organic carbon. AOC is a subset of organic carbon that is easily taken up by microbes as a food source. AOC typically makes up only 0.1 to 9.0 percent of TOC concentrations (Escobar and Randall 2001). AOC has been used as an indicator of potential regrowth (e.g., is well correlated with HPC) and is taken up by bacteria up to very low concentrations (0.010 g/L) (van der Kooij 1992). Ozonation has been shown to significantly increase AOC (Escobar and Randall 2001) and therefore is usually followed by biologically active filtration so that AOC can be reduced before entering the distribution system. However, Escobar and Randall (2001) propose that AOC be used together with analysis of biologically degradable organic carbon (BDOC) as complementary measurements of bacterial regrowth potential. Even though AOC is a fraction of BDOC, measuring only one of these parameters can potentially under- or overestimate the water's bacterial regrowth potential.

Microbial nutrients. Common microbial nutrients in drinking water include ammonia, sulfate, nitrate, and phosphorus. Nitrification occurs when free ammonia is present; therefore, residual free ammonia entering the distribution system should be minimized following chloramination. Chloramine decay can result in release of free ammonia in the distribution system. Nitrification is discussed in detail in Chapter 6.

Sulfate can serve as an electron acceptor for sulfate-reducing bacteria, producing sulfide in the process. Many corrosion inhibitors are phosphate based, such as orthophosphate. One side effect of phosphate-based inhibitors is that they serve as a nutrient source and can increase distribution system regrowth.

Nitrite. When nitrification occurs, ammonia is oxidized to nitrite by ammonia-oxidizing bacteria and then nitrite is subsequently oxidized to nitrate by nitrite-oxidizing bacteria. Nitrate is a product of nitrification but is also present in most aerobic source waters, making it difficult to use as an indicator of nitrification. Nitrite, however, is quickly oxidized to nitrate under aerobic conditions; therefore, any nitrite detected was likely formed recently prior to sampling. This makes nitrite a more direct indication of nitrification than more general growth metrics such as HPC or adenosine triphosphate (ATP). Typically, nitrite in excess of 0.05 mg/L would be considered a potential indicator of nitrification; however, it is a function of source water, so it must be determined on a site-specific basis.

Adenosine triphosphate. ATP is the basic energy carrier of almost every living cell. ATP (ASTM 2009) has gained popularity for rapid determination of biomass in biofilters (Pharand et al. 2014) and is also being used increasingly to make measurements in distribution systems. A distinct advantage of ATP tests is that they can be performed rapidly. An ATP test takes only about 15 minutes compared with 2 to 3 days for HPC culture. ATP methods are also not limited to bacteria that will grow on R2A agar, as is the case with HPC.

Characterizing the DSWQ Challenge: Water Quality Monitoring Drivers

- Applicable monitoring parameters
- Corrosion indices
- Microbial parameters

CORROSION CONTROL METHODS

Corrosion can be reduced using one of two general approaches: form a precipitate in the potable water supply that deposits onto the pipe wall to create a protective coating and cause the distribution system pipe material and the water supply to interact and form less-soluble metal compounds on the pipe surface (USEPA 1984). The former approach is not commonly practiced for corrosion control. The latter one, often called passivation, occurs where pipe material and existing scale deposits form a barrier film of less-soluble metal carbonates, phosphate, or silicate compounds on the inner pipe surface. This barrier layer of metal compounds isolates the plumbing materials from the water supply and minimizes dissolution.

In many cases, the easiest way to reduce corrosion is to modify the water quality at the treatment plant or well house. Chemical treatment options include pH adjustment, DIC adjustment, and application of corrosion inhibitors. USEPA recommendations (USEPA 2016) include several sets of flowcharts to help in the selection of viable chemical treatment options that are based on water quality characteristics. More than one chemically viable treatment option may be available to a water system.

It is important to note that the selection of a corrosion control technique may not be a one-time event. Utilities are encouraged to reevaluate corrosion control treatment whenever changes are encountered in the source water or when contemplating treatment changes such as changes to coagulant type or secondary disinfectant. AWWA Manual M58 (AWWA 2017) further outlines the impact on corrosion from changes to source of supply, water treatment, and distribution system activities.

Adjustment of pH and/or Alkalinity

The adjustment of pH is the most common method for reducing corrosion in distribution systems. pH is the primary factor that determines the solubility of most pipe materials and the films that form from corrosion by-products. Most materials in the distribution system dissolve faster at lower pH.

The relationships between pH, DIC, alkalinity, carbon dioxide, and ionic strength govern the solubility of metal carbonates, which are commonly involved in protective scale formation on interior pipe surfaces. The type of alkalinity present and the pH of the water are dependent on each other.

The solubility of metals typically decreases as pH increases. Increased pH may, however, prove troublesome for water that contains elevated levels of iron and manganese, which could cause discolored water, and elevated hardness, which can cause excessive calcium and magnesium precipitation. Excessive calcium and magnesium can cause increased pipe friction and pipe clogging.

Caustic soda (sodium hydroxide), quicklime, potassium hydroxide, soda ash, limestone contactors (calcite filters), and aeration/air stripping (where sufficient dissolved carbon dioxide exists in groundwater) are the principal methods for increasing pH. Soda ash, potash, and limestone contactors also increase DIC; aeration decreases DIC in the process of increasing pH. AWWA Manual M58 (AWWA 2017) and USEPA guidelines (USEPA 2016) provide extensive information comparing the relative use of chemical additives for pH and/or alkalinity adjustment.

Chemical Inhibitors

Corrosion control chemicals are commonly used in the drinking water industry. However, each chemical has different properties and may not truly inhibit corrosion. Some chemicals simply mask the effects of corrosion to prevent aesthetic problems such as red or black water. Currently, the most common inhibitors used in the drinking water industry are orthophosphates, polyphosphates, blended phosphates, and silicates.

Orthophosphates. Orthophosphate is most commonly used for lead and copper control. Yet, it has also been shown to reduce the corrosion rates of iron (AwwaRF and TZW 1996). Orthophosphate is significantly less expensive than all other inhibitors, and is available as phosphoric acid. There are also neutral orthophosphate products such as potassium and zinc orthophosphate.

The optimal pH range is between 7.0 to 8.0 for controlling lead and copper. Pipe loop studies have demonstrated an immediate increase in lead concentrations whenever the pH falls out of this range. At pH values around 8.0 where copper release is diminished because of the pH increase, orthophosphate addition shows no additional benefit (AwwaRF and TZW 1996; Schock and Sandvig 2006). Orthophosphate addition has been shown to be an effective copper control measure even in challenging distribution systems with seasonally varying blends of multiple source waters such as groundwater, surface water, and desalinated seawater (Taylor et al. 2008). Yet, in some cases, orthophosphate may be effective for controlling corrosion at pH as high as 9 (USEPA 2016).

Typical dosages of orthophosphate range from 0.5 to 3 mg/L as PO_4. In some cases, higher orthophosphate concentrations (>3.0 mg/L) have been found to form a precipitate when in contact with aluminum, calcium, and iron that can cause complaints of milky water at the customer tap (Tesfai et al. 2006; Fromell 2004). In addition, higher phosphate doses may cause issues with wastewater discharge limits.

Polyphosphates. In general, polyphosphates are used as a sequestering agent for iron and manganese in source waters and are not generally used for corrosion control. Polyphosphates are best used to sequester metals in order to prevent red or black water where lead and copper are not already significant. Low doses of polyphosphate (2 to 4 mg/L) are typically used and will help reduce the red water as dissolved iron and manganese are bound as a complex molecule. Corrosion rates, however, are not reduced and may actually increase (Cantor et al. 2000). Polyphosphates should only be considered when it is known that lead and copper levels will not substantially increase (Clement 2002).

Blended phosphates. Blended phosphates are proprietary mixtures of ortho- and polyphosphate. The fraction of orthophosphate in blended formulas ranges from 5 to 70 percent. While it is possible for blended phosphates to sequester metals and reduce metals release, the exact mechanism of protection is not well understood and blends should be used with caution (USEPA 2016).

Silicates. The protective mechanism of silicates appears to be the formation of a thin layer over the corroded metal layer. The films are self-limiting and do not build up in thicker layers with increased dosages. The films also break down relatively quickly if there is an interruption in feed.

Silicates are expensive and typically cost more than phosphates. In general, silicates have been found to be less effective for corrosion control than orthophosphate. The use of phosphate–silicate blends has increased in recent years. The blends appear to be as good as phosphate products alone but are used at lower concentrations. It is important to note that more research on the use of silicates for control corrosion is required. The high costs associated with silicates and the time required to lessen corrosion in the distribution system have eliminated the use of silicates as a possible corrosion treatment for many water systems.

Chlorine Oxidation

Chlorine oxidation has a positive impact on lead scales. In a high oxidation state (caused by the addition of high chlorine), lead can form an oxide called plattnerite. This compound is highly insoluble in water and can form the desired barrier of solids on metal surfaces.

The phenomenon has been seen to greatly slow the uniform corrosion of lead, with resultant low lead concentrations in drinking water (Lytle and Schock 2005). However, a high oxidation caused by chlorine is a major factor in shifting young copper hydroxide scales into a soluble phase (Schock and Lytle 2014). With this, researchers conclude that any chlorinated groundwater system is prone to excessive copper release by uniform corrosion. Additionally, "questions remain as to how systems and primacy agencies can ensure that disinfectant residuals required for the formation and maintenance of [these] scales are maintained within lead service lines throughout the distribution system and to the customer's taps" (USEPA 2016). Therefore, the use of chlorine oxidation as a corrosion control treatment technique is not recommended at this time.

Responding to the DSWQ Challenge:

- Adjustment of pH and/or alkalinity
- Addition of chemical inhibitors
  - Orthophosphates
  - Polyphosphates
  - Blended phosphates
  - Silicates

BEST PRACTICES FOR CORROSION CONTROL THROUGH DISTRIBUTION SYSTEM DESIGN, OPERATION, AND MAINTENANCE

Implementing and maintaining optimum corrosion control treatment requires appropriate quality of water leaving the treatment plant and the ability to maintain that quality throughout the distribution system. Many factors can alter water quality, making the system susceptible to corrosion. These factors include long detention times, tuberculated pipe scales that can provide a haven for microbes that can alter water chemistry, and velocity that is too high or too low. The most common best practice techniques to minimize corrosion include:

- Distribution system design considerations,
- Developing a corrosion control action plan
- Routine distribution system operations and maintenance

Distribution System Design Considerations

A water system can minimize distribution system pipe corrosion by selecting the correct distribution system materials and having a good engineering design. The reality is, however, that most water systems are forced to deal with existing distribution systems that were designed based on priorities other than corrosion, such as sufficient fire protection. Nonetheless, water systems can take advantage of system design for corrosion control during system upgrades, planned system materials replacement, and expansion of the distribution system.

System and pipe materials. Piping material should be selected based on the system's water chemistry and cost. The piping material's mechanical properties will also influence material selection (see Table 7-3). Comparable materials should be used throughout the system. Galvanic corrosion should be minimized by not having two materials with different electrical potentials, such as copper and steel pipe, come in direct contact. The placement of dielectric (insulating) couplings between dissimilar metals reduces galvanic corrosion.

System layout. The design and layout of pipes and structures in the distribution system can be as important as the selection of materials. Some important design-related steps that can be taken include the following:

- Avoid dead ends and stagnant areas as these can increase particulate accumulation and biological activity. Provide a looped design in order to reduce stagnation and particulate accumulation. See Chapter 2 for discussion of water age issues associated with dead ends.
- Plan for asset types and locations that support effective main flushing and cleaning. Refer to Appendix B for more information on flushing techniques.
- Consider hydraulic modeling to reduce mechanical stresses, such as flexing of pipes, water hammer, and change in the direction of flow, as these can disrupt pipe scales. Chapter 9 discusses pressure management in more detail.
- Reduce sharp turns and elbows to avoid erosion corrosion.

Developing a Corrosion Control Action Plan

The LCR requires water quality parameter monitoring in the distribution system. The purpose of this monitoring is to ensure that distribution system water quality is conducive to optimal corrosion control. However, monitoring of only pH, alkalinity, temperature, and silicate or orthophosphate (if adding a corrosion inhibitor) is required. As previously mentioned in this chapter, many additional parameters can impact corrosion, and these parameters should be monitored to ensure that uniform and nonuniform corrosion is optimized. USEPA recommendations provide additional parameters for systems to consider (USEPA 2016). In response to water quality parameters that fall outside of target ranges, water systems may develop standard operating procedures that can be followed when performing maintenance activities.

Selecting water quality parameters. To ensure optimal corrosion control, a monitoring strategy must be developed. This strategy begins by grouping mechanisms of metal release into the following three categories of factors that affect corrosion in distribution systems:

- Those related to by-products of uniform corrosion of metals,
- Those related to the biostability and microbiological activity in the water, and
- Those related to metal release by accumulation and transport.

Water quality characteristics that can be used to quantify and track trends in these categories are listed in Table 7-6. Some water quality parameters are helpful in describing more than one category of water quality–shaping factor.

Table 7-6 Water quality characteristics related to corrosion factors

By-Product of Uniform Corrosion of Metals	Biostability	Other Inorganics That Can Accumulate and Transport Lead and Copper
Total alkalinity	Free chlorine (if used)	Turbidity
Total hardness	Monochloramine (if used)	Conductivity
Total dissolved solids	Total chlorine	Total and dissolved iron
Conductivity	Assimilable organic carbon and/or total and dissolved organic carbon	Total and dissolved manganese
pH	Ammonia, nitrate, nitrite, total nitrogen	Total and dissolved aluminum
Temperature	Total phosphorus	Metals scan for total and dissolved metals (optional); list will vary based on laboratory and water system characteristics
Dissolved oxygen	Heterotrophic plate count	Color
Chloride	Adenosine triphosphate	Total and dissolved lead
Sulfate		Total and dissolved copper
Oxidation-reduction potential		
Orthophosphate		
Silicate		

Water quality parameters selected from Table 7-6 for a specific water system's monitoring plan should be justified based on the known source water characteristics and treatment chemicals used at the treatment plant. Some water systems may require other parameters not listed here. The objective is to prepare a list of possible water quality parameters that will aid in describing the three main categories of factors that shape a specific distribution system's water quality.

Sampling locations. Water systems may currently have distribution sampling points where they collect coliform bacteria samples and monitor disinfectant residuals. These are typically set as representative of the water system geographically and hydraulically.

Water systems may also use or add sampling sites where extremes of water quality may be found. A water distribution system changes over geographical location and over time; it can become difficult to obtain water samples that will be representative of the system's true nature. On one extreme, the freshest water with the highest concentrations of treatment chemicals can be found at the entry points to the distribution system. The effectiveness of the treatment chemicals is assumed to be at a maximum at these points. Opposite to the entry points of a distribution system are locations of high water age. These locations occur at the farthest reaches of a distribution system, dead ends, and in areas of low water usage. Many water quality issues in the distribution system, for example, lead and copper release, disinfectant by-product formation, microbial growth, and discolored water, occur by the time water reaches high water age locations. A strategic monitoring plan can be used to address these issues because interrelationships exist among these them.

Multilevel water quality monitoring. Some US water systems have incorporated routine corrosion monitoring at coliform sampling sites. This strategy involves measuring some key corrosion parameters, including pH, turbidity or apparent color, and orthophosphate (if adding orthophosphate), at the same time and location where coliform samples are collected for compliance with the RTCR. These parameters would also be measured routinely at the entry point to verify uniform corrosion treatment. Theoretically, entry point monitoring ensures control of chemical corrosion treatment at the plant, while distribution system monitoring identifies localized corrosion issues.

This technique does not require extra travel time, as the water quality technician is already collecting total coliform samples; however, this technique does lengthen the amount of time the sample collector must spend at each site. If the corrosion parameters fall outside of target ranges, systems may consider follow-up monitoring with more decisive and system-specific parameters. Some of the parameters that have the most impact on the distribution system are listed in Table 7-6. If two parameters are outside their range during follow-up monitoring, systems may wish to take follow-up action(s). For example, low pH, elevated nitrite, and ATP are indications of MIC, and elevated iron and color are indications of metal accumulation in the main. Table 7-7 lists initial and system-specific triggered water quality parameters, chosen from Table 7-6, for a large surface water system that adds alum, orthophosphate, and chloramine and also has hard water and a high percentage of cast-iron mains.

Depending on the issue, follow-up actions could involve localized unidirectional flushing or valve investigations (to determine if a valve was inadvertently closed). If parameters continue to be outside target ranges, further actions such as replacing and rehabilitating mains, installing chlorine booster stations, minimizing water age, and ensuring proper water velocities could be considered.

Pipe loop and coupon distribution monitoring stations. Mini pipe loops (Figure 7-8) placed at the entry point and in the distribution system can be used to assess metals release. Pipe loops at the entry point are used to assess corrosion treatment, while pipe loops in the distribution system are used to assess localized distribution corrosion. Additional metal coupons, such as brass and iron, can be placed inside the loops to determine the corrosion rates of these metals.

Table 7-7 Multilevel water quality parameters for a water system that uses chloramine and a phosphate-based inhibitor

Initial	Triggered
pH	pH
Orthophosphate (total)	Orthophosphate (total)
Total chlorine	Orthophosphate (dissolved)*
Color (apparent) or turbidity	Total chlorine
	Color (apparent)
	Iron (total)
	Calcium precipitate†
	Aluminum (total)
	Nitrite
	Adenosine triphosphate

* Use of a 0.45-micron filter.

† Calcium precipitation = total calcium hardness − dissolved calcium hardness.

Figure 7-8 Mini pipe loops

Old pipes harvested from water distribution systems as well as new pipes can be used in pipe loops. However, there are important limitations that must be considered when evaluating data from these apparatuses. When using new pipes, the role of pipe films and scales on existing system piping is not captured. However, when harvested pipes are used, there is no control over how representative each pipe may or may not be of the distribution system. In addition, metal particulates from the debris on the old pipe walls is disturbed and may interfere with the metal concentration data for a long period of time (AWWA 2017). Pipe loops should have the capability to simulate water movement and stagnation, similar to that in a typical home, and should contain enough metallic pipe material to draw a 1-L sample from the pipe.

Corrosion coupon (metal strips or plates) monitoring stations are more condensed and easier to use then pipe loops and can provide a better assessment of biofilm formation (Figure 7-9).

Process control charts. Another approach that takes advantage of RTCR sampling sites (and other distribution system regulatory sites) is to plot current and historic disinfection data using process control charts, also known as Shewhart charts. The data can be plotted over time for each RTCR site. Locations and times when disinfection concentrations are low indicate times when disinfection could be taken up by interaction with metal corrosion debris and with microbiological populations. Some sites indicate historically low disinfection, while others indicate historically high disinfection. Process control charts can be used to identify the norm and variability of disinfection concentrations.

Figure 7-9 Metal plates inside and stacked in open test chambers

Figure 7-10 shows process control charts for disinfection concentrations at two RTCR sites in one water system. Site D1 has a disinfection concentration that is sufficiently high with little variation; site D2 experienced a period when the disinfection concentration was inadequate but then suddenly increased to a high level. If water system personnel can determine why the disinfection concentrations vary at each site, water system operations can be changed to achieve a desired average concentration with little variation. Any action to remedy low disinfection concentrations can be tracked for effectiveness by continuing to study the disinfection trends at each site. This method can also be used with other water quality parameters such as pH and orthophosphate, if taken routinely. Details on how to create a water system–based process control chart for corrosion control can be found in chapter 6 of AWWA Manual (AWWA2017). Additionally, as part of a Water Research Foundation project, a tool was developed to assist systems in creating Shewhart control charts (Cantor et al. 2012).

Distribution System Operations and Maintenance

Routine system maintenance and actions in response to monitoring triggers can reduce corrosion and/or remove the by-products of corrosion. These actions include maintaining disinfectant residuals, controlling hydraulic detention times, cleaning pipe, and rehabilitating pipe.

Maintaining disinfectant residuals. Disinfectant residual concentrations decay as water passes through the distribution system because of reactions with materials present in the bulk water and at the pipe wall. At the treatment plant, long-lasting disinfectant residual can be promoted by removing chlorine-demanding material, particularly natural organic matter. Good control of pH and alkalinity can also result in low bulk decay for disinfectant residual. Once the finished water has left the treatment plant, decay is inevitable. Disinfection is discussed in depth in Chapter 8.

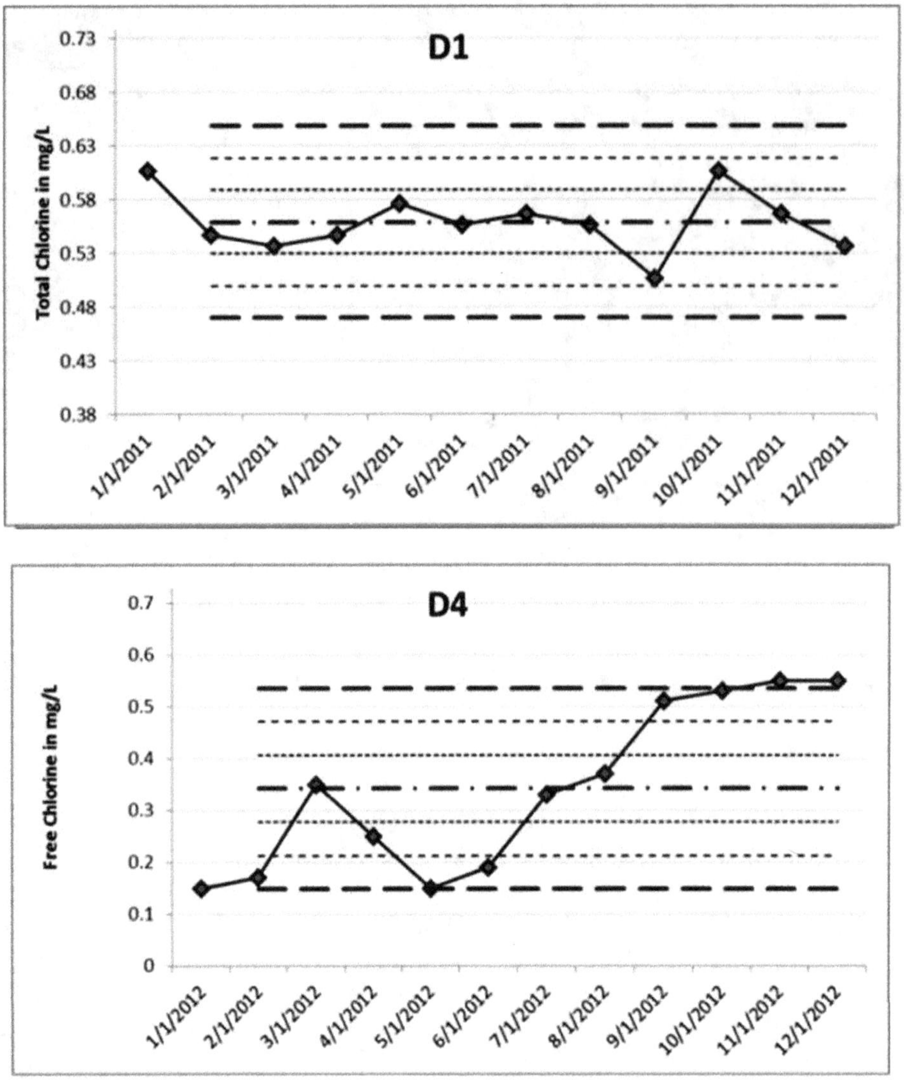

Figure 7-10 Process control charts for historical disinfection data at two Revised Total Coliform Rule sampling sites

Disinfectant-demanding materials present in the pipes include corrosion products and biofilm materials. In a synergistic way, cleaning out the accumulation of corrosion materials and biofilms will help to maintain disinfectant residuals, which in turn will help to manage corrosion and biofilm growth. While it is preferable to remove as much chlorine-demanding material from the water and pipe as possible, it is sometimes the case that disinfectant booster stations are needed to maintain adequate disinfectant at the farthest ends of the distribution system.

Controlling hydraulic detention times. Control of hydraulic detention time will help maintain disinfectant residual and reduce corrosion rates. Several operational strategies can be implemented to reduce hydraulic detention time. One of the most important operational issues to address is storage tank turnover. Most distribution systems include a large quantity of stored water, and many customers receive water that has passed through several storage tanks. Good tank turnover will refresh the volume of water and ensure that stagnation does not occur. Tank turnover is especially important when water temperature increases above 60ºF, which favors accelerated microbial activity (LeChevallier et al. 2010). In addition to volumetric cycling, alterations to inlet and outlet geometry as well as installation of active mixers can reduce stagnation in storage tanks. Chapter 2 discusses water age and capacity.

Hydraulic detention times are highly site specific; therefore, hydraulic analyses of the distribution system is recommended to identify areas of low tank turnover and low velocity such as dead ends. Stagnant areas can also be created within a distribution system at the boundary created by pumping or flow from different pressure zone entry points. In such a case, modifications to the zone supply strategies in order to alternate sources or variations of flow can help to reduce the extent of stagnation.

Pipe cleaning. The benefits of maintaining clean pipes include enhanced disinfectant residual stability, reduced bacterial levels, and improved aesthetic water quality as reflected by customer reactions. These outcomes are typically cited by water systems as the primary drivers for cleaning main. However, recent research that focused on pipe–water interactions, both within the distribution system and in premise plumbing systems, has shed additional light on the wide-ranging impacts posed by deposits. The research also demonstrated that effective routine main cleaning plays a vital role in controlling internal corrosion in premise plumbing, limiting metals release, and protecting the integrity of system infrastructure (Hill et al. 2014; Schock et al. 2014). Care should be taken, however, as pipe cleaning may have unwanted side effects, as described in Chapter 4.

Pipe rehabilitation. One technique to keep corrosive water away from pipe walls is to line the wall with a protective coating. With new pipe, these linings are usually applied during the pipe manufacturing process but can also be applied in the field when the pipe is installed. Some linings can be applied even after the pipe is in service.

Linings can also be installed to provide structural integrity or to improve a pipeline that has been in service for a long of time. Linings are categorized using structural classifications I through IV, with class I linings being considered nonstructural coatings and class IV linings as structurally independent. These classifications are further defined in AWWA Manual M28, appendix A (AWWA 2014).

The most common pipe linings are coal–tar enamels, epoxy paint, cement–mortar, and polyethylene. It is common practice to reline cleaned pipe to protect the new exposed pipeline material (Figures 7-11 and Figure 7-12).

SUMMARY

Corrosion control in distribution systems can be complex. Many factors influence uniform corrosion (throughout the entire distribution system) and nonuniform (localized) corrosion. It is extremely important to select the proper corrosion treatment based on the source water quality and distribution pipe material. It is just as important to maintain the quality stability of water leaving the treatment plant and throughout the entire distribution system. Water systems are encouraged to reevaluate corrosion control treatment when considering changes in source water and/or treatment to ensure distribution system water quality.

Table 7-8 provides a summary of best practices described in this chapter. The actions are based on the following two levels of service:

- Basic: actions water systems should perform if they are experiencing water quality issues due to corrosion and
- Advanced: actions water systems should perform if they wish to optimize and/or develop a comprehensive program related to corrosion control.

For a more complete guide on how to select corrosion treatment, assess corrosion control, and maintain corrosion control throughout the distribution system, refer to AWWA Manual M58 (AWWA 2017).

Source: USACERL Technical Report 99/39.

Figure 7-11 Example of pipe before and after cleaning and lining with cement–mortar

Figure 7-12 Steel pipe lined with epoxy

Table 7-8 Recommended practices to address corrosion issues

Best Practice	Level of Service
Distribution system design considerations	
• System and pipe materials	Basic
• System layout	Basic/Advanced
Corrosion control action plan	
• Water quality parameters	Basic
• Sampling locations	Basic/Advanced
• Multilevel-based water quality monitoring	Advanced
• Pipe loop and coupon distribution monitoring stations	Advanced
• Process control charts	Basic/Advanced
System distribution operations and maintenance	
• Maintaining disinfectant residuals	Basic
• Controlling hydraulic detention times	Basic
• Cleaning pipe	Basic/Advanced
• Rehabilitating pipe	Advanced

REFERENCES

ASTM International. 2009. ASTM D4012-81, *Standard Test Method for Adenosine Triphosphate (ATP) Content of Microorganisms in Water.* West Conshohocken, PA: ASTM International.

AWWA (American Water Works Association). 2005. *Managing Change and Unintended Consequences: Lead and Copper Rule Corrosion Control Treatment.* Denver, CO: AWWA.

———. 2014. M28, *Rehabilitation of Water Mains.* Denver, CO: AWWA.

———. 2017. M58, *Internal Corrosion Control in Water Distribution Systems.* Denver, CO: AWWA.

Clement, J., M. Hayes, P. Sarin, W.M. Kriven, J. Bebee, K. Jim, M. Becket, V.L. Snoeyink, G.J. Kirmeyer, and G. Pierson. 2002. Development of Red Water Control Strategies. Denver, Colo.: AwwaRF and AWWA.

AwwaRF and TZW (American Water Works Association Research Foundation and DVGW-Technologiezentrum Wasser. 1996 (2nd ed.). *Internal Corrosion of Water Distribution Systems.* Denver, CO: AwwaRF and AWWA.

Boyd, G.R.; P. Shetty; A.M. Sandvig; & G.L. Pierson. 2004. Pb in Tap Water Following Simulated Partial Lead Pipe Replacements. *Journal of Environmental Engineering,* 130:10:1188–1197.

Bremer, P.J.; B.J. Webster; & D.B. Wells. 2001. Biocorrosion of Copper in Potable Water. *Journal of the American Water Works Association,* 93:8:82–91.

Cantor, A.F. 2006. Diagnosing Corrosion Problems Through Differentiation of Metal Fractions. *Journal of the American Water Works Association,* 98:1:117–126.

Cantor, A.F.; D. Denig-Chakroff; R.R. Vela; M.G. Oleinik; & D.L. Lynch. 2000. Use of Polyphosphate in Corrosion Control. *Journal of the American Water Works Association*, 92:2:95.

Cantor, A.F.; E. Klefer; K. Little; A. Jacque; A. Degnan; B. Maynard; D. Mast; & J. Cantor. 2012. *Distribution System Water Quality Control Demonstration.* Jointly sponsored by the Water Research Foundation, Denver, CO, and North Shore Water Commission, Glendale, WI.

Carlson, K.; S. Via; B. Bellamy; & M. Carlson. 2000. Secondary Effects of Enhanced Coagulation and Softening. *Journal of the American Water Works Association*, 92:6:63–75.

Clement, J.; M. Hayes; P. Sarin; V. Snoeyink; W. Kriven; J. Bebee; K. Jim; M. Beckett; G. Kirmeyer; & G. Pierson. 2002. *Development of Red Water Control Strategies*. Denver, CO: AwwaRF.

Cohen, J.M.; Kamphake, L.J.; Harris, E.K.; & Woodward, Richard L.. 1960. Taste Threshold Concentrations of Metals in Drinking Water. *Journal of the American Water Works Association*, 52:5:660–670.

Copeland, A. & D.A. Lytle. 2014. Measuring the Oxidation-Reduction Potential of Important Oxidants in Drinking Water. *Journal of the American Water Works Association,*, 106:1:10–20.

Cuppett, J.D.; S.E. Duncan; & A.M. Dietrich. 2006. Evaluation of Copper Speciation and Water Quality Factors That Affect Aqueous Copper Tasting Response. *Chemical Senses*, 31:7:689–97.

Dangel, R.A. 1975. Study of Corrosion Products in the Seattle Water Department Distribution System. Report No. 670/2-75-036. Cincinnati, OH: USEPA.

Edwards, M. & A. Dudi. 2004. Role of Chlorine and Chloramine in Corrosion of Lead-Bearing Plumbing Materials. *Journal of the American Water Works Association*, 96:10:69–81.

Edwards, M.; S. Jacobs; & D. Dodrill. 1999. Desktop Guidance for Mitigating Pb and Cu Corrosion By-Products. *Journal of the American Water Works Association*, 91:5:66–77.

Edwards, M.; T. Meyer; & J. Rehring. 1994. Effect of Selected Anions on Copper Corrosion Rates. *Journal of the American Water Works Association*, 86:12:73–81.

Escobar, I.C. & A.A. Randall. 2001. Case Study: Ozonation and Distribution System Biostability. *Journal of the American Water Works Association*, 93:10:77–89.

European Union Council. 1998. *Council Directive 98/83/EC of 3 November 1998 on the Quality of Water Intended for Human Consumption.* Accessed May 16, 2016. http://ec.europa.eu/environment/water/water-drink/legislation_en.html

Federal Register. 1989a. National Primary Drinking Water Regulations; Filtration, Disinfection; Turbidity, *Giardia lamblia*, Viruses, *Legionella*, and Heterotrophic Bacteria, Final Rule. 54:124:27486–27567.

———. 1989b. National Primary Drinking Water Regulations; Total Coliforms (Including Fecal Coliform and *E. coli*); Final Rule. 54:124::27544–27568.

———. 1991. National Primary Drinking Water Regulations; Lead and Copper; Final Rule. 56:110:26460.

———. 2007. National Primary Drinking Water Regulations for Lead and Copper: Short-Term Regulatory Revisions and Clarifications. 72:195::57781–57820.

Friedman, M.; A. Hill; S. Booth; M. Hallett; L. McNeill; J. McLean; D. Stevens; D. Sorensen; T. Hammer; W. Kent; M. De Haan; K. MacArthur; & K. Mitchell. 2016. *Metals Accumulation and Release within the Distribution System: Evaluation and Mitigation*. Denver, CO: Water Research Foundation.

Frommell, D.M.; C.M. Feld; V.L. Snoeyink; B. Melcher; & C. Feizoulof. 2004. Aluminum Residual Control Using Orthophosphate. *Journal of the American Water Works Association*, 96:9:99–109.

Health Canada. 1992. Guidelines for Canadian Drinking Water Quality: Guideline Technical Document—Copper. Ottawa, Ontario: Water Quality and Health Bureau, Healthy Environments and Consumer Safety Branch.

———. 2009. Guidance on Controlling Corrosion in Drinking Water Distribution Systems. Ottawa, Ontario: Water, Air and Climate Change Bureau, Healthy Environments and Consumer Safety Branch.

———.2013. Risk Management Strategy for Lead. www.hc-sc.gc.ca/ewh-semt/pubs /contaminants/prms_lead- psgr_plomb/index-eng.php. Accessed May 25, 2017.

———. 2014. Guidelines for Canadian Drinking Water Quality—Summary Table. Ottawa, Ontario: Water and Air Quality Bureau, Healthy Environments and Consumer Safety Branch.

———. 2015. Guidelines for Canadian Drinking Water Quality—Summary Table. www.hc-sc.gc.ca/ewh-semt/pubs/water-eau/sum_guide-res_recom/index-engphp#t Accessed May 25, 2017.

Hill, A.; M. Johnson; T. Brewer; & R. Moore. 2014. Recognizing the Distribution System Asset/Water Quality Management Nexus. *Opflow*, 40:12:10.

Imran, S.A.; J.D. Dietz; G. Mutoti; J.S. Taylor; A.A. Randall; & C.D. Cooper. 2005a. Red Water Release in Drinking Water Distribution Systems. *Journal of the American Water Works Association*, 97:9:93–100.

———. 2005b. Modified Larsons Ratio Incorporating Temperature, Water Age, and Electroneutrality Effects on Red Water Release. *Journal of Environmental Engineering*, 131:11:1514–1520.

Jacobs, S.; S. Reiber; & M. Edwards. 1998. Sulfide-Induced Copper Corrosion. *Journal of the American Water Works Association*, 90:7:62–73.

JMM (J.M. Montgomery, Consulting Engineering Inc.). 1982. *Internal Corrosion Mitigation Study*. Final Report, prepared for the Portland, Oregon Water Bureau.

Khiari, D.; S. Barrett; R. Chinn; A. Bruchet; P. Piriou; L. Matia; F. Ventura; I. Suffet; T. Gittelman; & P. Leutweiler. 2002. *Distribution Generated Taste and Odor Phenomena*. Denver, CO: AwwaRF and AWWA.

Kirmeyer, G.J.; M. Friedman; & J. Clement. 2000. *Guidance Manual for Maintaining Distribution System Water Quality*. Denver, CO: AwwaRF and AWWA.

Langelier, W.F. 1939. The Analytical Control of Anticorrosion Water Treatment. *Journal of the American Water Works Association*, 28:10:1500–1521.

Larson, T.E. & Skold, R.V. 1957. Corrosion and Tuberculation of Cast Iron. *Journal of the American Water Works Association*, 49:10:1294–1302.

LeChevallier M.W.; M.C. Besner; M. Friedman; & V.L. Speight. 2010 (6th ed.). Microbiological Quality Control in Distribution Systems. In *Water Quality and Treatment* (J.K. Edzwald, editor). Denver, CO: AWWA.

LeChevallier, M.W; C.D. Cawthon; & R.G. Lee. 1988. Inactivation of Biofilm Bacteria. *Applied and Environmental Microbiology*, 54:10:2492–2499.

Lee, S.H.; J.T. O'Conner; & S.K. Banerji. 1980. Biologically Mediated Corrosion and Its Effects on Water Quality in the Distribution System. *Journal of the American Water Works Association*, 72:11:636.

Lytle, D.A.; T. Gerke; & J.B Maynard. 2005. Effect of Bacterial Sulfate Reduction on Iron-Corrosion Scales. *Journal of the American Water Works Association*, 97:10:109–120.

Lytle, D.A.; P. Sarin; & V.L. Snoeyink. 2003. The Effect of Chloride and Orthophosphate on the Release of Iron From Drinking Water Distribution System Cast Iron Pipe. Proceedings AWWA Water Quality Technology Conference, city, state Philadelphia, PA

Lytle, D.A. & M.R. Schock. 2005. Formation of Pb(IV) Oxides in Chlorinated Water. *Journal of the American Water Works Association*, 97:11:102–114.

McNeill, L.S. & M. Edwards. 2001. Iron Pipe Corrosion in Distribution Systems. *Journal of the American Water Works Association*, 93:7:88–100.

NHMRC (National Health and Medical Research Council). 2011. *Australian Drinking Water Guidelines*. Paper 6,National Water Quality Management Strategy. Commonwealth of Australia, Canberra: National Health and Medical Research Council, National Resource Management Ministerial Council.

NIH (National Institute of Health) 2014. *Wilson Disease*. National Digestive Diseases Information Clearinghouse, National Institute of Diabetes and Digestive and Kidney Diseases, Washington, DC: National Institutes of Health. https://www.niddk.nih.gov/health-information/health-topics/digestive-diseases/wilson-disease/Pages/facts.aspx

Pharand, L.; M.I. Van Dyke; W.B. Anderson; & P.M. Huck. 2014. Assessment of Biomass in Drinking Water Biofilters by Adenosine Triphosphate. *Journal of the American Water Works Association*, 106:10:E433–E444.

Pisigan, R.A. & J.E. Singley. 1987. Influence of Buffer Capacity, Chlorine Residual, and Flow Rate on Corrosion of Mild Steel and Copper. *Journal of the American Water Works Association*, 79:2:62–70.

Reasoner, D.J. & E.E. Geldreich. 1985. A New Medium for the Enumeration of Subculture of Bacteria from Potable Water. *Applied and Environmental Microbiology*, 49:1:1–7.

Sarin, P.; V.L. Snoeyink; J. Bebee; K.K. Jim; M.A. Beckett; W.M. Kriven; & J.A. Clement. 2004. Iron Release From Corroded Iron Pipes in Drinking Water Distribution Systems: Effect of Dissolved Oxygen. *Water Research*, 38:5:1259–1269.

Schock, M.R. 1999 (5th ed.). Internal Corrosion and Deposition Control. In *Water Quality and Treatment*. New York: McGraw-Hill.

Schock, M.R.; A.F. Cantor; S. Triantafyllidou; M.K. Desantis; & K.G. Scheckel. 2014. Importance of Pipe Deposits to Lead and Copper Rule Compliance. *Journal of the American Water Works Association*, 106:7:E366–E349.

Schock, M.R. & D.A. Lytle. 2014. Demystifying Copper Corrosion. New England Water Works Association Conference. Rockport, ME

Schock, M.R.; D.A. Lytle; & J.A. Clement. 1995. *Effect of pH, DIC, Orthophosphate and Sulfate on Drinking Water Cuprosolvency*. EPA/600/R-95/085. Cincinnati, OH: USEPA, Office of Research and Development.

Schock, M.R. & A.M. Sandvig. 2006. Long-Term Impacts of Orthophosphate Treatment on Copper Levels. Proceedings AWWA Annual Conference and Exposition. San Antonio, TX

Singley, E.J. 1981. The Search for a Corrosion Index. *Journal of the American Water Works Association*, 73:11:579–582.

Snoeyink, V.L. & I. Wagner. 1996 (2nd ed.). Principles of Corrosion of Water Distribution Systems. In *Internal Corrosion of Water Distribution Systems*. Denver, CO: AwwaRF and AWWA.

Stone, A.; D. Spyridakis; M. Benjamin; J. Ferguson; S. Reiber; & S. Osterhus. 1987. The Effects of Short Term Changes in Water Quality on Copper and Zinc Corrosion Rates. *Journal of the American Water Works Association*, 79:2:75–82.

Taylor, J.S.; J.D. Dietz; A.A. Randall; S.K. Hong; C.D. Norris; L.A. Mulford; J.M. Arevalo; S. Imran; M. Le Puil; S. Liu; I. Mutoti; J. Tang; W. Xiao; C. Cullen; R. Heaviside; A. Mehta; M. Patel; F. Vasquez; & D. Webb. 2005. *Effects of Blending on Distribution System Water Quality*. Denver, CO: AwwaRF.

Taylor, J.S.; J.D. Dietz; A.A. Randall; C.D. Norris; A. Alshehri; J. Arevalo; X. Guan; P. Lintereur; D. MacNevin; E. Stone; R. Vaidya; B. Zhao; S. Glatthorn; & A. Shekhar. 2008. *Control of Distribution System Water Quality Using Inhibitors*. Denver, CO: Water Research Foundation.

Tesfai, F.; P. Constant; S. Reiber; R. Giani; & M. Donnelly. 2006. Precipitate Formation in the Distribution System Following Orthophosphate Addition. Proceedings AWWA Water Quality Technology Conference, Denver, CO

Triantafyllidou, S. & M.A. Edwards. 2006. Effect of Coagulant Selection on Lead Leaching: Importance of the Chloride to Sulfate Mass Ratio. Proceedings AWWA Water Quality Technology Conference.

USEPA (US Environmental Protection Agency). 1984. *Corrosion Manual for Internal Corrosion of Water Distribution Systems*. EPA/570/9-84-001. Washington, DC: USEPA, Office of Water.

———. 2004. Lead and Copper Rule: Quick Reference Guide. EPA 816-F-04-009. Washington, DC: USEPA, Office of Water.

———. 2006. Lead in Drinking Water: Basic Information. Washington, DC: USEPA, Office of Water. http://www.epa.gov/safewater/lead/ basicinformation.html

———. 2016. Optimal Corrosion Control Treatment Evaluation Technical Recommendations for Primacy Agencies and Public Water Systems. EPA 816-B-16-003. Washington, DC: USEPA, Office of Water.

van der Kooij, D. 1992. Assimilable Organic Carbon as an Indicator of Bacterial Regrowth. 1992. *Journal of the American Water Works Association*, 84:2:57–65.

WHO (World Health Organization). 2011 (4th ed.). *Guidelines for Drinking-Water Quality*. Geneva, Switzerland: WHO.

Chapter 8

Disinfectants and Disinfection By-products

Hélène Baribeau, AQUAlity Engineering, Arcadia, Calif.
Yvonne Mazza, CH2M, Lakewood, Colo.
Susan Rivera, Central Washington University, Ellensburg, Wash.
Caroline Russell, Carollo Engineers, Austin, Texas
Rebecca Slabaugh, Arcadis, Indianapolis, Ind.
Raj Vaidya, Chastain Skillman, Lakeland, Fla.

In the developed world, the use of water supply disinfection as a public health measure has been credited with a major reduction in waterborne illnesses. However, chemical disinfectants react with organic and inorganic compounds to varying degrees and can form disinfection by-products (DBPs). Research shows that human exposure to these by-products may have adverse health effects (USEPA 2011). A major challenge for water systems is to provide adequate protection from microbial pathogens while simultaneously minimizing health risks from DBPs.

DSWQ Challenge:

A major challenge for water systems is balancing the risks from microbial pathogens and disinfection by-products (DBPs) with providing safe drinking water. It is important to ensure public health protection from microbial pathogens while simultaneously minimizing health risks from long-term exposure to DBPs.

While Chapter 3 provides detailed information on microbial pathogens and waterborne illnesses, this chapter focuses on the disinfectants that are most commonly used in the water industry and that affect distribution system water quality, including their associated DBPs. This chapter provides guidance to water suppliers and their consultants on approaches for providing adequate disinfection to mitigate exposure to microbial pathogens while minimizing DBP formation.

Several factors contribute to the fate of disinfectants and DBPs in the distribution system. Some factors are associated with source water quality and treatment processes, while others pertain to the distribution system. These factors are summarized in Table 8-1.

REGULATIONS

Countries throughout the world have developed regulations or guidelines directed specifically at the use of disinfectants and the formation of DBPs. In general, the goal of these standards is to provide an adequate level of disinfection to reduce the risk of human exposure to microbial contaminants while minimizing the formation of potentially harmful DBPs.

The US Environmental Protection Agency (USEPA) has developed a number of rules under the Safe Drinking Water Act and subsequent amendments related to the use of disinfectants and the formation of DBPs. A summary of key disinfection-related requirements from each of these rules is provided below. In the United States, individual primacy entities at the state level may apply more stringent rules and criteria than those mandated by the USEPA. This section also includes guidelines and regulatory limits from other countries and the World Health Organization (WHO 2011) for comparison purposes.

Regulations for Microbial Disinfection

As previously mentioned, the primary goal of drinking water regulations is to protect public health. Recognizing that microbial contaminants can present acute health effects, ensuring an adequate level of disinfection to reduce the risk of human exposure to these contaminants is of prime importance. This section presents the rules that have been promulgated by the USEPA.

Surface Water Treatment Rule. The Surface Water Treatment Rule (SWTR) was established in 1989 to prevent waterborne diseases caused by microbes present in water supplies. The rule requires all water systems that use surface water or groundwater under the direct influence of surface water to disinfect and, in most cases, also filter the water to an adequate level.

The primary requirements related to disinfection are as follows:

- Systems must provide at least 99.9 percent (3-log) inactivation and/or removal of *Giardia* cysts and 99.99 percent (4-log) inactivation and/or removal of viruses. Eligible disinfectants include chlorine, chloramines, chlorine dioxide, ozone, and ultraviolet light (UV).
- The concentration of the disinfectant residual entering the distribution system must not be less than 0.2 mg/L as Cl_2 for more than 4 hours within a 24-hour period.
- The residual disinfectant concentration in the distribution system must be measured at the same point in the distribution system and at the same time that total coliforms are sampled. At least 95 percent of the samples in a month, for any two consecutive months, must measure a detectable concentration.

Table 8-1 Factors that affect disinfectants and the fate of disinfection by-products in distribution systems

Factors Associated With Source Water Quality and Treatment Processes	Factors Associated With the Distribution System
• Type of primary and/or secondary disinfectant • Disinfectant doses used during water treatment and/or at the entry points to the distribution system • Physical and chemical characteristics of the water (e.g., temperature, pH, dissolved oxygen concentrations, particulates, and turbidity), including concentration and composition of organic and certain inorganic materials (e.g., bromide, iodide, sulfur-based compounds, iron, manganese) • Biological characteristics of the water (e.g., presence and type of bacteria, algae, protozoa) • Seasonal variations and blending of source waters	• Size, age, and configuration of the distribution network • Hydraulic conditions, including water velocity, pressure, turnover/mixing, and residence time in distribution system structures, including clearwells, pipes, reservoirs, and tanks • Variation in water demands, which are responsible for hydraulic conditions • Disinfectant types and doses used at booster chlorination and chloramination stations • Presence and density of suspended and fixed biomass in the distribution system • Pipe diameter, which characterizes the wetted surface area • Distribution system materials, including pipes, tanks, gaskets, seals, valves, fittings, meters, and similar components • Nature and intensity of corrosion of distribution system materials and corrosion control methods used • Storage tank size and conditions, including potential deterioration and water aeration • Presence of sediments in distribution system pipes, storage tanks, and reservoirs • Distribution system operation and maintenance practices, such as pipe flushing/cleaning and valve exercising • Potential distribution system contamination from intrusion of stormwater, groundwater, cross-connections, pipe breaks, backflows, and other events

The Interim Enhanced Surface Water Treatment Rule (IESWTR), Long Term 1 Enhanced Surface Water Treatment Rule (LT1ESWTR), and Long Term 2 Enhanced Surface Water Treatment Rule (LT2ESWTR) build on the requirements of the SWTR to include *Cryptosporidium*. Systems that are required to filter must provide a minimum of 3-log, 4-log, 5-log, or 5.5-log removal or inactivation of *Cryptosporidium* depending on influent *Cryptosporidium* concentrations. Under the LT2ESWTR, inactivation of *Cryptosporidium* can be achieved via ozone, chlorine dioxide, or UV. The LT2ESWTR also requires systems that store finished water in an open reservoir to either cover the reservoir or provide treatment to achieve 4-log virus, 3-log *Giardia*, and 2-log *Cryptosporidium* inactivation and/or removal.

Systems considering changes in disinfectant practices, such as a change in the type of disinfectant or moving the point of disinfection, may be required to conduct disinfection profiling and benchmarking. A disinfection profile consists of daily *Giardia lamblia* and/or virus inactivation achieved over a consecutive 12-month period or longer. The profile is used to generate a benchmark for microbial protection, which is used to design any treatment modifications.

Total Coliform Rule. The USEPA promulgated the Total Coliform Rule (TCR) in 1989. It establishes sampling and maximum contaminant levels (MCLs) for coliform bacteria in distribution systems. The objectives of the TCR are to evaluate the effectiveness of the treatment strategy, assess the integrity of the distribution system, and signal the potential presence of fecal contamination. Compliance is based on the number or percentage of samples that test positive during a given month. The rule implies that a secondary disinfectant (i.e., disinfection used after filtration or at the clearwell inlet) must be added and a residual must be maintained throughout the distribution system.

Revised Total Coliform Rule. As part of the Second Six-Year Review (Six-Year Review 2), the USEPA promulgated the Revised Total Coliform Rule to provide greater public health protection against waterborne pathogens in drinking water distribution systems. The rule became effective for water systems on April 1, 2016, and includes the following key provisions:

- Establishes an MCL for *Escherichia coli* and eliminates the MCL for total coliform,
- Establishes criteria for systems to qualify for reduced monitoring and criteria for continuing reduced monitoring following qualification, and
- Requires assessment and corrective action for systems vulnerable to fecal contamination.

Ground Water Rule. The Ground Water Rule was published in 2006 and extends the protection provided by the TCR to groundwater users. The rule establishes a risk-targeted approach to identify groundwater systems that are susceptible to fecal contamination. Disinfection is not required; however, systems are subject to triggered source water monitoring if the system does not provide 4-log virus inactivation, similar to the SWTRs. Additionally, several US states, including Colorado, Florida, and Pennsylvania, have added provisions mandating disinfection of groundwater systems.

Regulated Disinfection By-products

The USEPA Disinfectants and Disinfection Byproducts rules (DBPRs) establish maximum residual disinfectant level goals and levels for chemical disinfectants, as shown in Table 8-2. Under the most recent Stage 2 DBPR, compliance with the MCLs for the four regulated trihalomethanes (which have been referred to as total trihalomethanes or TTHMs) and five haloacetic acids (HAA5) is based on a locational running annual average at each compliance site in the distribution system. Chlorite and bromate were regulated for the first time under the DBPRs, with the MCLs listed in Table 8-2.

The Stage 2 DBPR also requires systems to calculate operational evaluation levels (OELs) at each monitoring location during each sampling period. If the OEL for TTHM or HAA5 exceeds their respective MCL, systems must evaluate treatment and distribution system practices to identify how the potential violation may be avoided. As part of the Six-Year Review 3, the USEPA reviewed the Stage 1 and 2 DBPRs and identified TTHMs and HAAs as candidates for revision (USEPA 2016).

The WHO has established guidelines for disinfectants and several DBPs, which are also shown in Table 8-2. Additional information about THMs, HAAs, chlorite, and bromate is presented later in this chapter.

The *Guidelines for Canadian Drinking Water Quality* (GCDWQ; Health Canada 2013, 2014) as well as guideline technical documents that are published by Health Canada recommend that a chlorine residual be maintained in distribution systems within an operational range from detectable to 5 mg/L Cl_2. Recommended residuals are between 0.04 and 2.0 mg/L Cl_2 for free chlorine. Free chlorine and chlorine dioxide do not have health-based maximum acceptable concentrations (MACs) in the GCDWQ. Chloramines are given an MAC of 3.0 mg/L Cl_2 based on health effects of monochloramine. The GCDWQ recommends MACs of 0.1 mg/L for the four THMs, 0.08 mg/L for HAA5, 1 mg/L for chlorite, and 0.01 mg/L for bromate.

Table 8-2 Regulated disinfectants and disinfection by-products

Disinfectant Residual	US DBPR Maximum Residual Disinfectant Level Goals, mg/L	US DBPR Maximum Residual Disinfectant Level, mg/L	WHO Guideline, mg/L
Chlorine	4.0 (as Cl_2)	4.0 (as Cl_2)	5.0 (as Cl_2)
Monochloramine (measured as total chlorine)	4.0 (as Cl_2)	4.0 (as Cl_2)	3.0 (as Cl_2)
Chlorine dioxide	0.8 (as ClO_2)	0.8 (as ClO_2)	N/A
Disinfection By-products	**Maximum Contaminant Level Goal, mg/L**	**Maximum Contaminant Level, mg/L**	**WHO Guideline, mg/L**
Regulated trihalomethanes (TTHM)	N/A	0.080	N/A
Chloroform	0.07		0.3
Bromodichloromethane	zero		0.06
Dibromochloromethane	0.06		0.1
Bromoform	zero		0.1
Five haloacetic acids (HAA5)	N/A	0.060	N/A
Monochloroacetic acid	0.07		0.2
Dichloroacetic acid	zero		0.05
Trichloroacetic acid	0.2		0.2
Bromoacetic acid	N/A		N/A
Dibromoacetic acid	N/A		N/A
Chlorite	0.8	1.0	0.7
Bromate	zero	0.010	0.01

Abbreviation: DBPR, Disinfection By-products Rule.

In Australia, the National Health and Medical Research Council (NHMRC) and Natural Resource Management Ministerial Council (NRMMC) (2011) limit free chlorine residuals to 5 mg/L measured as total chlorine and 3 mg/L for chloramines (or 4.1 mg/L as Cl_2). These limits are based on health considerations. Chlorine dioxide should not exceed 0.4 mg/L based on aesthetic considerations. As for DBPs, NHMRC and NRMMC (2011) have established the following limits based on health considerations:

- 0.25 mg/L for TTHM, either individually or in total (with a note that actions to reduce THMs are encouraged but must not compromise disinfection),
- 0.15 mg/L for monochloroacetic acid,
- 0.1 mg/L for dichloroacetic acid,
- 0.1 mg/L for trichloroacetic acid,
- 0.8 mg/L for chlorite, and
- 0.02 mg/L for bromate.

In comparison, the Council of the European Union (1998) recommends a maximum concentration of 100 μg/L for the four THMs and 10 μg/L for bromate; no other DBPs are discussed in Council Directive 98/83/EC.

Potential Future Regulations

The USEPA is also considering or is in the process of developing regulations for several currently unregulated contaminants. These are described below.

Chlorate. Chlorate is an impurity in hypochlorite solutions that results from the decay of hypochlorite with time and elevated temperature. Chlorate is also a degradation product of chlorine dioxide (see Inorganic By-products, Chlorite and Chlorate section below).

The USEPA has calculated a health reference level of 0.21 mg/L and has stated that there is an opportunity for addressing concerns with chlorate as part of the Six-Year Review 3 determination associated with Stage 1 and 2 DBPs (USEPA 2016). In addition, California has established a notification level (NL) for chlorate at 0.8 mg/L.

The WHO (2011) has set a guideline for chlorate in drinking water at 0.7 mg/L, and the GCDWQ proposes an MAC of 1 mg/L based on potential health effects.

Perchlorate. Perchlorate is a manufactured chemical that is used to produce rocket fuel, fireworks, flares, and explosives. It is also another impurity present in hypochlorite solutions, which explains why it is discussed in this chapter, and can also be naturally occurring (see Inorganic By-products, Chlorite and Chlorate section below).

The USEPA was expected to release a draft regulation on perchlorate in 2013; however, they expressed a need for additional data and analysis before selecting an MCL. In the meantime, some states have established limits or guidelines for perchlorate in the range of 0.0002 to 0.051 mg/L. Massachusetts and California, for instance, have adopted MCLs of 0.002 mg/L and 0.006 mg/L, respectively. The USEPA has also established a non-enforceable interim health advisory of 0.015 mg/L.

Nitrosamines. Nitrosamines are DBPs that form during chlorine or chloramine disinfection, typically in areas with high water age, such as storage tanks and within remote areas of the system or dead ends (see Inorganic By-products, Nitrosamines section below).

The USEPA stated that there is an opportunity to address concerns with nitrosamines as part of the Six-Year Review 3 determination associated with Stage 1 and 2 D/DBP candidates for revision (USEPA 2016). California has already established an NL of 0.00001 mg/L for the following three nitrosamines compounds: N-nitrosodiethylamine, N-nitrosodimethylamine (NDMA), and N-nitrosodi-n-propylamine.

The WHO has set a guideline value of 0.0001 mg/L for NDMA in drinking water (WHO 2011). Health Canada (2013, 2014) proposes an MAC of 0.00004 mg/L based on potential health effects. The Australian government represented by NHMRC and NRMMC (2011) has established a limit of 0.0001 mg/L, although it states that actions to reduce NDMA are encouraged but must not compromise disinfection.

Brominated haloacetic acids. HAAs are formed when chlorine reacts with organics and bromide and can be distinguished by the term "bromo" included in the particular species name, such as monobromoacetic acid. At the time this manual was written, the USEPA was considering including brominated HAAs in the proposed 4th Unregulated Contaminant Monitoring Rule (UCMR4) monitoring program starting in 2018. Based on results obtained during the monitoring period, these constituents may be included as part of future regulations in the United States and other countries. There are currently two brominated HAAs that are regulated as part of the US Stage 1 and 2 DBPRs. The remaining three species are not related to bromide and make up the HAA5 that are regulated as a sum total, instead of by individual species. There are an additional four brominated species that are unregulated, sometimes referred to as HAA6Br. The term "HAA9" describes all currently regulated and nonregulated HAAs. Table 8-3 provides a crosswalk of how HAAs are typically grouped. UCMR4 includes provisions for systems to monitor all three groups, including regulated HAA5, in order to gain a better understanding of co-occurrence with the unregulated HAAs.

Table 8-3 Haloacetic acids currently regulated and considered for regulations

Status	HAA5	HAA6Br	HAA9
US regulated	Monochloroacetic acid, dichloroacetic acid, trichloroacetic acid, monobromoacetic acid, dibromoacetic acid	Monobromoacetic acid, dibromoacetic acid	Monochloroacetic acid, dichloroacetic acid, trichloroacetic acid, monobromoacetic acid, dibromoacetic acid
Unregulated/ potential future US regulation		Bromochloroacetic acid, bromodichloroacetic acid, dibromochloroacetic acid, tribromoacetic acid	Bromochloroacetic acid, bromodichloroacetic acid, dibromochloroacetic acid, tribromoacetic acid

DISINFECTANTS

This section describes the disinfectants that are most commonly used in the water industry and that affect distribution system water quality, their advantages and drawbacks, and control strategies. Of the agents that are used to leave a residual disinfectant, free chlorine and chlorine dioxide are good disinfectants and oxidants, while monochloramine tends to be a weaker disinfectant at comparable concentrations. For purposes of this manual, disinfection refers to the selective inactivation or destruction of microorganisms, whereas oxidation focuses on the removal or breakdown of chemicals. A number of disinfectants/oxidants are strictly used during water treatment and do not create a residual in the distribution system; these agents include ozone, permanganate, and UV. Discussion of these agents in this manual is limited to their potential impacts on distribution system water quality.

In natural waters, all disinfectants show decreases in concentrations over time due to decay and demand. Disinfectant decay refers to the natural decomposition of the chemical itself, whereas disinfectant demand results from the reaction between disinfectants and water constituents such as organic and inorganic materials.

Free Chlorine

Free chlorine is by far the most common disinfectant used in the United States and around the world. Chlorine is a highly efficient disinfectant and oxidant; it is added to water systems to kill disease-causing pathogens such as bacteria, viruses, and protozoans that commonly grow in source waters or on the wetted surfaces of water mains and storage tanks. Thus, it is also used to lower the risk of contamination in distribution systems.

Traditional use of chlorination has been to meet primary disinfection requirements (i.e., disinfectant used at the plant influent or during the first stages of treatment, with the main goal of fulfilling CT requirements, i.e., sufficient disinfectant concentration and contact time) and provide a residual disinfectant in the distribution system. Chlorine, being a strong oxidant, undergoes both substitutive and oxidation reactions with organics and inorganics in water (Connell 1996; Krasner 1999). Therefore, it will form regulated DBPs such as THMs and HAAs. It is also consumed rapidly, especially in the presence of reducing compounds such as sulfides, iron, and manganese (reduced form) as well as a host of organics, especially aromatic compounds. Exposed iron and steel pipes and tank walls create a high initial chlorine demand and cause rapid initial corrosion in distribution systems. However, scale formation and continued pipe corrosion due to free chlorine are typically slow at the doses commonly used for disinfection. Additional information about internal corrosion is provided in Chapter 7.

The advantages and disadvantages of free chlorine are summarized in Table 8-4, and additional information about target residuals is presented later in this chapter.

In addition to its natural decay, many factors are responsible for decreases in free chlorine residuals in distribution systems; the main reasons include reactions between free chlorine and pipe wall materials, organic and inorganic substances, biofilms, and corrosion by-products. Water temperature also plays an important role in determining chlorine decay and demand; as temperature increases, the rate of chemical and biological reactions increases, biological activity increases, and organic material characteristics vary. Additionally, the longer the water is within the distribution system, the more disinfectant decay that will be observed due to the reactions described above. Booster chlorination stations positioned at strategic locations within distribution systems are used to prevent chlorine residual from decreasing below regulatory levels.

Several factors impact disinfectant residual maintenance as water temperatures increase, including higher rates of reaction, increases in biological activity, and variations in dissolved organic matter concentrations during the warm seasons.

Chloramines

Chloramines are produced by combining free chlorine with ammonia, with the intent to form monochloramine. Monochloramine is a weaker disinfectant and oxidant than free chlorine, but this enables it to persist further into the distribution system and more efficiently inactivate certain biofilm microorganisms. Chloramines form lower concentrations of THMs and HAAs due to their weak oxidative substitution capability when compared to that of free chlorine and, hence, have been used by many utilities to meet the regulatory requirements of the Stage 1 and Stage 2 DBPRs.

Table 8-4 Advantages and disadvantages of free chlorine

Advantage	Disadvantage
• Can be used both as an oxidant and as a disinfectant • Used as a preoxidant and as a distribution system residual disinfectant • Strong oxidant toward many constituents, including ammonia, iron, manganese, sulfide, nitrite • Good disinfectant to inactivate bulk fluid microorganisms • Reacts with taste- and odor-causing compounds • Relatively inexpensive and readily available	• Reacts with organic materials to form regulated DBPs, that is, trihalomethanes and haloacetic acids • Reacts with certain compounds (e.g., phenols, sulfur-based compounds) to cause taste- and odor-causing compounds • Can increase the fraction of biodegradable or assimilable organic matter (e.g., assimilable organic carbon and biodegradable dissolved organic carbon) if the source water contains oxidizable organic materials; the resulting organic material can be responsible for microbial (re)growth in distribution systems • Reacts with bromide to form hypobromous acid and hypobromite, which can form brominated DBPs • Relatively less effective for disinfection of turbid waters and clumped or biofilm microorganisms • Photosensitive

Abbreviations: DBP, disinfection by-product.

Monochloramine is the desired and typically predominant chloramine species at chlorine-to-ammonia-N mass ratios of 5:1 or less and pH values between 7.0 and 9.0. A chlorine-to-ammonia-N mass ratio that is higher than 5:1 and a pH value that is lower than 6 favor the formation of dichloramine, while a pH value lower than 4.0 is required for trichloramine to be the dominant species. Di- and trichloramine species are considered nuisance species and can result in taste and odor complaints. Mono-, di-, and trichloramines are referred to as inorganic chloramines. Detailed information about the chloramination process can be found in AWWA Manual M20 (AWWA 2006).

Water temperature, contact time between chlorine and ammonia, mixing or lack thereof, and the presence of interfering compounds such as organic material and bromide can also influence the chloramine speciation. In the presence of certain organic compounds (i.e., organic amines), organic chloramines can be formed. These compounds have much lower disinfection efficacy than inorganic chloramines, they exhibit lower degradation rates than inorganic chloramines (Baribeau et al. 2005), and they can interfere with measurement methods used to monitor for chlorine.

In the presence of bromide, bromamines can be formed in a complex chemistry that leads to poor disinfection capabilities and chloramine instability. Bromamines are the brominated counterparts of chloramines, and their speciation is a function of pH and the bromide-to-ammonia ratio (refer to Speitel 1999 for additional information about bromamines).

Chloramines are unstable at neutral pH values (which are normally found in drinking water distribution systems) due to autodecomposition reactions that ultimately result in oxidation of ammonia and reduction of the combined residual chlorine. This autodecomposition primarily depends on pH, chlorine-to-ammonia ratio, temperature, and carbonate species (alkalinity). Inorganic compounds such as nitrite were observed to react with monochloramine and exert a demand (Valentine, et al. 2000; Valentine 1985; Margerum et al., 1994), while bromide was observed to catalytically increase its autodecomposition (Valentine et al. 2000; Galal-Gorchev and Morris, 1965; Bousher et al., 1989). Chloramine decay and demand in distribution systems lead to nitrification, which is described in detail in Chapter 6.

The advantages and disadvantages of chloramines are summarized in Table 8-5, and additional information about target residuals is presented later in this chapter.

Chlorine Dioxide

Chlorine dioxide exhibits chemical and physical properties that are very different from those of other chemical disinfectants. Because of its limited ability to leave a residual disinfectant (it is highly volatile), chlorine dioxide is typically used as a preoxidant (in untreated water) or intermediate oxidant (during or after sedimentation, before filtration) in the United States. Free chlorine or monochloramine is typically added at the entry point to the distribution system to leave a residual disinfectant. In the United States, the use of chlorine dioxide as a distribution system residual disinfectant has been limited to source waters of good quality (i.e., low in disinfectant demand) and small distribution systems.

When reacting with organic and inorganic materials, chlorine dioxide initially degrades into chlorite (50 to 70 percent of the chlorine dioxide dosed is found as chlorite), chlorate (less than or equal to 20 percent of the chlorine dioxide dose), and chloride. Unlike chlorine, which reacts by addition, substitution, or oxidation, chlorine dioxide is much more specific and typically reacts by oxidation only. If a variety of organic by-products can be formed (many are similar to those produced by ozone), they are present at concentrations that are much lower than those measured in chlorine- or ozone-treated waters. The key by-products formed include HAAs (mainly dihaloacetic acids, and at much lower concentrations than those formed by free chlorine), aldehydes, carboxylic acids, ketones, bromate under certain conditions, and halogenated furanones (Mofidi et al. 2004; Reckhow and Singer 2011; Richardson et al. 1994, 2003, 2009; Weinberg et al. 2002; Zhang et al. 2000). Additional information about other nonregulated halogenated and nonhalogenated DBPs is provided in the Other By-product section below.

In distribution systems, chlorine dioxide can follow different trends based on the secondary disinfectant used, that is, whether chlorine dioxide was used as a pre- or intermediate oxidant/disinfectant or as a final disinfectant. When used as a distribution system residual disinfectant, chlorine dioxide tends to decrease in distribution systems with increasing water age (Baribeau et al. 2002; Bergeon et al. 1992; Français et al. 1994; Lafrance et al. 1992; Olivieri et al. 1986; Ring et al. 2004; Volk et al. 2002). When chlorine dioxide is used as a pre- or intermediate oxidant with free chlorine as the distribution system residual disinfectant (in the absence of a residual chlorine dioxide), it can reform in the distribution system as a result of the reaction between chlorine and chlorite (Ellenberger et al. 1998; Gallagher, Hoehn, and Dietrich 1994; Hautman and Bolyard 1991, 1992; Hoehn et al. 2003). Currently, limited information is available regarding the fate of chlorine dioxide when it is followed by the addition of monochloramine (in the absence of a residual chlorine dioxide), but it appears that chlorine dioxide residuals decrease with increasing water age (Thompson et al. 1988; Briggs et al. 2007). Unlike free chlorine, chloramines cannot oxidize chlorite into chlorine dioxide.

Table 8-5 Advantages and disadvantages of chloramines

Advantage	Disadvantage
• Do not form regulated DBPs such as trihalomethanes and haloacetic acids as much as chlorine • More stable residual than free chlorine • Better at inactivating certain biofilm microorganisms, most likely due to better penetration of biofilms • Monochloramine presents less chlorinous flavors (lower taste and odor thresholds) than free chlorine	• Form nonregulated DBPs (e.g., nitrosamines, cyanogen halides) • Not as effective as other disinfectants at inactivating microorganisms; not a good primary disinfectant • Require additional oversight and monitoring during treatment and distribution when compared to free chlorine • Responsible for nitrification in distribution systems, particularly in areas with higher water age and water temperature • More toxic to aquatic life; dechloramination is necessary when flushing • Present risks to sensitive populations (e.g., dialysis patients, fish) • Reacts with bromide to form hypobromous acid, hypobromite, and bromamines • Attacks most rubber materials, resulting in faster deterioration of pipe seals and gaskets • Di- and trichloramines present more offensive tastes and odors than free chlorine and decay rapidly once formed

Abbreviation: DBP, disinfection by-product.

From a regulatory perspective, it should be noted that chlorine dioxide is considered an acute contaminant, in contrast to other disinfectants that are considered chronic contaminants. This makes the use of chlorine dioxide more risky from a compliance standpoint because a single sample above the MCL may trigger a violation and the need for public notification.

The advantages and disadvantages of chlorine dioxide are provided in Table 8-6.

Ozone

In water treatment, ozone is used for both disinfection to remove microorganisms and chemical oxidation of organic and inorganic compounds. Because ozone decomposes spontaneously in water, it has little to no ability to leave a disinfectant residual. Thus, it is only used as a preoxidant/disinfectant (in untreated water) or as a intermediate oxidant (during or after sedimentation, before filtration) in combination with free chlorine, monochloramine, or chlorine dioxide to leave a residual disinfectant in the distribution system.

Ozone oxidizes organic materials to form a variety of organic oxidation by-products such as aldehydes, organic acids (such as carboxylic acids), and ketoacids (also called aldoacids or oxoacids) (Reckhow 1999; Reckhow and Singer 2011; Xie 2004). These by-products are low-molecular-weight organics that are much more amenable to biodegradation. They can be captured by bioassays such as the assimilable organic carbon (AOC) and the biodegradable dissolved organic carbon (BDOC) measurements. If not removed prior to chlorination and water distribution, these by-products can consume chlorine and chloramines, form halogenated by-products such as THMs and HAAs, and be associated with biological (re)growth in distribution systems (a phenomenon referred to as "biological instability"). Biological instability can be expressed as increases in coliform counts, heterotrophic plate counts, or biofilm density.

Table 8-6 Advantages and disadvantages of chlorine dioxide

Advantage	Disadvantage
• Chlorine-free chlorine dioxide forms far fewer halogenated DBPs than chlorine; it does not form trihalomethanes • Highly selective and does not react with bromide or ammonia • Stronger disinfectant than free chlorine and chloramines against protozoa, bacteria, viruses, spores, and algae • Requires lower concentrations and shorter contact times for equal microbial inactivation • Chlorite may control nitrification • Oxidizes inorganic chemicals (e.g., iron, manganese, and hydrogen sulfide • Controls certain compounds responsible for taste, odor, and color • Less corrosive than other disinfectants • Highly effective over a broad pH range	• Volatile above 50°F, with a strong chlorinous odor • Unstable, must be generated onsite (cannot be stored) • Limited ability to leave a residual disinfectant in the distribution system • Photosensitive; destroyed by ultraviolet light • Forms halogenated DBPs such as chlorite and chlorate, which are regulated or considered for regulations • May create odors

Abbreviation: DBP, disinfection by-product.

When bromide is present, ozone can oxidize it through a series of reactions to form hypobromous acid and hypobromite (OBr^-). Further oxidation of OBr- leads to the formation of bromate. Brominated organics and bromamines (the brominated counterparts of chloramines) can also be formed based on the chemical composition of the water being treated (Figure 8-1). Of these, bromate is of significant concern because it is regulated by the Stage 1 DBPR (see Regulations section above).

Factors that influence ozone oxidation include ozone dose, pH, alkalinity, the nature of the organic and inorganic materials, and contact time. DBP precursor destruction tends to predominate when ozonation is performed at low pH. Ozonation pH is typically less than 7.5 for most humic substances (USEPA 1999a), although a pH range of 5.5 to 6.5 is generally adopted during water treatment (the exact values are site and treatment specific). When bromide is present, less bromate is also formed at low pH because OBr^- is responsible for the formation of bromate, but higher brominated organic by-products are formed. Ozonation in the presence of ammonia also decreases bromate and brominated organics because bromamines are formed instead.

The advantages and disadvantages of ozone are summarized in Table 8-7.

Ultraviolet Light and Advanced Oxidation Processes

UV light has been used as a multibarrier treatment solution to provide protection against specific pathogens (e.g., *Cryptosporidium* and *Giardia*) and help minimize DBPs. It inactivates microorganisms by affecting their DNA or RNA and creating thymine dimers. This genotoxic effect does not necessarily destroy the microbial cells but prevents microorganisms from replicating. The key advantage and disadvantage of UV are that it does not generate DBPs nor does it provide a residual. Additional advantages and disadvantages are presented in Table 8-8.

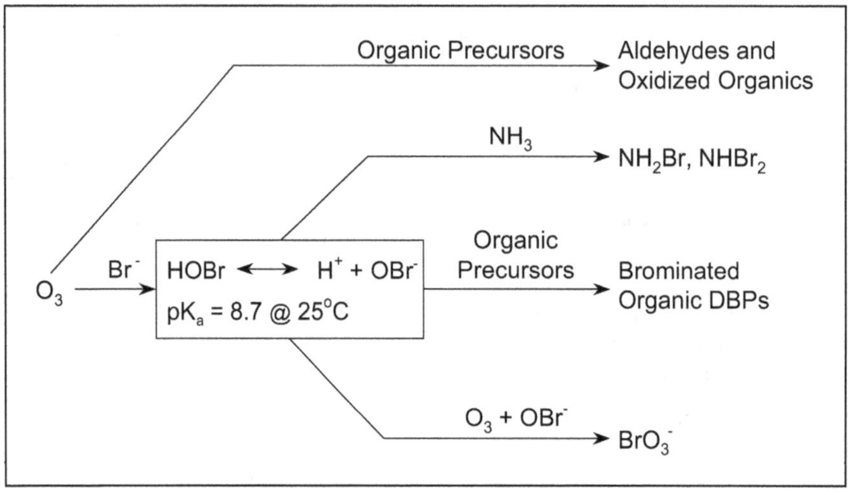

Source: USEPA 1999a.

Figure 8-1 Main pathways involved in the formation of ozone by-products

Table 8-7 Advantages and disadvantages of ozone

Advantage	Disadvantage
• One of the strongest oxidants and disinfectants • Does not form halogenated by-products, unless bromide is present • Stronger disinfectant than free chlorine, chloramines, and chlorine dioxide against protozoa, bacteria, viruses, spores, and algae • Its biocidal activity is not influenced by pH, though DBP formation potential can be influenced by pH • Improves coagulation, flocculation, and filtration • Oxidizes organic materials and micropollutants • When combined with biofiltration, can decrease disinfectant demand, remove some DBP precursors, and increase biological stability in distribution systems • Does not react with ammonia unless bromide is present • Oxidizes inorganic chemicals (e.g., iron, manganese, sulfide) (Langlais, Reckhow, and Brink 1991) • Can remove color-, taste-, and odor-causing compounds (Suffet, Anselme, and Mallevialle 1986), particularly in combination with activated carbon or hydrogen peroxide • End-product of decomposition is oxygen	• Unstable, must be generated onsite • Does not leave a residual disinfectant in the distribution system • Chemical reactions are highly sensitive to temperature and pH • Can form bromate, which is a regulated contaminant, when bromide is present • Biofiltration is typically required after ozonation to obtain biologically stable water • High energy requirement for ozone generation • Highly corrosive and requires specialized equipment and operations • Quenching of the excess dose may be required

Abbreviation: DBP, disinfection by-product.

Table 8-8 Advantages and disadvantages of ultraviolet light

Advantage	Disadvantage
• Very effective disinfectant against microorganisms, including resistant pathogens such as protozoa • Does not alter water quality (i.e., does not change pH, organic material, corrosiveness, or other water characteristics) • Does not form disinfection by-products as much as chlorine • Fast reaction time (i.e., seconds), limiting reactor footprint	• Some viruses are resistant to UV • Following UV exposure, microorganism reactivation is possible, including reactivation of *Cryptosporidium* • Does not leave a disinfectant residual in the distribution system • Turbidity may shield microorganisms from UV • Higher UV doses are required in the presence of organic material because it absorbs UV light • Organic material can foul lamp sleeves (biofilms) • Inorganic compounds can form scale on the quartz sleeves of the UV lamps • Certain UV sensors present inconsistent reliability/accuracy • Mercury mitigation strategies are required in case of lamp breakage • UV lamps need constant, continuous electric power supply

Abbreviation: UV, ultraviolet light.

Advanced oxidation processes (AOPs) form highly reactive oxidizing intermediates (e.g., hydroxyl radical) that can oxidize and degrade contaminants of concern. In addition to treating a broad range of contaminants, AOPs can provide high levels of disinfection. AOPs encompass a wide range of treatment technologies such as:

- Ozone/peroxide,
- UV/peroxide,
- Ozone/UV,
- Ozone/UV/peroxide,
- Titanium dioxide-catalyzed UV,
- Fenton's reactions, and
- UV/chlorine.

For municipal-scale applications, ozone/peroxide and UV/peroxide are the most common AOP technologies. UV/chlorine is an emerging technology with potential benefits over UV/peroxide; however, the number of installations is limited. AOPs have distinct advantages and disadvantages, including the ability to treat many contaminants that cannot be effectively removed with other technologies (e.g., 1,4-dioxane, NDMA), with the potential for simultaneous disinfection credit. Additional advantages and disadvantages are presented in Table 8-9.

Ozone- and UV-based AOPs share many of the same advantages and disadvantages as ozone and UV disinfection applications, as previously discussed (e.g., they must be generated onsite, they must be followed by other disinfectant to leave a residual in the distribution system, they require high energy demand). In general, AOPs can achieve similar or higher levels of disinfection, but the combination of agents used needs to be closely evaluated. For example, if peroxide is added to an ozone system, the peroxide residual will quench the ozone residual and stop the disinfection process.

Table 8-9 Advantages and disadvantages of advanced oxidation processes

Advantage	Disadvantage
• One of the strongest oxidation potential • Oxidation of a broad range of potential contaminants • Can provide high levels of disinfection depending on design and operating conditions • Can remove contaminants of concern such as nitrosodimethylamine, pharmaceuticals and personal care products, and endocrine disrupters, particularly when UV is used with hydrogen peroxide • Fast reactions with short residence times, which allows smaller reactor footprints	• Because the radicals are unstable, they must be generated onsite • Does not leave a residual disinfectant in the distribution system • Formation of by-products, which may require downstream treatment • Some advanced oxidation processes (e.g., ozone/peroxide and UV/chlorine) are highly sensitive to temperature and pH • High energy requirement

Abbreviation: UV, ultraviolet light.

AOPs have the potential to form by-products during and after treatment. AOP by-products depend on the selected AOP technology and can include the following:

- THMs and HAAs: AOPs have been shown to transform organic material, which can result in DBP formation, including THMs and HAAs, after free chlorine addition (Zheng, Andrews, and Bolton 1999; Dotson et al. 2010; Andrews 2009; Wang et al. 2015);
- AOC or BDOC when ozone-based AOPs are used;
- Bromate when ozone-based AOPs or UV/chlorine are used (Huang, Gao, and Deng 2008; Wang et al., 2015);
- Nitrite when UV-based AOPs with medium-pressure UV lamps are used. Although it is unlikely that nitrite formation will exceed the MCL of 1 mg/L N (Sharpless and Linden 2001), the addition of chlorine for secondary disinfection will oxidize nitrite back to nitrate and consume some chlorine (see Chapter 6); and
- Chlorate when UV/chlorine is used (Wang et al. 2015; Wetterau et al. 2015).

Characterizing the DSWQ Challenge: Disinfectants Versus Disinfection By-products

Characteristic	Disinfectant			
	Free Chlorine	Chloramines	Chlorine Dioxide	Ozone
Disinfection	Effective for primary, intermediate, and secondary oxidation and disinfection	Effective secondary disinfectant	Typically used for primary and intermediate disinfection; can be used for secondary disinfection	Effective for primary and intermediate oxidation and disinfection*
Persistence	Moderate	High	Moderate	None
Formation of regulated disinfection by-products	High	Low†	Moderate	Low to moderate
Control of taste- and odor-causing compounds	Moderate to high	Low	High	High
Contribution to tastes and odors	Moderate to high	Low to moderate	Moderate to high‡	Low
Ease of implementation and relative complexity of technology	Easy	Moderate¶	Moderate to difficult	Moderate to difficult
Implementation and operations and maintenance costs	Low	Medium	Medium to high	High

* Can form nonregulated disinfection by-products such as nitrosamines, cyanogen chloride, and bromide.

† Because it is typically used at low concentration, it may not lead to strong flavors, but strong odors can be formed if it reacts with certain precursors.

‡ Ozonation can stimulate biofilm growth if not followed by biological filtration to control the low-molecular-weight organic compounds that may be formed by ozonation.

¶ In itself, implementing chloramination is relatively simple. But nitrification may complicate the chloramination process, making it a complex technology.

**Responding to the DSWQ Challenge:
Strategies to Prevent Disinfectant Decay and Demand in Distribution Systems**

Factors Associated with Water Quality and Treatment Processes	Factors Associated with the Distribution System
• Ensure that most of the disinfectant demand is met before water enters the distribution system • Limit the presence of organic material prior to the first disinfectant application point, especially when dosing free chlorine • Limit the presence of inorganic constituents that can consume disinfectants (e.g., bromide, iodide, sulfur-base compounds, iron, manganese)	• Limit water age by eliminating dead ends and oversized pipes and by limiting the number of pressure zones (see Chapter 2) • Control hydraulic conditions in storage tanks and reservoirs by providing adequate mixing, preventing water stagnation and short-circuiting • Install booster chlorination or chloramination stations as needed • Limit the growth of suspended and fixed biomass in the distribution system (see Chapter 3) • Limit nitrification (see Chapter 6) • Limit corrosion of distribution system materials (see Chapter 7) • Limit the presence of sediments in distribution system pipes and storage tanks/reservoirs by conducting routine flushing and reservoir cleaning (see Chapter 2) • Prevent distribution system contamination from intrusion of stormwater, groundwater, cross-connections, backflows, and other events (see Chapter 10)

NOTE: These strategies can be viewed as the counterparts of the factors presented in Table 8-1.

CONTROL STRATEGIES

DBPs are chemicals that are formed when disinfectants/oxidants react with organic and inorganic substances present in water. More than 600 compounds recognized as DBPs have been identified, and it is likely that many more exist, though not uncharacterized yet (Richardson et al. 2007).

Factors Responsible for the Formation and Fate of DBPs

Disinfectant/oxidant type and dose. The effect of disinfectant/oxidant dose on DBP formation varies based on the nature of the disinfectant and the DBPs. DBPs that are end-products of reactions (e.g., THMs) increase in concentration as disinfectant dose increases. Intermediates of reactions (e.g., many of the HAAs) may behave differently, showing increases in concentration when moderate doses of disinfectants are used, but decreases in concentration at high disinfectant doses, as they are transformed into other products.

Reaction time with disinfectants. Reaction time includes contact time at the water treatment plant and water age in the distribution system. DBP concentrations that are end-products of reactions typically increase with reaction time, whereas DBPs that are intermediates of reactions may increase at the beginning of the reaction process and then decrease as they are transformed into other products.

Water quality characteristics. Water temperature and pH influence not only the concentrations of the DBPs formed but also their speciation. The presence of substances that may compete with disinfectants (e.g., sulfur, iron, manganese, nitrite) also influences DBP concentrations and speciation.

Nature and concentration of precursors. Precursors include organic compounds (e.g., natural organic matter) and inorganic substances (e.g., bromide, iodide, nitrite).

Microbial activity. Certain cellular materials can serve as DBP precursors (e.g., algae), thereby increasing DBP formation potential. On the other hand, a number of DBPs (including HAAs) are biodegradable.

Trihalomethanes and Haloacetic Acids

THMs and HAAs form when chlorine reacts with precursors such as organic material, bromide, and iodide. If bromide is present in the source water or is introduced during treatment, it will become oxidized by aqueous chlorine and then react with organic material to form brominated THMs and HAAs. Studies have shown that the rate of DBP formation is higher in water with increased concentrations of bromide (Krasner, S.W. 1999). When compared to THMs, HAAs typically form at a faster rate.

As water travels in the distribution system piping and storage tanks, THM concentrations increase as long as a disinfectant residual is present. This is depicted qualitatively in Figure 8-2; actual concentrations of residual disinfectant and THMs are system specific. Chapter 2 discusses water age related to demand issues in detail. At high water ages, the brominated species may cease to form and the chlorinated species (i.e., chloroform) will continue to form due to the reaction of chlorine and existing organic matter.

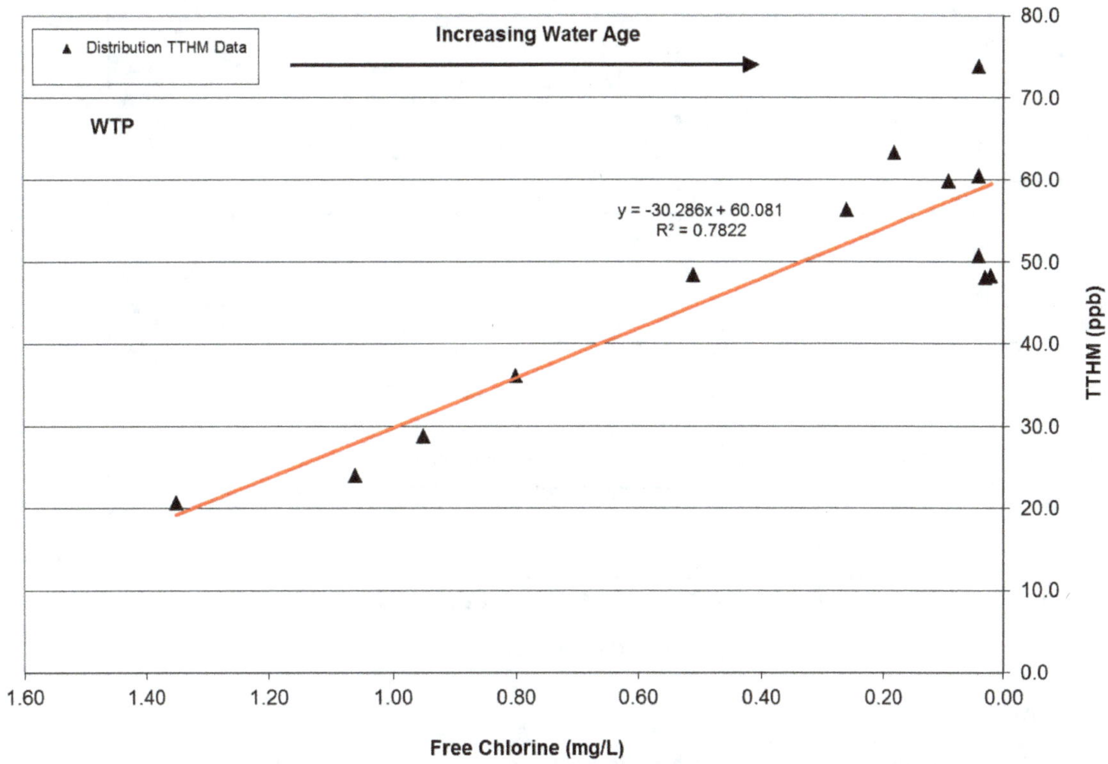

Source: USEPA 2013.

Figure 8-2 Example of the correlations between water age, trihalomethanes, and disinfectant residual in distribution systems

Although not currently regulated, iodinated THMs (I-THMs) can be formed when iodide is oxidized to hypoiodous acid by disinfectants (mainly monochloramine but also free chlorine and ozone) and then reacts with organic materials. I-THMs are believed to be more toxic than their chlorinated or brominated counterparts (Kargalioglu et al. 2002; Plewa et al. 2002; Woo et al., 2002) and can impart a strong medicinal odor to potable water, as explained in Chapter 5.

HAAs, in comparison to THMs, react differently within the distribution system. HAAs are typically highest in average residence time areas of the distribution system due to potential biodegradation at remote locations. Biological degradation is commonly found in areas with low disinfectant residuals, warmer water temperatures, and higher water age, which typically equate to elevated biogrowth with the potential presence of HAA-degrading bacteria. The typical effects of biodegradation in relation to disinfectant residuals, HAA concentrations, and water age are depicted qualitatively in Figure 8-3; actual concentrations of residual disinfectant and HAAs are system specific.

HAA biodegradation potential is dependent on disinfectant residual and water age throughout the distribution system. If elevated disinfectant residuals and low biomass concentrations are maintained throughout the system and water age is low, the highest HAA concentrations are likely to be found at the maximum residence locations, not the average residence locations.

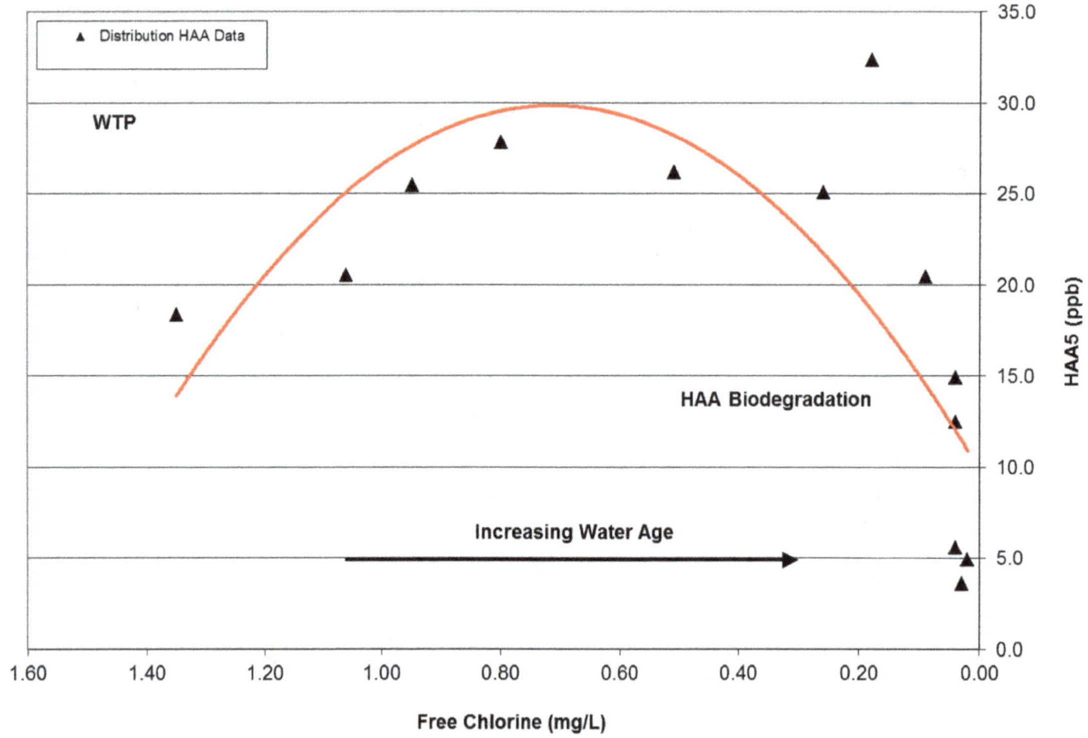

Source: USEPA 2013.

Figure 8-3 Example of correlations between water age, haloacetic acids, and disinfectant residual in distribution systems

Strategies to Control THMs and HAAs

Source water protection. Source water protection is not only one of the best methods to control DBP formation by limiting the introduction of organic material in raw water sources but it also limits disinfectant demand and the presence of pathogenic microorganisms that require disinfection. In addition, source water protection may help preserve use of more pristine water sources, limiting the need for sophisticated and/or costly treatment strategies.

Treatment plant controls. The control of THM and HAA precursors is most effective when it is accomplished at the treatment facility, where preoxidation with chemicals other than chlorine is used to not only enhance organic material removal but also minimize DBP formation within the plant. As an example, preoxidation with ozone can break down organic materials into simpler compounds that can be removed during subsequent biofiltration, thereby decreasing THM and HAA formation potential when the final disinfectant is applied. Additionally, enhanced coagulation and use of granular activated carbon (GAC) are recognized as best available technologies to remove organic materials prior to the addition of free chlorine and chloramines. These strategies focus on precursor removal prior to the addition of disinfectants. The GCDWQ (Health Canada 2014) recommends the removal of DBP precursors as the best strategy to limit THMs and HAAs (as discussed here), rather than any action that would compromise disinfection.

Alternatively, biologically active filters may remove HAAs that are already formed through biodegradation. Lou et al. (2014) showed that a pilot-scale biological filter significantly decreased the concentrations of HAAs, including the five regulated HAAs.

Alternative distribution system residual disinfectant. Switching to monochloramine, which has a lower formation potential of THMs and HAAs than free chlorine, as secondary disinfectant is another strategy that can be used at the treatment plant. Care must be taken when switching to chloramines as additional monitoring and formation control strategies must be used to ensure monochloramine is being formed and maintained within the distribution system. Distribution system monitoring that targets nitrification is also highly recommended and often required by state regulatory agencies when monochloramine is used as a distribution system residual disinfectant (see Chapter 6).

Reduce water age. The amount of time that water travels within the distribution system is referred to as "water age," or the number of hours or days that water is stored and conveyed prior to use or consumption (see Chapter 2). Minimizing water age through operational controls can be key to not only lowering THM and HAA formation but also to decreasing the overall disinfectant decay in the distribution system. Typical operational management controls include:

- Improving tank turnover and instituting adequate tank mixing to eliminate stratification and variations in water temperature and disinfectant residual;
- Eliminating oversized pipes and limiting dead ends through pipe looping or by operating valves to prevent the formation of stagnant zones;
- When possible, limiting the number of pressure zones in order to significantly decrease the overall water age in a distribution system; and
- Instituting flushing programs in low-use areas (see Chapter 2 more information on flushing).

Air Stripping THMs. In clearwells and storage tanks, air stripping can be instituted to remove volatile DBPs such as THMs. Fine-bubble diffusion, spray aeration, and surface aeration have all been documented as effective air-stripping methods for removing volatile DBPs (Brooke and Collins 2011; Ghosh et al. 2015).

The ability to remove volatile compounds by air stripping depends on the type of compounds, their relative concentration in water and air, variation of hydraulic residence time in the clearwell or storage tank, formation of THMs within the clearwell or tank as water ages, dilution of aerated water with fresh water, and contact time. Ghosh et al. (2015) reported that of these factors, contact time was the most influential in determining overall THM removal.

Air stripping is most effective on the chloroform species of THM. Bromoform species can also be removed by spray aeration at nearly the same rate as chloroform. An example of the effect of spray aeration on THM concentrations in a clearwell is illustrated in Figure 8-4; as illustrated, a 50-percent THM reduction was consistently achieved. Bubble-aeration removal rates for brominated species is much slower than for chloroform due to the differences in Henry's constants between these species (Ghosh et al. 2015).

Another point to consider is that aeration has been shown to have little to no effect on free chlorine and chloramine concentrations within the tank during the air-stripping process. THM aeration can also potentially have unintended negative consequences on distribution system water quality. During the aeration process, the water will quickly come to equilibrium with atmospheric carbon dioxide, which can affect the pH and the carbonate system. This may influence the aggressiveness of the water and trigger corrosion of distribution system structures (Ghosh et al. 2015).

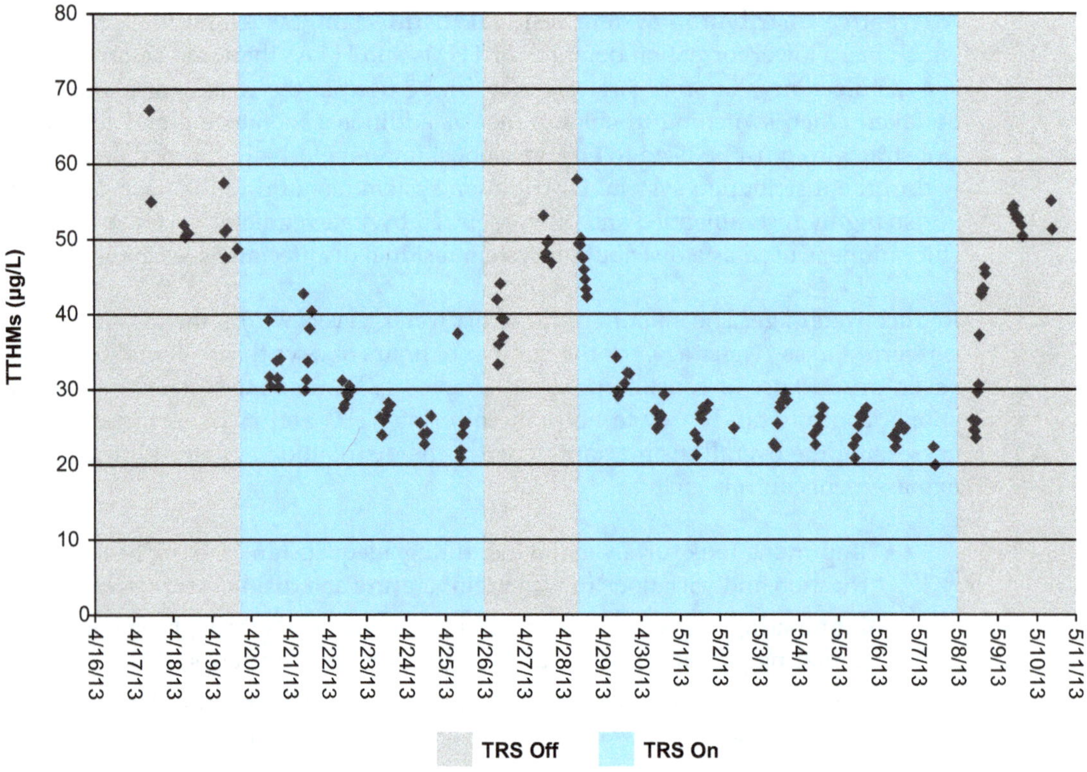

Source: PAX Water Technologies.

Figure 8-4 Effect of cycling spray nozzle aeration (on/off) on trihalomethane level in a 0.5-mil gal clearwell in Madison, North Carolina

Inorganic By-products

The inorganic DBPs that are discussed in this section can be generated by chlorine dioxide and ozone (such as chlorite, chlorate, and bromate) and also by free chlorine and chloramines (e.g., perchlorate, nitrosamines).

Chlorite and chlorate. Chlorite and chlorate are the main by-products of chlorine dioxide. Their occurrence is also related to the technique used to generate chlorine dioxide, whether hypochlorite salts are used instead of gaseous chlorine, the level of control over the generators, and exposure to light. Chlorate can also result from the decomposition of hypochlorite stock solutions (e.g., liquid bleach, calcium hypochlorite), which may lead to chlorate concentrations that are similar to those that result from the use of chlorine dioxide.

Chlorite decreases slightly with increasing water age in distribution systems where chlorine dioxide is used as a distribution system residual disinfectant, whereas chlorate remains more or less stable (Baribeau et al. 2002; Bergeon et al. 1992; Français et al. 1994; Lafrance et al. 1992; Olivieri et al. 1986; Ring et al. 2004; Volk et al. 2002). When chlorine dioxide is used as a pre- or intermediate oxidant with free chlorine as the distribution system residual disinfectant (in the absence of residual chlorine dioxide), chlorite concentrations tend to decrease in distribution systems, while variable trends were observed for chlorate concentrations. In certain cases, increases in chlorate concentrations were observed, resulting from the reaction between hypochlorite and chlorite (Gallagher et al. 1994; Gordon, Adam, and Bubnis 1995; Hautman and Bolyard 1991, 1992; Hoehn et al. 2003). When chlorine dioxide was followed by the addition of monochloramine (in the absence of a residual chlorine dioxide), decreases in chlorite concentrations with increasing water age were observed as well as fairly stable chlorate concentrations (Thompson et al. 1988; Briggs et al. 2007).

Control strategies to decrease chlorite after its formation include the use of reduced-sulfur compounds such as sulfur dioxide and sodium sulfite (which may lead to the formation of chlorate), activated carbon (granular or powder), and reduced-iron salts such as ferrous chloride and ferrous sulfate (which may lead to the production of chloride) (Gallagher, Hoehn, and Dietrich 1994; Iatrou and Knocke 1992; Singer and Reckhow 2011; Xie 2004).

Because there are no standardized treatment processes to remove chlorate once it is formed, chlorate is generally controlled by minimizing its formation during chlorine dioxide generation or storage of hypochlorite solutions (Alfredo et al. 2014; Singer and Reckhow 2011; USEPA 1999b). This being said, adsorption on GAC, ion-exchange processes, membrane biofilm reactors, and reverse osmosis (RO) have shown promising results (e.g., Alfredo et al. 2014).

Perchlorate. Perchlorate can be naturally occurring, but it is generally a manufactured chemical that is used to produce rocket fuel, fireworks, flares, and explosives. It is listed here because it is also an impurity present in hypochlorite solutions and because the USEPA was preparing a draft regulation when this manual was published.

Ion-exchange has been the preferred option for treatment of perchlorate in drinking water. The main disadvantage of this strategy is the production of a perchlorate-laden brine from the resin-regeneration process that needs to be treated or discharged, unless a single-pass process is used. Biological processes has been used successfully to treat high concentrations of perchlorate in industrial wastewater and is now gaining popularity for the removal of perchlorate from drinking water as well. It offers the advantage of destroying perchlorate without producing a waste brine. Modified GAC (metal doped or polymer doped) has also been tested. Filtration on an RO or nanofiltration membrane and electrodialysis reversal are also effective methods for removing perchlorate. Many of these treatment techniques have the added advantage of removing nitrate, which often co-occurs with perchlorate. Additional strategies have been tested, including chemical reduction, E-beam, and high-energy systems, but none have been implemented at full scale.

Bromate. Bromate is a well-known by-product of the reaction between ozone and hypobromite (OBr^-). It may also form from sunlight-catalyzed reactions between chlorine dioxide and bromide (Gordon, Emmert, and Bubnis 1995; Thompson 1989; Zhang et al. 2000; Zika et al. 1985).

Control strategies to decrease bromate formation include ozonation at lower pH, multistage ozonation (i.e., application of lower ozone doses at multiple points), use of ammonia to produce bromamines, and addition of hydrogen peroxide (Reckhow and Singer 2011; Singer and Reckhow 2011; Xie 2004). Once formed, bromate is somewhat difficult to remove, but chemical reduction using reduced sulfur compounds such as bisulfite or ferrous iron is possible. Adsorption on GAC, biodegradation, and UV treatment (with the formation of bromide) can also remove bromate (Singer and Reckhow 2011; Xie 2004).

Nitrosamines. Nitrosamines may be present in source waters from upstream industrial or wastewater discharge or they can be formed from reaction between disinfectants and precursors within the treatment plant or distribution system. Two possible NDMA formation mechanisms have been proposed (Mitch et al., 2003):

- NDMA can be formed through a nitrosation reaction, where organic nitrogen compounds react with nitrite to form NDMA and other n-nitroso compounds (Mirvish 1975).
- NDMA can form from the reaction of organic nitrogen compounds with chloramines, primarily dichloramine (Mitch and Sedlak 2002; Choi and Valentine, 2002; Schreiber and Mitch 2006).

Nitrosamine precursors may be naturally occurring but typically originate from wastewater effluents or from certain polymers or anion-exchange resins used for water treatment.

Conventional treatment processes (e.g., lime softening, coagulation with metal salts) that are effective at reducing the formation of regulated DBPs (i.e., TTHM and HAA5) show limited to no ability to remove NDMA precursors (Sacher et al. 2008). However, riverbank filtration has shown significant removals (90 percent removal of NDMA and 76 percent removal of an NDMA precursor, dimethylamine) (Schmidt et al. 2006). Other soil treatment processes such as aquifer recharge and recovery (ARR) and soil aquifer treatment have also been found to remove nitrosamines precursors (Sacher et al. 2008), with ARR resulting in an 88 percent reduction in NDMA formation potential (FP) and a 63 percent reduction in n-nitrosopyrrolidine (nitrosamine species) FP.

Oxidation with ozone (Sacher et al. 2008;), chlorine dioxide (Lee et al. 2007), or chlorine (Charrois and Hrudey 2007) can also be used to reduce NDMA formation in chloraminated water. Recent research has shown that chlorine and ozone can be more effective oxidants than chlorine dioxide (Shah et al. 2011;); however, the results vary depending on site-specific conditions.

UV photolysis and oxidation (i.e., AOP) have been applied to successfully degrade nitrosamines (99 percent removal of NDMA) in contaminated water at groundwater treatment plants. However, a high fluence (>400 mJ/cm^2) is required, which may be cost prohibitive (Sharpless and Linden 2003; Swaim et al. 2007). Several studies indicate a possibility for NDMA reformation after UV treatment (Mofidi et al. 2002; Liang et al., 2003), However, the observed increase in NDMA concentration from reformation is small (two-fold increase) compared to the amount of NDMA that is removed through UV photolysis.

A recent study suggests that UV photolysis can also destroy NDMA precursors at similar fluence as required for 1-log NDMA removal (Shah et al. 2011). Additional research is needed to confirm the applicability of UV (or UV/AOP) for precursor removal.

Nanofiltration has been found to reject more than 95 percent of NDMA precursors (Miyashita et al. 2009). However, rejection of nitrosamines varies from 9 to 75 percent depending on the membrane type. Similarly, RO is effective at removing NDMA precursors but shows mixed results for removing NDMA itself (Farré et al. 2011).

Additional strategies to control nitrosamine concentrations at the water treatment plant include optimizing disinfection procedures to provide free chlorine contact time (Sacher et al. 2008; Charrois and Hrudey 2007) and to minimize the formation of dichloramines (Schreiber and Mitch 2006) and reducing the dose of certain cationic polymers (Najm et al. 2004; Park et al. 2009). Within the distribution system, strategies include maintaining an adequate chloramine residual and chlorine to ammonia-N mass ratio to minimize dichloramine formation, reducing water age, and replacing or specifying alternate distribution system materials.

Other By-products

Beyond the regulated DBPs, a variety of other halogenated and nonhalogenated by-products can be formed during water disinfection and treatment. Some of the most common are briefly described here. For most, little to no information is available about their fate in drinking water distribution systems.

MX and E-MX in drinking water have been identified among the by-products that present the highest mutagenic effects. MX is 3-chloro-4-(dichloromethyl)-5-hydroxy-2(5H)-furanone, and E-MX is 2-chloro-3-(dichloromethyl)-4-oxobutenoic acid (Xie 2004). They are formed during chlorination of waters that contain organic materials.

In chlorinated waters, common halogenated DBPs include those that form when chlorine reacts with acetaldehydes (typically found in ozonated water) to produce trihaloacetaldehydes (trichloro- and tribromo-acetaldehydes are also known as chloral hydrate and bromal hydrate, respectively). In a national occurrence study on DBPs, Weinberg et al. (2002) reported that these compounds were widely present and at significant concentrations (low micrograms per liter levels). Mono- and dihaloacetaldehydes can also be formed in chlorinated waters. However, because they are typically oxidized into trihaloacetaldehydes, their occurrence and dominance in drinking water is limited (Xie 2004). The Australian government, represented by NHMRC and NRMMC (2011), proposes a maximum concentration of 0.02 mg/L for trichloroacetaldehyde (chloral hydrate) based on health considerations.

Halopropanones (also called haloacetones or haloketones) can be formed when chlorine reacts with propanone, which is a by-product of ozonation (Richardson et al. 1994, 2009; Weinberg et al. 2002; Xie 2004; Zhang et al. 2000). Halopropanones are common in chlorinated drinking waters but they are found at very low concentrations. They have also been detected in chlorine dioxide–treated waters but at even lower concentrations than in chlorinated waters. The most common forms are the dihalopropanones and the trihalopropanones.

Haloacetonitriles are commonly detected in chlorinated and chloraminated waters but have not been detected in significant concentrations in waters treated with chlorine dioxide. The common forms are the dihaloacetonitriles and trihaloacetonitriles; the monohaloacetonitriles are rarely found (Richardson et al. 2003, 2009; Weinberg et al. 2002; Xie 2004; Zhang et al. 2000). The WHO (2011) has established guidelines for dibromoacetonitrile and dichloroacetonitrile in drinking water at 0.07 mg/L and 0.02 mg/L, respectively.

In addition to haloacetonitriles, halonitromethanes and cyanogen halides may also form in chloraminated waters. These compounds were considered high-priority DBPs in the national occurrence study conducted by Weinberg et al. (2002). Trichloronitromethane is a common halonitromethane that is also known as chloropicrin. Monohalonitromethane and dihalonitromethane can also be formed but are not commonly detected in drinking waters. Cyanogen halides include cyanogen chloride and cyanogen bromide. The Australian government, represented by NHMRC and NRMMC (2011), has established a health-based limit of 0.08 mg/L for cyanogen chloride.

The most common nonhalogenated by-products are formed when ozone reacts with organic materials. These include aldehydes (the most common are formaldehyde, acetaldehyde, glyoxal, and methyl glyoxal), ketoacids (also called aldoacids or oxoacids), and carboxylic acids (which include oxalic, formic, and acetic acids). These compounds are low-molecular-weight polar organics that are much more amenable to biodegradation and, therefore, may promote microbial (re)growth in distribution systems.

Responding to the DSWQ Challenge: Minimizing Disinfection By-products

- Preserve source water quality and remove DBP precursors
- Control treatment processes at the plant
- Consider advanced oxidation processes
- Use alternative disinfectants for distribution system residual
- Reduce water age in the distribution system
- Consider air stripping to control THMs

BEST PRACTICES

The importance of maintaining disinfectant stability in distribution systems cannot be overstated. In fact, disinfectant residual monitoring and recording are required as part of Phase II Baseline and Annual Data Collection of the Distribution System Optimization Program of AWWA's Partnership for Safe Water (January 2011), which is a nonmandatory program comprising many participating water providers. As part of this program, the disinfectant residual database forms the primary element upon which systems conduct their comprehensive assessment of their system operations, which constitutes phase III of the program, Self-Assessment, and ultimately, Optimized Performance of the System (phase IV).

According to the AWWA Distribution System Optimization Program (January 2011), there are three main components of overall system integrity:

- Water quality preservation, for which disinfectant residual is the primary performance indicator;
- Hydraulic reliability, for which pressure is the primary performance indicator; and
- Physical security, for which main break frequency is the primary performance indicator.

The Partnership for Safe Water requires that all member systems use secondary disinfection. Systems are considered optimized when they meet the following disinfectant residual goals in at least 95 percent of their routine readings each month (January 2011):

- ≥0.20 mg/L and ≤4.0 mg/L Cl_2 for free chlorine,
- ≥0.50 mg/L and ≤4.0 mg/L Cl_2 for total chlorine (chloramines), and
- ≥0.20 mg/L and ≤0.80 mg/L ClO_2 for chlorine dioxide.

For a system to be considered optimized, sites that do not meet these goals cannot be remain below them in consecutive months within a distribution system.

The Partnership for Safe Water also addresses DBPs and states that THMs and HAA5 shall not exceed regulatory limits for each sample tested. Also, methods should be used to maintain or reduce levels even though disinfectant residuals may be increased at individual sites (January 2011).

The natural instability of DBPs may provide some removal as water flows in the distribution system, although this process generally contributes low DBP removals only. A variety of methods can be used to control DBPs, and these methods can be separated into four broad categories (Reckhow and Singer 2011; Xie 2004):

- Removal of precursors before disinfection or oxidation (e.g., enhanced coagulation, filtration, carbon adsorption, biodegradation, membrane filtration),
- Selection of disinfectants that form fewer DBPs (e.g., chloramine, chlorine dioxide, ozone),
- Adoption of disinfecting conditions that form fewer DBPs (e.g., lower disinfectant dose, shorter contact time, pH modification, alternative chlorination point), and
- Removal of DBPs after their formation (e.g., adsorption on GAC, biofiltration, aeration of the most volatile compounds, and potentially membrane filtration).

Table 8-10 summarizes the entities that can react with disinfectants and/or influence DBP formation and that can be included in a distribution system monitoring program. Many other parameters can influence disinfectant decay/demand and DBP formation in distribution systems, including water age, distribution system pipe material, and hydraulic conditions, but only the parameters that can be sampled and captured during routine monitoring are listed here. Table 8-10 should be used with caution. Monitoring programs are system specific and should be carefully tailored to the objectives of the program, system characteristics, system needs, and specific issues that may need to be considered.

Table 8-10 Entities that affect disinfectants and disinfection by-products in distribution systems and that could be included in a monitoring program

- Disinfectant residual
- Water temperature
- pH
- Organic material (total organic carbon, dissolved organic carbon, ultraviolet absorbance at 254 nm, specific ultraviolet absorbance, assimilable organic carbon, biodegradable organic carbon)
- Bromide, iodide
- Ammonia
- Nitrification indicators in chloraminated systems (see Chapter 6)
- Sulfur-based compounds
- Certain metals such as iron and manganese (see Chapter 7)
- Particulates (i.e., turbidity)
- Microbial biomass (see Chapter 3)

Regarding sampling sites, systems are considered optimized, per the Partnership for Safe Water, if sampling is performed at the following locations (January 2011):

- All finished water entry points and metering points to other systems,
- Stage 1 and/or Stage 2 DBP sites,
- If completed as part of the Stage 2 Disinfectant Byproduct Rule, Initial Distribution System Evaluation (IDSE) sites,
- Downstream of storage facilities (when water is being withdrawn),
- Upstream and downstream of booster disinfection stations,
- Areas served by older unlined cast iron mains,
- Low-flow areas, and
- Known high water age areas or other areas of concern.

Table 8-11 summarizes the best practices that can be used to preserve disinfectant stability and minimize DBPs. Many of these topics are discussed in other chapters of this manual; their effect on disinfectant residuals and DBPs are summarized here. Regarding the levels of service, basic practices are those actions that should be performed by all water providers that experience decreases in disinfectant residuals or increases in DBP concentrations. Basic practices also include actions that can be performed easily or are required by the regulatory agencies. Moderate practices offer additional options to limit disinfectant decay and DBP formation but may require resources that are not be available to all water providers. Advanced practices are more involved and require technical knowledge and a significant effort from water providers.

SUMMARY

The introduction of chemical disinfectants in water treatment is considered one of the greatest advancement to protect public health. However, disinfectants can react with organic and inorganic substances to form DBPs, some of which present health risks. Thus, water providers are facing the challenges of preserving disinfectant residuals in distribution systems to ensure that potable water remains free of microbial pathogens all the way to the customers' taps and also minimizing DBP formation.

The concentrations of chemical disinfectants generally decrease as water travels in the distribution system, and DBP concentrations may increase or decrease based on numerous factors. A thorough monitoring program is important to capture the magnitude of disinfectant decay and changes in DBP concentrations in distribution systems, identify the sites where the greatest changes occur, and determine the reasons for the changes observed. Any monitoring program is system specific and should be tailored to system characteristics, needs, and objectives.

Strategies that can be used to preserve disinfectant residuals in the distribution system are centered on water quality and the distribution system itself. Water quality factors include limiting the presence of organic and inorganic substances that can interfere with disinfectants and ensuring microbial water quality. This can be achieved either through source water protection or through the selection of adequate water treatment processes. Many issues discussed in other chapters of this manual help preserve disinfectant stability and limit the formation of DBPs, including controlling water age in pipes and storage facilities, limiting microbial growth and nitrification, limiting corrosion of materials and the presence of sediments in pipes and reservoirs, and preventing distribution system contamination.

Table 8-11 Best practices to preserve disinfectant stability and minimize disinfection by-products

Strategy	Effect on Disinfectant Residual	Effect on Disinfection By-Products	Level of Service
Source water protection	++ Help limit organic material and other constituents that contribute to disinfectant demand	++ Help decrease DBP formation	Moderate
Modify treatment strategy to decrease organic material	+++ Help preserve disinfectant stability	+++ Help decrease DBP formation	Moderate to advanced
Boost disinfectant residual in the distribution system	+++ Preserve disinfectant residual	+/- May form DBPs If boosting is used along with a lower disinfectant dose at the plant, DBP concentrations may be lower overall	Basic to moderate
Switch to monochloramine, either continuously or only in the warmer months	+++ More stable disinfectant residual	+++ Decrease the formation of regulated DBPs	Basic
Minimize water age in the distribution system	+++ Preserve disinfectant residual	+ Limit formation of most DBPs	Moderate to advanced
Minimize microbial growth and biofilms by limiting food sources of microorganisms (i.e., assimilable organic carbon, biodegradable organic carbon, ammonia, nitrite, and others)	++ Help preserve disinfectant residual	+ May limit the formation of the DBPs for which microbial biomass and products serve as precursor; it may limit the possibility of microbial degradation of certain DBPs	Advanced
Minimize release and/or accumulation of sediments and particulates by conducting routine flushing and reservoir cleaning	+ Help preserve disinfectant residual	+ May help limit DBP formation if particles harbor DBP precursors	Moderate
Control corrosion and corrosion scale formation	++ Preserve disinfectant residual	+/- Effect may vary based on the degree of pipe tuberculation and hydraulic conditions	Moderate
Prevent contamination (e.g., backflow, cross-connection)	++ Preserve disinfectant residual	+ May help limit DBP formation by limiting organic material that can serve as DBP precursors	Moderate

Abbreviation: DBP, disinfection by-product.

+++ Very effective or efficient.

++ Effective or efficient.

+ Somewhat effective or efficient, or in certain conditions only.

- Not effective or efficient.

REFERENCES

Alfredo, K.; C. Adams; A. Eaton; J.A. Roberson; & B. Stanford. 2014. *The Potential Regulatory Implications of Chlorate*. Briefing paper. Denver: CO: AWWA.

American Public Health Association (APHA); American Water Works Association (AWWA); & and Water Environment Federation (WEF). 2006 (20th ed.). *Standard Methods for the Examination of Water and Wastewater*. Washington, DC: APHA, AWWA, WPCF.

AWWA (American Water Works Association). 2013. *Hypochlorite Assessment Tool*. http://www.awwa.org/resources-tools/water-utility-management/hypochlorite-assessment-model.aspx (accessed Nov. 23, 2016).

———. 2006 (2nd ed.). M20, *Water Chlorination/Chloramination Practices and Principles*. Denver, CO: AWWA

Andrews, S. 2009. Canadian AOP Research/Practice for Treatment of Taste & Odour and EDC/PPCPs. *Americana*, March 17.

Baribeau, H.; N. Bazzurro; & G. Ziglio. 2009. Distribution Systems. In: *State of Science of Chlorine Dioxide in Drinking Water*. (D. Gates; G. Ziglio; & K. Ozekin, editors). Denver, CO: Water Research Foundation, and Genova, Italy: Fondazione AMGA.

Baribeau, H.; N.L. Pozos; L. Boulos; G.F. Crozes; G.A. Gagnon; S. Rutledge; D. Skinner; Z. Hu; R. Hofmann; R.C. Andrews; L. Wojcicka; Z. Alam; C. Chauret; S.A. Andrews; R. Dumancic; & E. Warn. 2005. *Impact of Changes in Distribution System Water Quality on Disinfection Efficacy*. Denver, CO: Water Research Foundation and AWWA.

Baribeau, H.; M. Prévost; R. Desjardins; P. Lafrance; & D.J. Gates. 2002. Chlorite and Chlorate Ion Variability in Distribution Systems. *Journal of the American Water Works Association*, 94(7):96–105.

Bergeon, T.; D. Cleret; J. Zagoury; M. Dernat; & M. Pouillot. 1992. Post-désinfection de l'eau potable au dioxyde de chlore. Qualité de l'eau distribuée et sous-produits d'oxydation. *L'eau, l'industrie et les nuisances*, 158:85–89.

Bolyard, M.; P.S. Fair; & D.P. Hautman. 1993. Sources of Chlorate Ion in U.S. Drinking Water. *Journal of the American Water Works Association*, 85:9:81–88.

Bousher, A., P. Brimblecombe, D. Midgley. (1989) Kinetics of Reactions in Solutions Containing Monochloramine and Bromide. *Water Research*, 23(8):1049-1058.

Briggs, D.A.; S. Weddle; A. Flores; H. Baribeau; E.A. Garvey; G.F. Crozes; T.P. Patten; R. Hofmann; K. Mowat; N. Paradis; M. Templeton; R.C. Andrews; E.G. Means III; J. Loveland; J.R. Leserman; M.J. McGuire; G.L. Amy; P. Schwer; C. Chauret; S.A. Andrews; and J.I. Musser. 2007. *Advanced Water Treatment of Estuarine Water Supplies*. Denver, CO: Water Research Foundation.

Brooke, E. & Collins, M.R., 2011. Posttreatment Aeration to Reduce THMs. *Journal of the American Water Works Association*, 103:10:84–96.

Charrois, J.W.A. and Hrudey, S.E. (2007). Breakpoint Chlorination and Free-Chlorine Contact Time: Implications for Drinking Water *N*-nitrosodimethylamine Concentrations. *Water Research,* 41(3), 674-682.

Choi, J. and R.L. Valentine. (2002). Formation of N-Nitrosodimethylamine (NDMA) from Reaction of Monochloramine: a New Disinfection By-Product. *Water Research,* 36:817-824.

Collins, J.; C. Cotton; B. Heeke; D. Dahl; & A. Royce. 2008. UV Advanced Oxidation Process for Taste and Odor Treatment: Evaluation of Assimilable Organic Formation Potential at an Indiana WTP. Proceedings AWWA Water Quality Technology Conference, Cincinnati, Ohio.

Connell, G.F. 1996. *The Chlorination/Chloramination Handbook.* Denver, CO: AWWA.

Dotson, A.D.; V.S. Keen; D. Metz; & K.G. Linden. 2010. UV/H2O2 Treatment of Drinking Water Increases Post-Chlorination DBP Formation. *Water Research,* 44:3707–3713.

Ellenberger, C.S.; R.C. Hoehn; D.L. Gallagher; W.R. Knocke; C.E. Via; E.V. Wiseman; R.W. Benninger; & A. Rosenblatt. 1998. Water Quality Impacts of Pure Chlorine Dioxide Pretreatment at the Roanoke County (Virginia) Water Treatment Plant. Proceedings AWWA Annual Conference and Exposition. Dallas, TX

Environmental Protection Agency, Office of Environmental Enforcement. 2011. *Water Treatment Manual: Disinfection.* Johnstown Castle, Co. Wexford, Ireland

European Union Council. 1998. *Council Directive 98/83/EC of 3 November 1998 on the quality of water intended for human consumption.* Accessed May 16, 2016. http://ec.europa.eu /environment/water/water-drink/legislation_en.html.

Farré MJ, Keller J, Holling N, Poussade Y, Gernjak W. (2011). Occurrence of N-nitrosodimethylamine precursors in wastewater treatment plant effluent and their fate during ultrafiltration-reverse osmosis membrane treatment. *Water Sci Technol.* 2011:63(4):605-12.

Français, T.; H. Suty; F. Mantisi; J.P. Gautier; M. Pouillot; & P. Hartemann. 1994. Sous-produits d'oxydation du dioxide de chlore: formation et devenir en réseau. Proceedings First International Research Symposium on Water Treatment By-products, Volume 1. Groupe de Recherche Universitaire sur les Techniques de Traitement et d'Épuration des Eaux, Poitiers, France. September 29–30, Poitiers, France.

Health Canada. 2013. *Guidance on the Use of the Microbiological Drinking Water Quality Guidelines.* Ottawa, Ontario: Water and Air Quality Bureau, Healthy Environments and Consumer Safety Branch, Health Canada, Ottawa, Ontario (catalogue no. H144-12/2013E-PDF).

———. 2014. Guidelines for Canadian Drinking Water Quality—Summary Table. Ottawa, Ontario: Water and Air Quality Bureau, Healthy Environments and Consumer Safety Branch.

Huang, X.; N. Gao; & Y. Deng. 2008. Bromate Ion Formation in Dark Chlorination and Ultraviolet/Chlorination Processes for Bromide-Containing Water. *Journal of Environmental Science*, 20:246–251.

Galal-Gorchev, H. and J.C. Morris. (1965) Formation and Stability of Bromamide, Bromimide, and Nitrogen Tribromide in Aqueous Solution. Inorganic Chemistry, 4:899–905.

Gallagher, D.L.; R.C. Hoehn; & A.M. Dietrich. 1994. *Sources, Occurrence, and Control of Chlorine Dioxide By-product Residuals in Drinking Water.* Denver, CO: Water Research Foundation and AWWA.

Gordon, G.; L. Adam; & B. Bubnis. 1995. Minimizing Chlorate Ion Formation. *Journal of the American Water Works Association*, 87;6:97–106.

Gordon, G.; G.L. Emmert; & B. Bubnis. 1995. Bromate Ion Formation in Water When Chlorine Dioxide is Photolyzed in the Presence of Biomide Ion. Proceedings AWWA Water Quality Technology Conference. New Orleans, LA

Ghosh, A.; C. Seidel; E. Townsend; R. Pacheco; & C. Corwin. 2015. *Reducing Volatile Disinfection Byproducts in Treated Drinking Water Using Aeration Technologies*. Denver, CO: Water Research Foundation, Water Environment Research Foundation, and USEPA.

Hautman, D.P. & M. Bolyard. 1991. Analysis of Inorganic Disinfection By-products Using Ion Chromatography. Proceedings AWWA Water Quality Technology Conference. Orlando, FL

Hautman, D.P. & M. Bolyard. 1992. Using Ion Chromatography to Analyze Inorganic Disinfection By- products. *Journal of the American Water Works Association*, 84:11:88–93.

Hoehn, R.C.; C.S. Ellenberger; D.L. Gallagher; E.V. Wiseman Jr.; R.W. Binninger; & A. Rosenblatt. 2003. ClO2 and By-product Persistence in a Drinking Water System. *Journal of the American Water Works Association*, 95:4:141–150.

Iatrou, A. & W.R. Knocke. 1992. Removing Chlorite by the Addition of Ferrous Iron. *Journal of the American Water Works Association*, 84:11:63–68.

Kargalioglu, Y.; B.J. McMillan; R.A. Minear; & M.J. Plewa. 2002. Analysis of the Cytotoxicity and Mutagenicity of Drinking Water Disinfection By-products in *Salmonella typhimurium. Teratogenesis, Carcinogenesis, and Mutagenesis*, 22:2:113–128.

Krasner, S.W. 1999. Chemistry of Disinfection By-product Formation. In: *Formation and Control of Disinfection By-products in Drinking Water.* (P.C. Singer, editor). Denver, CO: AWWA; pp. 27–52.

Lafrance, P.; D. Duchesne; N. Arcouette; & M. Prévost. 1992. The Use of ClO2: Case Study of Laval. Proceedings Second International Symposium on Chlorine Dioxide: Drinking Water Issues. Denver, CO: Water Research Foundation.

Langlais, B.; D.A. Reckhow; & D.R. Brink (editors). 1991. *Ozone in Drinking Water Treatment: Application and Engineering.* Boca Raton, FL: Water Research Foundation and Lewis Publishers.

Lasagna, C.; R. Binetti; & B. Bubnis. 2009. The Measurement of Residuals and Monitoring Strategies. In: *State of Science of Chlorine Dioxide in Drinking Water* (D. Gates; G. Ziglio; & K. Ozekin(editors). Denver, CO: Water Research Foundation, and Genova, Italy: Fondazione AMGA.

Lee, C., Schmidt, C., Yoon, J., and von Gunten, U. (2007). Oxidation of N-Nitrosodimethylamine (NDMA) Precursors with Ozone and Chlorine Dioxide: Kinetics and Effect on NDMA Formation Potential. *Environ. Sci. Technol.*, 2007, 41 (6), pp 2056–2063.

Li, H.; G.R. Finch; D.W. Smith; & M. Belosevic. 2001. Chlorine Dioxide Inactivation of *Cryptosporidium parvum* in Oxidant Demand-free Phosphate Buffer. *Journal of Environmental Engineering*, 127: 593–603.

Liang, S., H.M. Joon, M.K. Davis, J.F. Green, and D.S. Remer. 2003. Use of pulsed-UV processes to destroy NDMA. Journal of the American Water Works Association. 95(9): 121-131.

Lou, J.C.; H.Y. Chan; C.Y. Yang; W.B. Tseng; & J.Y. Han. 2014. Reducing and Verifying Haloacetic Acids in Treated Drinking Water Using a Biological Filter System. *Journal of Environmental Science and Health, Part A: Toxic/Hazardous Substances and Environmental Engineering*, 49:14:1693–1700.

Margerum, D.W., L.M. Schurter, J. Hobson, and E.E. Moore. (1994). Water Chlorination Chemistry: Nonmetal Redox Kinetics of Chloramine and Nitrite Ion. *Environmental Science & Technology*. 28(2):331–337

Mirvish, S. S. (1975). Formation of *N*-Nitroso Compounds: Chemistry, Kinetics, and *In-Vivo* Occurrence. *Toxicol. Appl. Pharmac.* 31:325–351.

Mitch, W.A., Gerecke, A.C., and Sedlak, D.L. (2003). A N-Nitrosodimethylamine (NDMA) precursor analysis for chlorination of water and wastewater. *Water Research*. 2003 Sep;37(15):3733–41.

Mitch, W.A., Sedlak, D.L., (2002). Formation of N-nitrosodimethylamine (NDMA) from Dimethylamine during Chlorination. *Environmental Science & Technology*. 36 (4), 588–595.

Miyashita, Y., Sang-Hyuck, P., Hyung, H., Huang, C., Kim, J. (2009). Removal of *N*-Nitrosamines and Their Precursors by Nanofiltration and Reverse Osmosis Membranes. *Journal of Environmental Engineering*. Vol. 135, Issue 9. September 2009.

Mofidi, A.A., Min, J.H., Palencia, L.S., Coffey, B.M., Liang, S., and Green, J.F. (2002). Advanced Oxidation Processes and UV Photolysis for Treatment of Drinking Water. Metropolitan Water District of Southern California. La Verne, California. Submitted as part of a report to the California Energy Commission. January 2002.

Mofidi, A.A.: T.I. Yun; S.W. Krasner; S. Liang; and B.M. Coffey. 2004. Formation and Removal of BOM Produced by Chlorine Dioxide Treatment: A Pilot Study. Proceedings AWWA Water Quality Technology Conference San Antonio, TX.

Najm, I; Teefy, S.; Barrett, S.E. (2004). NDMA Formation from Water Treatment Polymers. Proceedings, American Water Works Association Annual Conference, Orlando, Florida.

NHMRC (National Health and Medical Research Council) & NRMMC (Natural Resource Management Ministerial Council). 2011. *Australian Drinking Water Guidelines.* Paper 6, National Water Quality Management Strategy. Commonwealth of Australia, Canberra: National Health and Medical Research Council, National Resource Management Ministerial Council.

Olivieri, V.P.; M.C. Snead; C.W. Krusé; & K. Kawata. 1986. Stability and Effectiveness of Chlorine Disinfectants in Water Distribution Systems, *Environmental Health Perspectives,* 69:15–29.

Park, S.H., S. Wei, B. Mizaikoff, A.E. Taylor, C. Favero, and C.H. Huang. (2009). Degradation of Amine-Based Water Treatment Polymers during Chloramination as N-Nitrosodimethylamine (NDMA) Precursors. Environ. Sci. Technol.,43(5):1360-1366.

Plewa, M.J.; Y. Kargalioglu; D. Vankerk; R.A. Minear; & E.D. Wagner. 2002. Mammalian Cell Cytotoxicity and Genotoxicity Analysis of Drinking Water Disinfection Byproducts. *Environmental and Molecular Mutagenesis,* 40:2:134–142.

Reckhow, D.A. 1999. Control of Disinfection By-product Formation Using Ozone. In: *Formation and Control of Disinfection By-products in Drinking Water* (P.C. Singer, editor). Denver, CO: AWWA; pp. 179–204.

Reckhow, D.A. & Singer, P.C. 2011 (6th ed.). Formation and Control of Disinfection Byproducts. In: *Water Quality and Treatment; A Handbook on Drinking Water* (J.K. Edzwald, editor). Denver, CO: AWWA.

Richardson, S.D.; C. Rav-Acha; & G.D. Simpson. 2009. Chlorine Dioxide Chemistry, Reactions, and Disinfection By-Products. In: *State of Science of Chlorine Dioxide in Drinking Water.* (D. Gates; G. Ziglio; & K. Ozekin, editors). Denver, CO: Water Research Foundation, and Genova, Italy: Fondazione AMGA.

Richardson, S.D.; A.D. Thruston Jr.; T.W. Collette; K.S. Patterson; B.W. Lykins Jr.; G. Majetich; & Y. Zhang. 1994. Multispectral Identification of Chlorine Dioxide Disinfection Byproducts in Drinking Water. *Environmental Science and Technology,* 28:592–599.

Richardson, S.D.; A.D. Thruston Jr.; C. Rav-Acha: L. Groisman; I. Popilevsky; V. Glezer; A.B. McKague; M.J. Plewa; & E.D. Wagner. 2003. Tribromopyrrole and Other DBPs Produced by the Disinfection of Drinking Water Rich in Bromide. *Environmental Science and Technology,* 37:17:3782–3793.

Richardson, S.D.; M.J. Plewa; E.D. Wagner; R. Schoeny; & D.M. DeMarini. 2007. Occurrence, Genotoxicity, and Carcinogenicity of Regulated and Emerging Disinfection By-products in Drinking Water: A Review and Roadmap for Research. *Mutation Research,* 636:178–242.

Ring, E.; M. Samblebe; M. Leak; & S. Gray. 2004. NOM Removal for Extension of Chlorine Dioxide Residuals and Lower Biological Regrowth Potentials. *Water Science and Technology: Water Supply,* 4:4:251–254.

Sacher, F., C.K. Schmidt, C. Lee, and U. von Gunten. (2008). Strategies for Minimizing Nitrosamine Formation During Disinfection. Denver, Colo.: AwwaRF.

Schmidt, C.K., F. Sacher, and H. Brauch. (2006) Strategies for minimizing formation of NDMA and other nitrosamines during disinfection of drinking water. Water Quality Technology Conference 2006 Proceedings. Denver, Colo.

Schreiber, I.M.; Mitch, W.A. (2006). Nitrosamine formation pathway revisited: The importance of chloramine speciation and dissolved oxygen. Environmental Science & Technology, 40(19), 6007-6014.

Shah, A.D., A.D. Dotson, K.G. Linden, and W.A. Mitch. (2011). Impact of UV Disinfection Combined with Chlorination/Chloramination on the Formation of Halonitromethanes and Haloacetonitriles in Drinking Water. Environ. Sci. Technol., 45(8):3657-3664.

Sharpless, C.M & K.G. Linden. 2001. UV Photolysis of Nitrate: Effects of Natural Organic Matter and Dissolved Inorganic Carbon and Implications for UV Water Disinfection, *Environmental Science & Technology*, 35:14:2949–2955.

Singer, P.C. and Reckhow, D.A., 2011 (6th ed.). Chemical Oxidation. In: *Water Quality and Treatment; A Handbook on Drinking Water* (J.K. Edzwald, editor). Denver, CO: AWWA.

Speitel, G.E. 1999. Control of Disinfection By-product Formation Using Chloramines. In: *Formation and Control of Disinfection By-products in Drinking Water.* (P.C. Singer, editor). Denver, CO: AWWA; pp. 161–177.

Suffet, I.H.; C. Anselme; & J. Mallevialle. 1986. Removal of Tastes and Odors by Ozonation. Proceedings AWWA Seminar on Ozonation: Recent Advances and Research Needs. Denver, CO, AWWA.

Swaim, P.; A. Royce; T. Smiths; T. Maloney; D. Ehlen; & B. Carter. 2007. Effectiveness of UV Advanced Oxidation for Destruction of Micro-Pollutants. Proceedings 4th IUVA World Congress. Florence, Ken.: International Ultraviolet Association (8459-F US 42 #245. 4).

Thompson, A.L. 1989. Practical Considerations for Applications of Chlorine Dioxide in a Municipal Water System. Proceedings First International Chlorine Dioxide Symposium: Scientific, Regulatory and Application Issues. Denver, CO: Water Research Foundation.

Thompson, A.L.; N.O. Matthews; A.J. Mittl; & D.M. Owen. 1988. Chlorite and Chlorate Residuals in the Distribution System. Proceedings AWWA Water Quality Technology Conference. St. Louis, MO

USEPA (US Environmental Protection Agency). 1999a. Ozone. In: *Alternative Disinfectants and Oxidants Guidance Manual*. Washington, DC: USEPA, Office of Water (815-R-99-014).

———. 1999b. Chlorine Dioxide. In: *Alternative Disinfectants and Oxidants Guidance Manual.* Washington, DC: USEPA, Office of Water (815-R-99-014).

———. 2013. *Distribution System Comprehensive Evaluation Report.* Cincinnati, Ohio: USEPA Technical Support Center.

———. 2016. Six-Year Review 3 of Drinking Water Standards. https://www.epa.gov/dwsixyearreview/six-year-review-3-drinking-water-standards (accessed March 10, 2017).

Valentine, R.L. (1985). Disappearance of Monochloramine in the Presence of Nitrite. In Water Chlorination: Chemistry, Environmental Impact and Health. Effects Volume 5. Edited by R.L. Jolley et al. Chelsea, Mich.: Lewis Publishers.

Valentine, R.L., P.J. Vikesland, B.C. Angermand, S.A. Hackett, M. Shoup and S. Slattenow. (2000). The Role of the Pipe-Water Interface in DBP Formation and Disinfectant Loss. Denver, Colo: AwwaRF and AWWA.

Volk, C.J.; R. Hofmann; C. Chauret; G.A. Gagnon; G. Ranger; & R.C. Andrews. 2002. Implementation of Chlorine Dioxide Disinfection: Effects of the Treatment Change on Drinking Water Quality in a Full-Scale Distribution System. *Journal of Environmental Engineering and Science,* 1:323–330.

Wang, D.; J.R. Bolton; S.A. Andrews; & R. Hofmann. 2015. Formation of Disinfection By-Products in the Ultraviolet/Chlorine Advanced Oxidation Process. *Science of The Total Environment,* 518–519:49–57.

Weinberg, H.S.; S.W. Krasner; S.D. Richardson; & A.D. Thruston Jr. 2002. *The Occurrence of Disinfection By-products (DBPs) of Health Concern in Drinking Water: Results of a Nationwide DBP Occurrence Study.* EPA/600/R-02/068. Athens, GA: National Exposure Research Laboratory and USEPA.

Wetterau, G.; P. Fu; C. Chang; & B. Chalmers. 2015. Full-scale Testing of Alternative UV Advanced Oxidation Processes for the Vander Lans Water Treatment Facility. Proceedings WateReuse California Annual Conference, Los Angeles, Calif. Alexandria, VA: WateReuse Association.

Woo, Y.T.; D. Lai; J.L. McLain; M.K. Manibusan; & V. Dellarco. 2002. Use of Mechanism-Based Structure–Activity Relationships Analysis in Carcinogenic Potential Ranking for Drinking Water Disinfection By-Products. *Environmental Health Perspectives,* 110(Supplement 1):75–87.

WHO (World Health Organization). 2011 (4th ed.). *Guidelines for Drinking-Water Quality.* Geneva, Switzerland: WHO.

Xie, Y.F. 2004. *Disinfection Byproducts in Drinking Water; Formation, Analysis and Control.* Lewis Publishers: Boca Raton, FL.

Zhang, X.; S. Echigo; R.A. Minear; & M.J. Plewa. 2000. Characterization and Comparison of Disinfection By-products of Four Major Disinfectants. In: *Natural Organic Matter and Disinfection By- products: Characterization and Control in Drinking Water* (S.E. Barrett; S.W. Krasner; & G.L. Amy, editors). Washington, DC: American Chemical Society.

Zheng, M.; S.A. Andrews; & J.R. Bolton. 1999. Impacts of Medium-Pressure UV and UV/H2O2 Treatments on Disinfection Byproduct Formation. Proceedings AWWA Annual Conference, Chicago.

Zika, R.G.; C.A. Moore; L.T. Gidel; & W.J. Cooper. 1985. Sunlight-Induced Photodecomposition of Chlorine Dioxide. In: *Water Chlorination: Chemistry, Environmental Impact, and Health Effects,* Vol. 5. (R.L. Jolley; W.A. Brungs; R.B. Cumming; J.S. Mattice; & V.A. Jacobs, editors). Chelsea, MI. Lewis Publishers.

Chapter 9

Management of Low Pressure

Jian Yang, American Water–Engineering, Voorhees, N.J.

Mark LeChevallier, American Water, Voorhees, N.J.

Thomas M. Walski, Bentley Systems, Nanticoke, Pa.

Water systems must provide continuous and adequate pressure to ensure that contaminants do not migrate into their distribution systems. Water utilities are expected to provide potable water 24 hours a day, 7 days a week (24/7). If the goal is to provide potable water, it is essential for those systems to work toward a 24/7 water supply since it is impossible to ensure water quality when a system is not continuously pressurized. In the United States, for example, this is usually expressed as a requirement to provide at least 20 psi at ground elevations under all conditions, although higher minimum pressures, for example, 35 psi, may be targeted during normal operations.

Loss of system pressure or depressurization usually does not occur during normal operation in well-designed and well-operated systems. Low or negative pressures are the result of some exceptional occurrence that may last for an extended period (hours) or just a few seconds (e.g., hydraulic transients). This chapter describes low or negative pressures under transient conditions as well as depressurization for an extended period.

DSWQ Challenge:

- Water utilities must provide continuous and adequate pressure to ensure that contaminants do not migrate into their distribution systems.
- Low or negative pressures under transient conditions, as well as depressurization for an extended period, can result in contamination through leaks, cross-connections, or air valves (i.e., potential points for intrusion).

Hydraulic transients, sometimes referred to as "water hammer" or "surge," are caused by rapid changes in the momentum of flow and can introduce large pressure forces to a distribution system. Low or negative pressure transients may create a temporary breach of hydraulic integrity, allowing chemical and microbial contaminants to enter the distribution system through leaks, cross-connections, or air valves (i.e., potential points for intrusion). Additionally, high pressure transients may result in pump failures, joint separation, and an increase in the number of water main breaks. Persistent high and low pressure fluctuations can also contribute to stresses and strains that slowly fatigue and weaken distribution system piping, which eventually leads to pipe leaks/failures, pressure loss, and/or potential contamination. Low pressure transients may vaporize water, form bubbles in the pipeline, and cause water-column separation, which could lead to catastrophic pipe failures after collision of the two water columns.

An understanding of low pressure and hydraulic transients is essential for water system professionals who are involved in planning, designing, and operating safe drinking water distribution systems. This chapter covers the fundamentals of low pressures, hydraulic transients, potential for contamination, pressure monitoring, hydraulic and transient modeling, devices, guidelines, and procedures to minimize the likelihood of contamination due to low pressure. Pressure management is only one part of the multibarrier approach to ensuring safe drinking water. Other best practices and management strategies related to preventing depressurization-caused water quality problems are provided in Chapters 2, 4, 8, and 10.

PRESSURE STANDARDS AND GOALS

There are no universally established pressure standards for water distribution systems. Ghorbanian, Karney, and Guo (2016) summarized water distribution pressure standards based on a review of guidelines and regulations from around the world (Asia, Australia, Europe, and North and South America). In the United Kingdom, Brazil, and Hong Kong, the minimum pressure standards under all flow conditions are 14 psi, 22 psi, and 29 psi, respectively. In Australia, the minimum pressure standard is 29 psi during fire flow conditions. In the United States and Canada, many states and provinces have a minimal pressure requirement of 20 psi in water distribution systems under all flow or fire flow conditions, as listed in the Ten States Standards (2012).

Under normal flow and operating conditions, minimum and maximum pressure requirements are often either nonexistent or highly variable among water systems. Local regulatory agencies generally have pressure standards or regulations in place for planning, regulatory approval, or as operating guidelines. Water system professionals are often advised to carefully review regulatory requirements and to work closely with regulators when establishing pressure standards and goals. Beyond regulatory requirements, the AWWA Partnership for Safe Water (PSW 2011) previously recommended the following pressure performance goals:

- Above 20 psi under maximum day demand and fire flow conditions (excluding emergencies),
- Above 35 psi under normal conditions,
- Emergency conditions such as main breaks and power outages require implementation of preapproved procedures to protect public health, and
- The maximum pressure range (difference between the minimum and maximum within a pressure zone) should not be greater than a predetermined system-specific goal.

The first low pressure goal of 20 psi and the preapproved emergency condition procedures are related to the adequacy of a distribution system to protect public health. Systems that do not meet these two goals usually need specific operation mitigations or capital improvements such as providing additional storage or pipe improvements. The second goal of maintaining pressure above 35 psi under normal conditions is related to both public health protection and system efficiency. This goal may vary significantly among different systems. It results in greater customer satisfaction and also provides systems a cushion for pressure fluctuation during fire flow and emergency conditions. Areas with high elevations or low pressures from undersized mains might be more susceptible to negative pressure events and risk of intrusion. Conversely, it may not always be efficient to maintain the minimum normal pressures that is too high at the expense of increasing energy usage, main breaks, and leakage.

The pressure goal for the maximum pressure range is related to distribution system efficiency (to reduce unnecessary water loss, main breaks, and/or energy usage). Many water systems do not have specific goals to control high pressure or pressure variation at a particular location. In a system with pressures greater than 100 psi, a pressure-reduction project may not be cost effective. Similarly, there may be a few spots in the system where fluctuations are normal and acceptable, such as at the fringe of a system with limited capacity. Also, some systems intentionally turn down the pressure at night to reduce leakage (LeChevallier et al. 2014). The cost and benefit of planned system modifications should be evaluated to determine if the modifications will be economical. Systems that meet the predetermined system-specific goal may still have options for further optimization. For example, a system equipped with tight pressure control might consider further reducing system pressures to reduce unnecessary leakage, main breaks, or energy costs.

Finally, when changes are made in the system, the history of pressure delivery to the customer should be considered. For example, an industry might be fully adjusted to a very high pressure in order to perform some of its water-related functions, and a large pressure drop might impact their business. In practice, water system professionals should jointly consider regulatory requirements, system-specific operations and conditions, the cost of implementation, customers' expectations of service pressure, and anticipated benefits from the reduced pressures.

CAUSES OF DEPRESSURIZATION AND INTRUSION

Pressure in water distribution systems is usually positive, and contaminants are not likely to enter the system. However, in some situations (e.g., hydraulic transients, power outages, main breaks and large demand events) pressure can drop to the point where nonpotable water enters the water distribution system.

Hydraulic Transients

Figure 9-1 shows a pressure transient observed in a distribution system with pressure as low as −10 psi or −69 kPa (LeChevallier et al. 2003). The duration of negative pressure was observed to be longer than 16 s.

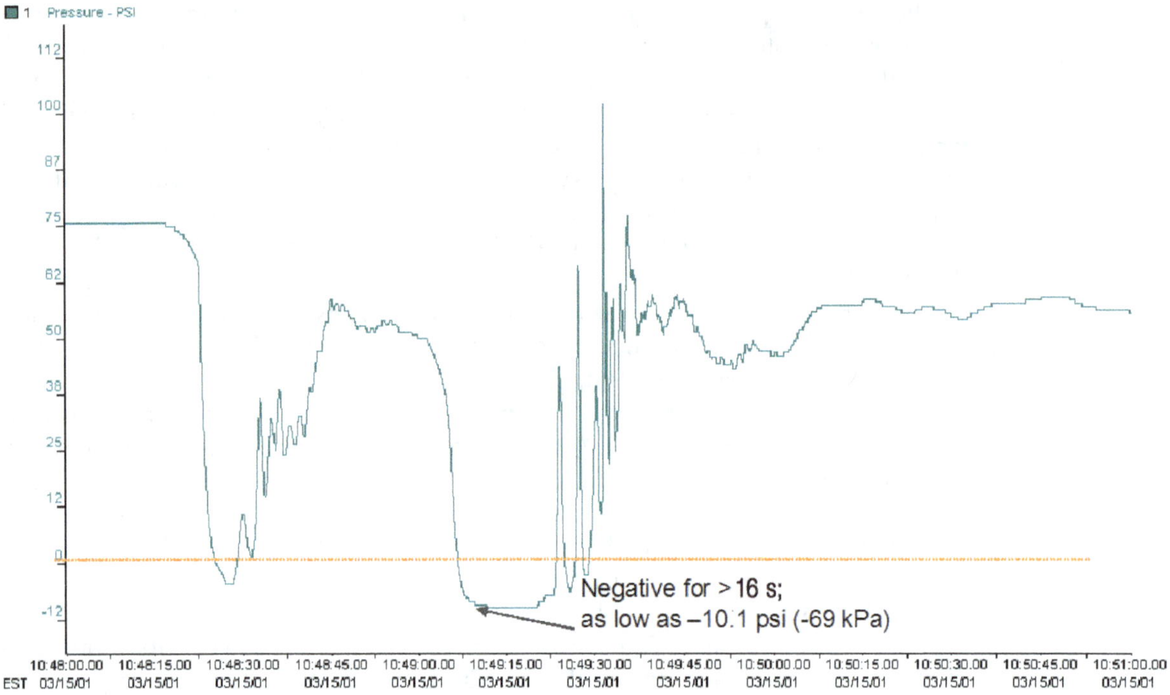

Source: LeChevallier et al. 2003.

Figure 9-1 Example of a pressure transient

For a quick and conservative estimation, every 1 ft/s of velocity change can generate a pressure transient with a magnitude of about 100 ft, or 43 psi. The actual magnitude of a pressure transient can be influenced by construction materials, pipe characteristics, and water velocity changes. Because pressure transients can propagate throughout a distribution system, the magnitude can be further complicated by distribution system characteristics and operational activities. For example, the presence of small quantities of entrapped air can significantly reduce the wave propagation speed in the pipeline. Other factors such as placement of air relief valves and other transient control devices, pipe friction, network water demands, and leaks also help reduce the magnitude of the pressure wave generated (Karney and Filion 2003).

The occurrence of pressure transients may be common, as these transients can be caused by customer demand variations and routine distribution system operations and maintenance (O&M) activities such as valve closures, hydrant flushing, hydrant flow testing, and pump start and shutoff. The duration of pressure transients is influenced by the size of the pressure zone, flow control operations, and system characteristics such as the presence of storage tanks, entrained air, and pipe leaks. While transients can be gradually dampened in systems due to friction effects, the period of greatest concern is the duration of negative or low pressures (e.g., <20 psi). Gullick et al. (2005) showed that negative pressures lasted from about 1 s to more than 160 s in four of eight distribution systems studied. The volume of intrusion can range from milliliters to thousands of liters depending on the effective size of the pipe break, external pressure, and the nature of the transient event (Kirmeyer et al. 2001). In pilot-scale investigations, estimated intrusion volumes were shown to be up to 50 mL and 127 mL through ⅛-in. and ¼-in. orifices, respectively, when 132 gpm of flow was brought to a stop with the sudden closure (less than 1 s) of a 2 ½-in. ball valve (Boyd et al. 2004a, 2004b).

Power Outage

The most severe low or negative pressure transients are often caused by unplanned events, for example, a power outage. Figure 9-2 shows the number of times pump stations lost power per year in more than 30 water systems surveyed (LeChevallier et al. 2011, 2014). The average number of power outages for these systems was about 3.6 to 4.7 per year. This is comparable to a water industry average of 5.0 power outages per year reported by the Electric Power Research Institute (Grebe, Sabin, and McGranaghan 1996).

Main Breaks, Firefighting, Flushing, and Large Demand Events

More than 700 water main breaks and repairs are estimated to occur in the United States every day, many of which may be small and repaired with a clamp while pressure is maintained in the pipe. In these instances, there is little chance for contaminants to enter the distribution network. However, at the other end of the spectrum, some breaks are much larger, even catastrophic and may result in widespread depressurization. Such events are likely to involve removal and replacement of sections of pipe and valves. This type of break has the potential to allow entry of microbial (and chemical) contaminants, both at the repair site and in the depressurized areas of the distribution system farther away from the break area. Similarly, firefighting, flushing, and other large-demand events (e.g., irrigation) may cause localized or regional low pressures and potential intrusion.

Cross-Connections

Cross-connections are actual or potential connections with a nonpotable source, where it is possible for a contaminant to enter the water distribution system. Chapter 10 discusses cross-connections in more depth. If a cross-connection exists, backflow may occur if the pressure in the distribution system drops below the external pressure (i.e., backsiphonage) or if the external pressure increases above the system pressure (i.e., backpressure). In a survey of more than 700 North American distribution systems, Lee et al. (2003) found that 65 percent of the systems had cross-connections that were susceptible to backflow via backsiphonage, while the remaining 35 percent had cross-connections that were susceptible to backflow via both backsiphonage and backpressure. This indicates that all surveyed systems had locations susceptible to backflow of nonpotable water during a low or negative pressure event.

PUBLIC HEALTH IMPACT OF DEPRESSURIZATION-RELATED WATER QUALITY PROBLEMS

Pathogen and indicator microorganisms were detected in undisturbed soil and water samples immediately near distribution system pipelines (Karim et al. 2003). During pressure-loss events, even those lasting a few seconds, there are many potential entry portals in a distribution system for contaminants to enter, including leakage points in water mains, areas where the water table is high, submerged air valves, faulty seals and joints, and cross-connections.

Contaminants Near Drinking Water Pipelines

Survey results (Kirmeyer et al. 2001) showed that at least 20 percent of systems had pipes below the water table. About 20 percent of systems reported between 25 and 75 percent of meter boxes being flooded (depending on the season). Figure 9-3 shows examples of potential intrusion sites such as a broken water main near sewage or storm pipes and a flooded vault. Standing water in vaults or meter pits may enter the distribution system through leaks, air inlets, or faulty seals and joints.

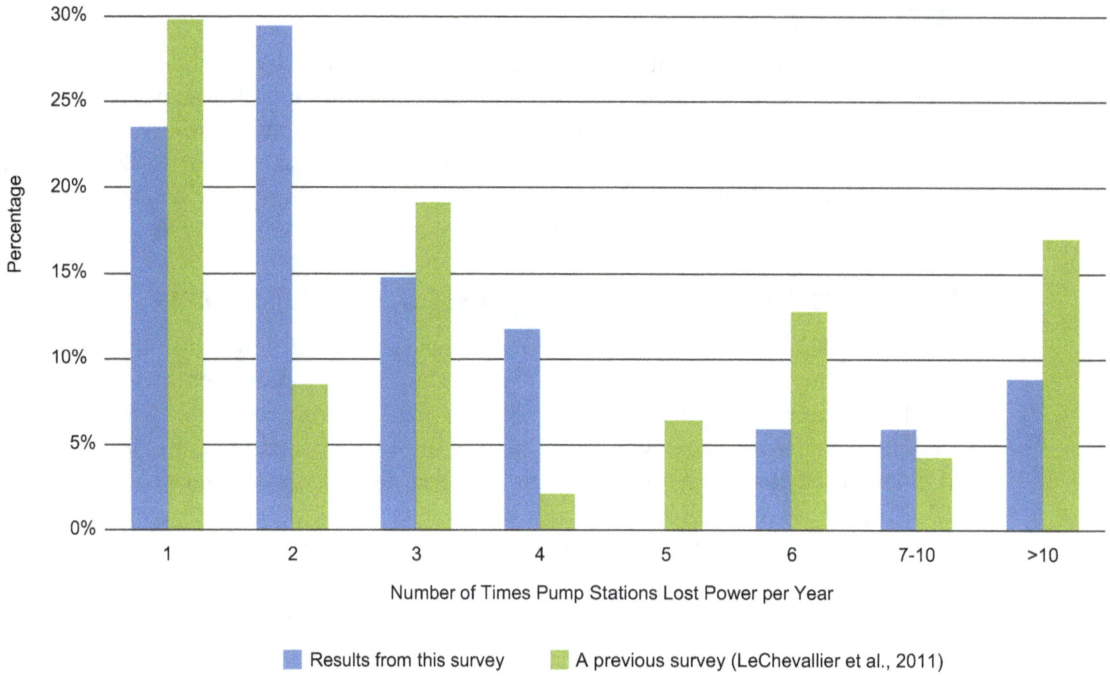

Source: LeChevallier et al. 2014.

Figure 9-2 Water system power outages per year

Source: LeChevallier et al. 2011.

Figure 9-3 Examples of potential intrusion sites: (a) broken main near sewer pipe, (b) broken main near storm pipe, and (c) flooded meter vault

Potential chemical contaminants that exist near distribution systems include pesticides, petroleum products, fertilizers, solvents, detergents, and pharmaceuticals. If sufficient concentration or volume of chemical intrudes during a pressure-loss event (e.g., via cross-connection), it may cause acute toxicity.

The intrusion of microbial contaminants is of greater concern because even with dilution, some microbes (e.g., noroviruses) can cause an infection at very low numbers

(Teunis et al. 2008; Messner et al. 2014). Microbial contamination could pose a high risk if distribution water has no or inadequate disinfectant residuals. Karim and colleagues (2003) examined 66 soil and water samples immediately adjacent to drinking water pipes from eight water systems in six US states. Total coliform and fecal coliform were detected in about half of the water and soil samples, indicating the potential presence of fecal contamination. About 56 percent of the samples were positive for human enteric viruses. In addition, fecal coliform levels in some soil samples were greater than 1.6×10^4 cfu/100 g of soil, suggesting the sampling locations were under the influence of leaking sewage.

Public Health Risks

Pathogens and indicator microorganisms are often detected in undisturbed soil and water samples immediately adjacent to distribution system pipelines. Consequently, low or negative pressure events may create pathways for direct intrusion from outside sources such as pipe leaks, faulty joints, cross-connections, and water main installation and repair sites. Negative pressure transients occur frequently in some distribution systems due to routine distribution system O&M activities. In addition, intrusion can occur downstream from the point of intrusion (Walski and Lutes 1994; Friedman et al. 2004).

Some associations can be made between disease outbreaks and the occurrence of low and negative pressure events. Nygård et al. (2007) conducted a cohort study of drinking water consumers in Norway and found that households exposed to water downstream of a main break or maintenance were 1.6 times more likely to report gastrointestinal illness than control households that were not exposed to such distribution system events.

Epidemiological studies (Payment et al. 1997; Tinker et al. 2009) suggest that the distribution system was at least partially responsible for increased levels of gastrointestinal illnesses. In the absence of other plausible explanations, Payment et al. speculated that low disinfectant residuals and negative pressure transients in the distribution system contributed to the viral-like etiology of the observed illnesses. Borchardt et al. (2007) reported increased levels of gastrointestinal viruses (primarily adenovirus, enterovirus, and norovirus) in eight groundwater systems that were primarily due to main breaks and low-pressure events. In April 2002, a *Giardia* outbreak occurred at a trailer park in New York state, causing six residents to become seriously ill (Blackburn et al. 2004). The contamination was attributed to a power outage that created a negative pressure transient in the distribution system. The transient allowed water to enter the system through either a cross-connection inside a mobile home or through a leaking underground pipe that was near sewer crossings. Hunter et al. (2005), in a subanalysis of controls from a case-control study, determined that there was a strong association between diarrheal illnesses in the controls and low water pressure at their home faucets.

Characterizing the DSWQ Challenge:

- Field measurements methods: supervisory control and data acquisition monitoring, hydrant flow tests, and pressure monitors
- Customer complaint tracking
- Online pressure monitors
- Online water quality monitors
- Hydraulic modeling

TRACKING DEPRESSURIZATION-RELATED WATER QUALITY CHALLENGES

The common approaches to tracking low pressures include supervisory control and data acquisition (SCADA) monitoring, hydrant flow tests, and portable pressure recorders installed at specific locations. Sometimes operator experience and field pressure measurements at specific locations for low pressure complaints may serve as indicators of operating pressures in distribution systems. Tracking main breaks and leaks spatially and temporally may provide additional information for tracking depressurization events.

Because continuous pressure monitoring at every service connection is not always feasible, hydraulic modeling and/or engineering studies are used to estimate operating pressures in distribution systems. Water systems should also track customer complaints related to low pressure and analyze that data for patterns and to delineate where events are occurring and possibly reoccurring. The operational guide to AWWA Standard G200 provides information on how systems should respond to low pressure complaints (Oberoi 2009).

Pressure Monitoring

To prevent contamination, a water distribution system needs to be pressurized. Monitoring should be performed continuously throughout the system to determine if system pressures are within the targeted ranges.

Limitation of existing practices. For pressure monitoring in distribution systems, the Ten States Standards (2012) do not provide detailed information on how to monitor pressure, such as location, frequency, or minimum time interval. Many US primacy agencies do not require water systems to submit monitoring data of low pressure events for regulatory compliance. In a recent survey, most systems (89 percent) did not have any state or local regulatory requirements for pressure monitoring (LeChevallier et al. 2014).

Typical pressure monitoring locations are often at convenient sites such as pump stations and storage facilities (for system operation) or at service connections that need certain pressures, such as wholesale and industrial customers. Additionally, low or negative transients often last seconds or less than a minute. A practical question to ask is whether it is unacceptable for pressures to fall below 20 psi for 1 s. Regardless of the answer, even when monitoring continuously, many systems record pressure readings only once every few minutes or even less frequently. Although some systems capture data at a sufficiently high rate to identify pressure transients, many uncertainties remain when it comes to the best practices for pressure monitoring.

Lack of monitoring at low pressure locations. Locations with low pressure are generally more vulnerable to negative pressure transients due to lack of a pressure "cushion" for the system's internal pressure drop. It is not uncommon for distribution systems to have locations with normal operating pressures of less than 35 psi; this is primarily due to high elevations. These locations may serve as pressure monitoring locations. Pressure monitoring at low-pressure areas was found to be uncommon, and not all systems have backflow prevention devices installed to prevent potential intrusion due to hydraulic transients (LeChevallier et al. 2014). Therefore, it is difficult to fully characterize the extent of low pressure and its role in system contamination.

In addition to occurrence under normal conditions, low or negative pressures may also occur during fire and main break and repair events. In general, the vicinity where the event occurred, the highest elevation in the pressure zone of the event, and areas far from floating storage tanks are most vulnerable to low or negative pressures. Low or negative pressures in these areas could go undetected due to lack of pressure monitoring.

During a fire event, many distribution systems rely on SCADA and telemetry monitoring. While typical SCADA pressure monitoring occurs at utilities (e.g., tanks or pumps), it is uncommon to place pressure monitors near fire areas. Many systems rely on the local fire department to monitor pressures on-site during fires. Since pressure monitoring by fire crew personnel may only be conducted adjacent to the fire flow withdrawal, low or negative pressures that occur elsewhere in the system could go undetected.

Similarly, few systems monitor pressures on-site during water main break and repair events. Often, these systems rely on SCADA monitoring at tanks or pumps and on customer complaints in order to estimate the extent of depressurization. It is uncommon to conduct hydraulic model simulation to determine the extent of depressurization.

Best practices for pressure monitoring. Pressure monitoring throughout a distribution system may not be feasible. The AWWA Partnership for Safe Water Distribution Program (PSW 2011) recommends pressure monitoring at a minimum of two critical locations in each pressure zone, representing the highest and lowest pressure in the zone. For very small pressure zones with little change in elevation and for small systems with limited resources, a single location for low pressure could be used. Data from these critical locations should supplement routine SCADA pressure monitoring that is installed for operational control (e.g., at storage tanks, pump stations, and valve chambers). For large pressure zones, two pressure monitoring locations may not be adequate (LeChevallier et al. 2014). Some systems have a single pressure zone with more than 1,000 mi of pipe.

Some low or negative pressure events are localized and often not captured by routine monitoring or pressure monitoring at two critical locations. Therefore, a water system may increase the number of permanent pressure monitoring locations for large pressure zones (e.g., more than 100 mi of pipe), considering available resources, existing monitoring locations, zone sizes, and topography. An alternative is to use metering and metering transmission systems that detect reverse flow over a 1-hour period or less. Negative pressure may cause reverse flow events that can be detected by a reverse-flow sensing meters. It should be noted that reverse flow does not always indicate a negative pressure event.

Monitoring locations should be selected based on a review of historical distribution system pressure data, the existing hydraulic model, operational experience, and customer complaints about pressure (LeChevallier et al. 2014). The monitoring location should not be in the vicinity of water storage tanks or pumps and should not be subject to large localized water demands. The nature and volume of regional water use should be considered as well. An area with a dense population has the potential to expose more customers to contamination than a semirural areas. Hydrants are convenient locations for pressure monitoring but not ideal for permanent installations because they may take fire protection out of normal use or could be affected by freezing conditions. Permanent monitoring sites can be located in underground valve pits, meter pits, blowoffs, or inside buildings where heat is available and the influence of domestic demands is not significant (e.g., upstream of the backflow preventer for a fire suppression system). Public buildings such as fire stations are good locations for monitoring, although the pressure sensor would need to be tapped into the main and not the service line to eliminate the effect of the pressure drop in the service line. All sites should be readily accessible by operators and have a low potential for vandalism. Figure 9-4 shows examples of pressure monitoring installations.

If a hydrant or blowoff is used for pressure monitoring under freezing conditions (e.g., pilot testing for permanent installation), insulation should be provided. For permanent installation of pressure monitors, telemetry and SCADA integration are recommended in order to continuously monitor and record data and alert operators of any extreme pressure events (LeChevallier et al. 2014). The higher initial cost of the telemetry system can often be justified by labor savings associated with the periodic data download required for portable pressure monitors.

(a) Hydrant (b) Underground meter pit (c) Premise plumbing inside a building

(d) Underground valve pit (e) Walk-in valve vault (f) Underground blow-off

Source: Adapted from LeChevallier et al. 2014.

Figure 9-4 Example field installations of pressure monitors in distribution systems

Electronic pressure monitors should be set with a tolerance of ±2 psi, properly zeroed, and calibrated at least annually according to the manufacturer's specifications. The convenience of mounting the loggers and protection from vandalism should be determined on a site-by-site basis. The manufacturer's guidance manual should be followed when installing the device, setting up pressure monitoring rate and tolerance settings, and downloading data. Depending on the monitor features, different low- and high-pressure alarms should be set with the monitors. Training should be provided to field operators for device setup, configuration, and data collection. If portable pressure loggers (often battery operated) are used, the logistics for monthly data download (before the stored data reaches capacity of the unit) should be planned as part of routine O&M activities.

A variety of pressure monitoring, logging, and communication equipment is commercially available to help operators make informed choices and evaluate potential tradeoffs. Some brands of pressure monitors are more difficult to use and may have higher tendencies for malfunction. Operators often have preferences for certain pressure monitoring equipment based on locations where housing, electricity, or telemetry are not be available. A partial list of pressure monitoring and recording equipment is provided by LeChevallier et al. (2014). The list provides comparisons in terms of cost, maintenance, ease of use, sampling frequency, capacity, power options, digital or chart recording, size constraints, working environment requirements, and communication options (telemetry integrated to SCADA).

Pressure fluctuations, hydraulic transients, and distribution system events such as flushing can persist for only a few minutes or less. Previous studies (LeChevallier et al. 2011) suggest that negative pressure transients that last for less than a few seconds do not pose significant risks of infection, while the risks are higher for negative pressure transients that last for more than a minute. To minimize monitoring data storage needs, pressure monitoring should be set with the recording interval in minutes (e.g., 3 to 5 min); the sampling interval in seconds, with minimum/maximum/average values recorded for each recording interval; and the impulse recording feature enabled to capture sudden pressure changes during distribution system events. Accelerated monitoring, also known as impulse recording, is a feature that some pressure monitors have to automatically record data more frequently when significant changes in pressure occur.

Pressure data assessment. For each pressure zone monitored, the pressure monitoring results should be analyzed and compared with the established pressure goals monthly or at least quarterly for small systems. The hourly pressure data can be analyzed using a previously developed spreadsheet program (PSW 2011). Figure 9-5 shows an example output of the spreadsheet program for 1 year of pressure monitoring. The program takes 96 hourly readings (four representative days) of data each month, summarizes descriptive statistics, and compares the data to the five optimized pressure performance goals.

Analyzing the hourly interval pressure data may be inadequate for a 20-psi or 0-psi low-pressure goal during distribution system fire flow (flushing) and emergency events (power outages). When the pressure monitor is set with a recording rate in minutes and the impulse reading features enabled, minute-interval pressure data can be further analyzed. Pressure sensors connected to a SCADA unit may be able to process multiple pressure readings during a polling interval such that at each polling interval, one can retrieve minimum, maximum, average, and instantaneous pressure. The minute, hour, and day information can be extracted from the time stamp of each pressure reading; this information is then placed into a pivot table and graphed in a spreadsheet for trending.

This process can show whether hourly pressure graphing omits pressure events or anomaly pressures (e.g., flushing or main breaks) that might be captured in the minute/second interval pressure data. Records of distribution system events, for example, flushing, breaks, sprinkler demand, and operation changes, should be collected and correlated with any confirmed pressure anomalies. If the pressures do not meet the goals (too low, too high, or too much fluctuation), the problem can be cross-validated within all monitored pressure data and with the hydraulic modeling (see additional details of hydraulic modeling later in this chapter). Data from the optimized pressure monitoring is then used to guide system improvement planning.

Online Water Quality Monitoring in Distribution Systems

Monitoring of both online pressure and water quality is needed to determine whether locations in the distribution system that are subject to periodic pressure loss are more likely to have poorer water quality than locations maintained at higher pressures. In other words, if backflow or intrusion of contaminants occurs when pressure is lost, will a change in water quality be detected, for example, chlorine residuals? If so, it would be important to identify the conditions and scope to which chlorine residual and pressure could potentially be correlated, at least for water systems that maintain a disinfectant residual.

Two ends of the risk scale are large-volume, low-frequency events and small-volume, high-frequency events. It is generally unlikely that a large volume of raw sewage would be introduced directly into the distribution system in the absence of a visible system breach. If a visible breach is present, the water system would respond with pipe isolation and flushing activities to remove the contaminated water. Therefore, low-level intrusion events such as through leaks and cracks in subsurface piping during negative pressure transients, which are likely to go undetected, are the focus when the efficacy of a disinfectant residual during intrusion events is being evaluated.

LeChevallier et al. (2011) conducted pressure and water quality monitoring at six locations for more than 40 days in a distribution system that served a hilly terrain (elevations ranged from 555 to 1,710 ft). The system had several areas where main breaks occurred frequently. No significant relationship between pressure and monitored water quality parameters, including free chlorine residual, conductivity, pH, and temperature, was observed at any monitoring location.

Pressure Zone Summary Table—Annual
Monitoring Site 1 (High Elevation)

Pressure Zone Name/ID:	**Main Zone**
Pressure Zone Monitor Name/ID:	**59th Street & Indian Bend**
Pressure Zone Monitor Elevation (ft msl):	
Pressure Zone Highest Elevation (ft msl):	
Pressure Zone Lowest Elevation (ft msl):	

Category/Criteria	**Month**												
	1	**2**	**3**	**4**	**5**	**6**	**7**	**8**	**9**	**10**	**11**	**12**	**Annual**
Summary Stats													
Number of Readings (#)	96	96	96	96	96	96	96	96	96	96	96	96	1,152
Minimum Number of Readings Satisfied (96 Readings)	Yes	Yes	Yes	Yes	Yes	Yes	Yes	Yes	Yes	Yes	Yes	Yes	Yes
Average Pressure Reading (psi)	44.3	44.3	44.0	44.2	43.9	44.6	44.5	44.4	44.5	44.4	44.4	44.3	44.3
Maximum Pressure Reading (psi)	44.9	45.1	44.8	45.4	45.3	45.1	45.3	45.0	45.3	45.2	45.5	45.4	45.5
Minimum Pressure Reading (psi)	43.4	43.0	40.9	40.7	37.9	40.9	40.7	43.7	44.0	43.9	43.8	43.4	37.9
Number of Readings >Average +10 psi (#)	0.0	0.0	0.0	0.0	0.0	0.0	0.0	0.0	0.0	0.0	0.0	0.0	0.0
Number of Readings <Average −10 psi (#)	0.0	0.0	0.0	0.0	0.0	0.0	0.0	0.0	0.0	0.0	0.0	0.0	0.0
Percent of Readings Exceeding ± 10 psi Range (%)	0.0	0.0	0.0	0.0	0.0	0.0	0.0	0.0	0.0	0.0	0.0	0.0	0.0
Standard Deviation (psi)	0.2	0.3	0.7	0.9	1.9	0.7	0.9	0.2	0.3	0.2	0.3	0.3	0.8
Readings NOT Meeting Criteria Below (#)													
1. (Not) Within ± 10 psi of average pressure	0	0	0	0	0	0	0	0	0	0	0	0	0
2. (Not) Less than 100 psi under normal conditions	0	0	0	0	0	0	0	0	0	0	0	0	0
3. (Not) Above 35 psi under normal conditions	0	0	0	0	0	0	0	0	0	0	0	0	0
4. (Not) Above 20 psi under peak and fire flow conditions	0	0	0	0	0	0	0	0	0	0	0	0	0
5. (Not) Above 0 psi during emergencies such as main breaks and power outages	0	0	0	0	0	0	0	0	0	0	0	0	0
Overall System Score	A	A	A	A	A	A	A	A	A	A	A	A	A

Source: LeChevallier et al. 2014.

Figure 9-5 Example output of the spreadsheet program for pressure monitoring

The chlorine demand of leaking sewage may be significantly reduced after it is transported to the location of drinking water pipes by dilution with groundwater, soil filtration, uptake for microbial growth, and similar processes. Additionally, near the location of water leaks and cracks, disinfectant residuals in the leaked drinking water, if not exhausted by the environment, may provide additional dilution and reduction of chlorine demand in the standing water.

In light of recent disinfectant demand and decay studies (Kirmeyer et al. 2014), the lack of any observable relationship between pressure fluctuation and loss of disinfectant residual can be better understood. It is uncertain if the background disinfectant residual can persist and be adequate after contaminants intrude during main breaks. Most disinfectant decay or demand is consumed instantaneously during initial mixing. In addition, disinfectant residuals are likely to remain in the distribution system after a low-level intrusion event, especially for systems maintained with chloramine residuals (a relatively weaker oxidant compared with chlorine). Unless intrusion of a large volume of water with high chlorine demand occurs, it is unlikely that intrusion would be detected by observing a noticeable loss of chlorine residual (LeChevallier et al. 2011).

Hydraulic Modeling

A water system may evaluate pressure deficiencies using a hydraulic model of the existing system. When developing the model and reviewing its results, model-predicted pressures should be evaluated under different water demands (e.g., average day, maximum day, peak hour), fire flow (if required), and other system-specific conditions. The model-predicted pressures can be compared with the pressure monitoring data. Pressures can be graphed and color-coded on distribution system maps. Elevation contours can be overlaid with pressure performance on the distribution maps, and areas that do not meet the established pressure goals usually can be readily identified.

The cost to develop and maintain a hydraulic model can be justified by its wide range of applications. A full-pipe hydraulic model is needed to evaluate pressure performance and potential deficiencies throughout the system. The hydraulic model should be calibrated to a reasonable accuracy level, one that resembles actual design and operation conditions (AWWA 2013). Models are usually calibrated using hydrant flow tests under steady-state conditions and/or verified with routine SCADA monitoring data (e.g., recorded flow, pressure, and tank levels) for extended period simulations (EPS). These two calibration methods are often referred to as fire flow calibration and EPS calibration, respectively (they are also referred to as microcalibration and macrocalibration, respectively). There are numerous references available about hydraulic model development, calibration, and applications, including AWWA Manual M32 (AWWA 2012), Speight et al. (2010), Walski et al. (2007), and Boulos, Lansey, and Karney (2006).

Transient Modeling

While pressure monitoring is the only way to document the occurrence of low pressures and negative transients, it has its limitations in that monitors are expensive and can be installed at only a few locations. In addition, monitors cannot be used to answer "what if" questions about transient control facility design or alternative operation. In order to analyze pressure transients, a transient analysis computer model is needed. With a transient model, it is possible to simulate past events, design alternative control devices, and implement operational modifications. Friedman et al. (2010) recommend transient modeling be conducted every 10 years for each major pump station or when any significant changes are made to the pump station, for example, installation of a larger pump or motor or addition of more pumps.

Pressure waves can either increase or decrease pressure. Increases can cause catastrophic pipe bursts. Unless the pressure drops below the vapor pressure of water to cause column separation and cavitation, the effects of low pressure may not be as dramatic but, as mentioned earlier, can result in intrusion of contaminants. The magnitude of the transient depends on the change in momentum; the movement and dampening of the wave depends on the pipe material, network layout, and existence of control devices. For a quick estimation, the magnitude of a pressure transient can be estimated using the Joukowsky equation (Thorley 2004).

$$\Delta H = \pm \frac{c}{g} \Delta V \tag{Eq. 9-1}$$

Where:

ΔH = the change in pressure head (ft)

c = the characteristic wave speed of the fluid (ft/s)

g = acceleration due to gravity (ft/s^2)

ΔV = the change in fluid velocity (ft/s)

The positive and negative signs represent a wave propagation traveling downstream and upstream, respectively. The wave speed (c) depends on the pipe material and amount of entrained air, among other factors. It can range from 4,000 ft/s to 1,500 ft/s or even lower for plastic pipes. Pressure waves eventually dampen out due to friction, turbulence, and reflection and attenuation of waves. Their effects often only last for seconds or minutes depending on system size and the magnitude of the transient wave. Chaudhry (1987), Almeida and Kolle (1992), Wylie and Streeter (1993), Walski et al. (2007), and Wood and colleagues (2005) provide additional background on the nature of transients.

Transient models can be used to improve understanding of how a distribution system will respond to a variety of events such as power outages, pump shut downs, valve operations, flushing, firefighting, main breaks, and other events that can create a rapid change in flow conditions. Key elements that should be evaluated include causes of transient events; magnitudes, durations, and locations of lowest pressures; and the transient pressure monitoring plan before and after transient control has been implemented. The primary reason for doing this is to prevent or mitigate "0 psi under emergency" conditions (e.g., power outage and main breaks) and/or to demonstrate sizing of the surge tank, vacuum breaker valves, and other measures to protect the system from negative pressure transients under emergency conditions. In some instances, a minimum pressure of 20 psi is used to size the surge protection tanks. In some case studies, a safety factor of 2 has been applied to account for uncertainty of key modeling parameters or for ease of tank maintenance. The redundancies are system specific and consider a combination of existing model accuracy, existing building footprint, and costs (LeChevallier et al. 2014).

Low or negative pressure transients are usually most severe near pump stations and control valves and in high elevation areas and locations with low static pressures or far from floating storage facilities (Boulos et al. 2005). Systems that do not have floating storage facilities are more susceptible to low and negative pressure transients (Fleming, Dugandzik, and LeChevallier 2006). In addition, locations with one or more of the following characteristics tend to be more susceptible to transient events:

- Near a pump station with downstream velocity greater than 3 ft/s,
- More than 1 mile away from elevated storage,
- With elevations greater than 40 to 50 ft above surroundings, and
- Near a distribution system dead end.

Several commercial computer tools for transient analysis are available. Since many systems already have a hydraulic model, it is not difficult to convert that model into a transient analysis model. If the hydraulic model is maintained in a different model format (e.g., EPANET), it needs to be converted into the format of the transient modeling tool. As a system develops transient models and identifies vulnerable areas, the vulnerable areas should be prioritized for pressure monitoring, disinfectant residual maintenance, surge control, leak detection and control, and cross-connection control. Similar to the calibration of hydraulic models for a distribution system, the accuracy and usefulness of uncalibrated surge models are limited. By calibrating the transient model using historical events or routine operations, the user will have confidence in the model and reasonable expectations for accuracy. Currently there are no industry standards and few guidelines on the calibration of distribution system transient models. Some general parameter adjustments for a transient model can be made if the observed transient data are available for comparison (Fleming, Dugandzik, and LeChevallier 2006; Walski et al. 2007), including the following:

- If the modeled transient period is significantly different from the observed, the wave speed needs adjustments.
- If the magnitudes of the modeled and observed transients are significantly different, adjustment should be made in the speed of the flow control event (e.g., pump shutdown or valve closure).
- In real systems, transients often die off faster than models predict. If the modeled transient dies off too quickly or too slowly, the system friction loss might need adjustments.
- If there are secondary transient peaks that are not predicted correctly, it may be due to a misrepresentation of system geometry (missing pipe loops or diversions).
- If the modeled and the observed baseline pressures are significantly different, the initial and boundary conditions may not be set correctly (e.g., pump status, tank water level, and system water demand).

Quantitative Microbial Risk Assessment

A quantitative microbial risk assessment (QMRA) for depressurization-related water quality problems can be used to evaluate operations and/or infrastructure improvements that are potentially important in reducing public health risk. For infectious microbial agents, the QMRA is still in the active research stage; it is used to estimate the risk of infection based on known or inferred pathogen levels at the point of water consumption. Key steps of a typical QMRA include hazard identification, exposure assessment, dose–response assessment, and risk characterization. This is similar to the chemical risk assessment framework, and more details can be found elsewhere (NRC 1983; ILSI 2000). The International Life Sciences Institute approach has been further clarified by the US Environmental Protection Agency and the US Department of Agriculture–Food Safety and Inspection Service (USEPA and USDA/FSIS 2012). The most challenging step in estimating microbial infection risk caused by pressure-related intrusion is the exposure assessment, which involves evaluating the characteristics of intrusion (location, volume, and duration) and the fate of intruded pathogens traveling within the distribution system. Viruses are often selected as the pathogen for risk modeling because they can cause acute risk of infection. The QMRA starts by assuming virus existence external to the distribution system and that a leak exists at locations where a pressure loss event occurs. This scenario is modeled as the baseline (or the worst case) and combines the four elements necessary to create risk:

source, pathway, exposure, and post-contamination defense failure. Hydraulic and transient modeling is coupled with Monte Carlo simulations to obtain risk estimates of the primary waterborne infections (LeChevallier et al. 2011).

Sufficient information is available from the literature and engineering calculations to estimate the infection risks of intrusion due to low or negative pressure events. LeChevallier et al. (2011) estimated intrusion due to a negative pressure transient to have a median infection risk of 2.1×10^{-4} for norovirus and 1.1×10^{-4} for rotavirus. These values are slightly higher than the USEPA target level of annual infection risk (1×10^{-4}). The primary value of a QMRA or risk model may not be to calculate exact risk levels but to provide a framework for organizing existing and known information, a tool for assessing data gaps and mitigation options, and a pathway for future risk assessment refinement. Most significant findings from a QMRA that addresses low pressure–related water quality problems can be summarized as follows (LeChevallier et al. 2011; Kirmeyer et al. 2014):

- Among all practices, including pressure management, leak detection and repair, ensuring separation distance from sewer mains, and similar actions, the most effective way to reduce risk from intrusion is to maintain adequate system pressures under all conditions, that is, to eliminate the driving force for intrusion. The risk of infection is most sensitive to the duration of pressure loss and the number of intrusion locations. Therefore, risk mitigation practices should be targeted to alleviate the magnitude of low or negative pressures, that is, improve pressure management.
- Maintaining adequate free chlorine residual (free chlorine of 0.2 mg/L or above) is the last defense against the risk of microbial infection due to low or negative pressures. Maintaining a chloramine residual may not significantly reduce the risk of infection, mainly due to its weaker disinfection effectiveness against virus. The contact time (CT) values for effective virus inactivation are on the order of magnitude of hundreds to thousands mg/L $Cl_2 \times$ min. So at 1 mg/L chloramine as total chlorine, significant virus inactivation would take hundreds of minutes to hours. For chloraminated systems, mitigation should rely on other practices such as pressure management.
- Separation distances from sewer mains should be ensured in order to reduce the potential overall public health risks and concerns from an aesthetics perspective. Leak detection and repair and cross-connection control practices should be prioritized in areas that are vulnerable to low pressures and negative transients, which may reduce the number of intrusion locations and reduce the overall risk of infection. Flushing may be the simplest and most effective way to remove the intruded contaminants if it is done in time (see Chapter 2 for additional details).

DSWQ Response Actions:

- Issue a boil-water advisory
- Review pump and tank cycling
- Locate and repair main breaks and leaks
- Maintain adequate disinfectant residual
- Investigate possible cross-connections
- Flush potentially contaminated water

PREVENTING AND MANAGING LOW PRESSURE

As discussed earlier, the microbial risk assessment suggested that of all mitigation practices, pressure management might be the most effective way to reduce the risk of microbial infection caused by low pressures (LeChevallier et al. 2011), which fits well in existing industry guidelines for distribution system operations. In the United States, the Ten States Standards (2012) recommend that 20 psi be maintained under all flow conditions, at ground level, and at all points in the distribution system. However, in reality, occasional short periods of pressure drops below 20 psi often occur undetected due to a variety of reasons.

First, pressure monitoring at locations vulnerable to low pressures and negative transients (e.g., high elevations) may not be conducted as frequently as at facility locations. Some water system are not equipped with high-speed electronic pressure loggers to detect low or negative pressure events of short duration. Some mechanical pen-and-chart recordings of system pressures are not able to capture negative transient pressures (Gullick et al. 2004).

Second, pressure deficiency analysis under transient conditions is uncommon in current hydraulic modeling practices. A typical pressure deficiency analysis involves evaluating pressures under steady-state conditions or for extended periods (hourly). For example, the minimum system pressure criterion of 20 psi is often interpreted as being met under the conditions of maximum day demand and a sustained fire flow. Transitory pressures caused by transients are often not considered.

Finally, lack of attention to low or negative pressure transients could be due to lack of regulatory standards and guidelines or a lack of quantitative indications of public health risks caused by negative pressure transients.

Improve and Optimize Pressure Management

Various terms have associated pressure management with distribution system improvement such as pressure reduction, flow modulated pressure control, and district meter areas (a hydraulically discrete part of a water distribution system, see definition in AWWA Manual M36 [2016]). To some water systems, "pressure management" means reducing pressure to reduce leakage; in Europe, it means to minimize pressures as a way to reduce leakage and possibly pipe breaks. Thornton and Lambert (2006) give a more formal definition of pressure management as "the practice of managing system pressures to an optimum level of service ensuring sufficient and efficient supply to legitimate uses and consumers, while eliminating or reducing pressure transients and variations, faulty level controls and reducing unnecessary or excess pressures, all of which cause the distribution system to leak and break unnecessarily." In this manual, pressure management involves a combination of planning (and modeling), monitoring measures, and mitigations used to ensure distribution system pressures. Based on the results from literature reviews, surveys, and case studies, practical issues that water systems face when improving or optimizing their pressure management programs include how to do it, costs, and benefits. These practices can be organized into the following four steps, and details can be found elsewhere (Friedman et al. 2010; PSW 2011):

1. Establish pressure performance goals and complete pressure monitoring and performance data assessment.
2. Conduct hydraulic modeling and transient modeling; identify pressure deficiencies and performance constraints.

3. Prepare a pressure improvement action plan; plan pressure control and/or reduction; evaluate cost and benefits associated with the optimized pressure management.
4. Document and track improvements in ongoing performance associated with pressure improvements and projects developed in step 3 (if implemented).

Plan pressure control and/or reduction. Steps 1 (establishing goals) and 2 (identifying deficiencies) have been discussed in detail in the previous sections. To start with step 3 (preparing improvement plans), a workgroup of multidisciplinary staff including engineers, modelers, operators, managers, and consultants should be convened to explore options for improving system pressure management. System hydraulic and/or transient models may be relied on to develop potential solutions. If pressure monitoring has been performed and anomalies have occurred, the model should be capable of simulating the event or point to a reason (e.g., closed valves, leaks) why field data and the model are not agreeing. Finding and correcting anomalies may be all that is required.

If the normal operating pressures are too high or low (e.g., greater than 35 to 100 psi), the solutions are typically to establish new hydraulic grade lines, create new or adjust existing pressure zone boundaries, or install pressure-regulating valves. If the system pressures are too low under fire flow conditions (less than 20 psi), solutions include upsizing water mains, looping dead ends, and installing new storage tanks or fire pumps, among others (see Chapter 2 for additional details). If pressure transients cause too much variation or negative pressures in the system, transient mitigations should be planned and implemented. Input from the multidisciplinary workgroup will form the basis for plans to optimize their system's pressure management. However, it is important to note that these plans need to be jointly considered with the impact on fire flow, water quality, leaks and breaks, and energy usage in the system. Cost estimates can be compiled for a range of mitigations for pressure deficiencies, for example:

- Check, realign, create new pressure zone boundary;
- Install pressure-regulating valve (PRV); change PRV or pump control settings;
- Change pumping configurations or make pump upgrades such as installing variable-frequency drive or slow-opening valves;
- Change the size of pipes; clean and line tuberculated pipes;
- Locate and repair leaks;
- Find unidentified closed valves that should be open;
- Provide surge protection including surge tanks (air chamber, hydropneumatic tank, open surge tank), pressure-relief valve, air release/vacuum valve, pump bypass; and
- Conduct capital improvement projects such as adding an elevated storage tank.

Cost–benefit estimate for pressure management. In addition to meeting regulatory requirements, the key reason for a system to consider optimizing its pressure monitoring and management is that the anticipated benefits or cost savings can significantly outweigh the implementation cost (LeChevallier et al. 2014). A cost comparison analysis can be conducted for various combinations of system upgrades. Benefits or cost savings are mainly from energy savings, reduced main break frequencies, and reduced system leakage. Data on system leakage, main breaks, energy usage, and other distribution system operational information should be collected to establish system-specific qualitative or quantitative (preferred) relationships between pressure and other performance indicators (e.g., water loss, breaks, energy usage). This is helpful in estimating anticipated cost savings or benefits in order to justify the pressure improvement action plans.

The average pressure–leakage rate relationship for large systems with mixed pipe materials is usually close to linear (Thornton 2003), that is, a 10 percent reduction in average system pressure should generally achieve about a 10 percent reduction in system leakage rate. For quick estimates of pressure impact on new break frequencies, a general 1-to-1 relationship (10 percent pressure reduction results in 10 percent reduction in new breaks) can be used and adjusted based on the system-specific pipe ages, deterioration conditions, and/or main break data (LeChevallier et al. 2014). For cost savings estimated from reduced leakage in the United States, for example, the unit water production cost is typically used because the unit water retail cost ($2.00 to $3.00/1,000 gal) is often up to 10 times the unit production cost. The unit main break cost and the unit insurance claim cost may be highly variable. The majority of labor and material costs for a single break or leak repair may be in the range of $400 to $8,000; some main breaks and insurance claims have been reported to have a cost or payment of more than $1 million (LeChevallier et al. 2014). Hydraulic modeling can be used to assess energy usage efficiencies of different pressure improvement plans and compare the output with energy billing data. Energy management and optimization tools are available in most commercial hydraulic modeling software. Case studies and more detailed examples of cost–benefit analysis for pressure management improvements can be found elsewhere (LeChevallier et al. 2014).

Practitioners should develop system-specific cost–benefit estimates for their pressure optimization action plan. Pressure reduction that results in significant savings from reduced break repair cost and energy expenditures in one system may not be cost effective in another system. The feasibility of project implementation can be further complicated by other factors such as legal, easement, and property acquisition; customer and community acceptance; regulatory permitting; and safety. The identified pressure improvements and capital projects should be placed in a pool together with other system improvement projects and prioritized for implementation. Empirically, a payback period of six to eight years or less is ideal for a typical municipal project. Projects with payback periods of more than 10 years have higher uncertainties and are likely to be prioritized at a lower level when compared with other system optimization projects (LeChevallier et al. 2014).

Pressure Transient Control

The strategies to control transient pressures are classified as direct actions and diversionary tactics (Thorley 2004). Direct action strategies are used to modify system operations that cause rapid flow rate changes. Examples include prolonging valve and pump status change operations and slowing the opening and closing of fire hydrants. When designing a new distribution systems, rerouting pipelines, using larger-diameter pipes or otherwise lowering the flow velocity, changing pipe material, and applying changes in system topology are all direct actions (Boulos et al. 2005). The diversionary tactics are used to control the magnitude of transients that may occur during unplanned events such as power failures. A brief overview of these control devices and methods is presented here:

- *Valve and hydrant operation.* In addition to power outages, fast operation of valves and hydrants can be a significant contributor to negative hydraulic transients. It is important that fire fighters and water distribution operators be trained to slowly operate valves and hydrants, especially for the first few turns on opening and the last few turns on closing. The effects vary depending on the magnitude of momentum changes and location of the operation. There are numerous warnings about the problems of quick operation (AWWA G200 [2004]; Walski 2009).

- *Pump controls on startup and shutdown.* Pumps can change the momentum of water fairly quickly. To control the effect of pump transients, pumps can be started and stopped using pump control valves, soft starts, or variable-frequency drives. Unfortunately, while these controls are effective for normal starts and stops, they do not help with transients caused by sudden power outages. Addition of devices such as a flywheel to provide additional momentum for a pump can help with this, although it is a costly option and needs to be justified.
- *Reduction in flow velocity.* If the velocity of flow is low, the change in velocity and hence the magnitude of the pressure change is low. Increasing pipe sizes solely to reduce transients is usually not justifiable. However, sizing pipes to avoid excessively high velocities (e.g., greater than 6 ft/s) is good practice.
- *Surge tanks.* Tanks located close to the cause of the transient can provide water to reduce the effect of a negative pressure wave. Open tanks on the discharge side of pump stations can be used; however, the height that is required to match the system hydraulic grade line often makes them too expensive or not feasible. Pressurized air chambers (hydropneumatic tanks) that use compressed air to force needed water into a pipe to minimize negative transients can be used. There are many variations in surge tank design.
- *Air-release and vacuum breaker valves.* Air-release valves at high points in the distribution system are useful for pipe filling but they also serve a valuable function in preventing negative pressures during transients. By opening when the pressure drops to zero, they can prevent water column separation and collapse. For water quality purposes, if pressure at high points is maintained at atmospheric pressure, the pressure should be positive for the rest of the system. For additional information on air-release and vacuum breaker valves, see ANSI/AWWA 2004a.
- *Surge control valves.* There are a variety of surge control valves including surge anticipator valves, pump bypass valves, pressure-relief valves, and automatic control valves. These are generally designed to reduce high pressures, but some designs can also affect low pressures. A reference for valve selection and installation in distribution systems is AWWA Manual M44 (2006a).

The strategies to control transient pressures may be system specific, and transient modeling may be conducted to determine transient control facility design or alternative operation. Figure 9-6 summarizes methods used by surveyed systems to control and manage pressure transients in their systems (LeChevallier et al. 2014). Elevated tanks, soft-start or variable-speed pumps, and surge tanks were reported to be the most common strategies in both the largest and the smallest pressure zones. Information about these surge protection devices together with rationale for choosing surge protection strategies (direct, diversionary, or both) can be found in standard engineering texts on pump design, water distribution system pipeline design, and transient analysis reports (Thorley 2004; Walski et al. 2007; Boulos et al. 2006).

The presence of storage tanks, especially those that float on the system (meaning the tank water level is the same as the system hydraulic grade line), is essential for systems with supply interruptions that last longer than a few seconds. Some systems can meet demands for hours to days in an emergency depending on storage volumes and demand management. Systems with multiple sources, interconnections with neighboring systems, and gravity-fed reservoirs usually do not need much emergency storage, except for equalization of peak hour water demand and fire storage. However, systems with limited source or treatment capacity, no interconnections with neighboring systems, and the need for pumping may need additional storage beyond that for equalization and fire protection (see Chapter 2 for more on this topic).

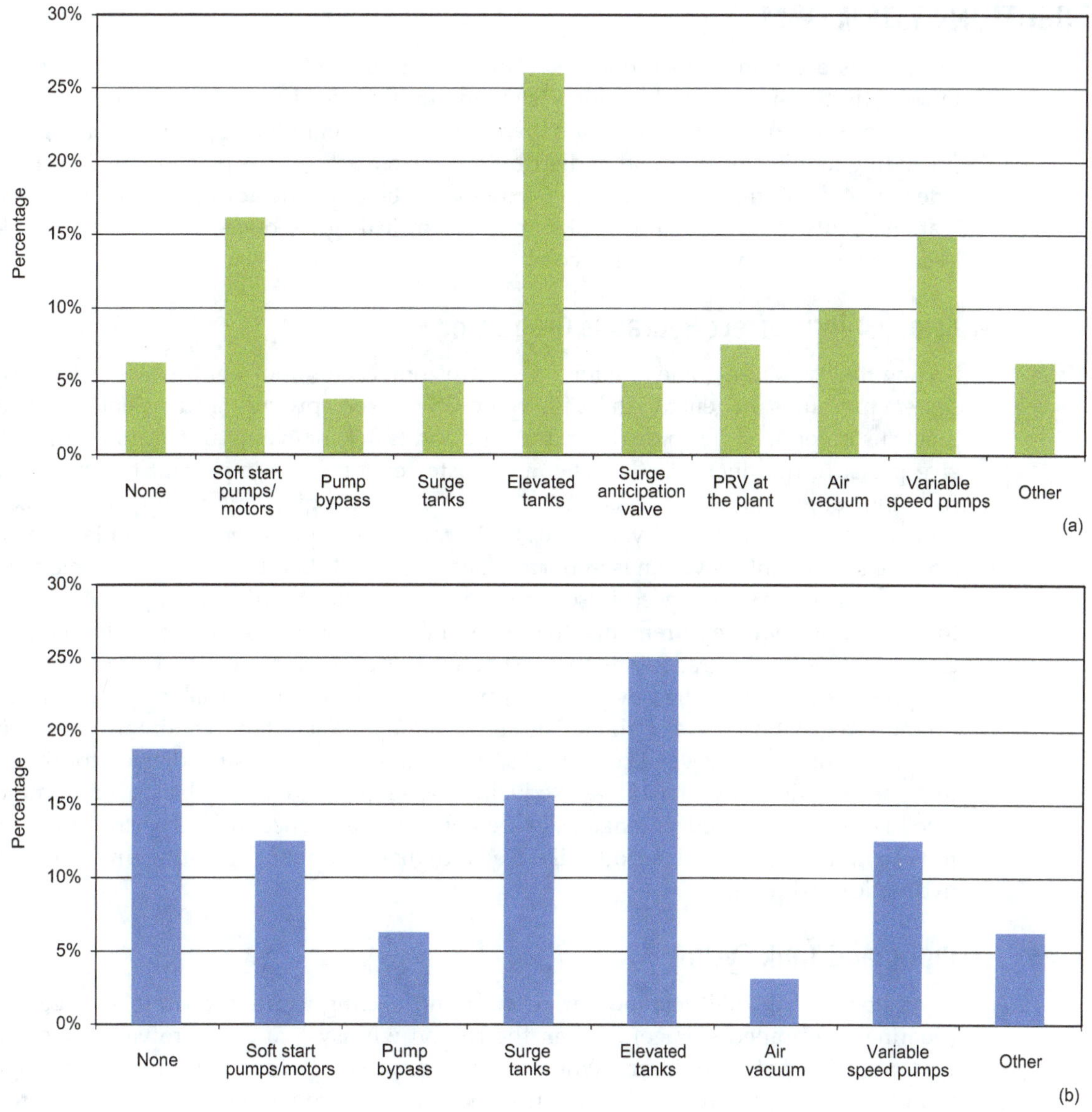

Source: Adapted from LeChevallier et al. 2014.

Figure 9-6 Pressure transient control and management in (a) large zones and (b) small zones of 36 surveyed systems

In pressure zones with floating storage or areas served from higher zones through the use of control valves, the loss of power or a mechanical failure at a pump station can instantly result in low or negative pressure depending on the elevation within the zone. Most systems with no storage in a pressure zone have standby generators to furnish power to pumps; however, these generators do not start instantly. In order to maintain adequate pressure, the control system must first detect the loss of power and begin ramping up the generator. Only when the generator reaches operating speed can the poor pressures be corrected. In most cases, some customers experience negative pressures. Pflanz and colleagues (2010) and Walski and colleagues (2011) showed that, in most cases, the generator is not able to start up quickly enough to prevent some negative pressures in the system. To mitigate, some storage (even a small hydropneumatic tank) is provided in order to prevent negative pressures while a generator starts up.

DISTRIBUTION SYSTEM O&M

Pressure is a common operational variable that is related to many distribution system O&M activities such as establishing system hydraulic grade line, cycling pumps to fill and drain tanks, flushing, repairing main breaks, etc. The integration of pressure management (including monitoring) with other distribution system activities is primarily to jointly consider the cost of implementation and anticipated benefits including reduced depressurization events, reduced main breaks and system leakage, reduced number of backflow events, or a reduction in customer complaints.

Establishing Correct Hydraulic Grade Line

Setting the hydraulic grade line for a pressure zone is an important process as it controls system pressures. In general, there is a trade-off between providing sufficient pressure to meet customer needs even during peak flow and transient events and eliminating excessive pressure that increases leakage and wastes energy. It is important to consider the full range of ground elevations that can be served from a given hydraulic grade. In zones with tanks that float on the system (i.e., tank water level equals hydraulic grade), the decision regarding tank elevation is important because it is difficult to modify tank elevations once the tank has been placed in service. A water system can use existing SCADA monitoring and/or field measurements, that is, portable pressure monitors, to help determine the correct hydraulic grade lines and system pressures. In zones without floating storage, the pressures are controlled by pump (in particular, those with variable-speed drives) or pressure-regulating valve settings. The settings must be selected to maintain pressure in the correct range. However, unlike tanks, these values can be easily adjusted for the long term, seasonally, or by time of day. Reducing the set point for a variable-speed pump can also save energy and reduce leakage. Preventative maintenance such as performing pump tests should be conducted periodically to ensure that pumps are operating under intended hydraulic conditions.

Pump and Tank Cycling

In addition to diurnal demand variation, pump cycling to fill and drain storage tanks (floating or pumped storage) is often the reason why system pressure variations occur under normal flow conditions. When pumps are on, discharge pressures can be higher than the average because of the pressure needed to overcome the system head loss to fill the tank. When pumping is reduced, the elevated tank augments supply back to the system, and the pressure in the area of the pump station is lower than the average. However, cases can be more extreme. For example, assuming a head loss of 25 ft each way in the pipe and a 10-ft tank water elevation change, the pressure variation at the pump station will be about 60 ft of head, or 26 psi.

Reducing tank turnover in order to stabilize pressures by modifying storage, booster pumps, or control valves may lead to increased water age and water quality issues (e.g., loss of chlorine residual and nitrification). Conversely, routinely maintaining a low tank water level or increasing storage facility fill–drain cycles to improve water mixing and turnover may reduce fire flow capacity, increase pressure fluctuations, and increase the potential for main break and backflow events. For a pressure zone with multiple storage tanks that control hydraulic grade lines, control of tank valves can be more complicated when preventing overflow, providing adequate turnover, allowing equalization usage of all tanks, and preventing overpressurization at the pumps. Therefore, a cross-disciplinary approach to improve pump cycling is recommended based on a combination of system hydraulics, variations in customer demand, fire flow capacities, water quality impact, energy management, and interzone operations. Hydraulic models can be used for such analysis, and additional details can be found in Chapter 2.

Pressure Monitoring for Flushing

Flushing is a common and routine maintenance activity conducted in water distribution systems; it can be planned as part of an annual program (either conventional or unidirectional flushing) or performed as sporadic spot flushing (in response to customer water quality complaints). In some instances, flushing is coordinated with fire flow tests to stress the system for fire flow analysis. Often, operators have no information regarding the possibility of a system pressure drop to less than 20 psi during flushing activities. Flushing may be throttled at the hydrant blowoff or nearby system valves or by using a smaller hydrant nozzle, which results in less effective pipe cleaning, less sediment removal, and inadequate pressure drops for hydraulic model calibration. Pressure monitoring (performed at critical locations in each pressure zone with remote sensing and access), supplemented with existing SCADA monitoring, can be used to guide planned or spot flushing activities. Operators in the field can maximize flushing effectiveness or gain higher pressure drops for fire flow tests (ideal for calibrating the hydraulic model) while preventing a system pressure drop to less than 20 psi. It should be noted that the critical location with the lowest pressure may change with flushing locations and flows in a large pressure zone, and a single monitoring station for lowest pressure may not be adequate. A system may increase the number of pressure monitoring locations for large pressure zones, considering available resources, existing monitoring locations, zone size, and topography.

Pressure Monitoring for Firefighting

Similar to flushing activities, using water for firefighting will draw down the pressure in the distribution system. Water systems should perform adequate pressure monitoring during firefighting events. The lowest pressures are usually at the location of the fire and should be monitored by the fire fighters. However, monitoring pressure is not their primary concern. Water system operators should also monitor pressure when they have high flows or weak portions of the distribution system. In hilly areas, the lowest pressures may not occur at the fire but at high elevations. Pressures should be monitored at suspected weak points; this may require portable pressure monitors if permanent pressure monitoring is not provided at these locations.

Main Breaks and Depressurization

Previous studies (Kirmeyer et al. 2014) have provided guidance on how to respond to main breaks in order to minimize the risk of depressurization. Main breaks are categorized into four types based on break severity, degree of depressurization, and impact area. Type 1 breaks are small breaks repaired under pressure. Type 2 breaks require a controlled shutdown and do not place the unpressurized leaking pipe in contact with the surrounding soil. Type 3 breaks involve localized depressurization and possible contamination; an emergency shutdown places the depressurized pipe in direct contact with adjacent soil. Mitigations such as flushing and disinfection are recommended to control the risk of infection. Type 4 breaks are the most severe. They require extraordinary measures to protect the system and consumers prior to return to service. Type 4 breaks involve loss of pressure at the break site and potentially widespread depressurization in the water system. Contamination is assumed because of the nature of the break and its severity. A boil-water advisory and bacteriological sampling are recommended as part of return-to-service procedures (see additional details later this chapter). There is a potential risk of contamination beyond the break repair area because loss of pressure allows reverse flow from customer supply lines and possible backsiphonage from unidentified losses of pipe integrity in the area affected.

How does an operator know if there is a regional or system-wide depressurization when at the break site? Customers can provide relevant feedback when they report the loss of pressure. Experience also teaches the distribution staff when loss of pressure beyond the shutdown area has occurred. An example of this is main breaks at low elevations, causing loss of pressure on a nearby hill. An opened hydrant–blowoff during the repair is often used to control and assess pressure levels in type 2 and 3 breaks. Pressure monitoring, performed at critical locations in each pressure zone with remote sensing and access, supplemented with existing routine SCADA monitoring can be used to guide quick determination of main break categories in the field. This is followed by mitigations and public notification.

Shutdown Management

Portions of a distribution system need to be shut down occasionally for routine or emergency maintenance. Although customers in the isolated segment will lose pressure, the effect of a shutdown is lower pressure downstream of the isolated segment. Whenever practicable, pressure should be monitored downstream of the shutdown. Hydraulic models can be used to determine the extent of impacts from a shutdown. Bottled water, water tankers, and/or emergency interconnections from hydrants to bypass the shutdown may be provided to customers during extended periods.

Valve Vault Flooding

Air-release vacuum-breaker valves are often used at high points in distribution systems to prevent column separation and minimize the impacts of transients. These valves are generally located in underground vaults. Even though these vaults are at high points, they can become flooded. As a result, when a low pressure event occurs, nonpotable water, instead of air, can be drawn into the pipe. Ebacher, Besner, and Prevost (2012, 2013) showed that it was possible to create intrusion in valve vaults during low pressure transients. This type of valve vault must be kept dry or, at a minimum, the inlet to the valve must be kept above the water level.

Boil-Water Advisory

Pressure loss or extended periods of low pressure can serve as a trigger to issue a boil-water recommendation or requirement, depending on the extent and pressure level. Many systems issue a boil-water advisory when a main break occurs and pressure drops to less than 20 psi, based on a recent survey (LeChevallier et al. 2011). Many systems also use customer complaints as an indicator to issue a boil-water advisory. The majority of surveyed systems use a combination of factors to determine whether to issue a boil-water advisory, including power outages, having parts of the system isolated or depressurized for repair, and/or regulatory requirements based on water quality testing (e.g., bacteria monitoring results).

Figure 9-7 summarizes the number of boil-water advisories reported by US water systems during the first half of 2010 (Raucher et al. 2014). The surveyed systems did not have consistent spatial patterns of issuing advisories during the year, suggesting various practices or triggers for issuing boil-water advisories. Since SCADA may not detect low or negative pressure within distribution systems, such as areas with high elevations or within the vicinity of a fire or main break, the number of systems that issue a boil-water advisory could be significantly higher if more pressure monitoring was performed in distribution systems.

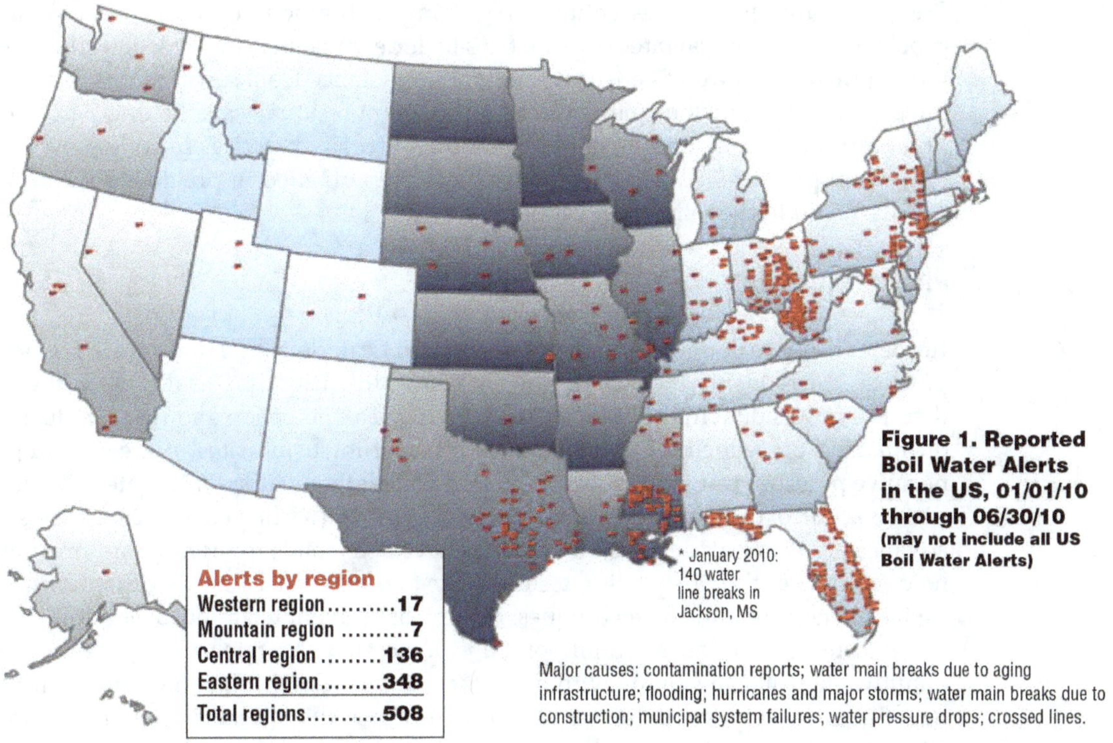

Source: Brigano and Burke 2010; Raucher et al. 2014

Figure 9-7 Location of boil-water advisories in the United States

Additionally, it is important to have high-speed recording in second or subsecond intervals to capture transients. Ordinary pressure monitoring using SCADA systems can often miss transient events entirely because of their long polling intervals (e.g., in minutes). This may present an additional challenge and motivate water systems to perform more extensive pressure monitoring since boil-water advisories are often negatively perceived from a public relations perspective. More guidance for water system professionals to consider in terms of reconciling the dilemma and communicating risks to customers effectively can be found elsewhere (CDC 2013).

THREE-INTEGRITY APPROACH

The National Research Council (2006) and PSW (2011) defined distribution system integrity as having three basic components:

- Physical integrity,
- Hydraulic integrity, and
- Water quality integrity.

As part of the multibarrier approach to protecting public health, maintaining the physical, hydraulic, and water quality integrity of water distribution systems is of paramount importance. Depressurization-related water quality problems represent loss of all three components: physical breach, pressure loss, and loss of disinfectant residual.

Pressure management is essential to regaining hydraulic integrity but cannot solve the problem alone. Physical integrity practices include implementing leak detection and main repair practices, as well as ensuring separation distance from sewer mains. Water quality integrity practices include maintaining disinfectant residual, establishing a cross-connection control program, and flushing. Hydraulic integrity practices include pressure monitoring and pressure management; these have been detailed in previous sections and are not discussed in the following section.

Physical Integrity

Implementing leak detection and main repair. Leak detection and repair are important practices for reducing the risk of infection from intrusion, especially when there is low disinfectant residual in the system. Additionally, leak detection and repair reduce the risk of infection by reducing the potential number of intrusion sites in areas susceptible to negative pressure transients. Leak detection is usually included in a water accountability and loss control program and is commonly performed (Kirmeyer et al. 2001). Considerable advancements in leak detection techniques have been made in recent years, and new technologies are continuously being tested. A system may start leak detection with conventional acoustic monitoring techniques, flow measurement techniques, or both.

Acoustic techniques can pinpoint a leak location accurately and quickly but cannot quantify the leakage volume with great accuracy. Although flow measurement does not have the capability to pinpoint a leak, it does allow the operator to quantify individual or multiple leakage flow rates that exist in different areas of the distribution system. The goal of leak detection for water loss control and for reducing the risk of viral infection from intrusion have different emphases. The purpose of the former is to reduce the water loss volume, while the purpose of the latter is to reduce the most number of leaks in areas susceptible to negative pressure transients. This difference may impact the selection and deployment of leak detection techniques.

Aging infrastructure. Excessive main breaks and leaks are signs of aging distribution system infrastructure, which has long been a concern for distribution system operators. As infrastructure deteriorates over time, it can create water quality issues such as corrosion by-products, changes to the system's chemical or hydraulic properties, or potential contamination during a failure event. It is important to understand the implications of an aging system in order to proactively manage them. For example, valves should be inventoried and operated regularly to ensure they are both operable and in the correct position. Valves that are difficult to operate can indicate tuberculation or sediment buildup in the line. Many systems find that a significant percentage of their system valves do not operate as intended (if at all). These valves are not locatable, not accessible, nonfunctional, or in the incorrect position. Not only does this affect system hydraulics, it also increases the time necessary to isolate a pipe during a break event.

The ultimate solution to aging infrastructure is replacement; however, this is often cost prohibitive. Instead, adopting a disciplined asset management approach that incorporates regular maintenance with condition assessment to determine appropriate rehabilitation activities is preferred. In short, small but timely investments will extend the service life of utility infrastructure and reduce water quality issues caused by corroding or failing pipelines. More details can be found in Chapter 4.

Ensuring separation distance from sewer mains. Ensuring adequate separation distance from sewer mains will reduce pathogen levels and chlorine demand of the leaked sewage into adjacent drinking water mains. Adequate free chlorine residual is effective in reducing the risks and severity of microbial infections, especially viral infections. Water mains submerged in leaking sewage, even with adequate pressures and no leaks, could cause significant concerns from an aesthetics perspective. In the United States, the Ten States Standards (2012) recommend a minimum separation distance between water mains and sewer mains of 10 ft horizontally for parallel installation and 18 in. vertically for crossings, with preference to the water main above the sewer. Exceptions may apply on a case-by-case basis.

Water Quality Integrity

Disinfectant residual maintenance. As detailed in Chapter 8, maintaining a disinfectant residual is the final barrier for protecting public health in the event of intrusion or backflow of a contaminant into distribution systems. In the United States, federal regulations (the Surface Water Treatment Rule) require large surface water systems to maintain a detectable free or combined residual in 95 percent of monthly samples or to demonstrate a heterotrophic plate count that is less than 500 cfu/mL. Smaller systems have a reduced monitoring requirement that provides minimal protection against microbial intrusion.

Some primacy entities in the United States have adopted more stringent minimum residuals for chlorinated and chloraminated distribution systems. For example, Texas requires 0.2 mg/L (as free chlorine) for chlorinated water and 0.5 mg/L (as total chlorine) for chloraminated water (TCEQ 2005). North Carolina requires 0.2 mg/L for chlorinated water and 2.0 mg/L for chloraminated water (NCDENR 2004). Industry guidelines also exist for disinfection residuals. For example, the Ten States Standards (2012) recommend a disinfectant residual of 0.2 mg/L for chlorinated water and 1.0 mg/L for chloraminated water. AWWA recommends a minimum free chlorine residual of 0.2 mg/L and a chloramine residual of 0.5 mg/L (AWWA 2006b).

LeChevallier et al. (2011) note that a free chlorine residual of 0.2 mg/L or greater is necessary to significantly reduce the risk of viral infection from intrusion. For systems that use chloramines as disinfectant residual, other preventive measures (e.g., pressure management and transient mitigation) should be used to reduce public health risks due to negative pressure transients. While chloramines have benefits for biofilm control (such as with *Legionella*), typical CT values of chloramine residuals in distribution systems are not effective for virus inactivation. Once a target disinfectant residual level is established in accordance with local regulatory requirements, especially in areas vulnerable to transients, disinfectant doses at the treatment plant can be adjusted to meet the targeted levels in the system. Another alternative is to install chlorine booster stations (see Chapter 8 for additional details). Loss of disinfectant residual can be reduced by treatment to minimize disinfectant demand in bulk water (e.g., enhanced coagulation or biofiltration, membrane processes, or source protection).

Distribution system operational practices are also effective in reducing the loss of disinfectant residuals (such as flushing or minimizing water age with turnover of stored water). A program for maintaining adequate disinfection residuals in a distribution system should be implemented as required by local regulations. If the regulatory standards do not exist, the program should be implemented according to AWWA (2006b) or the Ten States Standards (2012). Assuming compliance with the regulatory requirements that pertain to maintenance of a disinfectant residual, a general framework of the program can be implemented as described by the Partnership for Safe Water (2011), which summarizes criteria, metrics, and assessment approaches that water systems can use to optimize the disinfectant residual performance of their distribution systems.

Cross-connection control. Cross-connection control is the subject of Chapter 10 and is described in this chapter in the section titled Causes of Depressurization. In the context of water quality integrity being part of the three-integrity approach, cross-connection control is a best practice for ensuring water quality integrity.

The backsiphonage type of cross-connection is viewed as being similar to leaks that are influenced by sewage, that is, intrusion may occur when system pressures drop below external pressure (or the pressure of the cross-connection source). Reducing the number of cross-connections in an area susceptible to negative pressure transients (i.e., reducing the number of intrusion nodes) may significantly reduce the risk of infection. If a system does not have a cross-connection control program, it should prepare and implement a comprehensive cross-connection control and backflow prevention program as required by local regulations. If regulatory standards do not exist, the program should be implemented, at the very least, as stringently as those provided in AWWA Manual M14 (AWWA 2004). The program involves installing, testing, and maintaining backflow preventers in order to prevent potential contamination at various locations in the distribution system. A survey of 26 systems representing 17 US states showed that 16 states required cross-connection control programs and 14 required annual testing of backflow prevention devices (Kirmeyer et al. 2001).

Flushing for decontamination. Decontamination of pathogen intrusion due to low or negative pressures can be incorporated into emergency response planning. After intrusion due to a low pressure event, operators should consider boosting disinfectant residuals in the system, issuing a boil-water advisory, and/or flushing to remove the contaminants. Depending on the distribution system's response time, boosting disinfectant residual may not be effective if the intruded viruses have spread to customers. A boil-water advisory should be issued and broadcast as quickly as administratively possible. Flushing appears to be the simplest way to remove pathogens and can be effective if administered in time. Flushing can also be used for chemical contamination. Because the source and duration of contamination may not be known, it is advisable to take numerous samples during the contamination event to determine the extent of the contaminant plume. Before flushing, system operators should consider the following:

- The effectiveness of flushing to decontaminate depends on the promptness of flushing after the contamination event; the sooner flushing is performed, the more effective it will be. Generally, conventional flushing should be used instead of unidirectional flushing (UDF) because UDF requires significantly more time for valve operations and location of the contamination may be approximate.
- Flushing for contaminant removal should begin with the site(s) of intrusion and be limited to the estimated area of spread. Flushing far from the contamination source encourages intruded pathogens to spread out to more customers. For systems that experience frequent negative pressure transients, a map of potential intrusion locations, flushing locations, and flushing areas should be prepared in advance with the assistance of hydraulic and transient model simulations (Haxton and Walski 2009).
- Areas with contaminant intrusion are prone to low or negative pressures. Aggressive flushing can trigger more low or negative pressures and should be avoided. Flushing should be targeted to quickly replace the contaminated water with clean water rather than to achieve a scouring velocity using UDF.

Some intruded pathogens attach to the pipe surface and become resuspended later. This can create additional infection risks, even after the other intruded pathogens have been washed from the system. An active flushing program targeted to remove surface deposits and biofilms with effective scouring velocities (e.g., 5 ft/s for UDF) can reduce such risks. A system should prepare and implement such a flushing program after considering all cost–benefit factors (see Chapter 2 for more details).

CASE STUDIES

To improve management of low pressure for public health protection can be planned and designed in engineering studies (e.g., master planning and system specific studies) and implemented via capital improvement projects or improvements in system daily O&M activities. Two case studies are presented to demonstrate options of negative pressure transient control and integration of pressure monitoring with other distribution performance monitoring (DMA flow metering, continuous acoustic leak detection, and backflow-sensing metering).

Case Study 1—Pressure Transient Control Options

The following case study from Illinois (LeChevallier et al. 2011) demonstrates mitigation options for pressure transients. The distribution system was fed by surface water and had a relatively flat topology. The primary high lift pump station had a capacity of 30 mgd, and the hydraulic grade line leaving the plant varied from 889 to 906 ft (corresponding to 64 to 72 psi) under maximum day water demand conditions. The primary pump station did not have a stable power supply, and the distribution system had no floating storage facilities. Transient modeling was conducted for the system, assuming that a power outage occurred at the primary pump station. It was found that the system was vulnerable to negative pressure transients. As shown in Figure 9-8, the northwest portion of the system was estimated to draw negative pressures (highlighted in red) following the power outage at the primary pump station.

A number of transient mitigation options were evaluated in terms of the configuration and cost. As summarized in Table 9-1, the evaluated transient mitigation options included installation of hydropneumatic tanks, an uninterruptible power supply, an elevated storage tank, or 24/7 generators. It was noted that building a new 1-mil gal elevated storage tank in the system might provide not only surge protection benefit but also additional storage capacity for peak hour equalization and firefighting purposes. However, this option might add significantly more cost. The most effective and economical solution to mitigate the low or negative pressure transients in the system was found to be the installation of two 20,000-gal hydropneumatic tanks.

Case Study 2—Main Break, Low Pressure Event, and Intrusion

This case study (LeChevallier et al. 2014) illustrates that system depressurization might have caused backflow events and potential contamination from intrusion. Tracking depressurization-related water quality problems may require online pressure and/or water quality monitoring in the distribution system. Figure 9-9 illustrates the main break, system depressurization, pressure monitoring, and backflow events that occurred in a medium-sized surface water system (average day demand of 1.0 mgd) in Pennsylvania.

Two pressure loggers were installed at strategically selected locations to monitor system pressures. A major main break occurred on July 2, 2012. The break was sudden and severe. As shown in Figure 9-9, pressure dropped about 120 psi, from 140 to 20 psi, for 60 minutes at the high-pressure monitoring location. Pressure at the other monitoring location also dropped about 30 psi during the event. The hourly flow readings for water supply into the system were queried for three days: the break day, one day before, and one day after. The main break flow rate was estimated to be about 3,200 gpm (4.6 mgd) during the event. In addition to the flow increases, backflow events were also recorded by the smart meters installed as part of the advanced metering infrastructure system. The smart meters detect flow reversals (as little as 0.10 gal in 15 minutes). During the main break event, six backflow sensing flowmeters reported reverse flow, with one registering 150 gal.

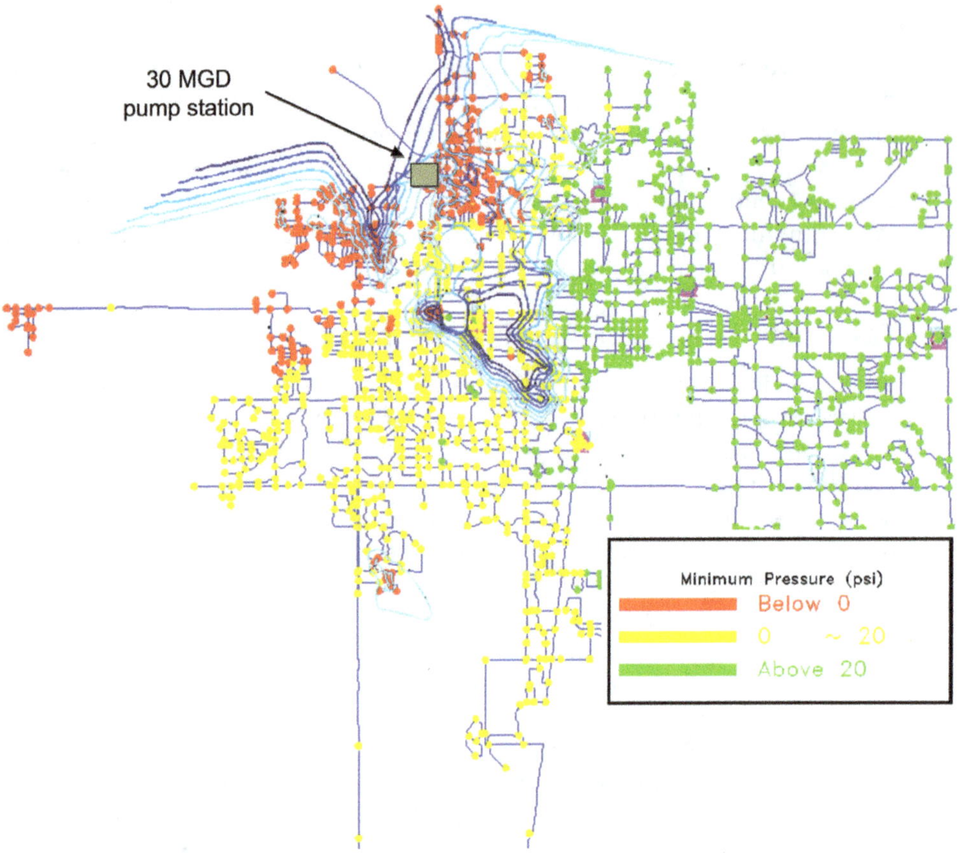

Source: LeChevallier et al. 2011.

Figure 9-8 Modeled minimum pressure in an Illinois water system during a power outage at the primary pump station

Table 9-1 Improvement planning for surge mitigation options

Option	Description	Cost
1	Two 20,000-gal hydropneumatic tanks	$0.3 million
2	Uninterruptible power supply sized to support nine pumps	$1.8 million
3	One 30,000-gal hydropneumatic tank and one 1-mil gal elevated tank at location A	$1.5 million
4	One 30,000-gal hydropneumatic tank and one 1-mil gal elevated tank at location B	$1.5 million
5	Do nothing, 24/7 generator	$30,000 per month

20,000-gal hydropneumatic tank

Location A and B

Source: LeChevallier et al. 2011.

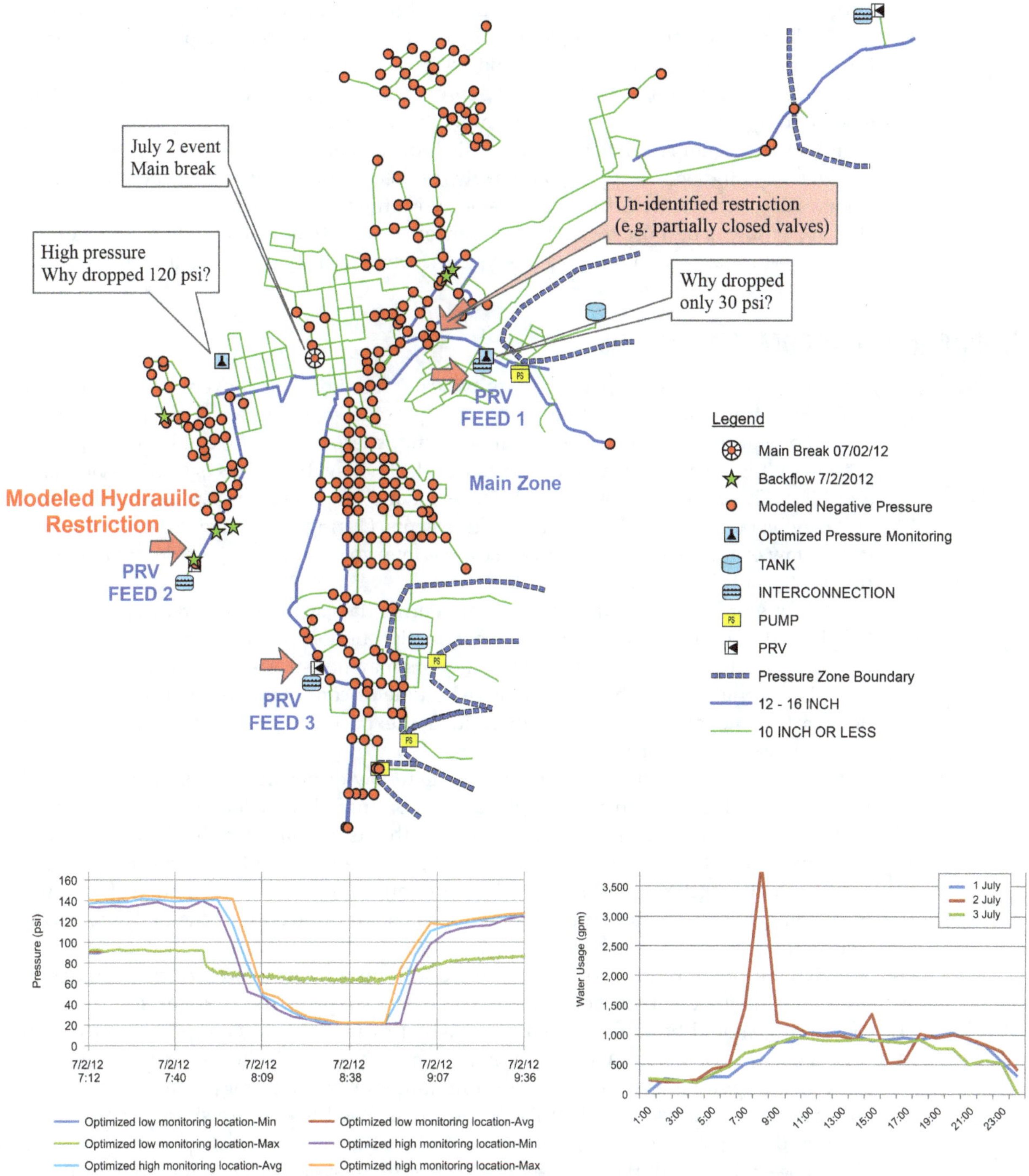

Source: Adapted from LeChevallier et al., 2014,

Figure 9-9 Backflows and low pressure event caused by main break: (A) broken main, backflow locations, and modeled negative pressures; (B) pressure drop at the two optimized pressure monitoring locations; and (C) water usage increase during the main break

Large reverse flow events often suggest a main break at a low elevation and loss of pressure at higher elevations. Using system hydraulic modeling to assemble all monitoring information, there appeared to be a weakness in the system (e.g., closed or partially closed valves, restricted PRVs), and the area of depressurization was estimated to be larger than evidenced by the backflow locations recorded by the smart meters. The approximate area of unknown system restriction was identified for further field investigation. Overall, the weaknesses could be investigated and likely resulted in increased system operational efficiency. Also, there were further discussions with the system operation and engineering staff about how to best use these multiple sensors to investigate the weakness, monitor the system performance, and improve the system's overall efficiency and resilience.

SUMMARY AND RECOMMENDATIONS

Water distribution system pressures are typically monitored at convenient locations, for example, pump stations, storage tanks, and pressure-regulating valves, to determine system operating pressures. Pressure monitoring is less commonly conducted at critical areas with high elevations and low pressures (locations where the system does not have a facility). These areas are often identified as having low pressures (e.g., less than 35 psi) and may be more vulnerable to pressure transients. During fire or main break and repair events, low or negative pressures in these areas may go undetected, considering that most transient events last less than a minute.

Online pressure and water quality monitoring data such as temperature, conductivity, pH, and free chlorine residual (if present) in distribution systems may not have significant correlations. Online water quality monitors may not successfully identify low-level intrusion events. In the absence of a large volume of contamination with high chlorine demand, maintenance of a disinfectant residual is expected to provide a protective barrier against low-level intrusions.

Microbial infection risks of intrusion due to low or negative pressure events could potentially be higher than the acceptable level of annual infection risk. Pressure management is an effective way to control the risk, that is, to maintain the distribution system at pressures that are higher than external pressures under all conditions. A pressure management program can be improved and/or optimized with a combination of pressure monitoring, hydraulic/transient modeling, transient pressure mitigations, and other O&M improvements.

Online pressure monitoring performed at a minimum of two critical locations in each pressure zone with remote sensing and access supplemented with existing SCADA monitoring should be implemented to tie actual distribution system performance with regulatory requirements. A pressure monitor can be set with the recording interval in minutes, sampling interval in seconds, minimum/maximum/average values recorded for each recording interval, and impulse recording enabled to capture short pressure fluctuations during distribution system events such as transients. Remote pressure monitors should be enabled with the impulse reading feature to avoid the need for large data storage and still capture pressure performance under fire flow and emergency conditions. Pressure monitoring should be integrated with other distribution monitoring such as tank levels, system flow, backflow, continuous leak detection, and water quality and with O&M activities such as flushing and main break repairs.

To reduce the risk of infection intrusion due to low or negative pressures, maintenance of an adequate residual of free chlorine (0.2 mg/L or above) is the last defense. Studies of microbial risk assessment of intrusion suggest that maintaining a chloramine residual might not significantly reduce the risk of infection, mainly due to its weaker disinfection effectiveness against viruses. For chloraminated systems, mitigation should rely on other practices such as pressure management. Separation distances from sewer mains should be maintained to reduce the potential of overall public health risks and to address concerns from an aesthetics perspective. Leak detection and repairs and cross-connection control practices should be prioritized in areas that are vulnerable to low and negative pressures, which may reduce the number of intrusion locations and reduce the overall risk of microbial infection. Flushing is the simplest and most effective approach to remove the contaminants if administered in time.

Table 9-2 provides a summary best practices for improving and optimizing pressure management based on the following two levels of service:

- Basic: all water systems perform this action as part of standard practice for ensuring pressure adequacy and
- Advanced: water systems with aims to improve pressure efficiency to reduce unnecessary breaks, leakage, and customer pressure complaints may need to perform this action, which is beyond standard practice of ensuring pressure adequacy.

Table 9-2 Best practices for management of low pressures

Best Practice	Level of Service
Improve and optimize pressure management	Advanced
Step 1. Establish pressure performance goals	Basic
Complete pressure monitoring	Intermediate
Assess performance data	Advanced
Step 2. Conduct hydraulic modeling, transient modeling	Basic/Advanced
Identify pressure deficiencies, if any, and performance constraints	Advanced
Step 3. Prepare pressure improvement action plan	Advanced
Plan pressure control and/or reduction	Advanced
Evaluate costs and benefits	Advanced
Step 4. Document and track improvements in ongoing performance associated with pressure improvements and projects developed in step 3 (if implemented)	Advanced
Implement transient control strategies	
Install pump control valves, motor soft starts, or variable-frequency drives	Intermediate Intermediate Advanced
Install air-release valves	Intermediate
Install elevated storage tanks and/or surge tanks	Advanced
Design pipe sizes to avoid velocities >6 ft/s	Basic
Train fire fighters and water distribution operators to slowly operate valves and hydrants	Basic
Install surge control valves	Advanced
Distribution system operations and maintenance	
Establish correct hydraulic grade line	Basic
Implement a cross-disciplinary approach to improve pump cycling	Intermediate
Guide planned or spot flushing activities based on pressure and supervisory control and data acquisition (SCADA) monitoring	Intermediate
Use hydraulic modeling and pressure monitoring to manage scheduled and emergency shutdowns	Advanced
Ensure valve vaults are kept dry	Basic
Implement leak detection program	Intermediate
Ensure adequate separation from sewer mains	Basic
Establish a cross-connection control program	Advanced

Recommended Reading List

Additional industrial guidelines and standards for depressurization-related water quality issues in distribution systems are available:

- LeChevallier et al. (2014) Water Research Foundation #4321—*"Pressure Management: Industry Practices And Monitoring Procedures"*
- Kirmeyer et al. (2014) Water Research Foundation #4307—*"Effective Microbial Control Strategies for Main Breaks and Depressurization"*
- LeChevallier et al. (2011) Water Research Foundation #4152—*"Managing Distribution System Low Transient Pressures for Water Quality"*
- Friedman et al. (2010) Water Research Foundation #4109—*"Criteria For Optimized Distribution Systems"*
- Fleming et al. (2006) Water Research Foundation #3008—*"Susceptibility of Distribution Systems to Negative Pressure Transients"*
- ANSI/AWWA (2004a) C512 (Standards for Air Release, Air/Vacuum, And Combination Air Valves for Waterworks Services);
- AWWA (2005) Manual M32 *Computer Modeling of Water Distribution Systems*
- AWWA (2006a) Manual M44 *Distribution Valves: Selection, Installation, Field Testing, and Maintenance*

REFERENCES

Almeida, A.B. & Kolle, E. 1992. *Fluid Transients in Pipe Networks.* Southampton, United Kingdom: Elsevier Applied Science.

AWWA (American Water Works Association). 2004a. Standard C512, *Air Release, Air/Vacuum, and Combination Air Valves for Waterworks Services.* Denver, CO: ANSI/AWWA.

———. 2004b. Standard G200, *Distribution System Operation and Management.* Denver, CO: ANSI/AWWA.

———. 2004. M14, *Recommended Practice for Backflow Prevention and Cross Connection Control.* Denver, CO: AWWA.

———. 2006a. M44, *Distribution Valves: Selection, Installation, Field Testing, and Maintenance.* Denver, CO: AWWA.

———. 2006b. M56, *Fundamentals and Control of Nitrification in Chloraminated Drinking Water Distribution Systems.* Denver, CO: AWWA.

———. 2012. M32, *Computer Modeling of Water Distribution Systems.* Denver, CO: AWWA.

———. 2016. M36, *Water Audits and Loss Control Programs.* Denver, CO: AWWA.

AWWA Water Distribution Model Calibration Subcommittee. 2013. Defining Model Calibration. *Journal of the American Water Works Association,* 105:7: 60–63.

Blackburn, B.G.; G.F. Craun; J.S. Yoder; V. Hill; R.L. Calderon; N. Chen; S.H. Lee; D.A. Levy; & M. J. Beach. 2004. Surveillance for Waterborne-Disease Outbreaks Associated With Drinking Water—United States, 2001–2002. Surveillance Summaries, October 22, 2004. *Morbidity and Mortality Weekly Report*, 53:SS-8:23–45.

Borchardt, M.A.; S.K. Spencer; G.W. Harrington; P.D. Bertz; M.J. Volenec; B.A. Jr. Kieke; E. Lambertini; & F.J. Loge. 2007. Inputs of Viruses into Municipal Drinking Water: Contaminated Groundwater Versus Distribution System Intrusions. American Society for Microbiology, Toronto, Ontario, May 21–25.

Boulos, P.; K. Lansey; & B. Karney. 2006. *Comprehensive Water Distribution Systems Analysis Handbook for Engineers and Planners*. Innovyze., Pasadena, Calif.: MWH Soft Inc.

Boulos, P.F.; B.W. Karney; D.J. Wood; & S. Lingireddy. 2005. Hydraulic Transient Guidelines for Protecting Water Distribution Systems. *Journal of the American Water Works Association*, 97:5:111–124.

Boyd, G.R.; H. Wang; M.D. Britton; D.C. Howie; D.J. Wood; J.E. Funk; & M.J. Friedman. 2004a. Intrusion Within a Simulated Water Distribution System Due to Hydraulic Transients. I: Description of Test Rig and Chemical Tracer Method. *Journal of Environmental Engineering.* 130:7(774).

Boyd, G.R., H. Wang, M.D. Britton, D.C. Howie, D.J. Wood, J.E. Funk, and M.J. Friedman. 2004b. Intrusion Within a Simulated Water Distribution System due to Hydraulic Transients. II: Volumetric Method and Comparison of Results. *Journal of Environmental Engineering*, 130:7:778.

CDC (Centers for Disease Control and Prevention). 2013. Drinking Water Advisory Communication Toolbox. http://www.cdc.gov/healthywater/emergency/dwa-comm-toolbox/index.html (access date may 25, 2017).

Chaudhry, H. 1987. *Applied Hydraulic Transients.* New York: Van Nostrand Reinhold.

Ebacher, G.; M.C. Besner; & M. Prevost. 2012. Field Validation of Intrusion Modeling Inputs: External Heads and Contamination Levels of Water in Valve Vaults and Groundwater. Water Distribution System Analysis Conference, Adelaide, Australia.

———. (2013). Submerged Appurtenances and Pipelines: An Assessment of Water Levels and Contaminant Occurrence. *Journal of the American Water Works Association*, 105:12:43–44.

Fleming, K.K.; J. Dugandzik; & M.W. LeChevallier. 2006. *Susceptibility of Distribution Systems to Negative Pressure Transients*. Denver, CO: AWWA Research Foundation.

Friedman, M.; L. Radder; S. Harrison; D. Howie; M. Britton; G. Boyd; H. Wang; R. Gullick; M. LeChevallier; D. Wood; & J. Funk. 2004. *Verification and Control of Low Pressure Transients in Distribution Systems*. Denver, CO: AWWA Research Foundation.

Friedman, M.; G. Kirmeyer; J. Lemieux; M. LeChevallier; S. Seidl; & J. Routt. 2010. *Criteria for Optimized Distribution Systems*. Denver, CO: Water Research Foundation.

Ghorbanian, V.; B. Karney; & Y. Guo. 2016. Pressure Standards in Water Distribution Systems: Reflection on Current Practice with Consideration of Some Unresolved Issues. *Journal of Water Resource Planning and Management, ASCE,* 10:1061.

Grebe, T.E.; D.D. Sabin; & M.F. McGranaghan. 1996. *An Assessment of Distribution System Power Quality*. Palo Alto, CA: Electric Power Research Institute.

Gullick, R.W.; M.W. LeChevallier; R.C. Svindland; & M.J. Friedman. 2004. Occurrence of Transient Low and Negative Pressures in Distribution Systems. *Journal of the American Water Works Association,* 96:11: 52–66.

Gullick, R.W.; M.W. LeChevallier; J. Case; D.J. Wood; J.E. Funk; & M.J. Friedman. 2005. Application of Pressure Monitoring and Modeling to Detect and Minimize Low Pressure Events in Distribution Systems. *Journal of Water Supply: Research and Technology–AQUA,* 54:2:65–81.

Haxton, T. & Walski, T. 2009. Modeling a Hydraulic Response to a Contamination Event. World Environmental and Water Resources Congress, pp. 1–9. Kansas City, MO.

Hunter, P.R.; R.M. Chalmers; S. Hughes; & Q. Syed. 2005. Self-Reported Diarrhea in a Control Group: A Strong Association With Reporting of Low-Pressure Events in Tap Water. *Clinical Infectious Diseases,* 40:e32–34.

ILSI (International Life Sciences Institute). 2000. Revised Framework for Microbial Risk Assessment. An ILSI Risk Science Institute Workshop Report. https://www.unc.edu/courses/2006spring/envr/133/001/mrabook_ILSI_EPA_MicroRisk.pdf. (access date May 25, 2017)

Karim, M.; M. Abbaszadegan; & M.W. LeChevallier. 2003. Potential for Pathogen Intrusion During Pressure Transients. *Journal of the American Water Works Association,* 95:5:134–146.

Karney, B.W. & Y.R. Filion. 2003. Energy Dissipation Mechanisms in Water Distribution Systems. Proceedings FEDSM'03, 4th ASME–JSME Joint Fluids Engineering Conference. Honolulu, HI.

Kirmeyer, G.J.; M. Friedman; K. Martel; D. Howie; M. LeChevallier; M. Abbaszadegan; M. Karim; J. Funk; & J. Harbour. 2001. *Pathogen Intrusion into the Distribution System.* Denver, CO: AWWA Research Foundation and AWWA.

Kirmeyer, G.J.; T.M. Thomure; R. Rahman; J.L. Marie; M.W. LeChevallier; J. Yang; D. Hughes; & O. Schneider. 2014. *Effective Microbial Control Strategies for Depressurization and Main Break Events*. Denver, CO: Water Research Foundation.

LeChevallier, M.W.; R.W. Gullick; M.R. Karim; M. Friedman; & J.E. Funk. 2003. The Potential for Health Risks From Intrusion of Contaminants Into the Distribution System From Pressure Transients. *Journal of Water and Health,* 1:1:3–14.

LeChevallier, M.W.; M. Xu; J. Yang; P. Teunis; & K. Fleming. 2011. *Managing Distribution System Low Transient Pressures for Water Quality*. Denver, CO: Water Research Foundation.

LeChevallier, M.W.; J. Yang; M. Xu; D. Hughes; & G. Kunkel. 2014. *Pressure Management: Industry Practices and Monitoring Procedures*. Denver, CO: Water Research Foundation.

Lee, J.J.; P. Schwartz; P. Sylvester; L. Crane; J. Haw; H. Chang; & H.J. Kwon. 2003. *Impacts of Cross-Connections in North American Water Supplies*. Denver, CO: AWWA Research Foundation.

Messner, M.J.; P. Berger; & S.P. Nappier. 2014. Fractional Poisson: A Simple Dose–Response Model for Human Norovirus. *Risk Analysis*, 34:10:1820. http://dx.doi.org/10.1111/risa.12207.

NRC (National Research Council). 1983. *Risk Assessment in the Federal Government: Managing the Process*. Washington, DC: National Academy Press.

NCDENR (North Carolina Department of Environment and Natural Resources). 2004. Title 15A, Subchapter 18C of the North Carolina Administrative Codes. Section 2000—Filtration and Disinfection. Raleigh, NC: NCDENR.

National Research Council. 2006. *Drinking Water Distribution Systems: Assessing and Reducing Risks*. Washington, DC: National Academy Press.

Nygård, K.; E. Wahl; T. Krogh; O.A. Tveit; E. Bøhleng; A. Tverdal; & P. Aavitsland. 2007. Breaks and Maintenance Work in the Water Distribution Systems and Gastrointestinal Illness: A Cohort Study. *International Journal of Epidemiology*, 36:873–880.

Oberoi, K. 2009. *Operation Guide to AWWA Standard G200: Distribution Systems Operation and Management*. Denver, CO: AWWA.

PSW (Partnership for Safe Water). 2011. *Distribution System Optimization Program (Overview)*. Denver, CO: AWWA (https://www.awwa.org/resources-tools/water-and-wastewater-utility-management/partnership-for-safe-water.aspx) (access date May 25, 2017)

Payment, P.; J. Siemiatycki; L. Richardson; G. Renaud; E. Franco; & M. Prevost. 1997. A Prospective Epidemiological Study of Gastrointestinal Health Effects Due to the Consumption of Drinking Water. *International Journal of Environmental Health Research*, 7:5–31.

Pflanz, M.; T. Walski; & M. Sharkey. 2010. Preventing Negative Pressure Due to Transients. Texas ASCE, 2010.

Raucher, R.; K. Raucher; J. Pratt; & L. Ragain. 2014. *Identifying the Gaps in Understanding the Benefits and Costs of Boil Water Advisories*. Denver, CO: AWWA Research Foundation.

Speight, V.; N. Khanal; D. Savic; Z. Kapelan; & P. Jonkergouw. 2010. *Guidelines for Developing, Calibrating, and Using Hydraulic Models*. Denver, CO: Water Research Foundation.

Ten States Standards (Water Supply Committee of the Great Lakes). 2012. Recommended Standards for Water Works. Great Lakes–Upper Mississippi River Board of State and Provincial Public Health and Environmental Managers. http://10statesstandards.com/index.html May 25, 2017

TCEQ (Texas Commission on Environmental Quality. 2005. 30 TAC290.110(b)(4) §290.110. Disinfectant Residuals, Austin TX.

Teunis, P.F.; C.L. Moe; P. Liu; S.E. Miller; L. Lindesmith; R.S. Baric; J. Le Pendu; & R.L. Calderon. 2008. Norwalk Virus: How Infectious Is It? *Journal of Medical Virology*, 80:8:1468–1476.

Thorley, A.R.D. 2004 (2nd ed.). *Fluid Transients in Pipeline Systems*. Herts, England: D.&L. George, Ltd.

Thornton, J. 2003. Managing Leakage by Managing Pressure: A Practical Approach, Water Loss IWA Task Force, *Water21* 54-55.

Thornton, J. & A. Lambert. 2006. Managing Pressures to Reduce New Breaks, Water Supply Networks, *Water21*, 8:6:24-26.

Tinker, S.C.; C.L. Moe; M. Klein; W.D. Flanders; J. Uber; A. Amirtharajah; P. Singer; & P.E. Tolbert. 2009. Drinking Water Residence Time in Distribution Networks and Emergency Department Visits for Gastrointestinal Illness in Metro Atlanta, Georgia. *Journal of Water and Health*, 7:2:332. http://dx.doi.org/10.2166/wh.2009.022.

USEPA (US Environmental Protection Agency) and USDA/FSIS. 2012. *Microbial Risk Assessment Guideline: Pathogenic Microorganisms with Focus on Food and Water.* Prepared by the Interagency Microbiological Risk Assessment Guideline Workgroup. EPA/100/J-12/001, Washington, DC.

Walski, T.M. & T.L. Lutes. 1994. Hydraulic Transients Cause Low-Pressure Problems. *Journal of the American Water Works Association*, 86:12:24–32.

Walski, T.M. 2009. Not So Fast—Close Hydrants Slowly, *Opflow*, 2:14.

Walski, T.M.; D.V. Chase; D.A. Savic; W.M. Grayman; S. Bechwith; & E. Koelle. 2007. *Advanced Water Distribution Modeling and Management.* Watertown, CT: Haestad Methods, Inc.

Walski, T.M.; M. Sharkey; & M. Pflanz. 2011. What Happens When You Lose Power. AWWA Distribution System Symposium, Nashville, TN.

Wood, D.; S. Lingireddy; & P. Boulos. 2005. *Pressure Wave Analysis of Transient Flow in Pipe Distribution Systems.* MWH Soft Inc., Pasadena, Calif..

Wylie, B. & V. Streeter. 1993 *Fluid Transients in Systems.* Englewood Cliffs, NJ: Prentice-Hall.

Chapter 10

Cross-Connection Control and Backflow Prevention

John Graham, California Water Service Co, Chico, Calif.
Patrick Cole, H2M Architects and Engineers, Brielle, N.J.
Peter Fiske, PAX Water Technologies, Oakland, Calif.
Adam Jacoby, Hatch Mott MacDonald, New York, N.Y.
Kira S. Smith, City of Houston Drinking Water Operations, Houston, Texas
Aneta King, Halton Regional Municipality, Oakville, Ont., Canada

A cross-connection is the physical connection of a safe or potable water supply with another water supply of unknown or contaminated quality or a connection such that the potable water could be contaminated or polluted. Backflow is a hydraulic condition that results from a difference in pressures, causing nonpotable water or other fluid to flow into a potable watersystem. The history of outbreaks and reported illnesses due to cross-connections and backflow incidents indicates that some level of public health risk is associated with these situations. Risk is a function of a variety of factors including cross-connection and backflow occurrence, type and amount of contaminants, and the contaminants' potential health effects.

The modern water distribution system is designed to provide potable water to customer connections and sometimes also for emergency fire flow, while maintaining positive system water pressure without interruption. In theory, this provides for safe water at the tap by ensuring forward flow in the distribution system at all times. In reality, many water systems experience flow rates that are higher than the designed rates due to fire department operations, pump failures, water main leaks, or water mains that are shut down for repair or replacement. These normal occurrences provide the conditions that can contribute to a backflow event.

Homes and businesses are required to comply with any adopted plumbing code for their jurisdiction. This should maintain a physical separation between the potable water piping and any hazardous substances. However, only the most rigorous code enforcement programs can detect small modifications to customer plumbing systems that are not in compliance with the plumbing code. Small, rural communities may not have or implement plumbing code requirements, even if they provide potable water. Home improvement stores and Internet shopping allow consumers to purchase plumbing fixtures and water-connected equipment and install them without permit or inspection. This can lead to cross-connections that provide a physical connection between the potable water piping and hazardous substances such as harmful microorganisms, pesticides, and herbicides.

Hydraulic fluctuations or pressure transients in the water distribution system are a common occurrence (Chapter 9 covers pressure transients in depth). Even severe adverse hydraulic conditions are often ignored because they are typically of short duration or have occurred for so long and with such frequency that they are treated as acceptable conditions. Cross-connections can remain unnoticed for long periods of time. Equipment may be improperly installed or plumbing changes made that do not conform to the plumbing code, and yet water still flows from the tap. Likewise, hazardous substances are common in everyday life, for example, detergents and cleaning products, so much so that proper handling is not always practiced.

Where cross-connections exist along with the presence of a hazardous substance and certain hydraulic conditions, backflow can and does occur. It is at this point that the water supplier loses control of drinking water quality and the safety of the potable water supply. The scope of contamination from a backflow event can range from one customer to an entire water distribution system.

It is only with a robust and active cross-connection control program that water suppliers can ensure their distribution system is truly safe and that water quality delivered at the tap is not affected by outside conditions. AWWA Manual M14 (2015) provides detailed guidance on cross-connection control. It is important for a water quality manager to understand that backflow is real, that it occurs in well-managed water systems, and that cross-connection control must be a central element of water system operations. Cross-connection control is truly the final barrier to water contamination.

DSWQ Challenge:

The history of outbreaks and reported illnesses due to cross-connections and backflow incidents indicates some level of public health risk is associated with these situations.

BACKFLOW AND HYDRAULIC PRINCIPLES

Backpressure is a condition that can cause water to backflow into the public water supply when a downstream water system is at a higher pressure than the public water system. To move fluid between two points, there must be two conditions: a pipe or conduit for a flow path and a difference in pressure between the two points. A fluid such as water flows from high pressure to low pressure.

Pressure is the force that is applied to a mass over an area, such as pounds per square inch (psi). In this manual, references are made to absolute, atmospheric, and gauge pressure. It is important to have an understanding of how these relate to each other. (In standard notation, atmospheric pressure is represented by psia, and gauge pressure is psig. When used, "ps" references a calculation-specific difference in pressure.) Following is an explanation of the different types of pressure:

- Atmospheric pressure—psia. This pressure is the force exerted on an area by the atmosphere surrounding the Earth. The value of atmospheric pressure changes with elevation or the thickness of the atmosphere's layer. At sea level, the atmospheric pressure is approximately 14.7 psi. At 1 mile higher than sea level, the atmospheric pressure is approximately 12.2 psi.
- Gauge pressure—psig. The pressure measured with a gauge is referred to as gauge pressure—the term is used to refer to pressure exerted within a piping system. The zero basis for gauge pressure is atmospheric pressure. If the pressure in a piping system is at atmospheric pressure, the gauge pressure is zero. For example, an empty piping system at sea level has a gauge pressure of 0 psi. The same system at 1 mile in elevation will have a gauge pressure of 0 psi. The term "head" is often used synonymously with psig.
- Absolute pressure—psi. Absolute pressure is defined as a combination of atmospheric pressure and gauge pressure. For example, if the atmospheric pressure is 14.7 psia and the pressure within a pipe is 30 psig, the absolute pressure within the pipe is 44.7 psi absolute. When gauge pressure is negative, the absolute reading is less than the atmospheric pressure. For example, if the gauge on a piping system reads −4.7 psi, the absolute pressure is 10 psi. (In Standard International (SI) units, for example, if the atmospheric pressure is 101 kPa and the pressure within a pipe is 207 kPa, the absolute pressure within the pipe is 308 kPa absolute. And, if the gauge read −32 kPa, then the absolute pressure would be 69 kPa.)

It is important to recognize that absolute pressure cannot have a negative value. An absolute pressure value of 0 psi absolute has absolutely no pressure—it is a total vacuum. Figure 10-1 provides a graphical representation of atmospheric, gauge, absolute, and vacuum pressures.

For example, a gauge value of 20 psig indicates that the pressure within the system is 20 psi greater than atmospheric pressure. At sea level, the absolute pressure within the system is 34.7 psi (14.7 psi atmospheric plus 20 psi gauge). (In SI units, for example. a gauge value of 138 kPa indicates that the pressure within the system is 138 kPa greater than atmospheric pressure. At sea level, the absolute pressure within the system is 239 kPa (101 kPa atmospheric plus 138 psi gauge.)

As another example, a gauge value of −14.7 psig is the lowest gauge pressure possible at sea level because the absolute pressure is zero. The water pressure measured within a distribution or plumbing system is measured as gauge pressure (psig).

As with the pressure that is exerted by the thickness of the atmosphere, water pressure also increases with its elevation above the gauge. A water column that is 2.31 ft high provides a pressure of 1 psig at the base. The water pressure increases at the rate of 0.433 psi per foot of elevation. An elevated pipe that is 100 ft high and filled with water provides a pressure of 43.3 psig at the base.

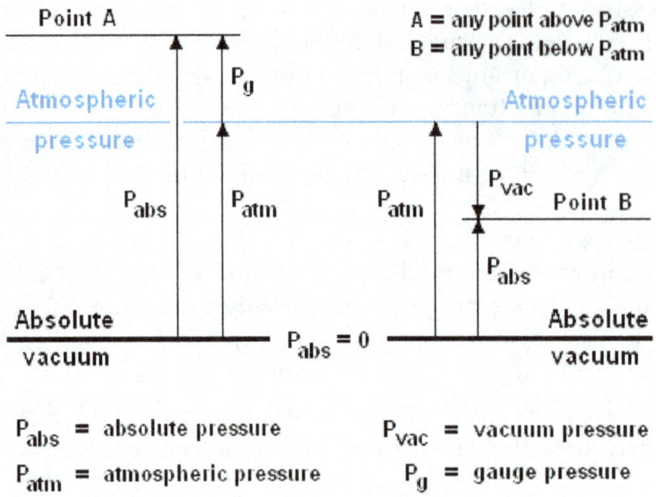

Source: Wikispaces, ChemEngineering, Pressure (https://chemengineering.wikispaces.com/Pressure).

Figure 10-1 Absolute, atmospheric, gauge, and vacuum pressures

Vacuum

As discussed above, a vacuum occurs within a piping system when the pressure drops below atmospheric pressure or becomes negative gauge pressure. Because of this low pressure condition, water within a piping system will flow to the low-pressure point where the vacuum exists. This can result from pressure transients in the distribution system, as described in Chapter 9.

Figure 10-2 shows another kind of vacuum, a barometric loop. A pump located above ground creates a suction. Water is then pushed upward from the atmospheric pressure that is applied at the water surface. At sea level, atmospheric pressure can push water up a column 33.9 ft high if a vacuum of −14.7 psi exists at the top of the column. The loop is elevated 35 ft above the highest water outlet, which prevents any siphon or return flow from that outlet. Some model plumbing codes allow barometric loops and others do not. Check with the local authority having jurisdiction to find out if it is allowed in your area.

Like the pressure exerted by the thickness of the atmosphere, water pressure will also increase with its elevation above the gauge. A water column that is 2.31 ft high will provide a pressure of 1.0 psi at the base. The water pressure will increase at the rate of 0.433 psi per foot of elevation. An elevated pipe that is 100 ft high and filled with water provides a pressure of 43.3 psi at the base, which is the height of the hydraulic grade line. The hydraulic grade line is a mathematically derived line, plotted to represent the system pressure or head at any point in the distribution system. The elevation of the line is determined by calculating the vertical rise of water at various points in the system. To adequately maintain system pressure, it is critical that all service connection water pressure demands fall below the line, under all flow conditions. If the hydraulic grade line falls below the demand, it is likely that a vacuum will be created, allowing a siphon to occur. This is referred to as backsiphonage, a form of backflow caused by a negative or subatmospheric pressure within a water system.

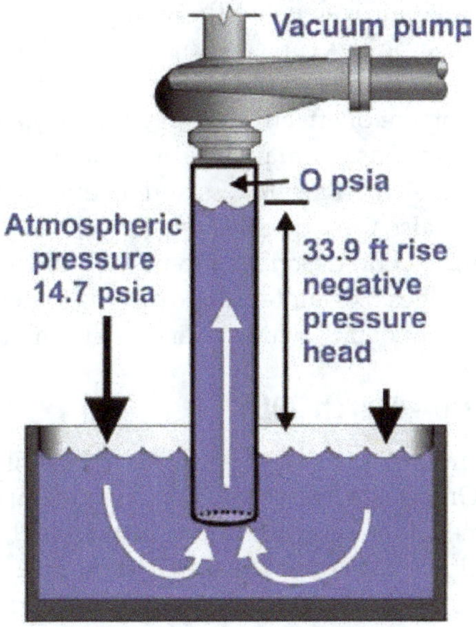

Figure 10-2 Example of a barometric loop

Like the pressure exerted by the thickness of the atmosphere, water pressure will also increase with its elevation above the gauge. A water column that is 2.31 ft high will provide a pressure of 1.0 psi at the base. The water pressure will increase at the rate of 0.433 psi per foot of elevation. An elevated pipe that is 100 ft high and filled with water provides a pressure of 43.3 psi at the base, which is the height of the hydraulic grade line. The hydraulic grade line is a mathematically derived line, plotted to represent the system pressure or head at any point in the distribution system. The elevation of the line is determined by calculating the vertical rise of water at various points in the system. To adequately maintain system pressure, it is critical that all service connection water pressure demands fall below the line, under all flow conditions. If the hydraulic grade line falls below the demand, it is likely that a vacuum will be created, allowing a siphon to occur. This is referred to as backsiphonage, a form of backflow caused by a negative or subatmospheric pressure within a water system.

Giovanni Venturi discovered another type of backsiphonage when fluid velocity increases through a piping restriction and a jet effect is created that can produce a negative gauge pressure. A piping connection at the point of the siphon may allow backflow to occur through an uncontrolled cross-connection. This principle, referred to as the Venturi effect, is frequently used to intentionally introduce chemicals such as chlorine into a water line. This principle also applies to a garden sprayer attached at the end of a hose in order to apply fertilizers and pesticides. Figure 10-3 illustrates the Venturi effect.

EXAMPLES OF BACKFLOW AND CROSS-CONNECTION INCIDENTS AND THEIR EFFECTS ON WATER QUALITY

Drinking water utilities occasionally experience some type of abnormal hydraulic event in their distribution systems. Many backflow incidents, which occur during system pressure variations, are not reported due to the normally short-term duration of the event. Customers are not aware of any changes in the quality of their water supply, making it difficult to confirm what caused the incident.

With continual improvement in the ability to record events (e.g., water meter register data-loggers), water suppliers can gain knowledge of backflow events in their distribution systems. This not only results in the ability to maintain the integrity of supplies but also leads to the development of effective emergency response plans.

Several organizations maintain lists of reported incidents that provide specific details and provide the reader with awareness and a better understanding of backflow. Internet news media sources also provide extensive information on "backflow incident reporting."

The following section describes recent backflow incidents that are representative of the everyday challenges of maintaining drinking water quality. An incident in Stratford, Ontario is discussed in detail, while others are summarized in Table 10-1.

Stratford, Ontario—March 2005

Sweet smelling, pink, soapy water was observed coming into toilet tanks and out of faucets in Stratford, Ontario, in March 2005. A city-wide drinking water advisory was issued, which soon became a boil-water advisory. Flushing of the water distribution system immediately began as city workers tracked down the source of the contamination.

Public Works staff, city officials, staff from various departments, and a Ministry of the Environment inspector joined forces to enact an emergency action plan. Samples were taken to a laboratory for analyses, as water was being flushed from the hydrants and diverted from the city-wide system. Through television and radio announcements, citizens were urged not to drink the water, while firefighters, postal carriers, city staff, and volunteers went door to door delivering flyers to provide notification that a drinking water advisory was in effect. Schools were closed, dialysis patients (hospital and at home) were sent to other treatment centers, and bottled water distribution depots were set up, with home delivery available to those unable to travel to the depots.

The investigation discovered that a coin-operated car wash was the source of the pink, soapy water. A car wash customer noticed that soap had come out of the clear water rinse nozzle, and a car wash employee discovered a soap container leaking into the rinse water. It was determined that during cleanup, a worker forgot to close a crucial valve. This allowed the release of 2-butoxy-ethanol detergent into the water distribution system. The cross-connection between the soap and rinse water had been installed without the city's knowledge. In addition, a pressure pump twice as powerful as indicated on municipal records was installed. The extra force generated by this pump (80 psi) had forced the chemical into the city pipes (65 psi).

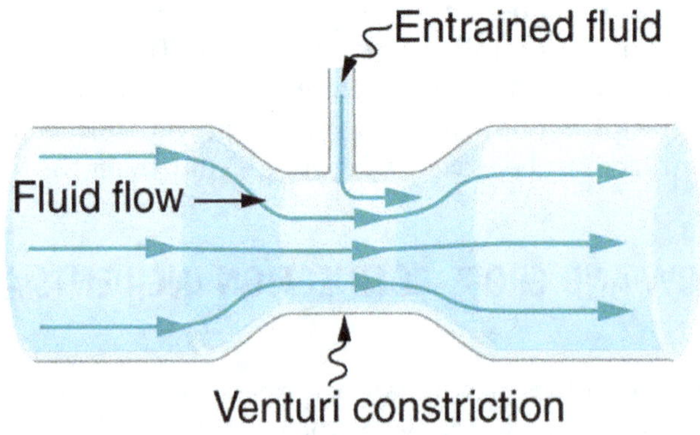

Source: philschatz, Physics Book (http://philschatz.com/physics-book/contents/m42206.html).

Figure 10-3 Diagram of the Venturi effect

The three components, as previously discussed, required for a backflow event were met:

- Cross-connection: valve turned on,
- Hazard: direct connection to soap, and
- Hydraulic condition: pump at higher pressure than in the water distribution system.

The car wash owner was fined $75,000 for "causing or permitting the discharge" of the chemical due to a combination of choices, that is, an unexpected leak and human error (Ministry of the Environment and Climate Change 2006).

INDICATORS OF A BACKFLOW INCIDENT

A clear indication that a backflow event has occurred is the presence of one of the following contaminants in the potable water system. This is based on information the US Environmental Protection Agency obtained from backflow incident records (USEPA 2001):

- Fecal coliform/*Escherichia coli*—from lawn irrigation systems, wastewater and septic systems, and auxiliary water supplies;
- Detergents—extensively used in many industries;
- Pesticides—chlordane, malathion, and others;
- Ethylene glycol—used as antifreeze in heating and cooling systems, plating plants, steam boiler plants, and dye plants;
- Propylene glycol—used as antifreeze and as a food additive;
- Freon—used in air conditioning and refrigeration;
- Copper—used in plumbing; contamination commonly caused by malfunctioning soft drink dispensers (carbonators);
- Chromium—used in the past in cooling towers as a rust and corrosion inhibitor; and
- Nitrate—used as an agricultural fertilizer.

Except for fecal coliform and *E. coli,* most contaminants listed are not routinely monitored in the distribution system. Therefore, and because there is little to no immediate warning that a backflow incident is occurring, it is difficult to detect a cross-connection or backflow event in a water distribution system. In some cases, it is not known for some time after an incident; in other cases, it is never discovered.

Possible reasons for water systems to monitor for cross-connections include customer complaints regarding water quality, drops in operating pressure, drops in disinfectant residual, water meters running in reverse, and coliform detections. More advanced monitoring and specific bacterial species also provide information, as discussed in Chapter 3. Cross-connections and contamination due to backflow events can occur in the absence of these indicators.

Complaints

Customer complaints of odor, discoloration, other aesthetic or health concerns, or direct physical harm from contact with the water can provide a warning to check for certain contaminants.

Table 10-1 Summary of reported cross-connections and backflow incidents

Event Location and Date	Description	Water Quality Impact(s)
Corpus Christi, Texas, July 2015 (Ellison 2015)	A private water well without proper backflow prevention may have contaminated a neighborhood	*Escherichia coli* contamination resulted in the city issuing a boil-water notice
Old River–Winfree, Texas, July 2015 (Barr 2015)	The water system had a power outage, and some equipment malfunctioned, likely causing depressurization	Several residents complained of worms flowing from their faucets
Cedar Hills, Utah, September 2012 (Warnock 2012)	An irrigation line had been cross-connected to a drinking water line, affecting 13 homes; the city provided bottled water to homes until they could fix the problem and sanitize the mains	Customers complained of discoloration and bad taste; bacteriological tests were positive for total coliform bacteria but negative for *Escherichia coli*
Boise, Idaho, June 2012 (Cotterell 2012)	A backflow preventer in an empty house in the Hobble Creek neighborhood failed to keep irrigation water out of the drinking water system; it was initially discovered when a water meter reading showed unusual readings and was flagged for further inspection	80 residents in one neighborhood claimed to have gotten sick as a result
Somerset, Massachusetts, May 2011 (Welker 2011)	An unauthorized connection to a hydrant without a backflow preventer permitted an unknown amount of hydro-seed material to enter the town's water system	Residents and businesses were told not to use the water for drinking, washing dishes, or preparing food for three days; no illnesses were known to have occurred
Dorval, Quebec, June 2011 (CTV Montreal 2011)	A private contractor mistakenly flushed out (back flowed) a massive underground tank from a hangar at Trudeau Airport that brought about 200,000 gal of contaminated water into the water distribution system	A drinking water alert was issued following contamination of Dorval's water supply where residents were told not to use tap water for drinking, brushing teeth, or rinsing food and washing dishes
Kilauea, Hawaii, May 2011 (The Garden Island 2011)	A cross-connection was created when a customer connected a surface water irrigation line to the water system without an approved backflow preventer; when the pressure in the irrigation line increased, irrigation water was pushed into the county water system	After receiving reports of dirty water, the Department of Water (DOW) obtained samples that tested positive for *Escherichia coli*; the DOW issued a "do not use" notice to customers
Bristow, Kentucky, October 2010 (*Bowling Green Daily News* 2010)	Glycol, a chemical used in a school's heating and cooling system, leaked into the school's drinking supply when a contracting company was working on the heating, ventilation, and air conditioning system; the contractor attempted to pump additional water from the school's supply into geothermal pipes; however, pressure in those pipes caused water to back flow, and a small amount of the chemical was pushed back into the drinking supply	Students were given bottled water and food had to be prepared off site, while a backflow-prevention system was installed
Minneapolis, Minnesota, October 2010 (Friedmann 2010)	A company twice contaminated the water system at Normandale Community College while working on a cooling tower ; they first attached a hose without using a backflow preventer, allowing ethylene glycol into the drinking water system; 17 days later, company workers let chemicals backflow into a water heater	No one got sick; the system was flushed both times
Queens, New York, February 2010 (Bailey 2010)	A contractor working on a recently installed central air-conditioning system may have caused propylene glycol to contaminate a school's potable water system	74 students were rushed to area hospitals with nausea and vomiting; water had to be delivered to the school

Drops in Operating Pressure

A drop in operating pressure creates the hydraulic condition necessary for backflow to occur. Restoration of normal system pressure cannot stop an event in progress or prevent an incident, unless the root cause is corrected. Continual monitoring for reduced pressure is recommended and can identify the likely area of pressure drop. See Chapter 9 for more discussion and guidance on pressure monitoring in distribution systems.

Drops in Disinfectant Residual

A drop in disinfectant residual may indicate the existence of foreign materials introduced into the distribution system through a backflow occurrence.

Water Meter Anomalies

During periods of reversed water flow, water meters can reverse their counters. Sporadic readings from month to month are also a good indicator of potential backflow situations.

Total Coliform Detections and Heterotrophic Plate Count Changes

A sudden spike in total coliform detections or a sudden change in heterotrophic plate count bacterial densities could indicate that contaminants have entered the distribution system. Persistent coliform contamination may indicate a long-standing cross-connection. Microbial issues and indicators are discussed in Chapter 3.

Characterizing the DSWQ Challenge:

Monitor for cross-connections and backflow incidents. Possible indicators include:

- Customer complaints
- Drops in operating pressure
- Drops in disinfectant residual
- Water meter anomalies
- Total coliform detections and heterotrophic plate count bacteria changes

---

RESPONDING TO BACKFLOW INCIDENTS

Backflow events are transient in nature. They can move slowly or quickly through the water distribution system. They can be readily obvious as sweet-smelling, pink, and foamy water or they can lend no color or odor to the water at all. The aesthetics of various contaminants are described in Chapter 5. At the basic level, a water provider should perform an investigation, sometimes referred to as a customer service inspection, in response to cross-connection and backflow incidents that include these steps:

1. Locate the source of the contamination (see potential indicators of backflow in previous section of this chapter).
2. Isolate that source to protect the water distribution system from further contamination.

3. Determine the extent of the spread of contamination through the distribution system and provide timely, appropriate notification to the public and to regulatory agencies.
4. Take corrective action to clean the contamination from the distribution system and eliminate the cross-connection.
5. Restore service to customers.

A post-incident investigation is typically a team effort. The investigation should be led, at least initially, by the cross-connection control program manager, or someone trained in cross-connection control and backflow prevention. Alternatively, the water quality manager can lead this effort. Additional team members include:

- Water distribution system manager and/or operators,
- Local health inspector,
- Local plumbing inspector,
- Property managers,
- On-site plumber or maintenance worker,
- Communication specialist,
- Utility manager, and
- Lab technicians and available certified laboratories to confirm the parameters of the A report should be generated detailing the backflow incident, and the corrective actions that were taken. This is often referred to as a backflow incident report. This report documents details of the investigation and subsequent corrective action taken for reportable backflow incidents. Some states and provinces require that details of backflow incidents be submitted to the regulatory authority. Many jurisdictions require the use of standardized backflow incident form.

DSWQ Response Actions:

- Check for indicators and typical cross-connection contaminants to locate the source
- Isolate the source to protect the water distribution system from further contamination
- Determine the extent of the contamination and provide notification to the public and to regulatory agencies
- Take corrective action to clean the contamination from the distribution system and eliminate any identified cross-connections
- Restore service to customers

BEST PRACTICES

It is the responsibility of the cross-connection control program manager or water quality manager to be knowledgeable about the water distribution system, for example:

- How it operates (the normal flow patterns of your system);
- Where valves are located;
- Where backflow-prevention devices and assemblies are installed (customers);
- Where industrial and residential areas are;
- Local impacts from seasonal weather, for example, in a cold climate, winterization of irrigation systems could increase the number of dirty-water calls;
- Locations with high hazards;
- Locations that have not complied with the cross-connection control program requirements;
- Locations of cooling towers, wells, ponds, auxiliary water supplies, and similar structures; and
- Projects—for example, new construction, water main replacements, flushing, fires, contractors permitted to use fire hydrants for construction water.

Ideally, this wealth of knowledge resides with one individual. In larger organizations, a team may be gathered to coordinate the detailed information needed.

Backflow Incident Response Plans

Generally, emergency response plans that cover a wide range of emergency conditions are beneficial to water suppliers when protecting water quality. A backflow incident response plan, as part of an overall emergency response plan, provides for a more efficient response by personnel trained in cross-connection control and basic water quality. The plan enables personnel to quickly respond to backflow incidents, prevent further contamination, and then investigate and document both the incident and utility response.

A backflow incident response plan includes defined roles and responsibilities, action plans, and corrective measures. A key component of the plan is the investigation of any water quality complaints that could indicate backflow contamination. According to AWWA Manual M14, a proper backflow incident response plan contains the following:

- Purpose and scope of the plan,
- Identification of roles and responsibilities,
  - Organizations and departments (e.g., water department, health department, communications department, mayor's office),
  - Teams and team members (water distribution manager and distribution operators),
- Definitions and common backflow indicators,
- Action plans and tasks,
- Notification protocols,
- Corrective measures,
- Reporting protocols and forms (e.g., backflow incident report), and
- Close-out procedures.

Cross-Connection Control and Backflow-Prevention Programs

The purpose of a cross-connection control and/or backflow-prevention program is to avoid contamination of the public water supply by preventing, eliminating, and/or controlling cross-connections. The water supplier should carry out a comprehensive and effective cross-connection control program to ensure that public health is protected, regulatory requirements are met, and the quality of the water supply is maintained.

An effective cross-connection control program should have a documented plan that includes the following components:

- Purpose;
- Definitions;
- Establishment of authority and legal requirements;
- Program administration and budgeting;
- Roles and responsibilities;
- Coordination with local authorities;
- Approved backflow-prevention methods and codes;
- Risk assessment;
- Public outreach and education; customer communication;
- Training of internal and external personnel;
- Certification of backflow testers, installers, repair personnel;
- Documentation; recordkeeping;
- Inventory and database of backflow-prevention assemblies;
- Backflow-prevention assembly testing;
- Enforcement;
- Site inspections; surveys;
- Backflow incidence response plan (as discussed in the previous section of this chapter); and
- Quality assurance procedures (see following subsection for details).

Quality assurance programs. A quality assurance program is important for all aspects of operating a public water system. For cross-connection control, the minimum quality assurance program should include a review of the performance of certified testers and monitoring of field test equipment.

Review of the performance of certified backflow-prevention assembly testers includes:

- Reviewing accepted backflow prevention assembly tester certification programs to ensure
  - A validated written examination that adequately measures the expected range of knowledge
  - A proctored hands-on performance component that simulates properly functioning and failed assemblies
  - Periodic recertification utilizing both written and hands-on elements
  - A clear separation between the trainer and the certification authority

- Spot-checking (auditing) the tester's work by observing the backflow-prevention assembly field test and/or by inspecting and retesting assemblies;
- Comparing field test data with manufacturer's data and previous field test reports;
- Checking on the proper completion of field test report forms;
- Verifying with the certification agency that the tester's certification is valid; and
  - Reviewing the field test results submitted for backflow preventers to determine if: the results are unsatisfactory (i.e., component[s] failed performance criteria);
  - replacement or repair is needed;
  - the backflow preventer has been replaced, relocated, repaired, modified, or removed without the water supplier's prior knowledge; and
  - the backflow preventer is improperly installed or in an improper application.

Monitoring of field test equipment to ensure that accuracy is within tolerances may be done by:

- Requiring that field test equipment has its calibration checked by an independent manufacturer certified or industry recognized laboratory and
- Having the water supplier check verification of the field test equipment annually, at a minimum.

Part of the program should also include recordkeeping and data management as well as a comprehensive inventory of backflow-prevention devices and assemblies and their locations.

Table 10-2 summarizes the best practices related to cross-connection control.

Table 10-2 Summary of best practices related to cross-connection control

Best Practice	Level of Service
Understand the normal flow patterns of the distribution system	Basic
Inventory valves	Basic
Inventory locations where backflow-prevention devices and assemblies are installed (customers)	Basic
Map industrial and residential areas	Basic
Track complaints seasonally, for example, winterized irrigation systems can increase the number of calls related to dirty water	Basic
Inventory locations with high and other hazards, e.g., cooling towers, wells, ponds, auxiliary water supplies	Basic/ Advanced
Track locations that are or potentially are out of compliance with the cross-connection control program	Basic/ Advanced
Track projects such as new construction, water main replacements, flushing, fires, contractors permitted to use fire hydrants for water	Basic/ Advanced
Train staff on how to identify the types of contaminants and pollutants; use a certified laboratory to confirm the parameters of the contaminant (remember you have to give them a starting point to test for)	Advanced
Develop a backflow incident response plan	Basic
Establish a cross-connection control program	Advanced

ADDITIONAL RESOURCES

The following resources provide guidelines for establishing effective cross-connection control and/or backflow prevention programs:

- Backflow Prevention and Cross-Connection Control—Recommended Practices M14 American Water Works Association
- Manual of Cross-Connection Control- University of Southern California—Foundation for Cross-Connection Control and Hydraulic Research
- InfraGuide's Methodologies for Setting a Cross-Connection Control Program (https://www.fcm.ca/Documents/reports/Infraguide/Methodology_for_Setting_a_Cross_Connection_Control_Program_EN.pdf)
- Ontario Ministry of the Environment and Climate Change, A Guide for Drinking Water System Owners Seeking to Undertake a Backflow Prevention Program (http://www.downloads.ene.gov.on.ca/envision/env_reg/er/documents/2014/012-1649.pdf)
- Canadian Standards Association, Selection and Installation of Backflow Preventers/Maintenance and Field Testing of Backflow Preventers (http://shop.csa.ca/en/canada/plumbing-products-and-materials/b6410-11b64101-11/invt/27013342011)
- Western Canada Section AWWA Canadian National Cross Connection Control Manual. http://www.wcsawwa.net/files/3814/8648/6617/CCC_Manual_Order_Form.pdf

THE FUTURE: WHAT EMERGING TECHNOLOGIES WILL HAVE AN EFFECT ON BACKFLOW AS A WATER QUALITY RISK FOR UTILITIES?

Water systems are moving into a more digital and automated operating environment. Automatic meters and online sensors may enable earlier warning and detection of backflow events, as discussed in the sections below.

Automatic Meter Reading and Advanced Metering Infrastructure

Across the United States and Canada water suppliers are replacing old direct or touch pad–read meters with new automatic meter reading (AMR) and/or advanced metering infrastructure (AMI) meters. This new technology provides the water supplier with increased meter readings, in some cases as often as every 15 seconds. The AMI systems use a meter data management system database that stores all the information per meter account. This information is reportable and can be analyzed to indicate occurrences such as leaks, meter tampering or reversal of the meter in line, and even backflow.

As discussed above, backflow is caused by increased pressure (booster pumps, auxiliary water pumps, high-rise buildings) on the downstream side of the meter that overpowers the water distribution system pressure, also known as back pressure. Backflow also occurs when the pressure drops in the water distribution system from such events as peak water demands, power outages, main breaks, and fire hydrant use. This is also referred to as backsiphonage.

These occurrences cause the water from the customers' piping to backflow into the water distribution system along with anything that may be directly connected to the internal plumbing. The AMR system can detect backflow through the meter when the meter reading is lower than the previous reading. Frequency of meter readings (every hour or every 15 minutes) also influences the accuracy of the information. A backflow incident can be captured after several gallons have been metered, but an event may be missed if the reading is not frequent enough. Some AMR/AMI systems can be set to show backflow flags, and some AMI systems can even provide immediate alerts of an event.

Online and Real-Time Water Quality Sensors

Data loggers, which are not new to water suppliers, are used to gather data in the water distribution system. New online technologies collect and upload data in real time. However, these sensors can detect a backflow event only if it is severe enough or if the sensor is near the event. A large number of sensors throughout the water distribution system would be needed to detect system anomalies. In addition, these sensors cannot detect precise location or identify contaminants. When a sensor does alert an operator of an anomaly, the operator would drive to the location, collect a sample, and analyze it. Unfortunately, backflow incidents are transient and can be gone within seconds without a trace.

As noted above, new technologies are not without some disadvantages and therefore cannot be used solely to detect backflow event. The need for proper backflow prevention devices and assemblies and a properly managed cross-connection control program are still the first line of defense when dealing with backflow prevention.

REFERENCES

AWWA. 2015. M14, *Backflow Prevention and Cross-Connection Control: Recommended Practices.* Denver, CO: AWWA.

Bailey, S. 2010. What Went Wrong: How a Queens School's Drinking Water Became Contaminated. *Popular Mechanics,* June 3, 2010. http://www.popularmechanics.com/home/how-to/a5823/glycol-in-school-drinking- water/. (access date May 25, 2017).

Barr, A. 2015. Worms Found in Neighborhood's Drinking Water. KHOU11. July 29, 2015. http://www.khou.com/news/local/worms-found-in-neighborhoods-drinking-water/148480891 (accessed Oct. 31, 2016).

Bowling Green Daily News. 2010. Water Contamination at Bristow School. Oct. 13, 2010. http://www.bgdailynews.com/news/water-contamination-at-bristow-school/article_de8aaa1c-a9c3-5db3-8e07-bad405af27e8.html (accessed Oct. 31, 2016).

Cotterell, A. 2012. Boise Residents Sickened by Bad Water Not Satisfied With Company's Response. Boise State Public Radio, Idaho. http://boisestatepublicradio.org/post/boise-residents-sickened-bad-water-not-satisfied-companys-response (accessed Oct. 31, 2016).

CTV Montreal. 2011. Safe to Drink Again: Dorval Water Warning Lifted. Montreal, Canada. June 5, 2011. http://montreal.ctvnews.ca/safe-to-drink-again-dorval-water-warning-lifted-1.652878 (accessed Oct. 31, 2016).

Ellison, A. 2015. City Inspecting Wells in Flour Bluff. KRISTV.COM. Corpus Christi, Texas. July 29, 2015. http://www.kristv.com/story/29663120/city-inspecting-water-wells- (accessed Oct. 31, 2016).

Friedmann, J.. 2010. Plumber Must Quit After Tainting Water. StarTribune Whistleblower Blog. Minnesota. Oct. 11, 2010. http://www.startribune.com/local/blogs/104712149.html (accessed Oct. 31, 2016).

Garden Island. 2011. Unlawful Connection Caused Water Warning. Kilauea, Hawaii. May 27, 2011. http://thegardenisland.com/news/local/unlawful- connection-caused-water-warning/article_1b1d4d3a-8902-11e0-9a48-001cc4c03286.html (accessed Oct. 31, 2016).

Ministry of the Environment and Climate Change. 2006. "Festival U Wash" and "Manager Fined total of $75,000 Stratford". Ontario News Room. June 23, 2006. http://news.ontario.ca/archive/en/2006/06/23/Festival-U-Wash039-and-manager-fined-total-of-75000-Stratford.html (accessed Oct. 31, 2016).

USEPA (US Environmental Protection Agency). 2001. Potential Contamination Due to Cross-Connections and Backflow and the Associated Health Risks. Washington, DC: USEPA, Office of Groundwater and Drinking Water, Standards and Risk Management Division.

Warnock, C. 2012. Cedar Hills Reveals Contaminated Drinking Water Investigation. *Daily Herald*. Provo, Utah. http://www.heraldextra.com/news/local/north/cedar-hills/cedar-hills-reveals-contaminated-drinking-water-investigation/article_c67bdc88-54c3-5a63-af03-96cdea00e179.html (accessed Sept. 13, 2016).

Welker, G. 2011. State Fines Hydroseeding Company Responsible for Somerset Water Ban. *The Herald News*. Fall River, Mass. http://www.heraldnews.com/news/x2084006061/State-fines-hydroseeding-company-responsible-for-Somerset-water-ban (accessed Oct. 31, 2016).

Chapter 11

Security Issues

John Dyksen, SUEZ Water NA, North Haledon, N.J.

Kevin Fisher, Southern Nevada Water District, Las Vegas, Nev.

Jack W. Moyer, AECOM Technical Services, Morrisville, N.C.

Andrew Ohrt, Arcadis, Minneapolis, Minn.

Carol Walczyk, Hatch Mott MacDonald, Iselin, N.J.

Security is a global concern for water suppliers. Since September 11, 2001, the United States has been especially concerned about national security. More recent terrorist attacks in the United States and around the world have resulted in more heightened security measures for a variety of facilities and systems, including the nation's water supply and water quality infrastructure. Of the country's entire inventory of infrastructure, few components are more critical than its municipal water systems. A contamination incident could pose a serious health risk to a large population over a vast geographic area. Even without a health risk, the economic impact of a contamination incident could be highly detrimental to a localized population by supply disruptions and/or a loss of consumer confidence in the water supply (Porco 2010).

This chapter provides water systems with the basic information needed to assess their distribution systems with respect to security as it relates to maintaining water quality in the distribution system. More detailed information on distribution system security can be found in various documents of the US Environmental Protection Agency (USEPA), US Department of Homeland Security (DHS), and other sources (this chapter includes a summary of other resources). While much of this chapter's content is based on larger water systems, most of the recommendations are scalable to smaller systems. Recommendations for smaller systems are included where appropriate.

DSWQ Challenge:

Contamination incidents in water distribution systems can pose serious health risks to large populations over vast geographic areas.

POTENTIAL THREATS AND PATHWAYS

The contamination of a potable water distribution system is considered a critical threat in the water sector. Real-world experience demonstrates that contamination with chemical or biological agents, intentional or accidental, poses a significant risk to health and has potentially severe economic effects (Porco 2010).

Threat Categories

A water system may be exposed to a range of threats, which fall under three potential categories:

- Natural,
- Unintentional, and
- Malevolent.

Natural threats are natural phenomenon such as hurricanes, earthquakes, toxin-producing algal blooms, and other extreme weather events. Unintentional (accidental) threats are caused by human activities but have consequences that are accidental or inadvertent. Examples include unauthorized use of a system's hydrants or valves by contractors, fire departments, or the public. Unintentional intrusion of contaminants can occur because of inadequate treatment, poor backflow prevention, and main repair practices that allow contaminants to enter the distribution system. Cross-connection control and backflow prevention are discussed in Chapter 10.

Malevolent (deliberate) threats are caused by human activities for the express purpose of damaging or destroying assets and terrorizing people or for the purpose of criminal enterprise. Historically, malevolent threats generally were limited to vandalism. However, over the past 15 years, intentional finished water contamination has become a more heightened malevolent threat that requires more stringent security considerations. Malevolent threats can occur because of terrorist activities or disgruntled consumers or employees. Water systems can use tools provided by DHS and other entities for their threat warning systems. These tools are described here.

DHS National Terrorism Advisory System. In April 2011, DHS established the National Terrorism Advisory System (NTAS), which provides notification about the degree and nature of terrorist threats. The two levels that were established are:

- Imminent—warns of a credible, specific, and impending terrorist threat against the United States and
- Elevated—warns of a credible terrorist threat against the United States.

These alerts will include a clear statement that there is an imminent threat or elevated threat. Using available information, the alerts will provide a concise summary of the potential threat, information about actions being taken to ensure public safety, and recommended steps governments can take to help prevent, mitigate, or respond to the threat. These threat levels should be used by the water sector to evaluate security procedures at all facilities and make appropriate adjustments. NTAS alerts are disseminated through the media and other avenues as appropriate. Water systems can sign up to receive NTAS alerts through e-mail, Facebook, or Twitter. More information is available at https://www.dhs.gov/how-do-i/check-national-terrorism-advisory-system-ntas.

Water Information Sharing and Analysis Center (Water ISAC). Water systems can also sign up for free or paid (Pro) membership in the Water Information Sharing and Analysis Center (Water ISAC). Water ISAC analysts collect and review infrastructure protection information from government and private sources to share with members. Pro members are alerted to all threats so they can take quick action to reduce or prevent damage or injuries. Water ISAC provides other resources such as security and emergency preparedness materials and vulnerability assessment tools. More information is available at https://portal.waterisac.org/web.

Intrusion Pathways

Contaminant intrusion into a distribution system can be chemical or microbial in nature. Intrusion occurrence has three main elements—presence of a contaminant, a pathway, and a mechanism. A pathway could be a point of leakage in a pipe or an open hatch in a storage tank. A mechanism could be a pressure differential or lack of disinfectant. Examples of potential pathways and mechanisms to the distribution system include the following:

- Microbiological contamination of closed storage not associated with a deliberate security breach;
- Contamination as a result of microbial regrowth (microbial issues are discussed in Chapter 3);
- Contamination of closed storage from foreign inanimate objects (e.g., dirt, mud, glass, rubbish) from a deliberate security breach;
- Contamination of closed storage as a result of deliberate sabotage (e.g., addition of chemical substance to stored water);
- Backflow event that causes contamination;
- Contamination caused by negative or low pressure transient resulting in intrusion of untreated water through submerged air valve, faulty seal, or point of leakage; and
- Contamination as a result of use of products, materials, or coatings that are not approved for contact with potable water, based on National Sanitation Foundation or Water Quality Association guidance and standards (NSF/ANSI 61 2013).

The most challenging aspect of security for a water system is most likely in the distribution system, since it is comprised of many potential points where a contaminant(s) could be introduced into the system. These points include, but are not limited to, the following locations:

- Hydrants,
- Valve boxes,
- Finished water storage reservoirs,
- Chemical feed points in pumping stations,
- Exposed pipeline appurtenances such as air release valves and sampling stations,

- Air vents, such as in water storage structures,
- Blowoffs,
- Sampling ports, and
- Interconnections.

Additionally, there are potential points of intrusion back into the public water supply from private plumbing systems (premise plumbing), including:

- Outside customer hose bibbs,
- Inside customer faucets or other fixtures,
- Service connections to the distribution system,
- Plumbing in abandoned buildings, and
- Backflow devices (cross-connection control and backflow prevention are discussed in Chapter 10).

CONTAMINANT DETECTION

Evidence exists that chemical and biological weapons have already been used or considered in plots to contaminate drinking water systems. The chronology of intentional water contamination events with chemical or biological hazards dates back thousands of years. While the actors, motives, tactics, and outcomes of chemical or biological attacks vary, the following is a partial list (Tucker 2000; Kroll 2010) of historical incidents that confirm the interest in such attacks (Van Leuven 2011):

- In North Carolina, a water reservoir was intentionally contaminated, resulting in denial of water to customers. Water had to be trucked in for residents (1977).
- New York City received anonymous threats of plutonium poisoning to the city's water supply. Subsequent testing for plutonium revealed 200 times the normal concentration levels but not enough to warrant public health concerns (1985).
- Al-Qaeda members were arrested in Rome while in the process of attacking a water distribution system with cyanide near the US Embassy. They had detailed plans and equipment but were thwarted at the last minute. While the compound turned out to be a benign cyanide derivative, it could have been a pre-event effort to trace the compound and the flow in the water system (2002).
- Two Al-Qaeda members in possession of documents about how to poison US water supplies were arrested in Denver (2002).

According to the USEPA, categories of contaminants that can be introduced into the distribution system are:

- Pathogens that can impact human health, such as *Escherichia coli*, *Cryptosporidium*, poliovirus, hantavirus, smallpox virus, and the microorganisms responsible for anthrax, bubonic plague, and cholera;
- Toxic metals with acute and/or chronic health effects, such as arsenic, cadmium, mercury, lead, and osmium;
- Toxic organic compounds such as petroleum products, pesticides, biotoxins (ricin), chlorinated compounds such as dioxin, and volatile organic compounds such as mustard gas;

- Radioactive materials such as radioactive isotopes used in hospitals, research labs, universities, and nuclear reactor fuels (USEPA 2004).

Roberson and Morley (2005) found that contaminants in a water distribution system can be detected by various indicators. These indicators are graphically depicted in Figure 11-1 and include the following categories:

- Customer complaints;
- Public health surveillance systems;
- Physical evidence;
- Routine water quality monitoring practices; and
- Real-time water quality monitoring.

Customer complaints. While the goal of all water systems is to detect contamination events before they affect customers, customer complaints are a major part of a contaminant detection system. Customer complaints should be promptly investigated, responded to, and tracked to detect potential contamination incidents. As noted in AWWA Standard G200 (AWWA 2015), water systems should have a system to document customer inquiries, including type, investigation results, resolution, and number.

Public health surveillance systems. In recent years, public health agencies have developed sophisticated systems, often called syndromic surveillance systems, for tracking public health outbreaks, including those potentially caused by water contamination, based on indicators that include:

- Doctor and hospital visits,
- Prescription and over-the-counter drug sales, and
- 9-1-1 and poison center hotline calls.

Figure 11-1 Contamination warning indicators

All water systems should develop and maintain good working relationships with their local public health departments and other drinking water regulatory agencies and be prepared to respond when health surveillance indicates a potential waterborne contamination incident.

Physical evidence. Physical evidence includes tangible material or changes left behind by intentional or accidental contamination acts, including:

- Cuts or other damage to fences;
- Cut or broken locks or latches on gates or hatches;
- Evidence of tampering, such as caps left off or hoses left connected, with fire hydrants, air release valves, sampling stations, blow-offs, or other system access points;
- Damage to vents in storage structures;
- Damage or modifications to backflow preventers;
- Liquid, powered, or granular material spilled on the ground or pavement;
- Discarded buckets or other containers;
- Discarded gloves or other protective clothing; and
- Foot prints or tire tracks.

Routine water quality sampling practices. All water systems should be prepared to respond to and address unusual conditions in their routine water quality sampling that may indicate potential contamination. For most potential points of intrusion, which are discussed in the next section, system pressure and residual disinfectant, if present, would have to be overcome to introduce a viable contaminant into the distribution system. Water systems should monitor their pressure and disinfectant residual as required by their primacy agency and, as possible, based on system size and available resources.

Real-time water quality monitoring. A serious threat to a water supply caused by either an accidental release or by an intentional act might not be identified during routine periodic monitoring at drinking water treatment plants and various distribution system locations. Real-time online water quality monitoring provides broad contaminant detection and can improve a system's understanding of day-to-day operations. Water quality monitoring can support regulatory compliance, process control, baseline water quality, and security monitoring (Porco 2010). However, it can be expensive, maintenance intensive, and difficult to site, and care should be taken during data analysis because of the possibility of false-positive readings. Real-time online water quality monitoring is not realistic for the smallest water system.

Online water quality monitoring includes instrumentation, data acquisition, data communication, and data management, as well as event detection to analyze and identify anomalies in water quality data. When evaluating costs for deployment of an online water quality monitoring instrument or instrumentation network, water systems should consider long-term operational costs in addition to the up-front capital costs. Recurring water quality monitoring instrument operational costs can include both water system and manufacturer staff and resources, depending on the operations and maintenance required.

Real-time water quality monitoring can be used as an indicator of drinking water contamination in order to provide broad coverage of many contaminant classes. The USEPA has defined 12 classes of potential water contaminants. Because of the large number of potential contaminants and the broad range in physical and chemical characteristics, many cannot be identified through conventional water quality monitoring (USEPA

2007). Thus, it is usually more practical to use sensors that measure indicator or surrogate parameters in order to detect abnormal water quality for possible contamination evaluation. The most prevalent surrogate parameters used are:

- Turbidity,
- pH,
- Conductivity,
- Total or free chlorine residual (for systems with chloramines, or free chlorine, respectively),
- Temperature, and
- Total organic carbon (TOC).

USEPA has shown that free chlorine residual, TOC, ultraviolet light, and conductivity are the most responsive to a wide range of contaminants in cases where free chlorine is the disinfectant. USEPA and others have demonstrated that the introduction of a contaminant into a water system impacts (increase or decrease) one or more of the following water quality parameters: turbidity, pH, conductivity, chlorine residual, or TOC. For example, the introduction of an acidic substance into the distribution system would be expected to result in a drop in pH and an increase in conductivity. The introduction of a basic substance would be expected to increase pH and conductivity. An organic substance might be expected to increase TOC and reduce total chlorine residual. A microbial substance might reduce total chlorine residual. The following resources can be used to determine the possible impacts on the water quality parameters:

- Water Contaminant Information Tool (WCIT, 2016): USEPA, available on the Internet at: http://www.epa.gov/waterlabnetwork/access-water-contaminant-information-tool;
- *Distribution System Water Quality Monitoring: Sensor Technology Evaluation Methodology and Results:* USEPA 2009;
- *Water Quality Event Detection Systems for Drinking Water Contamination Warning Systems—Development, Testing, and Application of CANARY:* USEPA 2010; and
- *Data Set: Water Quality Parameter Response Data for 5 Different Finished Drinking Water from 4 Large Municipalities:* USEPA 2013a.

In addition, the water pressure should be monitored at each monitoring location. System-wide monitoring should be established to provide a baseline of normal pressure fluctuations and to identify occurrences of low- and negative-pressure events. Pressure monitoring is discussed in Chapter 9.

When selecting monitoring locations, the following criteria should be considered:

- Proximity to critical facilities such as hospitals and schools, as well as sports venues where large groups of people gather;
- Water samples must be representative of the water being used at each location;
- Sufficient space must be available for installing the monitoring device;
- Power must be available to run the device(s);
- A drain line must be available for the wastewater from the device;
- Sufficient security must be provided; and
- A strong signal for cellular or radio communications must be available.

USEPA has developed a graphical interface tool, the Threat Ensemble Vulnerability Assessment–Sensor Placement Optimization Tool, to help systems with sensor placement. It is an interactive tool utilities can use to optimize the number and location of sensors and, in turn, minimize economic and/or public health consequences. Objectives are set by the number of people potentially exposed, the time to detection, and/or the length of pipe contaminated. Installation and maintenance costs can also be factored into the tool analysis (USEPA 2013b).

Event detection systems (EDSs) allow water systems to analyze data and alert users of deviations from baseline water quality conditions. Event detection can be performed on data streams locally and/or at a central location (where data from multiple data streams are analyzed together). EDSs can be as simple as user-defined set points for each parameter or use more sophisticated algorithms. Implementing an EDS requires training of staff, evaluating instruments, and optimizing sensitivity. The USEPA has developed a guidance document on how to design dashboards or human–machine interfaces for event detection systems (USEPA 2015).

Maintaining instrumentation requires varying degrees of skill that depend on the instruments deployed and the operating environment, among other factors. Instrument manufacturers can provide formal training classes for instrument technicians and engineers. Pilot testing of specific instruments should be considered prior to selection and purchase.

The real-time water quality monitoring instrumentation package needs to be procedurally interfaced with consumer complaint surveillance and public health surveillance information. The objective of the customer complaint surveillance component is to support early indication of potential contamination. This is done by enhancing the knowledge gained from the Philadelphia Water Department's demonstration project to develop a comprehensive contamination warning system. Fifteen white papers were developed to disseminate the knowledge gained as quickly as possible for use by the broader drinking water utility community (Philadelphia 2013). The Water Research Foundation also provides additional detail regarding the development of a contamination warning system (Hasit, Johnson, and Cummings 2013).

Characterizing the DSWQ Challenge:

- Characterize threats as natural, unintentional, or malevolent
- Determine intrusion pathway and mechanism of contamination
- Detect contaminant
    - Customer complaints
    - Public health surveillance systems
    - Routine monitoring: pressure, disinfectant residual
    - Real-time monitoring

RESPONDING TO CONTAMINATION THREATS AND EVENTS

The USEPA provides response guidelines for effectively and appropriately planning for and responding to contamination threats and incidents (USEPA 2003). There are three stages of action for the response to contamination incidents:

- Initial response,
- Intermediate response and assessment, and
- Restoration and recovery.

Initial response actions for a water contamination incident include:

- Isolate the contaminated water, if possible. Distribution models and geographic information systems (GIS) can be used to identify the affected area and source of contamination, as well as to develop a containment plan;
- Prevent further contamination, if possible (e.g., eliminate the source);
- Notify law enforcement if the contamination may be intentional;
- Notify and coordinate with the local health department, water primacy agency; and/or USEPA, as necessary depending on the nature of the contamination.

At the same time, consideration should be given to the following possibilities:

- Contamination incidents may involve concentrations of highly hazardous substances, especially at the point of entry/introduction. All water system personnel should use appropriate personal protective equipment and precautions,
- Intentional contamination incidents involve a crime scene. If water system personnel find that scene, they should exercise great caution not to disturb evidence such as cut fences, hoses, buckets, pumps, and gloves.

Intermediate response actions and considerations for a water contamination incident include:

- Identify regular and critical customers in the affected area. Models and GIS are useful tools to assist systems in this task;
- Perform sampling and analysis. Include a list of laboratories that can assist with sampling and/or analysis;
- Evaluate and implement, as appropriate, alternate chemical feeds to neutralize the contamination;
- Determine the need for and implement as determined water use restrictions (e.g., boil water, do not drink, do not use);
- Determine whether it is necessary to provide bottled water and/or bulk water for customers;
- Notify interconnected drinking water systems if there is the potential for the contamination to impact their water systems or to discuss the need for additional water from their system(s);
- Activate the media/crisis communications plan; and
- Flush, neutralize, or disinfect as necessary to decontaminate. Some contamination incidents, depending on the contaminant, require very substantial recovery actions.

Restoration and recovery actions and considerations for a water contamination incident include:

- Flush, neutralize, or disinfect as necessary to decontaminate. Note, precautions should be taken to ensure that flushing does not inadvertently cause contamination of the local waterway from the substance being flushed from the system;
- Undertake substantial recovery actions such as replacement of water distribution system components and/or building plumbing systems and appliances that may be required in some contamination incidents, depending on the contaminant; and
- Conduct or participate in additional investigation to determine the cause of the contamination and prevent its occurrence in the future.

Responding to the DSWQ Challenge:

- Initial response actions
  - Isolate and/or eliminate the source
  - Notify authorities and health agencies
- Intermediate response actions
  - Conduct sampling and analyses
  - Flush, neutralize, and/or disinfect to remove contamination
  - Identify alternate water sources
  - Notify customers
- Restoration and recovery actions
  - Decontaminate, as necessary
  - Repair and replace any damaged distribution system components
  - Determine prevention measures for the future

RISK ASSESSMENT AND PLANNING

The challenge for the water sector and water systems is to implement methods and measures that address security needs. This must be done in a manner that provides the appropriate level of security based on risk, available capital, and operations and maintenance considerations.

Vulnerability Assessments

Risk assessment techniques are used to identify and prioritize potential contaminant entry points that need to be monitored and controlled. The Public Health Security and Bioterrorism Preparedness Act of 2002 (Bioterrorism Act) requires all water systems serving populations of more than 3,300 persons to conduct a vulnerability assessment (VA) of the risk of experiencing "a terrorist attack or other intentional acts intended to substantially disrupt the ability of the system to provide a safe and reliable supply of drinking water." A VA is the process that identifies weaknesses in physical structures, personnel protection systems, policy and procedures, and any other area that may be exploited by terrorists. The process also identifies options and countermeasures for eliminating or mitigating those weaknesses.

The Bioterrorism Act also requires systems to develop or update emergency response plans (ERPs) by addressing the findings of the VAs. All VAs were to be submitted to the USEPA by June 30, 2004. There is currently no federal regulatory requirement to update the VAs, although many water and wastewater systems have continued to do this on their own. ANSI/AWWA Standard G430 (2014) recommends that VAs be updated no less than every five years. Although water systems that serve fewer than 3,300 persons were not required to conduct a VA or prepare an ERP, they should undertake a basic consideration of their vulnerabilities and prepare a simple plan of how they would respond to a terrorist attack.

ANSI/AWWA Standard J100 (2010) provides detailed guidance for conducting a VA. The standard encompasses an all-hazards risk and resilience management process for use specifically by water systems and publically owned wastewater treatment works. Some considerations unique to distribution system contamination threats include:

- Potential points of entry in highly isolated locations,
- Unsecured hatches, and
- Exposed appurtenances such as air relief valves.

In addition, the methodology for assessing risk has evolved into an all-hazards approach that incorporates multiple threats that can impact a water system, as defined in Standard J100. Figure 11-2 shows the seven steps in the J100 process. This standard is now the benchmark for how water and wastewater systems should execute VAs. The goal of a VA is to better enable a water system to do the following:

- Understand threats, consequences, and vulnerabilities that may negatively impact operations;
- Inform efficient resource allocations to maximize the risk mitigations and improve resiliency; and
- Use the risk and resilience management options selected for priority threats to update the ERP in order to enhance its effectiveness during an incident.

To support the application of Standard J100, water systems use in-house staff, various software packages, and/or consultant services. Other VA tools that have been developed for use by water systems are listed below:

- USEPA's Vulnerability Self-Assessment Tool available at http://water.epa.gov/infrastructure/watersecurity/techtools/vsat.cfm;
- DHS Risk Assessment Methodology for Water Systems, created by Sandia National Laboratories. More information is available at https://ip.sandia.gov/techpdfs/RAM%20Market%20Sheet%20SAND%202010%205794P.pdf; and
- SEMS Technologies' software program, which provides a complete asset inventory and condition assessment suite that combines the computer maintenance management system and asset management functionality into one easy-to-use program. More information is available at http://www.semstechnologies.com/.

Figure 11-2 Steps in the ANSI/AWWA Standard J100 risk assessment process

Emergency Response Plans

As stated previously, the Bioterrorism Act requires each US water systems serving more than 3,300 persons to have an ERP. ERPs are also required by US water primacy agencies, with differing requirements for submission and review. AWWA Manual M19 (AWWA 2001) offers guidance and tools that can be used to prepare for natural and human-caused emergencies. Also, as mentioned above, the USEPA Toolbox provides response guidelines for effectively and appropriately planning for and responding to contamination threats and incidents.

ANSI/AWWA Standard G440 (2011) recommends that all water systems prepare an ERP and update it every five years. As noted earlier, even water systems that serve fewer than 3,300 persons should prepare a basic ERP for their response to potential incidents. ERPs are used to:

- Standardize response and recovery protocols to prevent, minimize, and mitigate injury and damage from emergency situations;
- Describe the specific actions taken in the event of any emergency situation; and
- Act as a road map for a water system to use in the recovery process.

ERPs should include the following elements:

- Identification of responsible parties and their roles,
- Internal and external contact information,
- Emergency communication procedures and crisis communications plans,
- Identification of system priorities,
- Inventory of resources for the system,

- Identification of alternate water supplies,
- Plans for water use restrictions. and
- Procedures for responses to security incidents (e.g., vandalism, contamination event, cyber threat).

Security Strategy

Water systems can address many vulnerabilities found in their VA by developing a robust security strategy. Typical security methodologies are based on the concept of deter, detect, delay, and respond protocol. The intention of this protocol is to deter or prevent the attacker from attempting to attack the system, detect the attempt at intrusion/attack, delay the attacker as long as possible, and respond as quickly as possible. Water systems can, where applicable, harden specific facilities, thus deterring and delaying attackers. For example, water systems can install better locks and doors at various sites and install fencing and exterior lighting where applicable. Small water systems can take measures to harden the security of their facilities by improving padlocks and hardware.

Unfortunately, due to the configuration of water systems facilities, it is usually extremely difficult and expensive to prevent the entry of an attacker by hardening the entire distribution system, including the associated appurtenances. Therefore, water systems should optimize security expenditures by investing in security systems that best serve the individual water system and provide dual benefits in security and operations, whenever possible. Detection systems typically include closed-circuit TV cameras, security sensors, water quality monitoring sensors, and similar equipment. These detection systems serve as a line of defense behind the hardening strategies of locks, fences, lighting, and doors at each facility.

The response portion of the security strategy includes an ERP, business continuity plan, emergency response protocol to a water contamination event, and an emergency response protocol that addresses immediate threats to staff and infrastructure.

SUMMARY

Although a water system cannot completely control or prevent contamination threats in the distribution system, it can minimize the threat and prepare to manage the effect of a contamination incident. The first step is to assess and understand the vulnerability of the system. With knowledge of the vulnerabilities, the water system can then develop a security strategy that will mitigate many of the vulnerabilities identified in the VA. The next step is to develop a water quality monitoring program that provides detection of water quality changes throughout the distribution system. A contamination response plan is key to minimizing the extent of customers' exposure to a contaminant. A proper response plan includes methods for isolating the contaminated section of the system, procedures for handling the contaminated water, disposal methods, and coordination with local, state, and federal agencies. Table 11-1 provides a summary of best practices for water systems to follow with regard to security challenges.

Table 11-1 Recommended best practices to address security challenges

Best Practice	Level of Service
Contaminant detection	
• Customer complaints	Basic
• Public health surveillance systems	Basic
• Routine monitoring: pressure, disinfectant residual	Basic
• Real-time monitoring	Advanced
• Develop/update vulnerability assessments	Basic/Advanced
• Develop/update emergency response plans	Basic/Advanced
• Harden facilities: upgrade padlocks and doors, install fences and exterior lighting	Basic
• Install detection systems: closed-circuit TV cameras and other security sensors	Advanced

Resources

Following is a list of useful resources that all utilities should be familiar with when developing or enhancing their security profile:

- AWWA water security resources
  - Manual M19, *Emergency Planning for Water Utilities*
  - ANSI/AWWA Standard G440, *Emergency Preparedness Practices*
  - AWWA and ASME Innovative Technologies Institute's Standard J100, *Standard, Risk and Resilience Management for Water and Wastewater Systems*
- Water Information Sharing and Analysis Center
  - Security and preparedness information specific to the water sector
  - Basic and Pro membership with different levels of information (and vetting)
  - https://portal.waterisac.org/web
- HSIN Homeland Security Information Network
  - Information network based on communities of interest
  - Sign up at http://www.dhs.gov/files/programs/gc_1156888108137.shtm
  - Receive regular e-mail notifications
- USEPA water security resources
  - Guidance documents related to ERP VA and ERP requirements, water security, response protocol toolbox, and similar resources
  - http://water.epa.gov/infrastructure/watersecurity/
  - http://water.epa.gov/infrastructure/watersecurity/emerplan/index.cfm
  - Vulnerability self-assessment tool (http://water.epa.gov/infrastructure/watersecurity/techtools/vsat.cfm)

- Water/Wastewater Agency Response Network
  - Comprehensive mutual aid and assistance agreement among utilities
  - Identifies available resources (e.g., personnel, equipment, materials) and a framework for activation, reimbursement, liability concerns, and similar resources
  - https://www.awwa.org/resources-tools/water-knowledge/emergency-preparedness/water-wastewater-agency-response-network.aspx
- DHS Risk Assessment Methodology for Water Systems, created by Sandia National Laboratories. More information is available at https://ip.sandia.gov/techpdfs/RAM%20Market%20Sheet%20SAND%202010%205794P.pdf
- SEMS Technologies' SEMS software program, which provides a complete asset inventory and condition assessment suite that combines the computer maintenance management system and asset management functionality into one easy-to-use program. More information is available at http://www.semstechnologies.com/
- USEPA. 2007. *Water Security Initiative: Interim Guidance on Planning for Contamination Warning System Deployment.* Revision 1. July.
- *Anna M. Doro-on. 2011. Risk Assessment for Water Infrastructure Safety and Security.* London: Alliance House.
- Copeland, Claudia. December 2010. *Terrorism and Security Issues Facing the Water Infrastructure Sector.* Congressional Research Service.
- States, Stanley Ph.D. 2010. *Security and Emergency Planning for Water and Wastewater Utilities.* Denver, Co. AWWA.
- Gleick, Peter H. 2006. *Water and Terrorism.* Pacific Institute—Water Policy 8. Pages 481-503. July 2006.
- Murphy, Brian M., Radder, Lori L., and Kirmeyer, Gregory J. 2005. *Distribution System Security Primer for Water Utilities.* Water Research Foundation and EPA.
- USEPA. 2006. A Water Security Handbook: *Planning for and Responding to Drinking Water Contamination Threats and Incidents.* EPA 817/B-06/001. April 2006
- Danneels, Jeffrey J., and Finley, Ray E. *Assessing the Vulnerabilities of U.S. Drinking Water Systems.* Journal of Contemporary Water Research and Education. Issue 129, Pages 8–12, October 2004.

REFERENCES

The American National Standards Institute (ANSI) and AWWA. 2011. G440 Emergency Preparedness Practices. Denver, Colo.: AWWA.

AWWA (American Water Works Association). 2010. AWWA Standard J100, *Standard, Risk and Resilience Management for Water and Wastewater Systems.* Denver, CO: AWWA.

———. 2015. AWWA Standard G200, *Distribution System Operation and Management.* Denver, CO: AWWA.

———. 2014. AWWA Standard G430, *Security Practices for Operation and Management*. Denver, CO: AWWA.

———. 2001. M19, *Emergency Planning for Water Utilities*. Denver, CO. AWWA.

Hasit, Y.J.; R. Johnson; & E. Cummings. 2013. *Developing Surveillance and Response Systems for Water Distribution Systems*. Denver, CO: Water Research Foundation.

Kroll, D. 2010. *Securing Our Water Supply: Protecting a Vulnerable Resource*. Tulsa, OK: PennWell Publishers; pp. 1–25.

NSF/ANSI (National Scientific Foundation/American National Standards Institute). 2013. Standard 61, *Drinking Water System Components – Health Effects*. Ann Arbor, MI: NSF.

Philadelphia Water Department and CH2M HILL. 2013. *Philadelphia Water Department Contamination Warning System Demonstration Pilot Project: Customer Complaints Surveillance Guidance*. White paper submitted to USEPA as part of the Water Security Initiative grant awarded to Philadelphia Water Department. https://www.ch2m.com/newsroom/innovation/philadelphia-water-departments-surveillance-response-system (accessed March 1, 2017).

Porco, J.W. 2010. Municipal Water Distribution System Security Study: Recommendations for Science and Technology Investments. *Journal of the American Water Works Association*, 102:4:30–32.

Roberson, J. A. & K.M. Morley. 2005. *Contamination Warning Systems for Water: An Approach for Providing Actionable Information to Decision-Makers*. Denver, CO: AWWA.

Tucker, J.B. 2000. *Lessons from Case Studies. Toxic Terror: Assessing Terrorist Use of Chemical and Biological Weapons*. Cambridge, MA: MIT Press; pp. 250–251.2013a

USEPA (US Environmental Protection Agency). 2003. Response Protocol Toolbox: Planning for and Responding to Drinking Water Contamination Threats and Incidents. EPA 817/D-03/001. Washington, DC: USEPA.

———. 2004. Interim Final Response Protocol Toolbox: Planning for and Responding to Contamination Threats to Drinking Water Systems. EPA 817/D-04/001. Washington, DC: USEPA.

———. (2007). Water Security Initiative: Interim Guidance on Planning of Contamination Warning System Deployment. EPA 817-R-07-002. Washington, DC: USEPA. May 2007.

———. Water Contaminant Information Tool (WCIT). 2016 Washington, DC: USEPA. http://www.epa.gov/waterlabnetwork/access-water-contaminant-information-tool (accessed Mary 23, 2016).

———. 2009. Distribution System Water Quality Monitoring: Sensor Technology Evaluation Methodology and Results. EPA 600/R-09/076. Washington, DC: USEPA.

———. 2010. Water Quality Event Detection Systems for Drinking Water Contamination Warning Systems – Development, Testing, and Applications of CANARY. EPA 600/R-010/036. Washington, DC: USEPA.

———. 2013a. *Data Set: Water Quality Parameter Response Data for 5 Different Finished Drinking Water from 4 Large Municipalities.* EPA 600/R-13/179. Washington, DC: USEPA.

———. 2013b. Threat Ensemble Vulnerability Assessment—Sensor Placement Optimization Tool (TEVA-SPOT) Graphical User Interface. EPA/600/R-13/014. Washington, DC: USEPA https://cfpub.epa.gov/si/si_public_record_report.cfm?subject=Homeland%20Security%20Research&dirEntryId=257684 (accessed March 1, 2017).

———. 2015. Dashboard Design Guidance For Water Quality Surveillance and Response Systems. EPA 817-B-15-007. EPA Office of Water, Water Security Division. Cincinnati, OH 45268. https://www.epa.gov/sites/production/files/2015-12/documents/srs_dashboard_guidance_112015.pdf (accessed March 1, 2017).

Van Leuven, L. J. 2011. *Handbook of Water and Wastewater Systems Protection: Water/Wastewater Infrastructure Security: Threats and Vulnerabilities.* Berlin, Germany: Springer, 2:35.

This page intentionally blank.

Appendix A

Techniques to Characterize Microbial Communities

This appendix provides basic information on characterizing microbial communities. It includes an overview of microbial detection methods, information on detection of coliform and heterotrophic bacteria, selected culture-independent methods, as well as information on investigating other microbial communities.

Currently, microbial detection methods are primarily based on growth-dependent, or culture-dependent, strategies that track growth of the microbial groups that need to be determined. These methods are time consuming, show variable specificity, and typically recover only a fraction of the total population of the microorganisms present in a given water sample.

Culture-independent methods do offer the possibility of a rapid and more complete evaluation of drinking water microbiology (Douterelo et al. 2014; Law et al. 2015; Girones et al. 2010; Mendes Silva and Dominguez 2015). Direct methodologies such as flow cytometry total cell counts (FCM-TCC), viable cell counts (FCM-VCC), and related adenosine triphosphate (ATP) determination can be used for the overall evaluation of the microbial integrity or stability of drinking water distribution systems DWDS. Similarly, nucleic acid–based identification and/or quantification methods such as the polymerase chain reaction (PCR) and quantitative PCR (QPCR), also referred to as real-time PCR (RT-PCR), offer the possibility of a specific and complete determination of the drinking water's microbiology. These methods make use of the more stable genotypic characteristics of the targeted microorganisms (Botes, Kwaadsteniet, Cloete 2013; Dorigo, Volatier, and Humbert 2005). Conserved ribosomal RNA (rRNA) gene regions or other diagnostic genes in the microbial genome can be exploited as targets for genus or species taxonomic identifications, whereas genes that encode virulence factors or toxins can be used to determine the presence of potentially pathogenic or toxic organisms in the sample under investigation (Law et al. 2015; Maheux, Bissonnette, and Bergeron 2013). While PCR targets nucleic acid in dead or living cells, pretreatments can be used, such as propidium monoazide, that bind DNA and inhibit amplification from dead cells; short-lived rRNA or pre-rRNA within viable cells can also be targeted (Bae and Wuertz 2015; Pitkänen et al. 2013; Cangelosi and Meschke 2014).

The following sections describe in further detail growth-dependent methods, growth-independent methods, and methods for characterizing heterotrophic bacterial, coliform, and other indicator groups, as well as for biofilm assessment.

GROWTH-DEPENDENT METHODS

This section describes commonly-used growth-dependent methods that include coliform and heterotroph bacteria.

Coliform Detection Methods

Total coliforms have been defined according to the growth-dependent methodologies used for their detection and identification. Coliforms can be defined as aerobic or facultative anaerobic bacteria that have the biochemical or metabolic ability to ferment lactose to acid (previously defined to also produce gas) within 48 h at 35°C; can develop a red colony with a metallic sheen within 24 h at 35°C on an Endo-type medium containing lactose; or are positive for β-galactosidase activity using a suitable enzyme-substrate growth medium. As summarized below, numerous coliform detection and quantification methods are available. Enzyme-based media are used to detect only microorganisms that possess a specific enzyme(s) and therefore can be targeted, for example, to total coliforms or *Escherichia coli* specifically.

The following methods detect coliforms and require a confirmation phase:

- Presence–absence (P/A) assay,
- Multiple-tube fermentation assay, and
- Standard membrane filtration assay.

The following are defined substrate P/A or quantification tests that are performed in liquid media:

- Colilert: provides P/A and quantification,
- Colitag: provides P/A and quantification,
- E colite: provides P/A,
- Colisure: provides P/A, and
- Readycult: provides P/A.

The following are defined substrate quantification tests that are performed in combination with the membrane filtration technique:

- MI agar
- Chromocult coliform agar
- m-Coli Bule 24
- Coliscan

Heterotrophic Bacteria Detection Methods

There is no single growth, culture medium, or growth-dependent method that will recover or enumerate all culturable heterotrophic bacteria present in a given water sample (Allen, Edberg, and Reasoner 2004). Many heterotrophic bacteria present in water are not culturable either because of stress or injury or because they are in a physiologically viable but not culturable physiological state (Wilks, Gião, and Keevil 2013). Standard plate count (SPC) agar (tryptone glucose yeast agar), m-HPC (heterotrophic plate count) agar medium, and R2A agar are standard growth media for the HPC assay. Because different HPC media and growth conditions select for microorganisms able to grow under differing conditions, the growth medium, temperature, and incubation time have an impact on enumeration (Allen et al. 2004). The following three plating techniques have been described in *Standard Methods* (APHA, AWWA, and WEF 2012) for the HPC assay:

- Pour plate method (method 9215B): Warm media and 1-mL sample aliquot are mixed prior to incubation. This method can cause heat stress and a microaerobic environment for organisms embedded in the HPC agar; hence, it may not be optimal for growth but it does reduce surface-spreading colonies that can interfere with enumerations.
- Spread plate method (method 9215C): Sample aliquot is spread across the surface of solidified agar. Since the sample has to be absorbed into the HPC agar surface, only a small sample volume (0.1–0.5 mL) can be used. This method eliminates the possibility of heat shock stress and generally yields higher bacterial counts than the pour plate method.
- Membrane filter method (method 9215D): A water sample (typically 100–1,000 mL) is passed through a 0.45-μm filter that is then placed on top of the HPC solidified agar medium and incubated for microbial growth. This method eliminates the possibility of heat shock stress and is useful for the analysis of waters with expected very low levels of bacteria.

In a review of research on HPCs in drinking water, Chowdhury (2012) notes that, with respect to enumeration, the spread plate method typically yields the highest results followed by membrane filtration and pour plate methods, respectively. Selection of the appropriate method can depend on cost (pour plate is less expensive) and analytical needs (Chowdhury 2012).

While the most common HPC method in the United States relies on SPC agar, R2A agar (incubation at 20°C; 7 days) provides more sensitivity (Chowdhury 2012; Allen, Edberg, and Reasoner 2004; Reasoner 2004). Figure A-1 shows a comparison of HPC measurements from an undisinfected groundwater system using SPC agar versus R2A agar. The R2A results were generally two orders of magnitude greater than the SPC results. If only the SPC method was used, this utility may not have realized that they had a significant biofilm problem within the distribution system. Use of the more sensitive R2A method revealed significant differences between monitoring locations and provided a better baseline for assessing the effectiveness of subsequent mitigation strategies. Regardless of the method used, an important finding is a significant increase in HPC counts, suggesting possible biofilm, loss of chlorine, or intrusion concerns.

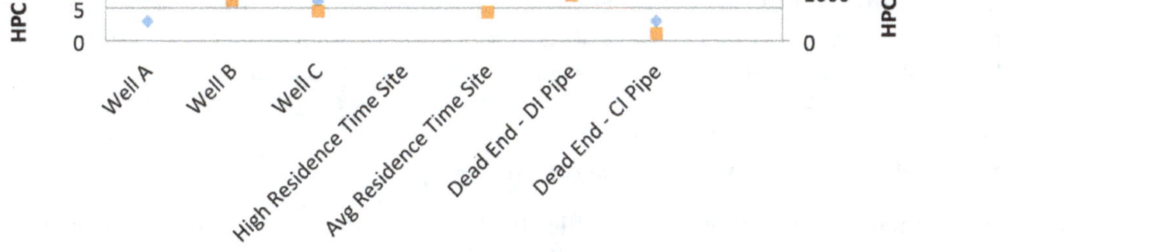

Source: Confluence Engineering Group, LLC Project Files.

Figure A-1 Importance of heterotrophic plate count method (plate count agar versus R2A)

GROWTH-INDEPENDENT METHODS FOR THE EVALUATION OF DRINKING WATER MICROBIAL STABILITY

In past years, growth-independent methods such as FCM-TCC, ATP determination assays, and fluorescence in situ hybridization probes have gained importance as quantitative, accurate, and rapid methodologies for determining biomass changes associated with drinking water treatment or distribution processes.

Flow Cytometry Total Cell Counts Assay

Flow cytometry (FCM) is a technique for analyzing individual cell particles by suspending the particles in a narrow flow stream that passes through an excitation light source, typically a laser beam, similar to an online particle-counting or turbidity-measuring device. When the light beam strikes a particle, part of the incident radiation is scattered. Since both biotic (microorganisms) and abiotic (e.g., crystals, dust) particles scatter light, microbial cells require labeling with some form of fluorochrome, from which fluorescence generated by the laser beam can be measured. Different fluorescent dyes are used that can bind to different cell molecules (nucleic acid, proteins), which can also be used to differentiate between living or active microorganisms from dead or inactive microorganisms. Nonspecific or universal fluorescent probes can be used for a total cell count that provides an easy, quick, complete count of cells, regardless of culturability (Siebel et al. 2008; Hammes et al. 2010; Prest et al. 2016).

ATP Assay

The ATP bioluminescence assay is described in *Standard Methods* (APHA, AWWA, and WEF 2012) as a rapid microbial detection test. Because ATP has a short half-life, dead cells rapidly lose their ATP content. Nonetheless, results obtained from ATP assays cannot be considered direct indicators of water quality (Liu et al. 2013). Instead, ATP should be used as a tool for drinking water treatment process control (e.g., ATP determination has been successfully used in the development, control, and maintenance of filters and biofilters) or for the rapid assessment of drinking water's general microbial stability (Magic-Krezev and van der Kooij 2004; Velten et al. 2007; van der Wielen and van der Kooij 2010; Vital et al. 2012).

Fluorescence In Situ Hybridization Molecular Probes

Fluorescence in situ hybridization (FISH) probes are short lengths of fluorescently labeled DNA used to identify specific microorganisms present in a sample. Typically, diagnostic regions of rRNA are targeted, as there are thousands of rRNA copies in active bacterial cells, making it possible to see the fluorescent signal from individually labeled bacterial cells with a microscope, with only a few hours of sample preparation (Juhna et al. 2007). The adaptation of a peptide nucleic acid probe has improved the sensitivity of this technique (Juhna et al. 2007). The FISH technique can require enrichment steps such as cultural pre-enrichment or a magnetic bead type of enrichment (Girones et al. 2010). However, the autofluorescence of some microorganisms can affect reliability of test results (Girones et al. 2010).

PCR and QPCR: Nucleic Acid-based Methods

As summarized by Douterelo et al. (2014), the PCR method is used to exponentially amplify a specific region of DNA, up to a million-fold after some 30 cycles of the process. Then, the amplicon generated is identified. Hence, PCR-based methods require the extraction of nucleic acids (DNA/RNA), If using reverse transcriptase to turn RNA into complementary

DNA, that requires the amplification of a target DNA region via PCR and post-PCR analysis of the product. The ability of PCR to produce extremely large numbers of copies of a specific nucleic acid segment allows for the rapid, sensitive, and specific detection of desired microorganisms in a water sample. Real-time or QPCR is a sensitive modification where the amplicon is identified at each cycle of amplification, and where cycle threshold is determined against a quantitative control to provide quantification of the original nucleic acid sample. Hence, the result is generally expressed as gene copies per sample.

Table A-1 summarizes characteristics for selected microbial detection methods including:

- Culture-dependent methods: coliforms, heterotrophic bacteria;
- Other selected microorganisms: iron-oxidizing bacteria, sulfate-reducing bacteria, *Pseudomonas aeruginosa,* aerobic and anaerobic growth assessment;
- Selected culture-independent methods; and
- Biofilm assessment.

Table A-1 Summary of microbiological methods

Method	Summary	Application
Culture-dependent methods for the detection and quantification of coliform and heterotrophic bacteria		
Membrane filtration*	• A sample is filtered to concentrate cells followed by incubation of the filter on a specific medium for a specified period of time • Diagnostic colonies that develop are then counted • Different media are used for different groups of target bacteria • Requires confirmation follow-up to ensure results	• Often used to detect coliforms and *Escherichia coli*
Multiple tube fermentation*	• The concentration of bacteria is estimated by inoculating a series of tubes containing liquid medium with 10-fold dilutions of the sample • Growth of bacteria is indicated by development of turbidity • Results can be expressed using an estimation of the average number of bacteria in the sample, known as the most probable number technique • Requires confirmation	• Often used to detect coliforms and *E. coli* • Particularly useful for enumeration when solids are in a sample
Commercial products for detecting coliform bacteria and *E. coli*	• Based on enzymatic reactions • The water sample is used to inoculate a medium containing specific enzyme substrate(s) that, when in contact with a particular group of active microorganisms, produces a quantifiable color change • P/A test, within agar media or in liquid media for MPN	• Examples include Colilert, Quanti-Tray, Colifast, Colitag, E-colite, Readycult, and MI-agar
BioLog	• Commercial method for identifying bacteria and fungi (molds and yeasts) • A plate with more than 100 wells containing various substrates and chemicals is inoculated with the sample • Colors develop in wells depending on microbial reactions carried out by the microbes in the sample • Software is used to identify the microbes present based on the pattern of colors produced by the microbial reactions	• Commercial method for identifying (speciating) bacteria, yeasts, and molds • Used in a wide range of industries including drinking water

(continued)

Table A-1 Summary of microbiological methods *(continued)*

Method	Summary	Application
Heterotrophic plate count†	• Measures culturable bacteria that develop in Petri dishes with agar grown under specific conditions include type of medium, incubation temperature, and time • R2A (7 days, 20°C) is the preferred agar, although standard plate count media is often used • Results reported as colony-forming units per milliliter	• Extensive historical use in water treatment and distribution system studies and regulation to identify unstable water or potential intrusions
Culture-independent methods		
Flow cytometry*	• Technique that uses fluorescent dyes to stain target microorganism in the water sample before analysis with a flow cytometer • Cells in solution pass through a capillary that is intersected by a laser beam • Interaction of the beam with the cells causes light to scatter and also excites the dye, and the florescence signal is quantified using specific filters and detectors	• Different fluorescent dyes/detectors can be used to estimate total bacterial counts, virus-like particles, *Cryptosporidium*, and *Giardia* oocysts in drinking water • Also used in medicine and many other fields
ATP*	• Measures active portion of bacterial community • Cellular ATP reacts with a luciferin–luciferase complex • Luminescence produced in this reaction is proportional to the concentration of ATP that is then correlated to the quantity of mass in the sample • Commercially available methods allow for quantification of several samples simultaneously using a microplate reader	• Numerous studies have successfully used this technique to assess microbial viability and biological stability of water
Fluorescent in situ hybridization*	• Extends epifluorescent microscopy, allowing for the fast detection and enumeration of specific microorganisms • Fluorescent-labeled oligonucleotide probes are used to bind specifically to microbial DNA in the sample, allowing visualization of the cells using an epifluorescent or CLSM	• Has been used to characterize numerous microorganisms in bulk water and biofilm samples in drinking water • CLSM in combination with different dyes can also be used to detect and quantify different biofilm components (a noninvasive microscopy technique that avoids need for biofilm extraction prior to analysis) • 4′6-diamidino-2-phenylindole is a dye often used for total cell counts using CLSM • Software is required to analyze the confocal images • Fluorescence in situ hybridization is not used as a standalone technique and is usually used in combination with other methods to characterize microbial communities

(continued)

Table A-1 Summary of microbiological methods *(continued)*

Method	Summary	Application
Nucleic acid–based methods: PCR and QPCR	• Extraction of DNA or RNA, then amplify target gene or genes using PCR and post-PCR analysis • Allows for rapid, sensitive, and specific detection of microorganisms	• Most sensitive and precise approach to enumerate specific pathogens or fecal markers; can be adapted to detect viable/infectious forms but with lesser accuracy, yet potentially greater sensitivity when rRNA is the molecular target
Next-generation sequencing	• Extracted sample DNA is sequenced (metagenomics) or specific genes, such as the 16 S rRNA gene, within a sample extract is sequenced and the generated "reads" are then reconstructed into partial sequences to allow for operational taxonomic units to be described • Allows for direct assessment of microbial diversity within a sample with the least amount of method bias	• Best way to describe the diversity of a sample, but only semiquantitative, hence can be coupled with qPCR to quantify specific member abundance
Functional genes*	• A molecular approach known as metatranscriptomics based on the study of actively transcribed ribosomal and messenger RNA • RNA is extracted from the sample and cDNA is synthesized by reverse transcription • Resulting cDNA can then be used to measure the expression of functional genes by real-time PCR or functional microarrays • Functional microarrays enable assessment of the overall gene expression of a microbial community	• Facilitates linking specific functions to certain members of a microbial community under investigation
Detection of other common bacterial groups		
Iron-oxidizing bacteria‡	• Several variations of MPN and P/A methods available that use iron sulfide as a substrate to grow these bacteria in a low-oxygen environment • Formation of a brown slime or foam indicates the presence of iron-oxidizing bacteria • Tests include P/A and determination of the MPN based on the amount of precipitate formed • Multiple dilution series make it possible to isolate pure cultures of iron-oxidizing bacteria • More advanced methods, such as molecular methods, have been used in research applications	• Several commercially available MPN and P/A methods are used in many industries • Major limitation in that only a small subset of the target group can typically be cultured
SRB¶	• Specific media are inoculated with the sample in order to culture SRB • Most available media contain lactate as the main carbon and energy source; the presence of SRB is evaluated by the formation of a black precipitate of ferrous sulfide • Tests include P/A and determination of the MPN • More advanced methods such as molecular methods and immunoassays for specific enzymes are used in research applications	• Several commercially available MPN and P/A methods are used in many industries • Major limitation is that only a small subset of the target group can typically be cultured

(continued)

Table A-1 Summary of microbiological methods *(continued)*

Method	Summary	Application
Pseudomonas aeruginosa[||]	• MTF, MF tests are available to detect these bacteria, often of fecal origin but also ubiquitous in the environment and increasingly of concern in healthcare settings with antibiotic-resistant strains • See Standard Method 9213 (APHA, AWWA, and WEF 2012) • Also commercial P/A products are available • PCR-based methods have been developed in research studies	• Some are interested in tracking *P. aeruginosa* as a water quality indicator, although this bacteria does not often cause infection in humans and does not tend to proliferate on normal human tissue • Indicator of contamination of recreational waters and pools/spas
Legionella pneumophila	• USEPA recommends collection of swab samples where water flows (e.g., showerheads) • Samples need to be the first flush to collect biofilm-associated material; generally 1-L samples require filtering for concentration • Cultured in a buffered charcoal yeast extract and differential glycine-vancomycin-polymyxin B media, but often unsuccessful, hence, qPCR for initial screening recommended • Plates incubated at 37°C for a minimum of five days	• Within hospital and healthcare premise plumbing system**
Aerobic and anaerobic growth assessment	• Sample is added to a test tube containing a thioglycollate broth • Aerobic bacteria tend to concentrate at the top of the liquid level where oxygen can diffuse into the broth • Anaerobes tend to concentrate at the bottom of the tube where oxygen levels are lowest and fermentation reactions can occur • Facultative anaerobes also tend to gather at the top of the tube since aerobic respiration is generally preferred	• Within distribution systems, can provide indication as to maturity of a biofilm • More mature biofilms tend to have more anaerobic growth
Methods applicable to biofilms and attached growth		
Cell adhesion*	• Several methods are used to study potential for bacterial adhesion to surfaces such as the inner wall of distribution system pipes, including cell hydrophobicity, electrostatic potential, and thermodynamic potential • Other methods include angle contact measurements and zeta potential to determine the cell surface charge	• Used in studies of biofilm formation and development

(continued)

Table A-1 Summary of microbiological methods *(continued)*

Method	Summary	Application
Coaggregation*	• Once the substratum is colonized by microorganisms, cells grow and produce extracellular polymeric substances (EPS), microcolonies develop, and cohesion and coagulation of different bacterial cells contribute to development of biofilm • Most commonly used method to study coagulation in biofilms is the visual coaggregation assay • Technique involves mixing, generally in pairs, of planktonic batch cultures of specific bacteria and assessing visually the degree of coaggregation in a semiquantitative fashion • Bacterial coaggregates can also be visualized using total cell counts and viable cell counts in combination with fluorescence microscopy	• Used to evaluate biofilms under laboratory conditions or at the bench scale • Not applied for the study of in situ biofilms

Source: Confluence Engineering Group, LLC and Nicholas Ashbolt.

Abbreviations: ATP, adenosine triphosphate; cDNA, complementary DNA; CLSM, confocal laser scanning microscope; MPN, most probable number; P/A, presence/absence; PCR, polymerase chain reaction; QPCR, quantitative polymerase chain reaction; SRB, sulfate-reducing bacteria.

* See Douterelo et al. 2014.

† See Means et al. 1981.

‡ See Emerson and De Vet 2015.

¶ See Vester and Ingvorsen 1998 and Lytle, Gerke, and Maynard. 2005.

|| See LaLancette et al. 2013.

** See Sidari et al. 2004.

REFERENCES

Allen, M. J., S. C. Edberg, and D. J. Reasoner. 2004. Heterotrophic plate count bacteria – what is their significance in drinking water? *International Journal of Food Microbiology,* 92(2004), 265–274.

APHA, AWWA, and WEF (American Public Health Association, American Water Works Association, and Water Environment Federation). 2012. *Standard Methods for the Examination of Water and Wastewater.* American Public Health Association, American Water Works Association, and Water Environment Federation.

Bae, S. and S. Wuertz. 2015. Decay of host-associated *Bacteroidales* cells and DNA in continuous-flow freshwater and seawater microcosms of identical experimental design and temperature as measured by PMA-qPCR and qPCR. *Water Res,* 70: 205–13.

Botes M., de Kwaadsteniet M. and Cloete T.E. 2013. Application of quantitative PCR for the detection of microorganisms in water. *Anal. Bioanal. Chem.,* 405(1): 91–108.

Cangelosi, G. A. and J. S. Meschke. 2014. Dead or alive: molecular assessment of microbial viability. *Appl Environ Microbiol,* 80 (19): 5884–91.

Chowdhury, S. 2012. Heterotrophic bacteria in drinking water distribution: A Review. *Environmental Monitoring and Assessment,* 184(10): 6087–6137.

Dorigo U., L. Volatier, and J. F. Humbert. 2005. Molecular approaches to the assessment of biodiversity in aquatic microbial communities. *Water Research,* 39(11): 2207–18.

Douterelo I., J. B. Boxall, P. Deines, R. Sekar, K. E. Fish, and C. A. Biggs. 2014a. Methodological approaches for studying the microbial ecology of drinking water distribution systems. *Water Research,* 65(2015): 134–56.

Emerson, D. and W. De Vet. 2015. The role of FeOB in engineered water ecosystems: a review. *JAWWA,* 107(1): 47–57.

Girones R., M. A. Ferrús, J. L. Alonso, J. Rodriguez-Manzano, B. Calgua, A. de Abreu Corrêa, A. Hundesa, A. Carratala, and S. Bofill-Mas. 2010. Molecular detection of pathogens in water—the pros and cons of molecular techniques. *Water Research,* 44(15):4325-39.

Hammes, F., C. Berger, O. Koöster, and T. Egli. 2010. Assessing biological stability of drinking water without disinfectant residuals in a full-scale water supply system. *Journal of Water Supply: Research and Technology—AQUA,* 59(1): 31–40.

Juhna, T., D. Birzniece, S. Larsson, D. Zulenkovs, A. Sharipo, N. F. Azevedo, F. Menard-Szczebara, S. Castagnet, C. Feliers, and C. W. Keevil. 2007. Detection of *Escherichia coli* in biofilms from pipe samples and coupons in drinking water distribution networks. *Applied and Environmental Microbiology, 73*(22):7456–7464.

Lalancette, C., E. Bedard, C. Laferriere, E. Deziel, and M. Prevot. 2013. *Do Drinking Water System Biofilms Act as Reservoirs for Hospital-Acquired Pseudomonas aeruginosa Infections?* AWWA Water Quality Technology Conference: Poster.

Law J.W., N. S. Ab Mutalib, K.-G. Chan, and L.-H. Lee. 2015. Rapid methods for the detection of foodborne bacterial pathogens: principles, applications, advantages and limitations. *Front. Microbiol.,* 12(5):770.

Liu, G.; E. J. Van der Mark; J. Q. J. C. Verberk; and J. C. Van Dijk. 2013. Flow cytometry total cell counts: a field study assessing microbiological water quality and growth in unchlorinated drinking water distribution systems. BioMed Research International. Published with open access at: http://dx.doi.org/10.1155/2013/595872. Accessed 9/21/2016.

Lytle, D. A., T. L. Gerke, and J. B. Maynard. 2005. Effect of bacterial sulfate reduction on iron corrosion scales. *Journal AWWA,* 97(10): 109–120.

Magic-Knezev, A. and D. van der Kooij. 2004. Optimisation and significance of ATP analysis for measuring active biomass in granular activated carbon filters used in water treatment. *Water Research,* 38(18): 3971–3979.

Maheux A.F., L. Bissonnette, and M. G. Bergeron. 2013. Rapid detection of the *Escherichia coli* genospecies in water by conventional and real-time PCR. *Methods Mol. Biol.* 943:289–305.

Means, E. G., L. Hanami, H. F. Ridgway, and B. H. Olson. 1981. Evaluating mediums and plating techniques for enumerating bacteria in water distribution systems. *Journal of the American Water Works Association,* 73(11): 585–590.

Mendes Silva, D. and L. Domingues. 2015. On the track for an efficient detection of *Escherichia coli* in water: A review on PCR-based methods. *Ecotoxicol. Environ. Saf.*, 113: 400–411.

Pitkänen, T., H. Ryu, M. Elk, A. M. Hokajärvi, S. Siponen, A. Vepsäläinen, P. Räsänen, and J. W. Santo Domingo. 2013. Detection of fecal bacteria and source tracking identifiers in environmental waters using rRNA-based RT-qPCR and rDNA-based qPCR assays. *Environ Sci Technol*, 47(23): 13611–13620.

Prest, E. I., F. Hammes, M. C. M. van Loosdrecht, and J. S. Vrouwenvelder. 2016. Biological stability of drinking water: controlling factors, methods, and challenges. *Frontiers in Microbiology*, 7(45).

Reasoner, D. J. 2004. Heterotrophic plate count methodology in the United States. *Int. J. Food Microbiol*, 92(3):307-315.

Sidari III, F. P., J. E. Stout, J. M. Vanbriesen, A. M. Bowman, D. Grubb, A. Neuner, M. M. Wagener, and V. Yu. 2004. Keeping *Legionella* out of water systems. *Journal of the American Water Works Association*, 96(1): 111-119.

Siebel, E., Y. Wang, T. Egli, and F. Hammes. 2008. Correlations between total cell concentration, total adenosine tri-phosphate concentration and heterotrophic plate counts during microbial monitoring of drinking water. *Drink. Water Eng. Sci.*, 1:1–6.

van der Wielen, P. and D. van der Kooij. 2010. Effect of water composition, distance and season on the adenosine triphosphate concentration in unchlorinated drinking water in the Netherlands. *Water Research*, 44(17): 4860–4867.

Velten, S., F. Hammes, M. Boller, and T. Egli. 2007. Rapid and direct estimation of active biomass on granular activated carbon through adenosine tri-phosphate (ATP) determination. *Water Research*, 41(9): 1973–1983.

Vester, F. and K. Ingvorsen. 1998. Improved most-probable-number method to detect sulfate-reducing bacteria with natural media and a radiotracer. *Applied and Environmental Microbiology*, 64(5): 1700–1707.

Vital, M., M. Dignum, A. Magic-Knezev, P. Ross, L. Rietveld and F. Hammes. 2012. Flow cytometry and adenosine tri-phosphate analysis: alternative possibilities to evaluate major bacteriological changes in drinking water treatment and distribution systems. *Water Research*, 46(15) 4665–4676.

Wilks, S. A. M. S. Gião, and C. W. Keevil. 2013. Comparison of methods for the detection of culturable and VBNC Escherichia coli O157:H7 in complex drinking water biofilms. In *Water Contamination Emergencies: Managing the Threats*. Edited by U. Borchers, J. Gray, and K. C. Thompson. The Royal Society of Chemistry: Cambridge UK.

This page intentionally blank.

Appendix B

Summary of Flushing Techniques, Likely Water Quality Responses, and Potential Applications

Flushing Technique	Likely Water Quality Response	Potential Application
Conventional spot flushing • Select hydrants are opened at low rate • Flow originates from multiple directions and pipe segments	• Replaces stagnant water through bulk water turnover • Little to no pipe cleaning benefit • Water quality improvements expected to be: localized, limited in magnitude (due to need to use low flow rate), and temporary • Conditions revert to preflush baseline, with exception of any bulk water contamination purged; rate of reversion typically on the order of hours to days • Higher flows can accelerate process but risk disturbing sediment and/or spreading contamination • Improvements may be observed at flowing hydrant, but impacts on surrounding area typically unknown	• Not a main cleaning strategy • Possible improvement in disinfectant residual, heterotrophic plate count, disinfection by-products, and other characteristics attributable to water age reduction, not pipe cleaning • In response to low chlorine residual when there is fresher water nearby • Purge water in response to a customer complaint • Purge water after main repair or other distribution system maintenance activities • To pull water into parts of the system, such as after a treatment change • Temporarily reduce water age • Other instances in which bulk water turnover is needed • Occurs with fire hydrant exercising • Reactive or proactive application NOTE: Conventional flushing should *not* be used at high flow rates that can stir up sediment, biofilm, or accumulated metals or spread contaminants. Velocities should be maintained at 2 fps or less.

(continued)

Flushing Technique	Likely Water Quality Response	Potential Application
Dead-end conventional flushing • Similar to conventional spot flushing, but dead-end location results in a single flow path for the local pipe segment	• Typically similar response as conventional spot flushing, although the unidirectional flow may provide some degree of local pipe cleaning and improve the post-flush response • Higher flows can accelerate process but risk disturbing sediment upstream • To truly purge water from a dead end, a blowoff must be present at the physical location of the dead end	• Similar to conventional spot flushing; see above • Automatic flushing devices, or "auto-flushers," may be installed in dead-end or low-flow areas to provide water turnover; however, this should include consideration for area drainage NOTE: Auto-flushers do not provide main cleaning. • With proper monitoring, presents an opportunity to evaluate water age versus pipe wall demand effects, given known pipe length and diameter (surface area)
Quasi-unidirectional flushing • Hybrid of conventional and unidirectional flushing • Specific mains (not part of a larger, complete sequence) are valved for high-rate unidirectional flow • Lacks a true upstream clean water source	• Specific main segments are cleaned • Pipe wall issues are addressed on a local basis, but the water introduced from upstream may contain sediment and not be truly "clean" • After approximately two pipe volume turnovers, additional improvements are primarily water age reduction • Post-flush water quality response would likely be comparable to conventional flushing, with improved new baseline depending on the length of upstream main cleaned • Use of high flow rates creates potential for upstream issues (disturbing sediment)	• See applications above for conventional flushing • Can remove some chlorine demand–causing sediments from system • In response to coliform positive if boundaries of impacted area(s) are known • Rapid response to nitrification until more robust responses can be implemented • Some level of proactive planning required • Still risk of stirring up sediments in upstream areas

(continued)

Flushing Technique	Likely Water Quality Response	Potential Application
Strict unidirectional flushing • Organized sequential main cleaning from a clean starting point • Requires extensive planning	• All local and upstream pipes are (at least partially) cleaned; also achieves >100 percent bulk water turnover • After flushing, age-related component of water quality degradation will return to prior condition • Overall water quality improvement depends on percent of initial degradation attributable to water age versus pipe effects and on effectiveness of cleaning • When done properly, should remove hydraulically mobile and loosely attached sediments from the system without disturbing surrounding areas • Use of high flow rate with controlled process results in least amount of water used and avoids risk of disturbing sediment without removal from system • Care should be taken to avoid velocities that dislodge tubercles from unlined cast-iron pipes • Typically, systems with harder, well-buffered waters can withstand higher velocities and more aggressive cleaning compared to systems with soft, poorly buffered water	• Proactive program • Too labor intensive for just bulk water turnover • Can actually remove some fraction of accumulated materials from water main but rarely all • Removes hydraulically mobile, loosely attached sediments; some fraction of scale; biofilm • Best response to contamination event (microbial, chemical, cross-connection) since contaminants will not be spread; program must already be developed • Least expensive method of reducing pipe wall chlorine demands (i.e., compared to swabbing, ice pigging) • Preventative water quality and distribution system maintenance practice

Source: Andrew Hill and Melinda Friedman, Confluence Engineering Group, LLC

This page intentionally blank.

Appendix C

Methods for Identifying and Monitoring Water Quality Aesthetics in Distribution Systems

This appendix provides details on sensory and physicochemical analysis methods, species that influence taste, and turbidity analyses.

SENSORY ANALYSIS: TASTE, ODOR, AND COLOR

Sensory methods include the threshold odor number test, flavor threshold test, flavor profile analysis, flavor rating analysis, and visual color comparison. Additionally, evolving methods include total intensity of odor, attribute rating test, discrimination tests, and particle identification using microscopes.

Threshold Odor Number Test

The threshold number (TON) test is based on preparing successive dilutions of a sample with odor-free water until no odor is detected at a controlled temperature. The test result is expressed in terms of the TON, which is the highest dilution that presents a perceptible odor (APHA 2012, method 2150-B). The current version of the TON method was introduced in the ninth edition of *Standard Methods* (APHA 1946). Water suppliers' experience with this classic method shows some important limitations and drawbacks (Dietrich and Burlingame 2015) such as lack of any descriptive information about the nature of the odor and the fact that it does not correlate with consumer perception and complaints. The development of new methods to replace TON should consider a utility-appropriate, syndromic surveillance system for statistically tracking and evaluating consumer feedback related to taste and odor and, when a certain threshold of complaints is exceeded, the utility should investigate. Another consideration is establishing specific guidelines for select odorants such as geosmin and 2-methylisoborneol (MIB).

In the United States there is a secondary standard for odor, and the maximum recommended TON is 3 at 60°C. In Canada, odor reduction is an aesthetic objective, and the only specification is that the odor must be "inoffensive." In the EC (1998), odor is an "indicator parameter" (secondary), and the requirement is that it must be "acceptable to consumers and no abnormal change." Because this condition is very diffuse and subjective, most European countries have restricted the compliance of this parameter including the TON. As in the United States, the maximum permitted value is also 3, but at 25°C.

Flavor Threshold Test

Some regulations have established the flavor threshold test (APHA 2012, method 2160 B) as the reference method for flavor testing. It is analogous to the TON method, but water is tasted instead of smelled. Although the term "taste" is used in some regulations, it is actually "flavor," which is the combination of taste and odor.

Taste (or flavor) is not considered for aesthetic guidance by the US Environmental Protection Agency and Canada Health. In Europe, taste requirements are identical to those described for odors, a dilution number of 3 at 25°C. The significance of the results for taste is even more uncertain than for odor because although some drinking waters can be practically odor free, all drinking water has a specific taste that is driven by the water's mineral content.

Flavor Profile Analysis

Flavor profile analysis (FPA) is a descriptive method (Krasner 1988; Suffet et al. 1995, 1999; APHA 2017). Its purpose is to define the sensory aspects of a drinking water, both qualitative and quantitative, by a trained panel. Taste is monitored at room temperature, or 25°C, and odor at 45°C. This standard method is a quantification of what is a naturally qualitative, or subjective, sensation. By gathering subjective assessments from a significant number of human participants, a quantitative value that is of use to the water system can be achieved; the value represents odors released when warm water is used for bathing or washing.

The core of this method is the extensive taste and odor wheel (see Chapter 5, Figure 5-1) and its more simplified version (see Chapter 5, Figure 5-2). These graphics describe the different attributes and descriptors for taste, odor, and mouthfeel. Mouthfeel describes the physical sensation felt on the tongue, for example, chalky. These wheels were adapted from the food industry and are available in a larger format in *Standard Methods* (APHA 2012). The simplified version shown in Figure 5-2 is useful for customer interactions, whereas the more complete and extensive version shown in Figure 5-1 is useful for a more in-depth investigation of system-specific taste, odor, or mouthfeel issues.

The taste and odor wheel includes the four traditional Western tastes: sweet, salty, sour, and bitter; the fifth basic taste used in Eastern global regions: umami; a variety of odor descriptors: earthy–musty, chlorinous, grassy, marshy, vegetable, fishy, medicinal, and chemical; and also mouthfeel: metallic, astringent, pungent, and similar descriptors. The wheel is continuously expanded with new compounds and terms so that water can be described accurately. For each descriptor, the wheel provides example characteristic compounds that are used for training taste and odor panels, and these correspond to taste and odor episodes reported in the literature.

In the FPA process, the perceived intensity is described using a scale from 0 (undetectable) to 12 (strong), with intermediate ratings of 4 (weak) and 8 (moderate). Individuals are train with standardized taste solutions for sweet (sucrose), salty (sodium chloride), sour (citric acid), and bitter (caffeine), as described in *Standard Method* (Method 2170; APHA 2012). A hexanal (grassy) odor standard has been proposed; currently its use is optional but encouraged (Ömür-Özbek and Dietrich 2008). After individual panelists assess the samples, the panel leader promotes discussion about the samples and a consensus is reached. The minimum number of panelists per session is five.

FPA has been used to resolve several taste and odor episodes in distribution networks worldwide and covering a variety of descriptors (Burlingame, Muldownery, and Maddrey

1992; Malleret and Bruchet 2002; Suffet, Schweitzer, and Khiari. 2004; Martin-Alonso et al. 2007; McGuire et al. 2014), most often along with chemical analysis. For example, complaints of musty odors were linked to trihaloanisoles when consumers complained of musty odors and geosmin or 2-MIB was not present. Chemical analysis and subsequent odor analysis identified trihaloanisole as the causative agent (Malleret and Bruchet 2002; Díaz, Ventura, and Galceran 2005). The main drawback to FPA, which has limited its widespread adoption, is that it requires a qualified, trained panel, which must be established and maintained through a continuous training and evaluation process.

Flavor Rating Analysis

Flavor rating analysis (FRA) (APHA, 2012 method 2160) is an affective analysis that uses a hedonic scale to rate the acceptability of a water flavor sample. The method is less widely used in the water industry than FPA. However, it is gaining popularity for its simplicity and its direct assessment of a water's overall acceptability. Four to seven panelists, who do not need special training or qualifications, taste water samples and rate them according to one of the following descriptions.

- I would be very happy to accept this water as my everyday drinking water.
- I would be happy to accept this water as my everyday drinking water.
- I am sure that I could accept this water as my everyday drinking water.
- I could accept this water as my everyday drinking water.
- Maybe I could accept this water as my everyday drinking water.
- I don't think I could accept this water as my everyday drinking water.
- I could not accept this water as my everyday drinking water.
- I could never drink this water.
- I can't stand this water in my mouth and I could never drink it.

Seattle Public Utilities, Washington, uses FRA. The utility found that if the water's rating falls between 1 and 5, its flavor will be acceptable to their consumers (Dietrich et al. 2004a). For the calendar year 2012, the Seattle Public Utilities successfully met its flavor goal for its drinking water; the water was rated a 3 or better on the FRA scale (Seattle Public Utilities 2012).

Color—Visual Comparison

Pure water is colorless, but natural waters and drinking waters can have color due to the presence of agents such as natural organic matter, minerals, organisms, and anthropogenic compounds. The terms "color" and "true color" are used synonymously to mean the color produced only by constituents in solution, as measured in water from which turbidity has been removed. The term "apparent color" includes both the color from substances in solution and from suspended solids. True color is analyzed unless it is specifically indicated that it should not be. Waters with noticeable suspended matter or turbidity must be filtered (usually through a 0.45-μm filter) or centrifuged before visual evaluation. If this is not done, the result will correspond to apparent color.

Even though several spectrophotometric methods have been proposed for the evaluation of color, a sensory method—the classic visual comparison method with platinum–cobalt (Pt–Co) standards (APHA, 2012 method 2120 B)—continues to be the method of choice. The color of the sample is determined by visual interpolation against color standards of the Pt–Co scale in Nessler tubes, both vertically and horizontally on a white surface. The method is adequate for measuring yellowish-brown colors derived from natural organic matter but it cannot be used in other cases.

One color (or Hazen) unit (CU) is equivalent to 1 mg/L Pt–Co. *Standard Methods* (Method 2120 B; APHA, 2012) describes how to prepare a concentrated stock solution of 500 CU with potassium chloroplatinate and cobalt (II) chloride; it is also available commercially. Working standards are prepared by dilution with reagent-grade water.

There is a nonenforceable standard for color in the United States, an aesthetic objective in Canada, and "indicator parameter" in the EC DWD (1988). Most people detect color above 15 CU in a glass of water (WHO 2011). The maximum recommended color in the United States and Canada is 15 CU. The ECDWD requirement is that the color must be "acceptable to consumers and no abnormal change." A number of European Union member states have introduced a maximum admissible value in the Pt–Co scale of between 15 and 20 units.

Field and online analysis. The conventional method of evaluating color using Nessler tubes is not convenient for field analysis. Several color cards and glass color disc comparators are commercially available. Color can also be monitored using online analyzers and based on photometric sensors (EC DWD 2013).

Evolving Methods for Sensory Analysis

The five methods described above are codified in *Standard Methods*. To meet the needs of the evolving drinking water industry and customer expectations, sensory evaluation methods commonly used in the food and beverage industries are being adopted for use in the water industry (Dietrich et al. 2003, 2004a; Desrochers 2008). The methods described here are newer methods that may some day be formalized and adopted into standard methods.

Total intensity of odor. Total intensity of odor (TIO) is a method that describes the perceived odor of a water sample. It was developed in the United States as a simplification of FPA because it results in a single numerical value for overall odor (Dale et al. 2011). In this process, the overall odor intensity of an undiluted water sample is determined at 25°C and compared to odor reference standards prepared from commercially available soy sauce at concentrations that correspond to values between 0 and 12 on the FPA scale. This method requires minimal training and does not require advanced laboratory techniques. TIO is currently under review to become a Standard Method.

Attribute rating test. The attribute rating test (ART) (Bae et al. 2002; Dietrich et al. 2004b) is a practical odor evaluation method that requires about 1 hour for training and implementation. The method rates the intensity of a known target problematic odorant such as geosmin, 2-MIB, or tribromoanisole in water samples in comparison to the odor in a known standard mixed from that known odorant. The concentration of the standard is selected based on what a water system does not want to exceed based on their site-specific determination of the level that would result in poor water quality and customer complaints. The ART is valuable for assessing drinking water samples and also for evaluating treatment, such as the appropriate dose of carbon for removing an odor, when the compound responsible for the unwanted odor is known and a standard is available.

Discrimination tests. Discrimination testing is easy to implement; water system personnel or customers answer the question, "which sample is different" or "which sample has more of a certain attribute." Discrimination testing is useful for comparing water from the distribution system or customer premise to the finished water leaving the plant.

Two commonly used discrimination tests include the triangle test and the 2-of-5 test (Dietrich et al. 2004a; WRF 2014). For the triangle test, three samples are presented; two are the same and one is different. A panelist is asked to choose the "different" sample.

Statistically, large numbers of samples, for example, 50 to 100, are needed for statistical significance when using the triangle test (Meilgaard, Carr, and Civille 2006). The 2-of-5 test (Dietrich et al. 2004a) compares two samples of one type of water to three samples of another water to determine sameness or difference. The advantage of the 2-of-5 test is that statistically fewer panelists and trials are required compared to the triangle test (Gallagher 2004).

The sensory tests described above can be used for multiple purposes to serve the needs of utilities and their customers. Table C-1 summarizes applications of the sensory test methods.

Particle identification using microscopes. Several types of particles can enter a distribution system and accumulate and/or be released from distribution system pipe walls, as summarized in Table C-2 and discussed in Chapter 7. Depending on their origin, these particles appear to impart color to the water, including red, black, yellow, white, or gray.

In some cases, these particles are identified using visual inspection with the naked eye. When this is not possible, use of a microscope is a viable option. Microscopes range in optical power depending on the needed application. Options include magnifying glass, stereomicroscope, bright-field compound microscope, and scanning electron microscope (SEM). Magnifying glasses, stereomicroscopes, and compound microscopes are used to identify microorganisms and particles approximately the size of a grain of sand, anthracite, or rust. An SEM is used to analyze smaller particles such as nanoparticles in suspension and may be coupled with an energy-dispersive X-ray spectrascope to identify their chemical composition. Booth and Brazos (2005) describe these microscopes and other qualitative identification methods in more detail and recommend procedures to use them for particle identification. In addition, digital microscopes that can take photos of the particles are available. These may either be stand-alone units that resemble digital cameras or attachments for compound microscopes.

Table C-1 Matching sensory testing methods to water quality and treatment objectives

Objective	Potential Test to Use
Bench-scale studies on the approval of materials for leaching and drinking water contact	FPA
Approval for the release of water in new mains or new and rehabilitated facilities	FPA, FRA, 2-of-5, ART, TIO
Screening source water for problems	FPA, TON, ART
Screening drinking water for problems	FPA, FRA, TON, ART
Customer complaint sample evaluation	FPA, FRA, ART, TIO
Odor or taste threshold determination	FPA, FTT, triangle test
Correlation to analytical test results	FPA, FRA, 2-of-5
Developing a baseline of the sensory characteristics of a water	FPA
Taste-and-odor event response	FPA, FRA, 2-of-5, ART, TIO
Evaluation of treatment effectiveness at bench scale or full scale	FPA, FRA, 2-of-5, ART, TIO
Routine drinking water quality control	FPA, FRA, 2-of-5, triangle test, TIO
Evaluating analysts' sensory abilities	Triangle test

Reprinted from Dietrich, Burlingame, and Hoehn (2003), with updates.

Abbreviations: 2-of-5, two-of-five test; ART, attribute rating test; FPA, flavor profile analysis; FRA, flavor rating assessment; FTT, flavor threshold test; TIO, total intensity of odor; TON, threshold odor number.

Table C-2 Particulate matter that may be found in distribution systems that will likely result in customer complaints

Particle Color	Particulate Matter
Black	Activated carbon, anthracite filter media, disintegrated rubber, bitumastic coating, manganese, lead-rich scale, rubber gasket chunk, grease, black copper carbonate
Brown	Sand, iron rust particles, zeolite resin, garnet sand, copper shavings
Biological	Particles with attached biofilm, bacterial clumps, nematodes, chironomids, crustacean, rotifers, *Asellus frasse*, algae, fungi, woody plant fibers, iron bacteria, pink slime
White or gray	Plastic chunks, minerals, air in water, oxidized zinc, lead carbonate, pipe joint materials, aluminum, cement pieces, natural (white-colored) garnet sand
Others	Copper corrosion (patina), pipe joint materials, artificial fibers, glass chips, non-white plastic

From Booth and Brazos (2005).

PHYSICOCHEMICAL ANALYSIS

Physicochemical analyses test for specific parameters or types of parameters, such as aldehydes, N-chloroaldimines, iodinated trihalomethanes, and halogenated phenols.

Aldehydes

There are several methods for evaluating aldehydes in water. A water system that needs to analyze aldehydes should ensure that the analysis is performed at a laboratory that routinely works with drinking water samples and ideally meets any local laboratory quality assurance regulations or guidelines.

USEPA methods 556 and 556.1 and method 6252B (APHA 2012) are based on a derivatization procedure that uses O-(2,3,4,5,6-pentafluorobenzyl) hydroxylamine hydrochloride (PFBHA). Small polar aldehydes and ketones and also dicarbonyl compounds are derivatized with PFBHA to form the corresponding oximes. Two geometrical isomers (*E* and *Z*) are formed for simple aldehydes, except for formaldehyde or symmetrical carbonyls. The oxime derivatives are then extracted with n-hexane and analyzed using gas chromatography–electron capture detection (GC/ECD) or GC-mass spectrometry (GC/MS; electron ionization [EI] mode). Method detection limits in the range of 0.09 to 0.62 μg/L are achieved using USEPA method 556.1. Others have used PFBHA derivatization, but the oximes are then extracted by liquid–liquid extraction, solid-phase extraction (SPE), purge and trap (PT), or solid-phase microextraction (SPME) procedures (Cancho and Ventura 2005).

N-chloroaldimines

A methodology based on headspace (HS) coupled to GC/MS allowed the determination of N-chloroaldimines at the microgram per liter level (Freuze et al. 2005). A further refinement using HS combined with a sorbent trap system and coupled to GC/MS allowed the determination of N-chloroaldimines at nanomolar levels (Brosillon et al. 2009).

Iodinated Trihalomethanes

Iodinated trihalomethanes (I-THMs) can be measured using methods that are similar to those for THMs (i.e., USEPA method 551.1). Cancho et al. (2000) showed that liquid–liquid extraction (LLE)-GC/ECD, with detection limits in the range of 0.01 to 0.03 μg/L, is more effective than HS-GC/ECD and PT-GC/ECD in determining I-THMs. The simultaneous analysis of 10 THMs can also be carried out using SPME-GC/MS (Allard et al. 2012). The carboxen/polydimethylsiloxane/divinylbenzene fiber was found to be the most suitable and allowed limits of detection ranging from 1 to 20 ng/L for iodoform and chloroform, respectively.

Halogenated Phenols

As noted in the literature, SPE is the preferred method for the enrichment of halogenated phenols together with stir bar sorptive extraction (SBSE), SPME, and LLE. They are analyzed mainly by derivatization of target compounds. Trimethylsilylation (i.e., N,O-bis(trimethylsilyl)trifluoroacetamide) and acetylation (acetic acid anhydride) are the most common ones (Faludi et al. 2015), although pentafluorobenzylbromide is also used (USEPA methods 528, 8041). Extracts are then analyzed using GC/MS, GC-MS/MS, or GC/ECD.

Faludi et al. (2015) recently reviewed the GC/MS/(MS) proposals for their analysis according to their chromatographic characteristics, derivatization conditions, acquisition protocols, and limit of detection/limit of quantification (LOD/LOQ) values. As examples, SBSE with in situ derivatization with acetic acid anhydride and thermal desorption followed by GC/MS has been used to determine trace amounts (1 to 2 ng/L levels) of chlorophenols in tap water (Kagawuchi et al. 2005). Similarly, a HS-SPME-GC/MS method with in situ acetylation was performed with LOD values below 0.1 μg/L using polydimethylsiloxane (appropriate for phenol, methylphenols, and low-chlorinated chlorophenols) and carboxen/polydimethylsiloxane fibers (for highly chlorinated phenols) (Llompart et al. 2002). SPE enrichment was used to analyze a wide range of substituted phenols at low nanogram per liter levels (0.29 to 14 ng/L) determined as their trimethylsilyl-ether derivatives by GC/MS–MS applying multiple-ion monitoring acquisition (Faludi et al. 2015).

SPECIES THAT INFLUENCE TASTE

Taste depends on minerals dissolved in water. This section focuses on analysis of the parameters and species that impact taste. Total dissolved solids (TDS) are an aggregate parameter that provides information about the overall level of mineralization. The ions that are most influential on taste are sodium, calcium, and magnesium as cations and bicarbonate, chloride, and sulfate as anions. Other species such as aluminum, copper, and iron can also play a significant role by conferring a metallic taste to waters. Recent research shows that this metallic sensation has a significant retronasal, that is, after ingestion, component from odorous compounds produced by biochemical reactions in the mouth (Ömür-Ozbek and Dietrich 2011; Ömür-Ozbek. 2012).

Inorganic Species That Influence Taste and Their Quantification Methods

TDS are determined by gravimetry, which is a measure of the salt residue obtained after the evaporation of a known volume of water (usually at 180°C), expressed in milligrams per liter. In general, this value is lower than the sum of the analytical results of cations and anions due to the partial decomposition of bicarbonates during analysis. TDS analysis is time consuming; for this reason, it is generally estimated using electrical conductivity. This parameter, usually expressed in microsiemens per centimeter, is corrected to the value at a reference temperature, typically 20 or 25°C. It is important to monitor the temperature because of the significant variation in conductivity, which is an increase of about 2 percent per degree centigrade.

Inductively coupled plasma (ICP) is a useful technique for the analysis of individual metals (and other elements) mainly due to its multielemental character. A sample is nebulized and atomized in an argon plasma at about 8,000 to 9,000 K. The measurement can be based on the emission of light at specific lines for each element (optical or atomic emission spectrometry, ICP-OES) or from the mass-to-charge ratio of the ions (mass spectrometry [ICP-MS]). The sensitivity of ICP-MS is much better than for the optical technique, and the mass interferences that seriously limited its performance have been almost totally solved with collision and reaction cells. However, ICP-OES easily reaches the levels that are needed to characterize the taste of waters or to solve any taste and odor problem. Other common techniques for these species include flame and graphite furnace atomic absorption spectrometry, flame emission photometry, complexometric titration, visible spectrophotometry, and cation chromatography (AWWA 2012; Hill 2006; Rouessac and Rouessac 2007; Fritz and Gjerde 2009).

Today ion chromatography is the most common analytical technique for anions due to recent developments such as highly selective stationary phases, conductivity suppression, and electrolytic eluent generation. Chloride, sulfate, and other anionic species are easily analyzed using this instrumental technique. Other classic methods (e.g., titrimetry, nephelometry) are still in use; however, they are time consuming and typically not considered for routine monitoring. Bicarbonate, sometimes expressed in terms of alkalinity as calcium carbonate, is commonly determined using potentiometric acid–base titrimetry (Rouessac and Rouessac 2007; Fritz and Gjerde 2009).

Organic Compounds That Cause Taste and Odor Problems

GC/MS is the most suitable technique for analyzing taste and odor compounds at trace levels (low nanograms per liter). Several extraction techniques are currently used to preconcentrate the target compounds before analysis. Some of the most common are closed-loop stripping analysis (CLSA), PT, LLE, SPME, SBSE, and simultaneous distillation extraction. Selected references are given in Table C-3.

Table C-3 below provides a comparison of the performance of different methods described in the literature for determining two relevant musty–earthy compounds, geosmin and MIB.

CLSA and LLE have potential disadvantages of being time consuming and requiring the use of solvent, respectively. PT and SPME are two solvent-free methods that integrate sampling, extraction, concentration, and sample introduction and require much smaller sample size. CLSA and SPME and further analysis by GC/MS are included among the Standard Methods (6040B and 6040D) with limits of detection of 2 ng/L and <5 ng/L for both compounds and methods, respectively. SBSE has the disadvantage of long enrichment times. However, several samples can be prepared simultaneously without additional labor, and additional stir bars can be stored after sampling.

Complementary information can be obtained by using gas chromatography/olfactometry (GC/O) that combines the odor evaluation of the individual GC peaks eluting from a chromatographic column at an olfactory detector outlet and the chemical identification of the odorous compounds by GC/MS. Applications of the GC-O technique in solving taste and odor episodes in water have been reported by Gallagher et al. (2015).

All analytical methods cited are suitable for identifying the compound(s) responsible for the taste and odor incidents. Method selection will depend on the availability of analytical instrumentation in the laboratory and familiarization with the specific techniques.

Disinfectant Flavor

Post-disinfection agents are a well-known source of characteristic odors and flavors in distributed water. There is a wide range of techniques (titrimetric, photometric, electrometric) for analyzing chlorine and chloramines, but DPD (N,N-Diethyl-p-phenylenediamine) colorimetry is by far the most common method for chlorine (Standard Method 4500-Cl G). The addition of potassium iodide allows the determination of free and total chlorine; combined chlorine (assimilated/equivalent to chloramines) is obtained by subtracting free

Table C-3 Comparison of different methods for evaluating geosmin and 2-methyl-isoborneol

				Limit of Detection, ng/L		Recovery, %	
Pre-extraction Method	Solvent Extraction	Extraction Time, min	Working Range, ng/L	GSM	MIB	GSM	MIB
Closed-loop stripping analysis*	Carbon disulfide	180	10–400	14.9	15.4	80–90	80–100
Liquid–liquid extraction†	Pentane	30	1–500	0.1	0.1	98–108	75–77
Purge and trap‡	Solvent free	11	0–100	4	5	85	104
Purge and trap¶	Solvent free	13	10–500	3.6	4.1	82–95	99–101
Headspace–solid-phase microextraction¶	Solvent free	40	10–500	3.0	3.1	92–96	93–103
Stir bar sorptive extraction||	Solvent free	120	0.1–10	0.5	1	83–101	89–110

Abbreviations: GM, geosmin; MIB, 2-methyl-isoborneol.
* Romero, Manero, and Laso (2007).
† Ma et al. (2007).
‡ Manickum and Wilson (2012).
¶ Yuan et al. (2013).
|| Benanou et al. (2003).

chlorine from total chlorine The simplicity of the method allows this parameter to be analyzed in the distribution system.

The DPD method with addition of glycine as a masking agent has been extensively used for the analysis of chlorine dioxide, but today it is not considered accurate enough and has been removed from *Standard Methods* (APHA2012). The colorimetric lissamine green method (Pepich et al. 2007), amperometry, and other alternatives are available (Gates, Ziglio, and Ozekin 2009).

Field and online analysis. Disinfectant residuals are usually analyzed in situ. DPD colorimetry is a common field and on-line analysis method used in the drinking water sector. In addition, disposable amperometric sensors are an emerging technology for measurement of disinfectant residuals. They are a good alternative to the classic DPD photometric analyzers because reagents are not needed and maintenance is minimal (USEPA 2009; McPherson 2012).

TURBIDITY

Turbidity is caused by suspended and colloidal matter that obstructs light transmission, making the water appear cloudy. Turbidity-causing particles in different types of waters originate from natural or anthropogenic sources and show a great variety in their nature: clays and silts that come from the weathering of rocks and soils; inorganic precipitates, such as hydrous oxides of iron and manganese, calcium carbonate, basic salts of aluminum and zinc; natural organic matter from soils (humic), plants, and animal debris; microorganisms (e.g., algae, cyanobacteria, protozoa, biofilm detachment); and others (WHO 2011). Turbidity is a recognized indicator of possible microbial contamination because microorganisms tend to attach to particles and colloids.

Turbidity has historically been measured using different visual and optical techniques, both field or laboratory methods. Today the most common method is nephelometry, which is based on measurement of light that is scattered by suspended particles (APHA 2012, method 2130). Scattered light depends on the number of particles and also on their characteristics and size distribution. It also depends on instrument parameters such as type and wavelength of the incident light and on the angle of observation. For this reason nephelometers cannot give reliable, absolute values, and they need to be properly calibrated. True color can also influence the measurement, although this interference is prevented if an instrument with more than one detector is used.

Formazin suspensions are the most common primary standard used for calibration. *Standard Methods* describes how to prepare a concentrated stock solution of 4,000 ntu with hydrazine sulfate and hexamethylenetetramine (method 2130-B). Working standards are prepared by dilution of reagent-grade water filtered through 0.1 to 0.45 μm. Because the preparation of these standards is time consuming and requires a degree of skill, commercial calibration standards are often used.

Field and Online Analysis

Turbidity must be analyzed as soon as possible because particles tend to flocculate or settle quickly when temperature and other sample conditions change. For this reason, field measurements for this parameter have been common. Today, a wide range of portable turbidimeters are available in different grades and specifications. Secondary standards based on liquid suspensions of latex or polymer beads are no longer recommended, but secondary stabilized formazin standards (sealed vials) are a good alternative (Standard Method 2130 B).

Turbidity was one of the first parameters to be continuously monitored. Today, online turbidity analyzers are very common in water systems that operate surface water treatment plants. Most instruments work by nephelometry with ratio measurement. Accurate results can be obtained over a wide measurement range. Special attention must be given to avoid the formation of bubbles, which leads to results that cannot be reproduced.

REFERENCES

Allard, S.; J.W.A. Charrois; C.A. Joll; & A. Heitz. 2012. Simultaneous Analysis of 10 Trihalomethanes at Nanogram per Liter Levels in Water Using Solid-Phase Microextraction and Gas Chromatography Mass-Spectrometry. *Journal of Chromatography A*, 1238:15–21.

APHA & AWWA. 1946 (9th ed.). *Standard Methods for the Examination of Water and Sewage.* Washington, DC: American Public Health Association.

Rice, E.W., Baird, R.B., Eaton, A.D., & Clesceri, L.S. (editors). 2012. *Standard Methods for the Examination of Water and Wastewater.* 22nd ed. ISBN: 9780875530130. American Public Health Association, American Water Works Association, & Water Environment Federation: Washington, D.C.

APHA, AWWA & WEF (American Public Health Association, American Water Works Association, & Water Environment Federation). 2011. *Standard Methods for the Examination of Water and Wastewater,* 6040 Constituent Concentration by Gas Extraction. 2011. Washington, DC: APHA, AWWA, WPCF.

– – –. 2012 (22nd ed.). *Standard Methods for the Examination of Water and Wastewater.* Washington, DC: American Public Health Association.

AWWA (American Water Works Association). 2012. Global Taste and Odor Survey of Water Utilities. Final report. Prepared by P. Ömur-Özbek. Denver, CO: AWWA.

Bae, B.-U.; Y.-I. Kim; D.W. Dugas; G.A. Burlingame; & A.M. Dietrich. 2002. Demonstration of New Sensory Methods for Drinking Water Taste-and-Odor Control. *Water Science and Technology: Water Supply*, 2:5–6:241–247.

Benanou, D.; F. Acobas; M.R. de Roubin; F. David; & P. Sandra. 2003. *Analysis of Off-Flavors in the Aquatic Environment by Stir Bar Sorptive Extraction-Thermal Desorption-Capillary GC/MS/Olfactometry. Analytical and Bioanalytical Chemistry*, 376:1:69–77.

Booth, S. & B. Brazos. 2005. *Qualitative Procedures for Identifying Particles in Drinking Water.* Denver, CO: Water Research Foundation.

Brosillon, S.; M. Lemasle; E. Renault; D. Tozza; V. Heim; & A. Laplanche. 2009. Analysis and Occurrence of Odorous Disinfection By-products From Chlorination of Amino Acids in Three Different Drinking Water Treatment Plants and Corresponding Distribution Networks. *Chemosphere*, 77:1035–1042.

Burlingame, G.A.; J.J. Muldowney; & R.E. Maddrey. 1992. Cucumber Flavor in Philadelphia's Water. *Journal of the American Water Works Association*, 84:8:92–97.

Cancho, B.; F. Ventura; M.T. Galceran; A. Diaz; & S. Ricart. 2000. Determination, Synthesis and Survey of Iodinated Trihalomethanes in Water Treatment Processes. *Water Research*, 34:13:3380–3390.

Cancho, B. & F. Ventura. 2005. Optimization of Methods for the Determination of DBPs. *Global NEST Journal*, 7:1:72–94.

Dale, M.S; R.D. Desrochers; S.M. Labernik; & M.S. Moylan. 2011. Threshold Odor Number Versus Flavor Profile Analysis: A Case Study in the Endeavor to Measure Consumer Acceptance. 9th IWA Symposium on Off-Flavours in the Aquatic Environment, Aberdeen, Scotland.

Desrochers, R. 2008. Sensory Analysis in the Water Industry. *Journal of the American Water Works Association*, 100:10:50–54.

Díaz, A.; C. Fabrellas; F. Ventura; & M.T. Galceran MT. 2005. Determination of the Odor Threshold Concentrations of Chlorobrominated Anisoles in Water. *Journal of Agricultural and Food Chemistry*, 53:2:383–387.

Dietrich, A.M. & G.A. Burlingame. 2015. Critical Review and Rethinking of USEPA Secondary Standards for Maintaining Consumer Acceptability of Organoleptic Quality of Drinking Water. *Environmental Sciences and Technology*, 49:2:708–720.

Dietrich, A.M.; G.A. Burlingame; & R.C. Hoehn. 2003. Strategies for Taste-and-Odor Testing Methods. *Opflow*, 29:10:10–14.

Dietrich, A.M.; D. Glindemann; F. Pizarro; V. Gidi; M. Olivares; M. Araya; A. Camper; S. Duncan; S. Dwyer; A.J. Whelton; T. Younos; S. Subramanian; G.A. Burlingame; D. Khiari; & M. Edwards. 2004. Health and Aesthetic Impacts of Copper Corrosion on Drinking Water. *Water Science and Technology*, 49:2:55–62.

Dietrich, A.M.; R.C. Hoehn; G.A. Burlingame; & T. Gittelman. 2004a. *Practical Taste-and-Odor Methods for Routine Operations: DECISION TREE*. Denver, CO: AWWA Research Foundation.

Dietrich, A.M.; A. Whelton,; R.C. Hoehn; R. Anderson; & M. Wille. 2004b. Attribute Rating Test for Sensory Analysis. *Water Science and Technology*, 49:9:61–67.

EC DWD (European Commission Drinking Water Directive). 1988. Council Directive 98/83/EC of 3 November, on the Quality of Water Intended for Human Consumption.

European Commission. 2013. Review of Sensors to Monitor Water Quality. ERNCIP Thematic Area, Chemical & Biological Risks in the Water Sector. European Commission. European Commission, J. Raich.

Faludi, T.; A. Vasanits-Zsigrai; G. Záray; & I. Molnár-Perl. 2015. Identification, Quantification and Distribution of Substituted Phenols in the Dissolved and Suspended Phases of Water Samples by Gas Chromatography Tandem Mass Spectrometry: Derivatization, Mass Fragmentation and Acquisition Studies. *Microchemical Journal*, 118:45–54.

Freuze, I.; S. Brosillon; A. Laplanche; D. Tozza; & J. Cavard. 2005. Effect of Chlorination on the Formation of Odorous Disinfection By-products. *Water Research*, 39:2636–2642.

Fritz, J & D.T. Gjerde. 2009 (4th ed.). *Ion Chromatography*. Wiley-VCH Verlag.

Gallagher, D.L. 2004. Statistical Comparison of the Triangle Test and the Two-of-Five Test for Taste and Odor Evaluation. *Water Science and Technology*, 49:9:107–114.

Gallagher, D.L.; K. Phetxumphou; E. Smiley; & A.M. Dietrich. 2015. Tale of Two Isomers: Complexities of Human Odor Perception for cis- and trans-4-Methylcyclohexane Methanol From the Chemical Spill in West Virginia. *Environmental Science and Technology*; DOI: 10.1021/es5049418.

Gates, D.; G. Ziglio; & K. Ozekin. 2009. *State of the Science of Chlorine Dioxide in Drinking Water*. Denver, CO: Water Research Foundation and Fondazione AMGA.

Hill, S.J (editor). 2006 (2nd ed.). *Inductively Coupled Plasma Spectrometry and Its Applications*. Chichester, United Kingdom: John Wiley and Sons.

Kawaguchi, M.; Y. Ishii; N. Sakui; N. Okanouchi; R. Ito; K. Saito; & H. Nakazawa. 2005. Stir Bar Sorptive Extraction With In Situ Derivatization and Thermal Desorption–Gas Chromatography–Mass Spectrometry for Determination of Chlorophenols in Water and Body Fluid Samples. *Analytica Chimica Acta*, 533:57–65.

Krasner, S.W. 1988. Flavor-Profile Analysis—An Objective Sensory Technique for the Identification and Treatment of Off-Flavors in Drinking Water. *Water Science and Technology*, 20:8–9:31–36.

Llompart, M.; M. Lourido; P. Landın; C. Garcıa-Jares; & R. Cela. 2002. Optimization of a Derivatization–Solid-Phase Microextraction Method for the Analysis of Thirty Phenolic Pollutants in Water Samples. *Journal of Chromatography A*, 963:137–148.

Ma, X.; N. Gao; B. Chen;; L. Qingsong; Z. Qiaoli; & G. Guofen . 2007. Detection of Geosmin and 2-Methylisoborneol by Liquid-Liquid Extraction-Gas Chromatograph Mass Spectrum (LLE-GCMS) and Solid Phase Extraction-Gas Chromatograph Mass Spectrum (SPE-GCMS). *Frontiers of Environmental Science & Engineering,* 1:286.

Malleret, L. & A. Bruchet. 2002. A Taste and Odor Episode Caused by 2,4,6 Tribromoanisole. *Journal of the American Water Works Association,* 94:7:84–95.

Manickum, T. & J. Wilson. 2012. Dehydration of 2-methylisoborneol to 2-methyl-2-bornene in the Trace Analysis of Taste-Odorants in Water by Purge and- Trap Sampling with Gas Chromatography (GC)-Mass Selective (MS) Detection. *Hydrol Current Res,* 3:127.

Martín-Alonso, J.; R. Devesa; A. Bernal; & L. Matía. 2007. Managing an Odour Episode in Barcelona's Water Supply: Strategies Adopted, the Causative Agent (Diacetyl) and Determination of Its Organoleptic Properties. *Water Science and Technology,* 55:5209–216.

McGuire, M.J.; J. Rosen; A.J. Whelton; & I.H. Suffet. 2014. An Unwanted Licorice Odor in a West Virginia Water Supply. *Journal of the American Water Works Association,* 106:6:72–82.

McPherson, L. 2012. Amperometric vs. Colorimetric Methods for On-line Measurement of Chlorine. *Waterworld,* 28:8.

Meilgaard, M.C.; B.T. Carr; & G.V. Civille. 2006 (4th ed.). *Sensory Evaluation Techniques.* Boca Raton, FL: CRC Press.

Ömür-Özbek, P. & A.M. Dietrich. 2008. Developing Hexanal as an Odor Reference Standard for Sensory Analysis of Drinking Water. *Water Research,* 42:2598–2604.

Ömür-Özbek, P. & A.M. Dietrich. 2011. Retronasal Perception and Flavor Thresholds of Iron and Copper in Drinking Water. *Journal of Water and Health,* 9:1:1–9.

Pepich, B.; T. Dattilio; P. Fair; D. Munch; & G. Gordon, G. 2007. An Improved Colorimetric Method for Chlorine Dioxide and Chlorite Ion in Drinking Water Using Lissamine Green B and Horse Radish Peroxidase. Washington, DC: US Environmental Protection Agency; paper 174.

Rice, E.W., Baird, R.B., Eaton, A.D., & Clesceri, L.S. (editors). 2012. *Standard Methods for the Examination of Water and Wastewater.* 22nd ed. ISBN: 9780875530130. American Public Health Association, American Water Works Association, & Water Environment Federation: Washington, D.C.

Romero, J.; I. Manero; & J. Laso. 2007. Validation of Geosmin and 2-Methyl-i-borneol Analysis by CLSA-GC-FID Method to Obtain ISO-17025 Accreditation. *Journal of Chromatographic Science,* 45:7:439–46.

Rouessac, F. & A. Rouessac, A. 2007 (2nd ed.). *Chemical Analysis: Modern Instrumentation Methods and Techniques.* Chichester, United Kingdom: John Wiley and Sons.

Seattle Public Utilities. 2012. Annual Report Card. http://www.seattle.gov/util/AboutUs/Management/AnnualReportCard/index.htm. (accessed May 23, 2017).

Suffet, I.H.; D. Khiari; & A. Bruchet. 1999. The Drinking Water Taste and Odor Wheel for the Millennium: Beyond Geosmin and Methylisoborneol. *Water Science & Technology*, 40:6:1–13.

Suffet, I.H.; J. Mallevialle; & E. Kawczynski (editors). 1995. *Advances in Taste-and-Odor Treatment and Control.* Denver, CO: AWWA Research Foundation and Lyonnaise des Eaux.

Suffet, I.H.; L. Schweitzer; & D. Khiari. 2004. Olfactory and Chemical Analysis of Taste and Odor Episodes in Drinking Water Supplies. *Reviews in Environmental Science and Biotechnology*, 3:3–13.

USEPA (US Environmental Protection Agency). 1996. Method 8041. Phenols by Gas Chromatography. Washington, DC: US Environmental Protection Agency.

– – –. 2000. Method 528. Determination of Phenols in Drinking Water by Solid Phase Extraction and Capillary Column Gas Chromatography/Mass Spectrometry (GC/MS), revision 1.0. Cincinnati, OH: National Exposure Research Laboratory, Office of Research and Development, US Environmental Protection Agency.

– – –. 2009. Method 334.0: Determination of Residual Chlorine in Drinking Water Using an On-Line Chlorine Analyzer. EPA 815-B-09-013, Washington, DC: Office of Research and Development.

WHO (World Health Organization). 2011 (4th ed.). *Guidelines for Drinking-Water Quality.* Geneva, Switzerland: WHO.

WRF (Water Research Foundation). 2014. *Advancing the Science of Water: WRF and Research on Taste and Odor in Drinking Water.* Denver, CO: Water Research Foundation.

Yuan, B.; F. Li: D. Xua; & M. Fub. 2013. Comparison of Two Methods for the Determination of Geosmin and 2-Methylisoborneol in Algae Samples by Stable Isotope Dilution Assay Through Purge-and Trap or Headspace Solid-Phase Microextraction Combined with GC/MS. *Analytical Methods*, 7:1631–1888.

Appendix D

Nitrification Monitoring Plan

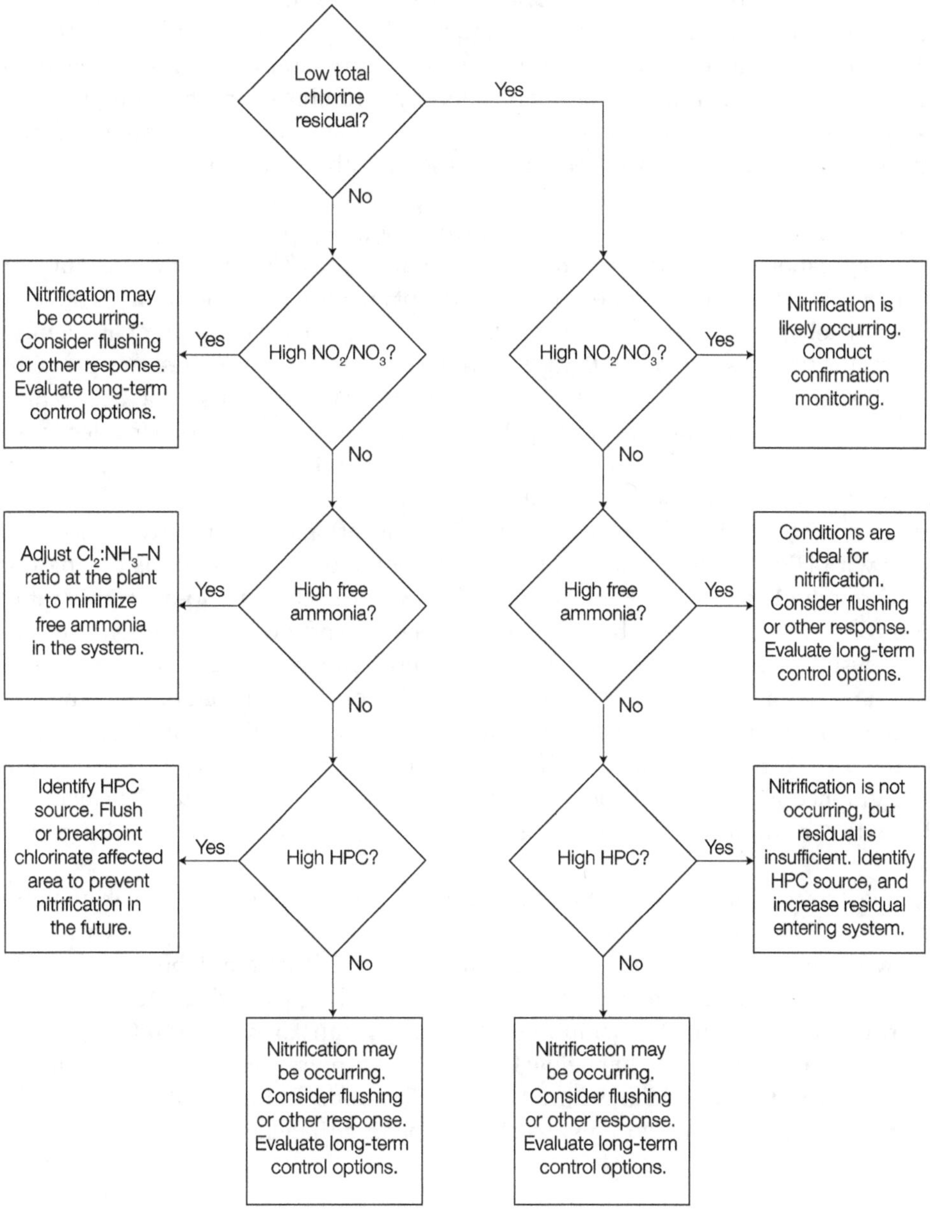

Source: AWWA M56 2013.

Figure D-1 Nitrification assessment flowchart

My Water Utility
Somewhere, USA
Date

Objectives: To reduce the impact of nitrification on water quality (e.g., disinfectant residual levels) and support the ability of the utility to meet process control and regulatory requirements.

System Details: My Water Utility has relied on chloramine to control DBP formation since 1978. In the late 1980s, we noticed that monthly monitoring indicated that one tank (5 mg standpipe at Lost Hope Road) routinely lost the chlorine residual in the month of August. The following year, we increased monitoring of this tank to weekly during the months of July, through October. We observed that the tank had low chlorine levels (or no chlorine) for six weeks in August and September. We later discovered that this was caused by nitrification, because the low chlorine residual level samples also had elevated nitrite levels. Since that time we have worked to control nitrification by a combination of cycling tanks over a greater range during the summer, and increasing the chloramine residual leaving the plant to 4.0 mg/L during the summer months. This program has been mostly successful with exception of the 5 MG standpipe at Lost Hope Road.

Our treatment process includes application of free chlorine (approximately 4.5 mg/L average) to the effluent of our coagulation basins, at pH ~6.8-7.2. At the end of the chlorine contact chamber (~ 30 minutes at typical summer flowrates), we add lime to increase the pH to ~8.5, and ammonia to convert the free chlorine to a predominately monochloramine residual. Our target is to leave the plant at 2.5 mg/L in the winter, and to adjust this up to as much as 4.0 mg/L as the water temperature increases past 60 °F. Maximum temperature in the summer is ~90 °F. At maximum temperature (mid-August), the 7-day chloramine decay (die-off) is approximately 30% (from 4.0 mg/L to 2.8 mg/L) according to our chlorine decay curve measurements (finished water stored in the dark at ambient water pH and temperature). The maximum water age is estimated at 5 days (at TCR sampling point on Last Chance Road).

Beginning in 1998, we have been monitoring 5 key locations for total, free, and monochloramine residuals, along with HPC, nitrites, nitrates and free ammonia on a weekly basis during the nitrification period (July 1 through October 1). We have closely followed the chlorine-to-ammonia ratio in our plant feeds to minimize free ammonia in the distribution system, and cycle our tanks to limit water age in the distribution system. As we gained experience with nitrification, we have reduced the monitoring for nitrites, nitrates, and free ammonia and rely primarily on the levels of total chlorine in the distribution system as an indicator of nitrification. We have also identified 5 key areas (3 elevated tanks and 2 standpipes) that provide adequate warning of when nitrification problems require changes in our operational plans. These data are graphed at the end of each month for analysis. For TCR data, we calculate the monthly average and monthly minimum values for all sites, and the percent of samples falling below 0.50 mg/L total chlorine.

Data from last year (2013) are presented in Figure 6-10. Weekly data from the key monitoring points indicate that except for the 5 mg standpipe at Lost Hope Road, the nitrification plan we have been working has been effective. The standpipe at DS 4 showed indications of nitrification, but increased chlorine levels leaving the plant and water age management through tank cycling was adequate to keep residuals above 1.0 mg/L. Chlorine residuals from TCR sampling were greater than 0.5 mg/L in 96% of all samples taken. Of the 240 TCR samples taken in 2013, 10 fell below 0.5 mg/L benchmark. Downstream flushing was adequate to restore chlorine levels to above 0.5 mg/L in 9 of these locations. In one location, a chronic low chlorine level was caused by an inadvertent closed valve. Upon opening the valve, chlorine levels were restored to acceptable levels.

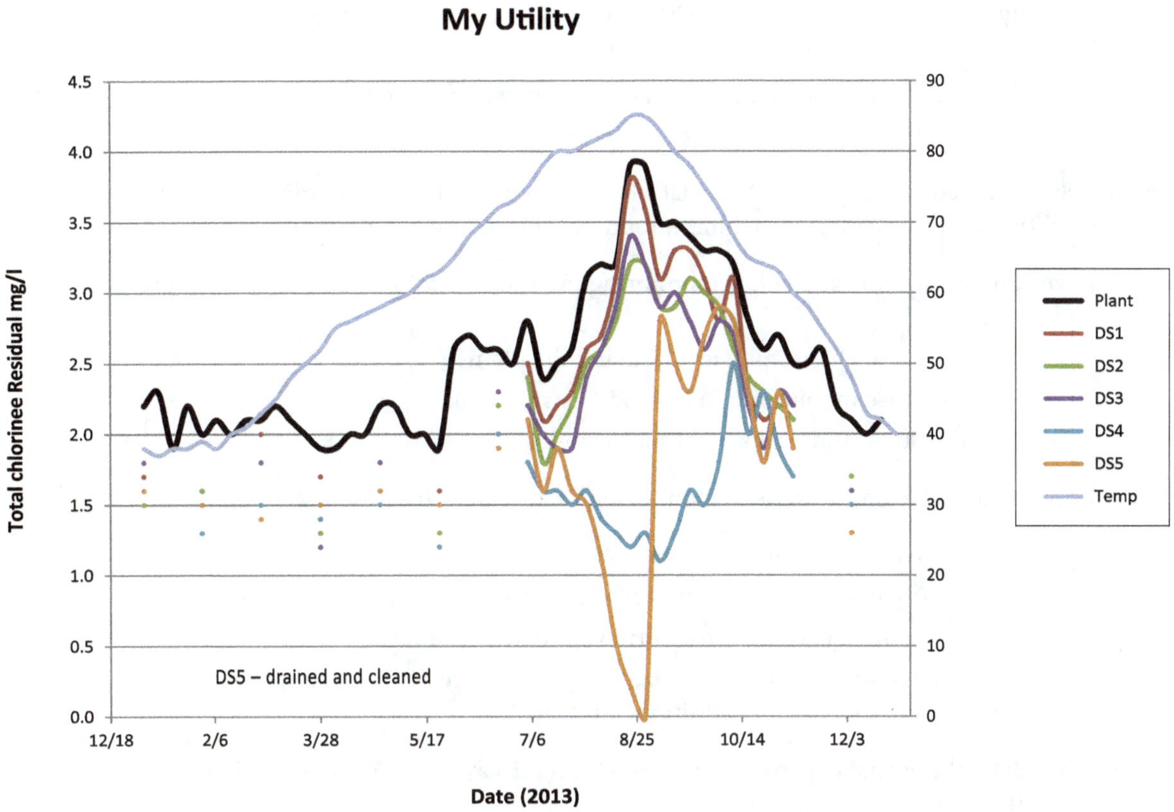

Figure D-2 Total chlorine residual as a function of time at various distribution system sampling locations

Strategies: The nitrification plan relies on the following strategies to help control nitrification:

- Minimizing available free ammonia in the distribution system
- Controlling water age through tank cycling and flushing
- Increasing chloramine levels leaving the plant during the nitrification season
- Reducing organic food sources (TOC, AOC) that indirectly support growth of nitrifying bacteria

Monitoring Protocol: For 2014, My Water Utility plans on maintaining the same monitoring protocol used in 2013, with the addition of daily monitoring of any location falling below 0.50 mg/L chlorine residual:

- Monthly monitoring at all tanks (DS1-DS5) November through June; weekly monitoring July through October (total and free chlorine, pH)
- Additional analysis for nitrite, nitrate, and free ammonia at locations where persistent low chlorines are found (3 consecutive samples less than 0.5 mg/L)
- Daily monitoring at any sampling location where total chlorine falls below 0.5 mg/L

Data Analysis: For 2014, My Water Utility plans to continue data analysis as was conducted in 2013. These data will be submitted monthly to the State.

- Monthly analysis of total chlorine data from key monitoring locations (as indicated in graphic above)

- Monthly summary of all TCR total chlorine data for monthly average, monthly minimum, and % below 0.50 mg/L
- Monthly submittal to State Regulatory Agency (in excel format provided by State Regulatory Agency)

Benchmarks and Action Plans: My Water Utility will continue to use the same benchmarks as in 2013, with the addition of the 0.50 mg/L benchmark for daily chlorine residual monitoring:

- **Level 1:** total chlorine at any key monitoring point **less than 1.50 mg/L** (or temp greater than 60°F):
  - Increase chloramine level leaving the plant
  - Increase monitoring in key locations to weekly
  - Evaluate and optimize chlorine/ammonia feed ratios
- **Level 2:** total chlorine at any key monitoring location **less than 1.0 mg/L:**
  - Increase tank turnover rate
  - Flush at localized hydrants as appropriate
  - Evaluate and optimize chlorine/ammonia feed ratios
  - Consider treatment optimization processes for TOC/AOC removal (enhanced coagulation or PAC addition to raw water)
- **Level 3:** total chlorine **approaching or less than 0.50 mg/L** at any key monitoring point **or TCR site:**
  - Daily monitoring for total chlorine
  - Localized hydrant flushing as appropriate
  - Consider adding analysis for nitrate, nitrite, free ammonia, and HPC
  - Consider drain/refill tank
  - Evaluate optimize chlorine/ammonia feed ratios
  - Consider booster chlorination as appropriate

At this time, My Water Utility does not consider conversion to free chlorine in the distribution system as an appropriate response to nitrification. Our experience has shown that nitrification can be managed without resorting to this if a staged response is executed at the earliest signs of nitrification. In our experience, this occurs when the total chlorine residual drops towards 1.5 mg/L.

Submitted by: ______________________________

Printed name and title/position: ______________________________

Date: ______________________________

Glossary

Absolute Pressure. The total pressure in a system, including both the pressure of the water and the pressure of the atmosphere (contrast with gauge pressure)

Absorption. The penetration of one substance into the structure of another substance. The penetration of a chemical through the skin, wall of the gastrointestinal tract, or the lung into the systemic circulation (i.e., blood). Contrast with adsorption.

Accelerated Monitoring. Also known as impulse recording. Is a feature that some pressure monitors have to automatically record data more frequently when significant changes in pressure occur.

Acidic. Water or soil that contains a sufficient amount of acidic substances (hydrogen [H^+] ions) to lower the pH to less than 7.0

Acoustic Monitoring. Technology that allows for more prompt identification of potential leaks.

Action Level (AL). A specified concentration of a contaminant in water that determines treatment requirements or actions that a water system is required to complete to control the contaminant (such as further treatment or further monitoring, or both) to comply with a drinking water regulation in the United States.

Activated Carbon. A form of particulate carbon (a crude form of graphite) with increased surface area to enhance adsorption of soluble contaminants. The primary source of activated carbon is bituminous coal, which is activated through a combustion process. Activated carbon is sometimes mistakenly called charcoal or just carbon.

Acute. Pertaining to an intense effect of a chemical or infectious agent. Acute effects are generally closely associated with high doses of the causative agent. They are not necessarily confined to brief exposure to the agent.

Acute Toxicity. Any severe poisonous effect resulting from a single short-term exposure to a toxic substance.

Adenosine Triphosphate (ATP). A compound found in all living cells in which energy is stored in high-energy phosphate bonds. Its components are the purine adenine, D-ribose, and three phosphoric acid groups.

Adsorbent. Any material that can be used to adsorb substances on its surface (e.g., activated carbon).

Adsorption. The attraction and adhesion of molecules of a gas, liquid, or dissolved substance to a surface. Adsorption is generally a passive and reversible process. Granular or powder activated carbon is often used as an adsorption medium.

Advanced Metering Infrastructure (AMI) Network. Enables two-way communication over a fixed network between the utility systems and the metering endpoints. AMI systems provide water systems with the ability to monitor the entire distribution system network continuously compared to Automatic Meter Reading (AMR) which provides one-way communication regardless of how the meter is read (walk-by, drive-by, or fixed network).

Advanced Oxidation Process (AOP). A process in which the oxidative capacity of a parent compound is modified to make oxidation-reduction reactions more complete. For example, hydrogen peroxide (H_2O_2) may be added to ozone (O_3), or hydrogen peroxide may be photolyzed by ultraviolet light, to generate hydroxyl radical, which are strong, nonspecific oxidizing agents. AOPs are feasible for full-scale use to destroy organic compounds because they generate hydroxyl radicals at ambient temperature and atmospheric pressure, whereas other processes that generate hydroxyl radicals require elevated temperatures and/or high pressures are mediated by free-radical chain reactions involving hydroxyl radicals.

Aeration. A gas transfer unit process of contacting a liquid with air by which a gas is transferred from one phase to another: either the gas phase to the liquid phase (gas absorption) or the liquid phase to the gas phase (gas stripping).

Aerobic. In reference to microorganisms, pertaining to the requirement for oxygen for growth. In reference to environmental conditions, pertaining to the presence of oxygen.

Aesthetic. Pertaining to the quality of water that is determined by the senses, e.g., color, taste, or odor.

Aggressive. Pertaining to corrosive water that will deteriorate material such as distribution piping. Aggressive water is water that has corrosive qualities.

Aggressiveness Index (AI). One of the indices available to estimate the general corrosivity of water. It is derived from pH, calcium hardness and total alkalinity. Compared to the Langelier Index (LI), where applicable, the AI is simpler and more convenient than the LI. Because AI does not include the effects of temperature or dissolved solids, it is less accurate as an analytical tool than the LI. An AI of 12 or above indicates nonaggressive (not corrosive) water. AI values below 10 indicate extremely aggressive (corrosive) conditions. Values of 10-11.9 suggest that the water is moderately aggressive.

Air Chamber. A closed pipe chamber installed on the discharge line of the reciprocating pump to take up irregularities in hydraulic conditions and relieve the pump of shocks caused by the pulsating flow.

Air Release Valve. *See* Air Relief Valve.

Air Relief Valve. An air valve placed at the summit of a pipeline to release air automatically and prevent air binding and pressure buildup or to allow air to enter a line if the internal pressure becomes less than that of the atmosphere.

Air Stripping. A process that removes volatile compounds from a liquid phase by passing air through the liquid. The process uses the principles of Henry's law to transfer volatile pollutants from a solution of high concentration into an airstream of lower concentration. Air can be diffused into a bulk liquid stream or can be introduced into packed towers by compressors, countercurrent to the water flow.

Air-And-Vacuum Relief Valve. An air valve that permits air to enter an empty pipe to counteract a vacuum and prevent accumulated air from escaping. This type of valve is also called a vacuum relief valve.

Aldehyde. A class of organic compounds that have a carbonyl (C=O) functional group on the first or last carbon atom in the chemical structure. Some aldehydes are created during the reactions of oxidants used as disinfectants, particularly ozone (O_3), with natural organic matter.

Algae. The simplest plants that contain chlorophyll and require sunlight; they vary from microscopic forms to giant seaweed. In drinking water sources, blooms of microscopic forms (phytoplankton) cause taste-and-odor problems.

Algorithm. An arrangement of logical steps to perform a certain operation within a computer program.

Alkalinity. A measure of the capacity of a water to neutralize strong acid; in natural waters this capacity is usually attributable to bases such as bicarbonate (HCO_3^-), carbonate (CO_3^{2-}), and hydroxide (OH^-) and to a lesser extent silicates, borates, ammonia , phosphates, and organic bases. It is expressed in milligrams of equivalent calcium carbonate per liter (mg/L as $CaCO_3$).

Aluminum (Al). A metallic element. Aluminum is the most abundant metal in the earth's crust; it does not occur free in nature.

Aluminum Sulfate (alum). ($Al_2(SO_4)_3$), an inorganic compound commonly used as a coagulant in water treatment. It contains waters of hydration, $Al_2(SO_4)_3 \times XH_2O$ (where X is a variable number). Aluminum sulfate is often called alum.

American Society of Testing and Materials (ASTM). An organization founded in 1898 that establishes standards for materials, products, systems, and services through technical committees. More than 9,000 standard test methods, specifications, classifications, definitions, and recommended practices are now in use.

Ammonia (NH_3). An inorganic gas commonly detected in wastewater. Ammonia may be present in drinking water that has been chloraminated, as well as in source waters.

Ammonia (free). Ammonia (NH_3) that is free in solution, not bound up as in chloramines.

Ammonia Oxidizing Bacteria (AOB). Bacteria that can be supported by the addition of ammonia to the distribution system associated with chloramine formation. These bacteria catalyze the first step in nitrification, oxidizing ammonia to nitrite. The growth of nitrifiers is typically accompanied by deterioration in water quality parameters such as an increase in heterotrophic plate count (HPC) bacteria, nitrite, and nitrate and a decrease in total chlorine residual.

Anion-Exchange. A process in which anion contaminants are removed from a liquid phase by contacting a synthetic, porous medium or resin that is coated with other anions. The anions on the medium are exchanged for the anion contaminants. When the medium is depleted of the exchanging anions, it is backwashed with a concentrated solution to restore the bed with the exchange anions and to flush the contaminant anions for subsequent disposal.

Anthropogenic. Pertaining to or involving the impact of humans.

Apparent Color. The color caused by substances in solution as well as that caused by suspended matter. Apparent color is determined on an original sample without filtration or centrifugation. In some waters, colloidal or suspended material may contribute to color. In such cases both true color and apparent color should be determined.

Arsenic (As). An inorganic contaminant that is found in water supplies primarily as a result of natural geologic formations and has been associated with adverse health effects.

Asbestos. A fibrous silicate mineral generally ranging from 0.03 to 0.10 micrometers in diameter and less than 10 micrometers in length. Asbestos fibers, from both natural and human made sources, have been associated with adverse health effects when inhaled. Very little information is available regarding the adverse health effects of ingesting asbestos fibers, and studies are under way.

Asbestos Cement Pipe (AC). A water main material made from a mixture of cement and asbestos fibers.

Assets. The entire resources of a government entity or person, or private business, both tangible and intangible.

Assimilable Organic Carbon (AOC). The fraction of organic carbon that can be used by specific microorganisms and converted to cell weight. AOC also represents a potential for biological regrowth in distribution systems. Ozone (O_3) can convert organic matter in water to assimilable organic carbon, whereas biological filtration can reduce the AOC level.

Atmospheric Pressure. The pressure exerted by the atmosphere at any point. It decreases as the elevation above sea level increases.

Automatic Meter Reading (AMR). The electronic reading of customers' meters from a remote central location. Data from meters are transmitted via telephone, cable television lines, or radio frequency networks to a central computer for billing purposes.

Average Day Demand. Annual water consumption divided by 365 days in a year; the average water demand a given water distribution system experiences over a one-day period. The average day demand applies to a specific time period over which the average is calculated.

Backflow. A hydraulic condition, caused by a difference in pressures that causes nonpotable water or other fluid to flow into a potable water system.

Backflow Preventer. A device for a water supply pipe to prevent the backflow of water into the water supply system from the connections on its outlet end.

Backflow Prevention. The use of a system or a device such as a double gate, double check assembly, or air gap to separate a potable water system from a system of water of unknown quality or source. The device or system prevents water from flowing back into a protected system when the pressure in the protected system drops lower than in the system of unknown quality.

Backflow Prevention Device. Any effective device, method, or construction used to prevent backflow into a potable water system.

Backpressure. A pressure that can cause water to backflow into the water supply when a user's water system is at a higher pressure than the public water system.

Backsiphonage. A form of backflow caused by a negative or sub atmospheric pressure within a water system.

Ball Valve. A valve with the closing and opening mechanism formed in the shape of a ball with a hole. The valve is opened by rotating the ball so the hole is parallel to the flow, allowing it to pass. The valve is closed when the hole is perpendicular to the flow.

Barium (Ba). A divalent cation and alkaline-earth metal. Barium can be found in naturally occurring mineral deposits and is used in various industrial applications. It has also been associated with adverse health effects..

Baseline. The condition of a natural community or environment at the starting point (e.g., prior to restoration activities) that will be used to track future changes.

Bicarbonate (HCO_3^-). An inorganic monovalent anion usually found in natural water.

Biodegradable Dissolved Organic Carbon (BDOC). A subset of biodegradable organic matter that is the portion of the organic carbon in water that can be mineralized by heterotrophic microorganisms. Ozone (O_3) can convert organic matter in water to biodegradable dissolved organic carbon, whereas biological filtration can reduce the biodegradable dissolved organic carbon level.

Biofilm. A layer of microorganisms held together by a matrix of organic polymers and found at the interface between water and a solid substrate in an aquatic environment. Biofilms in natural environments are heterogeneous and contain a range of microenvironments conducive to development of a diverse bacterial population. A biofilm can be detrimental to a process (say, by fouling a water treatment membrane) or beneficial (as in a biological treatment system).

Biofilter. Media, usually granular in material (e.g., activated carbon, anthracite coal, or sand) that has developed a biofilm capable of degrading organic matter, ammonia (NH_3), or both.

Biofiltration. The process of filtering water through a filter medium that has been allowed to develop a microbial biofilm that assists in the removal of fine particulate and dissolved organic materials.

Biofouling. A phenomenon in which the performance of a unit process is compromised by biological growth. For example, biofouling can occur in a membrane process when microorganisms grow on the membrane surface and result in a premature and excessive loss of flow rate, or flux, through the membrane. The clogging of a filtration medium as a result of microbiological growth. It is typically associated with (1) a loss of flux and potential filtrate quality degradation in membrane filters or (2) growth on a groundwater well casing.

Biological Regrowth. The phenomenon of biological growth in treated water as a result of an increase in bacterial density in treated distribution water caused by growth of bacteria released from the pipe wall biofilm and sediments, generally more apparent in areas where the disinfectant residual has been depleted. Biological regrowth indicates the presence of a persistent population of bacteria in a water distribution system. Such regrowth may lead to problems in the compliance monitoring of total coliform bacteria.

Biologically Active Filtration (BAF). The removal of biodegradable organic matter by passing water over a fixed medium covered with a biofilm.

Biomass. (1) The total weight of biological matter, including any attached extracellular polymeric materials. (2) A material that is or was a living organism or was excreted from a microorganism.

Black Water. An off-color condition that develops in drinking water from the presence of excess manganese that has been oxidized. Normally, less than 0.005 mg/L of manganese will not cause black water; however, as levels approach 0.10 mg/L, black water is more likely to occur.

Boil Water Advisory (BWA). A public notice issued by a federal, state, or local health department via broadcast media that informs users of a public water system that their drinking water is, or potentially is, unsafe microbiologically and that the drinking water should be boiled before use. To kill waterborne pathogens, water at sea level should be brought to a rolling boil, held in that condition for 1 minute, and then allowed to cool (protected from further contamination) before consumption.

Booster Station. (1) A pumping station in a water distribution system that is used to increase the pressure in the mains on the discharge side of the pumps. (2) A station that pumps water from low-level or ground storage to a distribution system.

Bottled Water. Water intended for human consumption that is sealed in bottles or other containers with no added ingredients except a disinfectant, if necessary, to ensure microbial quality. Bottled water is regulated in the United States as a food by the Food and Drug Administration.

Brass. A metal alloy of copper, zinc, and usually some lead. Brass is harder and stronger than copper because of its zinc content; lead contributes malleability and ductility.

Breakpoint. The point at which the chlorine dosage has satisfied the chlorine demand exerted by ammonia (NH_3).

Breakpoint Chlorination. A process by which ammonia (NH_3) is removed from water through the continuous addition of chlorine (Cl_2 or HOCl). As chlorine is added, monochloramine (NH_2Cl) is initially formed, followed by dichloramine ($NHCl_2$) and trichloramine (nitrogen trichloride—NCl_3). At the breakpoint, the nitrogen has been completely released and any additional chlorine added is measured as free, rather than combined, chlorine residual.

Bromate (BrO_3^-). The highest oxidation state of the bromide ion (Br^-). The bromate ion can be formed during the ozonation of bromide-containing waters.

Bromide (Br^-). An inorganic ion found in surface water and groundwater and caused by (1) seawater intrusion, (2) the impact of connate water, or (3) industrial and oil- field brine discharges. When oxidized by chlorine (Cl_2 or HOCl) or ozone (O_3), it can result in the formation of bromide-substituted disinfection by-products.

Bromine (Br_2). The oxidized form of the bromide ion (Br^-). In water, bromine is present as hypobromous acid (HOBr) and the hypobromite ion (OBr^-).

Buffer. A chemical substance that stabilizes the pH value of solutions.

Bulk Water. The potable water that is conveyed through distribution system piping. As it travels through the distribution system, it undergoes various chemical, physical and aesthetic transformations, impacting water quality.

Cadmium (Cd). A malleable, ductile, bivalent metallic element (atomic number 48). In nature, cadmium is found in the ores of greenockite. It has various industrial uses, especially in protective platings and bearing metals. It has also been associated with adverse health effects.

Calcium (Ca). One of the principal elements making up the earth's crust. The presence of calcium in water is a factor contributing to the formation of scale and insoluble soap curds that are a means of clearly identifying hard water.

Calcium Carbonate ($CaCO_3$). A colorless or white crystalline compound that occurs naturally as chalk, limestone, marble, and other forms. Pure calcium carbonate exists in two distinct crystalline forms: the trigonal solid, calcite; and the orthorhombic solid, aragonite. $CaCO_3$ is a sparingly soluble salt, the solubility of which decreases with increasing temperature. It has the potential to cause scaling if it is concentrated to supersaturation.

Calcium Carbonate Precipitation Potential (CCPP). The amount of solid calcium carbonate ($CaCO_3$) that will precipitate or dissolve as water equilibrates.

Calibrate. To adjust a measuring instrument so that it gives the correct result with a known concentration or sample.

Calibration. A procedure that adjusts or checks the accuracy of an instrument by comparison with a standard or reference.

Capital Improvement Plan. A plan updated or compiled annually by most utilities, which identifies facility requirements over an extended period, often 20 years or more. The capital improvement plan is often part of or stems from a water system master plan that combines water demand projections with supply alternative and facility requirements. It is different from an asset replacement program concentrating on smaller assets that are generally funded from current revenues or from funds established for replacement. The capital improvement plan, by contrast, is focused on larger expenditures that are often debt funded but are sometimes funded from capacity changes or current revenues.

Carbon Dioxide (CO_2). A colorless, odorless, incombustible gas that is a normal component of natural waters. It may enter surface water and groundwater by absorption from the atmosphere or biological oxidation of organic matter.

Carbonate (CO_3^{2-}). A divalent negatively charged anion.

Carboxylic Acid (R-COOH). Any of a broad array of organic acids composed chiefly of hydrocarbon groups, usually in a straight chain, terminating in a carboxyl group (COOH).They contribute cation exchange ability to some resins. Carboxylic acids are sometimes called weak acid cation exchangers.

Cast Iron Pipe. A pipe made from pig iron cast in a revolving, water-cooled mold, or in a stationary, cylindrical sand mold with a round, central core.

Cathodic Protection. Reduction or elimination of corrosion by making the metal a cathode by means of an impressed direct current or attachment to a sacrificial anode (usually magnesium, aluminum, or zinc).

Cationic Polymer. A polymeric substance with a net positive charge used in coagulation, flocculation, or filtration processes to improve the removal of negatively charged particles from natural waters. Cationic polymers can be used to destabilize negatively charged particles or to form a bridge between destabilized floc.

Caustic Soda (Sodium Hydroxide). Sodium hydroxide, a strongly alkaline chemical used for pH adjustment, water softening, anion exchange demineralizer regeneration, and other purposes. It is sometimes called caustic.

Cavitation. The formation and sudden collapse of vapor bubbles in a liquid, usually resulting from local low pressures, as on the trailing edge of a propeller. This phenomenon develops a momentary high local pressure that can mechanically destroy a portion of a surface on which the bubbles collapse. Cavitation can occur in pumps when the suction side has insufficient head for the current operating conditions..

Cement Lined Pipe. A pipe for which the interior has been lined with a smooth, dense, thin layer of cement mortar to reduce corrosion and increase smoothness..

Cement–Mortar Lining. A mixture of portland cement, sand, and water used to line pipe to prevent internal corrosion and maintain flow coefficients over the life of the system. The lining may be applied in the factory or in the field after cleaning in place. A seal coat of asphaltic material may be applied to the uncured lining at the purchaser's option.

Chloramination. The process of disinfecting water with chloramines.

Chloramine. Disinfectants produced from the mixing of chlorine (Cl_2) and ammonia (NH_3). The general formula is NH_xCl_y, where x can be 0, 1, or 2; y can be 1, 2, or 3. Typically, monochloramine (NH_2Cl) and a small percentage of dichloramine ($NHCl_2$) are formed, depending on the pH and the chlorine-to-ammonia ratio that reacts. Under certain circumstances, nitrogen trichloride (trichloramine, $NHCl_3$) can be formed. In the presence of the bromide ion, bromamines can be produced. In the presence of organic nitrogen, organic chloramines can also form; however, organic chloramines are not considered to be disinfectants.

Chlorate (ClO_{3-}). A disinfection by-product formed during the treatment of water with chlorine dioxide (ClO_2). The chlorate ion is also produced during the decomposition of bleach (NaOCl) in stock solutions. Furthermore, ozonation of water with a chlorine residual can yield chlorate ions..

Chloride (Cl^-). One of the major anions commonly found in water and wastewater. Its presence is often determined by ion chromatographic or volumetric analysis. Consumers who drink water with concentrations of chloride exceeding a secondary maximum contaminant level of 250 mg/L may notice a salty taste.

Chlorine. A chemical used as a disinfectant and oxidizing agent. Gaseous molecular chlorine (Cl_2), when introduced into water, is converted into hypochlorous acid (HOCl) and the hypochlorite ion (OCl^-); the ratio of the two substances is dependent on the pH of the solution ($HOCl \rightarrow OCl^- + H^+$). Chlorine is also commercially available in liquid form as a hypochlorite ion solution. Individual chlorine atoms are halogen atoms that can be substituted into organic molecules during chlorination, thus forming disinfection by-products. Chlorine atoms are also present in a number of pesticides and other synthetic organic chemicals.

Chlorine Demand. The quantity of chlorine consumed in a specified time period by reaction with substances present in water that exert an oxidant demand (e.g., natural organic matter, ammonia [NH_3], hydrogen sulfide [H_2S]). The chlorine demand for a given water varies with both contact time and temperature.

Chlorine Dioxide (ClO_2). A red-yellow gas that is very reactive and unstable. It is a strong oxidizing agent and is also used as a disinfectant. Chlorine dioxide decomposes in water to yield the chlorite ion (ClO_2^-) and, to a lesser extent, the chlorate ion (ClO_3^-).

Chlorine Residual. A concentration of chlorine species present in water after the oxidant demand has been satisfied. Chlorine residual can be determined and expressed in a number of ways. Commonly used analytical techniques include a colorimetric method using the reagent N, N-diethyl-p-phenylenediamine and a volumetric method using an amperometric titration. The concentration is often expressed in terms of free chlorine, total chlorine, or combined chlorine.

Chlorine to Ammonia Ratio. The ratio (typically on a weight basis) of the chlorine and ammonia doses during chloramination. During breakpoint chlorination of ammonia-containing waters, a theoretical Cl_2/NH_3–N ratio of 7.6:1 milligrams/milligram is required.

Chlorite (ClO_2^-). A disinfection by-product formed during treatment of water with chlorine dioxide.

Chloroform ($CHCl_3$). The common name for trichloromethane, a trihalomethane containing only chlorine atoms. Typically, this is the principal trihalomethane formed during chlorination, except for in waters containing moderate amounts of bromide.

Chromium (Cr). Chromium is a blue-white metallic element (atomic number 24) with various industrial uses, e.g., in alloys and electroplating. It has also been associated with adverse health effects.

Clearwell. A tank or vessel used for storing treated water. Typical examples of storage needs include (1) finished water storage to prevent the need to vary the rate of filtration with variations in distribution system demand, and (2) backwash water for filters. Clearwells are located on-site at a water treatment plant. A clearwell is also called a filtered-water reservoir.

Coagulant. A chemical added to water that has suspended and colloidal solids to destabilize particles, allowing subsequent floc formation and removal by sedimentation, filtration, or both (e.g., the use of alum or iron salts for removing turbidity in a water treatment process).

Coagulation. The process of destabilizing charges on particles in water by adding chemicals (coagulants). Natural particles in water have negative charges that repel other material and thereby keep it in suspension. In coagulation, positively charged chemicals are added to neutralize or destabilize these charges and allow the particles to accumulate and be removed by physical processes such as sedimentation or filtration. Commonly used coagulants include aluminum and iron salts and cationic polymers.

Coliform Bacteria. Bacteria of the family Enterobacteriaceae, commonly found in the intestinal tracts of warm-blooded animals. In sanitary bacteriology, these organisms are defined as all aerobic and facultative anaerobic, gram-negative, nonspore- forming, rod-shaped bacteria that ferment lactose with gas and acid formation within 48 hours at 95° Fahrenheit (35° Celsius). Coliform group bacteria are a group of related bacteria comprised of the genera Escherichia, Enterobacter, and Klebsiella. Commonly associated with the intestinal tracts of warm-blooded animals and shed in their fecal material.

Complexation. The inactivation of an ion by addition of a reagent that combines with it and, in effect, prevents it from participating in other reactions. Complexation is also called sequestration.

Concrete Pipe. A conduit manufactured from portland cement and aggregates combined with different forms of reinforcement, used to convey water.

Conductivity (Specific Conductance). A measure of the ability of a solution to conduct electrical current. Its value is inversely proportional to the solution's electrical resistance. The conductivity value is commonly used in water-desalting processes as a means to evaluate desalting efficiency and to estimate the total dissolved solids concentration; the conductivity value of a water sample is multiplied by an empirical factor representative of the typical total dissolved solids/conductivity ratio for the specific type of water. The units of conductivity are often reported as microhms per centimeter at 25° Celsius, but this is not an International System of Units measurement; multiplying such a value by 10–4 converts the value to units of siemens per meter.

Contaminant. Any physical, chemical, biological, or radiological substance or matter in water.

Contaminant Plume. The pattern formed as polluted water extends downstream from the pollution source (analogous to smoke drifting downwind from a smokestack). It contains undesirable concentrations of some pollutant. Water at the periphery of the mass may also contain the pollutant but at acceptable concentrations and so would not be considered part of the plume. Plumes are often elliptical or teardrop shaped, with the long axis aligned with the direction of flow.

Conventional Treatment. The use of coagulation, flocculation, sedimentation, filtration, and disinfection, together as sequential unit processes, in water treatment. This process is also called complete treatment.

Copper (Cu). A common reddish metallic element (atomic number 29). In nature, copper is found in the ores of azurite, azurmalachte, chalcocite, chalcopyrite, covellite, cuprite, and malachite. It is ductile and malleable and is used to manufacture pipe and other products used in water and treatment. Copper has various industrial uses and, as copper sulfate, is used to control algal growths. It has also been associated with adverse health effects.

Copper Plumbing. A conduit made from copper that has been heated and forced through dies to form a hollow cylindrical body of a required size for water services or plumbing pipes.

Corrosion. The gradual deterioration or destruction of a substance (usually a metal) or its properties as a result of a reaction with the substance's surroundings.

Corrosion Control. (1) Any water treatment method that keeps the metallic ions of a material, typically a conduit, from going into solution, e.g., increasing the pH of the water, removing free oxygen from the water, or controlling the carbonate (CO_3^-) balance of the water. (2) The sequestration of metallic ions and the formation of protective films on metal surfaces by chemical treatment.

Corrosion Inhibitor. A substance that slows corrosion by forming a protective film on the interior surface of pipes and tanks.

Corrosivity. An indication of the corrosiveness of water. The corrosiveness of water is described by the water's pH, alkalinity, hardness, temperature, total dissolved solids, dissolved oxygen concentration, and Langelier saturation index.

Cross Connection. The physical connection of a safe or potable water supply with another water supply of unknown or contaminated quality or such that the potable water could be contaminated or polluted.

Cross Connection Control. The enforcement of an ordinance or other legal statement regulating cross connections.

***Cryptosporidium*.** A widespread intestinal coccidian protozoan parasite about 3.5 micrometers in diameter, causing diarrhea and capable of infecting humans, birds, fish, and snakes. It is responsible for waterborne disease outbreaks. It causes cryptosporidiosis, a protozoan infection of the epithelial cells of the gastrointestinal, biliary, and respiratory tracts of humans and other vertebrates, including birds, reptiles, rodents, cats, dogs, sheep, and cattle. In humans, the average incubation period is 7 days and the infectious dose is low. The primary symptom is profuse, watery diarrhea with cramping abdominal pain, but infection is often asymptomatic. In immunologically healthy individuals, symptoms usually end in 30 days; however, immunocompromised persons, especially AIDS patients, may be unable to clear the infection and illness may be prolonged, contributing to death. Cryptosporidiosis occurs worldwide, with children, animal handlers, people who have close personal contact with infected persons and travelers most likely to be infected. Person-to-person transmission occurs by hand-to-mouth transfer of oocysts from feces of an infected person, especially in institutions and day-care centers. Animal-to-person contact and waterborne transmission are also important. In 1993, contamination of the Milwaukee, Wis., water system caused more than 400,000 cases of this disease; over 1,000 people were hospitalized.

CT. The product of disinfectant concentration (in mg/L) determined before or at the first customer and the corresponding disinfectant contact time (in minutes). It is also called the CT value. Units are mg-min/L.

Dead End. A section of a water distribution system that is not connected to another section of pipe by means of a connecting loop. Such portions of a distribution system can experience lower flows than surrounding portions, which can lead to water quality problems caused by somewhat stagnant water. Examples of problems include tastes or odors, bacteriological growth, loss of chlorine residual, or any combination of these.

Dechlorination. The process of removing chlorine (HOCl, OCl^-) from solution. Dechlorination may be required (1) when high dosages are provided to destroy tastes and odors or to reduce their concentration in the distribution system, (2) to reduce the potential toxic effect of a chlorinated discharge to aquatic life, (3) to protect downstream processes, such as zeolite softeners, from the oxidative effects of chlorine, or (4) to reduce chlorinous tastes in drinking water (particularly bottled water). Dechlorination is typically achieved through chemical addition of a reducing agent, the most common of which is sulfur dioxide (SO_2).

Decomposition. The conversion of chemically unstable materials to more stable forms by chemical or biological action. If organic matter decays when no oxygen is present (i.e., under anaerobic conditions or putrefaction), undesirable tastes and odors are produced. Decay of organic matter when oxygen is present (i.e., under aerobic conditions) tends to produce much less objectionable tastes and odors.

Degradation. (1) The loss of capacity, reduction of resin particle size, excessive swelling of resin particles, or any combination of these factors resulting in a lessening of the ion exchange capabilities of a resin. This may occur as a result of the type of service for which the resin was used, the solution concentrations used, heat, or aggressive operating conditions. (2) A decrease in the quality of water. (3) A reduction in performance of a unit process used to treat water. (4) The changing of an organic material or substrate into smaller, less complex compounds, principally by biological means.

Degradation Product. The result of changing of an organic material or substrate into smaller, less compounds, principally by biological means.

Deposition. (1) The material that collects on the inside surface of a distribution system pipe as a result of suspended material in the water. (2) The material that forms in a water heater or utensil when water is heated.

Detention Time. The average length of time a drop of water or a suspended particle remains in a tank or chamber. Mathematically, it is the volume of water in the tank divided by the flow rate through the tank.

Detergent. Any material with cleansing powers: soaps, synthetic detergents, artificial alkaline materials, solvents, and abrasives. In common domestic usage, the term is often used to refer to synthetic detergents.

Dezincification. The parting of zinc from an alloy. In some cases, zinc is lost, leaving a weak, brittle, porous, copper-rich residue behind.

Dibromoacetic Acid (DBAA) ($CHBr_2COOH$). A haloacetic acid containing two bromine atoms. It is formed to a significant extent during the chlorination of waters containing moderate amounts of bromide (Br^-). It may also be formed during the ozonation of bromide-containing waters.

Dibromochloroacetic Acid (DBCAA) ($CBr_2ClCOOH$). A haloacetic acid containing two bromine atoms and one chlorine atom.

Dichloramine ($NHCl_2$). A disinfectant produced from the mixing of chlorine and ammonia (NH_3).Typically, this chloramine species is produced to a lesser extent than monochloramine (NH_2Cl). However, chloramination at a low pH (e.g., 7), high chlorine-to-ammonia–nitrogen ratios (e.g., 7:1 by weight), or both results in higher levels of dichloramine.

Dichloroacetic Acid (DCAA) ($CHCl_2COOH$). A haloacetic acid containing two chlorine atoms. Typically, this substance and trichloroacetic acid are the principal haloacetic acids formed during chlorination, except for in waters containing moderate amounts of bromide (Br^-).

Diffusion. A process whereby molecules or particles move and intermix because of a concentration gradient driving force; the movement of a compound within a medium or from one medium to another. For example, longitudinal diffusion refers to the movement of a compound in a conduit at a speed either faster or slower than the mean velocity of the solution, whereas boundary layer diffusion refers to the movement of a compound to or from a solution through a bo–undary layer surrounding a particulate medium. Molecular diffusion is quantified by Fick's law.

Disinfectant. An agent that destroys or inactivates harmful microorganisms.

Disinfectants and Disinfection Byproducts Rule (DBPRs). A national primary drinking water regulation promulgated by the US Environmental Protection Agency to regulate drinking water disinfectants and by-products of disinfection.

Disinfection. (1) The process of destroying or inactivating pathogenic organisms (bacteria, viruses, fungi, and protozoa) by either chemical or physical means. (2) In water treatment, the process in which water is exposed to a chemical disinfectant—chlorine (HOCl, OCl^-), chloramines ($NHCl_2$ or NH_2Cl), chlorine dioxide (ClO_2), iodine, or ozone (O_3)—for a specified time period to kill pathogenic organisms.

Disinfection Byproduct Formation Potential. The amount of disinfection byproducts (DBPs) formed during a test in which a source or treated water is dosed with a relatively high amount of disinfectant (normally chlorine—HOCl, OCl^-) and is incubated (stored) under conditions that maximize DBP production (e.g., neutral to alkaline pH, warm water temperature, contact time of 4 to 7 days). This value is not a measure of the amount of DBPs that would form under normal drinking water treatment conditions, but rather an indirect measure of the amount of DBP precursors in a sample. If a water has a measurable level of DBPs prior to the formation potential test (e.g., in a prechlorinated sample), then the formation potential equals the terminal value measured at the end of the test minus the initial value.

Disinfection Byproduct Precursor. A substance that can be converted into a disinfection byproduct during disinfection. Typically, most of these precursors are constituents of natural organic matter. In addition, the bromide ion (Br^-) is a precursor material.

Disinfection Byproducts (DBPs). A chemical byproduct of the disinfection process. Disinfection byproducts are formed by the reaction of the disinfectant, natural organic matter, and the bromide ion (Br^-). Some disinfection byproducts are formed through halogen (e.g., chlorine or bromine) substitution reactions; i.e., halogen-substituted byproducts are produced. Other disinfection byproducts are oxidation byproducts of natural organic matter (e.g., aldehydes—RCHO).Concentrations are typically in the μg/L or ng/L range.

Dissolved Oxygen (DO). The concentration of oxygen in aqueous solution, which is often expressed in units of mg/L. It is usually determined by one of two methods: a dissolved oxygen probe or Winkler titration.

Dissolved Carbon Dioxide. The carbon dioxide (CO_2) that is dissolved in a liquid medium, typically expressed in mg/L. The saturation concentration is dependent on several factors, including partial pressure, temperature, and pH.

Dissolved Inorganic Carbon (DIC). The fraction of inorganic carbon (the carbonate (CO_3^{2-}), bicarbonate (HCO_3^-), and dissolved carbon dioxide (CO_2) in water that passes through a 0.45-micrometer pore-diameter filter.

Dissolved Organic Carbon (DOC). That portion of the organic carbon in water that passes through a 0.45-micrometer pore diameter filter. For most drinking water sources, the dissolved organic carbon fraction represents a very high percentage of the total organic carbon pool. It is composed of individual compounds as well as nonspecific humic material, although humic substances account for a large portion of dissolved organic matter in natural waters. Typically, the dissolved organic carbon level provides some indication of the amount of disinfection byproduct precursors in a water source. After filtration, dissolved organic carbon is determined in the same manner as total organic carbon. Organic carbon concentrations should be reported as dissolved organic carbon only if the sample has been filtered through a 0.45 micrometer pore diameter filter before analysis.

Distribution System. A system of conduits (laterals, distributaries, pipes, and their appurtenances)by which a primary water supply is distributed to consumers. The term applies particularly to the network of pipelines in the streets in a domestic water system.

Diurnal Demand. (1) Occurring during a 24-hour period, as in diurnal variation. (2) Occurring during the daytime (as opposed to nighttime).

Dose-Response Assessment. A component of risk assessment that describes the quantitative relationship between the amount of exposure to a substance and the extent of toxic injury or disease.

Downstream. The direction toward which a stream flows.

Drinking Water Advisory. A public notice issued by a federal, state, or local health department via broadcast media or through any other acceptable medium that informs users of a public water system that their drinking water is, or potentially is, unsafe and that specific actions (such boiling) are required before using the drinking water for consumption or other regular use. An advisory may also be issued during times of drought, in which drinking water may need to be conserved and restricted for use.

Ductile Iron Pipe (DIP). A water pressure pipe made from a material similar to cast iron, with the primary graphite in a nodular or spheroidal form as a result of the addition of an inoculant such as magnesium to molten iron of appropriate composition. These structural changes increase strength and flexibility.

E. coli (Escherichia coli). A gram-negative, facultatively anaerobic, nonspore-forming bacillus commonly found in the intestinal tracts of humans and other warm-blooded animals. In sanitary bacteriology, *Escherichia coli* is considered the primary indicator of recent fecal pollution.

Easement. An acquired legal right to the use of land owned by others. An easement is ordinarily more restricted in scope than a leasehold.

Effective Storage. The volume of water available for a designated purpose.

Electrolysis. Chemical changes caused in an electrolyte by an electrical current. The US Environmental Protection Agency discourages the use of this word to mean "corrosion by stray currents."

Elevated Storage Tank. A water storage facility located on and supported by a tower constructed at an elevation to provide useful storage and pressure for a water pressure plane, thereby allowing for controlled pumping rates, water for peak usage and fire protection, and minimized service interruptions during equipment maintenance requirements.

Emergency Action Plan (EAP). A plan developed by a water utility that defines actions that will be taken in case of an emergency.

Enhanced Coagulation. The addition of excess coagulant for improved removal of disinfection byproduct precursors by conventional coagulation–sedimentation–filtration treatment. In the Disinfectant/Disinfection Byproducts Rule, the removal of total organic carbon is used as a performance indicator for the removal of disinfection byproduct precursors. The Disinfectant/Disinfection Byproducts Rule requires no conversion to optimized coagulation practices, but rather enhancement of an existing process to remove specified levels of total organic carbon based on influent water quality. Enhanced coagulation can also be used to remove arsenic during the coagulation process.

Enteric. Of intestinal origin, especially with respect to wastes or bacteria.

Enteric Virus. Any virus that inhabits the alimentary and gastrointestinal tracts. Many such viruses are stable in the feces and wastewater and can be transmitted through contaminated water supplies.

Entry Point. A place where water enters the distribution system. Frequently, water is sampled at such points.

Equalization Storage. The quantity of water from storage that enters the system during peak demand periods to augment other sources of supply. Equalization storage is replenished during off-peak demand periods.

Erosion Corrosion. Deterioration of a surface by the abrasive action of moving fluids. This phenomenon is accelerated by the presence of solid particles or gas bubbles in suspension.

Escherichia coli (E. coli). A gram-negative, facultatively anaerobic, nonspore-forming bacillus commonly found in the intestinal tracts of humans and other warm-blooded animals. In sanitary bacteriology, *Escherichia coli* is considered the primary indicator of recent fecal pollution.

Exposure. (1) The condition of being subject to legal liability. (2) Contact with a chemical, physical, or microbial agent. Being precise when exposure to various agents is discussed is important because different routes of exposure can have greatly different health impacts. Exposure can best be expressed in terms of the agent's free concentration at the site in the body where the agent exerts its toxic effect. This is preferred because many factors modify the delivery of an external exposure to a chemical or microbial agent to the active site. Practical considerations frequently require exposure to be expressed in terms of units of agent administered per unit of body weight or unit of body surface area.

Exposure Assessment. The determination or estimation (in qualitative or quantitative terms) of the magnitude, frequency, duration, route, and extent (number of people) of exposure to a chemical.

Extended Period Simulations (EPS). A pipe network simulation comprising a series of steady-state runs that allow the user to view system operations over a period of time with defined intervals. EPS can model time-varying features, such as tank levels, head- or pressure-controlled valves, and the status of pumps and pipes.

Fecal Coliform. Members of the total coliform group of bacteria that are characterized by their ability to ferment lactose at 112.1° Fahrenheit (44.5° Celsius) and that are considered more specific indicators of fecal contamination than are coliforms that ferment lactose only at 95° Fahrenheit (35° Celsius). *Escherichia coli* and some *Klebsiella pneumoniae* strains are the principal fecal coliforms.

Fecal Contamination. Contamination of soil or water by feces from warm-blooded animals.

Ferric Chloride ($FeCl_3$). An iron salt used as a coagulant in water treatment. The iron has a valence of +3.

Ferric Iron (Fe^{+3}). Small solid iron particles containing trivalent iron, usually as gelatinous ferric hydroxide ($Fe(OH)_3$) or ferric oxide (Fe_2O_3), which are suspended in water and visible as "rusty water." Ferric iron can normally be removed by filtration. It is also called precipitated iron or iron +3.

Ferrous Chloride ($FeCl_2$). A soluble iron salt.

Ferrous Iron (Fe^{+2}). A reduced form of iron. When ferrous salts are added to water, the ferrous iron can act as a reducing agent. Ferrous iron is also referred to as iron +2.

Ferrous Sulfate ($Fe(SO_4)$). An iron salt used as a coagulant in water treatment. The iron has a valence of +2.

Fertilizer. A substance (manure or chemical mixture) used to make soil more fertile.

Filtration. The removal of suspended materials in a fluid stream by passage of the fluid through a filter medium.

Fire Flow. The rate of flow, usually expressed in gallons or liters per minute, which can be delivered from a water distribution system at a specified residual pressure for firefighting. When delivery is to fire department pumpers, the specified residual pressure is generally 20 pounds per square inch (138 kilopascals).

Fire Hydrant. A specifically designed fitting that serves as a connection point on a distribution system to provide water for fire protection directly through a fire hose or to a fire pumper.

Firm Capacity. Maximum day demand available with the largest pump out of service.

Flocculation. The water treatment process following coagulation that uses gentle stirring to bring suspended particles together so they will form larger, more settleable clumps called floc.

Flow Control. An in-line device, orifice fitting, or method implemented in a treatment device's piping that will regulate and control the flow of water or regenerant over a broad range of inlet water pressures. Some types are manually adjustable.

Flow Measurement. A measurement of the quantity of water flowing through a given point in a given amount of time, such as gallons or cubic meters per day.

Flushing. The act of running water through a distribution system or water main in order to remove debris, discolored water, or chemical solutions in order to clean the line or system.

Free Chlorine. The amount of chlorine available as dissolved gas (Cl_2), hypochlorous acid (HOCl), and hypochlorite ion (OCl^-), that is not combined with ammonia (NH_3) or other compounds in water.

Free Chlorine Residual. The free chlorine residual formed once all the chlorine demand has been satisfied. The chlorine is not combined with other constituents in the water.

Friction Loss. The head lost by water flowing in a stream or conduit as a result of the disturbances set up by the contact between the moving water and its containing conduit, as well as by intermolecular friction. In laminar flow, the head loss is approximately proportional to the first power of the velocity. In turbulent flow, friction head loss is proportional to a higher power, approximately the square of the velocity. Although, strictly speaking, head losses such as those caused by bends, expansions, obstructions, and impacts are not included in this term, the usual practice is to include all such head losses under this term. Friction loss is also called friction head or friction head loss.

Galvanic Corrosion. Corrosion that is increased because of the current caused by a galvanic cell. It is sometimes called couple action.

Gas Chromatography (GC). A technique commonly used in the analysis of organic compounds in water. Inmost cases, an extract of a water sample is injected into a gas chromatograph. Analytes are volatilized in an injector port and migrate as a gas through a chromatography column. The speed with which the analytes migrate depends on the relative affinity of the analyte for the stationary phase in the chromatography column. Compounds are identified based on their retention time in the column.

Gastrointestinal Illness. An illness affecting the gastrointestinal system. Such an illness is often caused by microbial infection, but it may also result from non-microbial causes.

Gauge Pressure. As measured by a gauge, the water pressure that is in excess of atmospheric pressure at the referenced elevation. Gauge pressure is not the total pressure. Total water pressure (absolute pressure) also includes the atmospheric pressure exerted on the water. Because atmospheric pressure is exerted everywhere, however (against the outside of the main as well as the inside, for example), it is generally not written into water system calculations. Therefore, the zero basis for gauge pressure is atmospheric pressure. Gauge pressure is reported in pounds per square inch or kilopascals.

Generator. A piece of equipment used to transform rotary motion (e.g., the output of a diesel engine) into electric current.

Geographic Information System (GIS). A computer system that stores and links nongraphic attributes and geographically referenced data with graphic map features to allow a wide range of information processing and display operations, as well as map production, analysis, and modeling.

Giardia. The genus name for a group of single-celled, flagellated, pathogenic protozoans found in a variety of vertebrates, including mammals, birds, and reptiles. These organisms exist either as trophozoites or as cysts, depending on the stage of the life cycle. The species of Giardia that is a common cause of human diarrheal disease Giardia lamblia. Protozoan infections, principally of the upper small intestine, with Giardia lamblia cause giardiasis. The median incubation period is 7 to 10 days, and the infectious dose is low—10 cysts have caused infection. Although infected persons are usually asymptomatic, infection may be associated with a variety of intestinal symptoms, such as abdominal cramps, chronic diarrhea, frequent loose and greasy stools, bloating, flatulence, fatigue, and weight loss. Malabsorption of fats or fat-soluble vitamins may occur. Occurrence is worldwide, with children found to be more frequently infected than adults. Person-to-person transmission occurs by hand-to-mouth transfer of cysts from the feces of an infected person, especially in institutions and daycare centers. Outbreaks often occur from ingestion of cysts in fecally contaminated water, especially unfiltered surface water, and less frequently from contaminated food. Reservoirs of infection include humans and wild and domestic animals.

Granular Activated Carbon (GAC). A form of particulate carbon manufactured with increased surface area per unit mass to enhance the adsorption of soluble contaminants. Granular activated carbon is used in fixed-bed contactors in water treatment and is removed and regenerated (reactivated) when the adsorption capacity is exhausted. In some applications, granular activated carbon can be used to support a biological population for stabilizing biodegradable organic material.

Gravimetry. Gravimetry includes all analytical methods in which the analytical signal is a measurement of mass or a change in mass.

Ground Water Rule (GWR). Ground Water Rule (GWR) was issued by EPA to improve drinking water quality and provide protection from disease-causing microorganisms. Water systems that have ground water sources may be susceptible to fecal contamination. In many cases, fecal contamination can contain disease causing pathogens. The purpose of the GWR is to reduce disease incidence associated with harmful microorganisms in drinking water. The GWR applies to public water systems that use ground water as a source of drinking water. The rule also applies to any system that delivers surface and ground water to consumers where the ground water is added to the distribution system without treatment. The GWR was published in the *Federal Register* on November 8, 2006. If significant deficiencies are identified or if the initial source sample or one of the five additional ground water source samples test positive for fecal contamination, ground water systems must take corrective actions.

Ground Water Under the Direct Influence of Surface Water (GWUDI). Water defined by EPA in the Surface Water Treatment Rule as any water beneath the surface of the ground that has (1) significant occurrence of insects or other macroorganisms, algae, organic debris, or large diameter pathogens such as Giardia lamblia, or (2) significant and relatively rapid shifts in water characteristics—such as turbidity, temperature, conductivity, or pH— that closely correlate with climatological or surface water conditions. The Enhanced Surface Water Treatment Rule amends the first item of this definition to include Cryptosporidium.

Groundwater Systems. Any water that is derived from beneath the Earth's surface. Groundwater issued by almost 50% of the public water supply systems in the United States; agriculture in the Midwestern United States also depends heavily on the use of groundwater for irrigation. Groundwaters generally have higher dissolved mineral concentrations than surface waters because of intimate contact between the CO_2-bearing water and rocks and soils in the ground and the length of time for dissolution. Additionally, CO_2 may be added to the water in the soil by the activities of soil microorganisms. Groundwaters usually require softening to make them acceptable for general domestic use and most industrial uses.

Groundwater Table. (1) The upper surface of a body of unconfined groundwater. (2) The elevation of depth below the ground surface of such a water surface. In a confined aquifer, this elevation is defined by the static levels in wells that draw from the aquifer.

Haloacetic Acid (HAA). (CX_3COOH, where X = Cl, Br, H in various combinations) A class of disinfection by-products formed primarily during the chlorination of water containing natural organic matter. When bromide (Br^-) is present, a total of nine chlorine-, bromine-and-chlorine-, or bromine-substituted species may be formed. Trihalomethanes and haloacetic acids are the two most prevalent classes of by-products formed during chlorination; and subject to regulation under the Disinfectant/ Disinfection By-Products Rule.

Halogen. Any of the chemical elements composing group VIIA of the periodic table: fluorine, chlorine, bromine, iodine, and astatine.

Hard Water. Generally, water that requires considerable amounts of soap to produce a foam or lather and that also produces scale in hot water pipes, heaters, boilers, and other units in which the temperature of water is increased materially. With respect to hardness, waters have been classified as follows: 0–25 milligrams per liter as calcium carbonate ($CaCO_3$), very soft; 25–75 milligrams per liter as $CaCO_3$, soft; 75–150 milligrams per liter as $CaCO_3$, moderately hard; 150–300 milligrams per liter as $CaCO_3$, hard; 300 milligrams per liter as $CaCO_3$ and up, very hard.

Hardness. A quality of water caused by divalent metallic cations and resulting in increased consumption of soap, deposition of scale in boilers, damage in some industrial processes, and sometimes objectionable taste. The principal hardness-causing cations are calcium, magnesium, strontium, ferrous iron, and manganese ions. Hardness may be determined by a standard laboratory titration procedure or computed from the amounts of calcium and magnesium expressed as equivalent calcium carbonate ($CaCO_3$).

Hazard Analysis Critical Control Point (HACCP). A science-based methodology and systematic system that identifies specific hazards and measures for their control to ensure safety and prevent harm to people.

Hazardous Substance. Any material that is dangerous and poses a threat to public health or safety.

Head. The energy per unit weight of a liquid. In practical terms, head is the pressure at any given point in a water system. It may also be called pressure head or velocity head. Head is calculated as the pressure exerted by a hypothetical column of water standing at the height to which the free surface of water would rise above any point in a hydraulic system. Head is often measured in pounds per square inch or kilopascals. Head is sometimes expressed as the height of a column of water in feet or meters that would produce the corresponding pressure; this measurement may be called hydrostatic head.

Head Loss. A reduction of water pressure (head) in a hydraulic or plumbing system. Head loss is a measure of (1) the resistance of a medium bed (or other water treatment system), a plumbing system, or both to the flow of the water through it, or (2) the amount of energy used by water in moving from one location to another. In water treatment technology, head loss is basically the same as pressure drop.

Health Effects. Modifications in the health of an individual as a result of exposure to a chemical, physical, or infectious agent. The meaning of this term is vague, but it is usually thought of in terms of adverse effects on health rather than beneficial effects. Even with this additional specificity, what is meant by the term is not very clear. The opinions of what constitutes an adverse health effect vary widely. Some professionals include small decreases in enzyme activities, whereas others require that clear-cut disease and observable physical pathology be present.

Health Reference Level (HRL). Risk derived concentrations against which to evaluate the occurrence data to determine if contaminants may occur at levels of public health concern. They are not the level of a contaminant in drinking water that must not be exceeded to protect any particular population (i.e., an HRL is not an MCL). There are two different approaches to the derivation of an HRL. One approach is used for chemicals that cause cancer and exhibit a linear response to dose. The other applies to non-carcinogens and carcinogens evaluated using a non-linear approach.

Hedonic Scale. A method used in which the stimuli are presented singly and are rated on a scale where the 9 categories range from "dislike extremely" to "like extremely."

Henry's Law. A relationship expressing the directly proportional relationship between the liquid-phase and gas-phase concentration of a substance at equilibrium and at a given temperature. Henry's law can be expressed in many forms. This expression is typically used in water treatment to evaluate the potential or extent to which (1) a contaminant will volatilize from the liquid phase to the gaseous phase or (2) a gas will dissolve into the liquid phase.

Heterotroph. An organism that metabolizes organic compounds as sources of carbon and energy.

Heterotrophic Plate Count (HPC). A bacterial enumeration procedure used to estimate bacterial density in an environmental sample, generally water. Other names for the procedure include total plate count, standard plate count, plate count, and aerobic plate count.

Humic Acid. A brown, polymeric constituent of soils, lignite, and peat. It is not a well- defined compound; rather, it is a mixture of polymers containing aromatic and heterocyclic structures, carboxyl groups, and nitrogen. Aquatic humic acids are major precursors of disinfection by-products.

Hydraulic Grade Line (HGL). A line (hydraulic profile) indicating the piezometric level of water at all points along a conduit, open channel, or stream. In a closed conduit under pressure, artesian aquifer, or groundwater basin, the line would join the elevations to which water would rise in pipes freely vented and under atmospheric pressure. In pipes under pressure, each point on the hydraulic profile is an elevation expressed as the sum of the height associated with the pipe elevation, the pipe pressure, and the velocity of the water in the pipe. In an open channel, the hydraulic grade line is the free water surface. Hydraulic profiles are commonly used to establish elevations through the processes that make up a treatment facility under maximum and minimum flow conditions.

Hydraulic Model. A flow system operated so that the characteristics of another similar system may be predicted. A model is generally a small-scale reproduction of the prototype, but it may be larger or geometrically distorted. A hydraulic model is used in studies of spillways, stilling basins, flood regulation, or river beds. It may also be called a hydraulic similitude model.

Hydraulic Residence Time (HRT). The theoretical amount of time required for a given flow to pass through a unit process. It is calculated by dividing the volume of a unit process by the applied flow rate. It is also called the hydraulic residence time.

Hydrogen Peroxide (H_2O_2). Strong oxidizing agent used for bleaching and, to some extent, as a disinfectant. Ozone (O_3) in combination with hydrogen peroxide (i.e., PEROXONE), or ultraviolet light in combination with hydrogen peroxide, is used in some advanced oxidation processes to produce hydroxyl radicals (^{-}OH).

Hydroxide. The ion formed by an oxygen and a hydrogen atom. This term is used to describe the anionic hydroxyl radical that is partially responsible for the alkalinity of a solution.

Hypobromite (OBr^-). The ionized form of hypobromous acid (HOBr). Ozonation of hypobromite ion produces bromate (BrO_3^-).

Hypobromous Acid (HOBr). An acid used as a bactericide. During the disinfection of drinking water with chlorine (HOCl) or ozone (O_3), bromide ion (Br^-) is oxidized to hypobromous acid. Hypobromous acid reacts with natural organic matter to form bromine- substituted disinfection by-products.

Hypochlorite (OCl-). The ionized form of hypochlorous acid (HOCl).

Indicator Microorganisms. An organism that can be detected with relative ease and specificity and whose presence (or absence) has been shown to be correlated with a specific condition of interest. The condition of interest may be as specific as the potable quality of treated and disinfected drinking water or as broad as the assessment of the general level of pollution in an ecological system. In the latter case, several indicator organisms may be used.

Inductively Coupled Plasma (ICP) Spectroscopy. An atomic emission technique that is used in the analysis of metals. Modern inductively coupled plasma instruments have four main components: an inductively coupled plasma source, a spectrometer, sample introduction hardware, and a computer. The inductively coupled plasma source is composed of an argon plasma confined in a radio frequency induction coil. The technique is sensitive for many metals and can allow for the rapid, even simultaneous, determination of metals.

Inorganic Compound. A substance that is not derived from hydrocarbons. Examples include sodium chloride (NaCl), ferric hydroxide ($Fe(OH)_3$), and aluminum sulfate ($Al_2(SO_4)_3$).

Interconnection. A physical connection between two water supply systems.

Iodine (I). A nonmetallic element that is the heaviest and least reactive of the naturally occurring halogens. It may be used for disinfection. In both its liquid and vapor forms, iodine is readily adsorbed by activated carbon.

Iron (Fe). An abundant element (Fe) found naturally in the earth. Dissolved iron is found in most water supplies. An iron concentration higher than 0.3 milligrams per liter causes red stains on plumbing fixtures and on other items in contact with the water. Dissolved iron can also be present in water as a result of corrosion of cast-iron or steel pipes, a frequent cause of red-water problems.

Irrigation System. A group of components in which there is the artificial application of water to land to meet the water needs of growing plants not met by rainfall.

Isolation Valve. A valve that is used in a water system to stop flow or isolate a portion of the water system until it is desirable to achieve flow downstream of the valve. The basic design requirement of isolation valves is to offer minimum resistance to flow in the fully opened portion and to exhibit tight shut-off characteristics when fully closed. A valve is classified as an isolation valve because of its intended function in a system, not because of the architecture of the valve itself. Therefore, many different types of valves can be classified as isolation valves. Gate, globe, ball, butterfly, plug, and diaphragm valves satisfy the above requirements in varying degrees and, therefore, are widely used in shut-off service. The actual type of valve selected is dictated by several parameters, including pressure drop, seat leakage, fluid properties, system leakage, actuation requirements, initial cost, and maintenance.

Ketoacids. A class of organic compounds that have both a ketone (R1–CO–R2) and carboxylic acid (COOH) functional group in the chemical structure. Some ketoacids are created during the reactions of oxidants used as disinfectants, particularly ozone (O_3), with natural organic matter.

Ketones. A class of organic compounds in which the carbonyl group (C=O) is attached to two alkyl (hydrocarbon) groups. In industry, ketones are used primarily as solvents. Some ketones are created during the reactions of oxidants used as disinfectants, particularly ozone (O_3), with natural organic matter.

Langelier Saturation Index (LSI). The most famous of the calcium carbonate ($CaCO_3$) saturation indexes, the formula for the Langelier saturation index is based on a comparison of the measured pH of a water (pHa) with the pH the water would have (pHs) if at saturation with $CaCO_3$ (calcite form) given the same calcium hardness and alkalinity for both pH cases. Many of the other indexes found in the water treatment and corrosion literature are merely approximations. Several versions of the LSI exist in the literature. The version presented here is the one developed by J.R. Rossum and D.T. Merrill for the quadratic equation form of the pHs equation solution; This material has been updated using newer equilibrium constant and activity coefficient data. (Note that a simplified approximation equation originally formulated by T.E. Larson and A.M. Buswell was presented in the National Interim Primary Drinking Water Regulations amendments.) The basic formula for the Langelier saturation index is: LSI = pHa − pHs. It is interpreted in the following way: If LSI > 0, water is potentially scale forming (supersaturated with respect to $CaCO_3$). If LSI = 0, water is in equilibrium with $CaCO_3$, not tending to dissolve or precipitate it. If LSI < 0, water will potentially dissolve existing $CaCO_3$ deposits (undersaturated with respect to $CaCO_3$). The quantity pHs is found from the computed hydrogen ion concentration at hypothetical saturation ($[H^+]_s$) by $pH_s = -\log_{10}([H+]s) - \log_{10} f_m$

Where:

f_m = the activity coefficient for monovalent ions.

The parameter [H+]s is computed by the following quadratic equation:
[H+]s = (−B±√(B2^2-4AC))/2A

Where:

A = $1 - [Ca^{2+}]K_2/K_s$

B = $K_2(2 - [Ca^{2+}]Alk/K_2)$

C = KwK_2/K_s $[Ca^{2+}][Ca^{2+}]$ = calcium concentration, in milligrams per liter as $CaCO_3$

Alk = alkalinity, in milligrams per liter as $CaCO_3$

K_2 = solubility constant

K_s = equilibrium constant

$K_w = [H^+][OH^-]$

Lead (Pb). A metallic element that has had various industrial uses, including in a gasoline additive, in solder and fusible alloys, and in radiation shielding. Lead is a toxic metal that can be harmful to human health even at low exposure levels.

Lead and Copper Rule (LCR). A rule promulgated by the US Environmental Protection Agency on June 7, 1991 (*Federal Register*, 56(110):26460–26564) that set National Primary Drinking Water Regulations for lead and copper.

Lead Service Line (LSL). A water line constructed of lead material that was used for the small water line running from the water main to the customer cutoff at the meter box or to the cutoff for the building served. Lead service lines are no longer permitted.

Lead-free. Meeting the conditions defined in Safe Drinking Water Act, Section 1417 for solder, flux, pipes, and fittings. When enacted, to be considered lead-free, solder and flux could contain no more than 0.2 percent lead; pipes and fittings no more than 8.0 percent. This was the definition of "lead-free" until January 4, 2011, when the Reduction of Lead in Drinking Water Act was enacted to reduce the amount of lead in contact with drinking water by lowering the maximum lead content of plumbing materials, including any pipes, fittings, fixtures, solders or flux, to 0.25 percent by weight.

Leak. An unintended hole or crack through which water escapes as from a pipe.

Leak Detection. The precise locating of underground water leaks in a water system by the use of sounding or sonic devices where leakage is known or suspected.

Leakage. (1) The uncontrolled loss of water from artificial structures as a result of hydrostatic pressure. (2) The uncontrolled loss of water from one aquifer to another. The leakage may be natural, as through a semi-impervious confining layer, or artificial, as through an uncased well. (3) The presence in treated effluent of the type of ions that the ion exchange process was supposed to remove. Ion removal may be incomplete because of incomplete resin regeneration, excessive service rates, low temperatures, high concentrations of sodium, interfering total dissolved solids in the water being treated, or other factors. This type of leakage is also referred to as slippage.

Lime Softening. The process of removing water hardness by adding lime to precipitate solids composed of metal carbonates and hydroxides. Clarification may or may not also occur.

Limestone Contactors. A treatment device consisting of a bed of limestone through which water is passed to dissolve calcium carbonate ($CaCO_3$). The addition of calcium carbonate to the water decreases corrosivity by increasing the pH, calcium concentration, and alkalinity of the water.

Localized Corrosion. Corrosion caused by action of local cells, i.e., galvanic cells arising because of non-uniformities between two adjacent areas at a metal surface exposed to an electrolyte.

Locational Running Annual Average (LRAA). The average of any four consecutive quarterly averages of monitoring data at any particular sampling location. The US Environmental Protection Agency Stage 2 Disinfectants and Disinfection By-Products Rule sets limits on locational running annual averages for total trihalomethanes and sum of five haloacetic acids at monitoring sites that are most likely to have high disinfection by-product (DBP) concentrations.

Log Removal. A shorthand term for $\log_{10}$ removal, used in reference to the Surface Water Treatment Rule (SWTR) and the physical-chemical treatment of water to remove, kill, or inactivate pathogenic organisms such as Giardia lamblia and viruses. A 1-log removal equals a 90 percent reduction in density of the target organism; a 2-log removal equals a 99 percent reduction; a 3-log removal equals a 99.9 percent reduction; and so on.

Long Term 1 Enhanced Surface Water Treatment Rule (LT1ESWTR). A rule promulgated January 14, 2002, by US Environmental Protection Agency (67 FR 1812-1844) that requires small public water systems using surface water to meet strengthened filtration requirements. The LT1ESWTR requires systems to calculated levels of microbial inactivation to ensure that microbial protection is not jeopardized if systems make changes to comply with the requirements of the Stage 1 Disinfectants and Disinfection By-Products Rule (Stage 1 DBPR). This rule, which addresses sub-part H systems serving fewer than 10,000 persons, is similar to the Interim Enhanced Surface Water Treatment Rule (IESWTR) established for larger systems.

Long Term 2 Enhanced Surface Water Treatment Rule (LT2ESWTR). A rule promulgated January 5, 2006, by the US Environmental Protection Agency (71 FR 653-702) to reduce the possibility of disease incidence associated with Cryptosporidium and other pathogenic microorganisms in drinking water. The LT2ESWTR supplements existing regulations by targeting additional Cryptosporidium treatment requirements to higher-risk systems. This regulation also contains provisions to mitigate risks from uncovered finished water storage facilities and to ensure that systems maintain microbial protection as they take steps to reduce the formation of disinfection by- products. The LT2ESWTR applies to all systems that use surface water or groundwater under the direct influence of surface water.

Magnesium (Mg). One of the elements that make up the Earth's crust as components of many rock-forming minerals, such as dolomite. Magnesium and calcium dissolved in water constitute hardness. The presence of magnesium in water contributes to the formation of scale and the insoluble soap curds that typify hard water.

Manganese (Mn). An abundant element found naturally in the earth. Dissolved manganese is found in many water supplies. At concentrations in excess of 0.05 milligrams per liter, it causes black stains to plumbing fixtures, laundry, and other items in contact with the water.

Manganese Dioxide (MnO_2). A dark brown or gray-black insoluble compound found in nature as pyrolusite or made synthetically. It is used as an oxidizing agent in water treatment and as a starting material for such permanganate compounds as potassium permanganate ($KMnO_4$).

Mass Spectrometry (MS). A method of chemical analysis in which compounds emerging from a gas chromatograph are fragmented and ionized by bombardment with a beam of electrons. An electromagnetic field separates the ions according to their individual mass-to-charge ratios into a characteristic mass spectrum for each molecule. An analog computer analyzes the spectra and makes it possible to identify molecules even in cases of poor separation on the chromatography column; hence the advantage of mass spectrometry compared to selective chromatograph detectors.

Maximum Contaminant Level (MCL). A value defined under the Safe Drinking Water Act Section 1401(3) as the maximum permissible level (concentration) of a contaminant in water delivered to any user of a public water system. Maximum contaminant levels are the legally enforced standards in the United States. There are primary and secondary maximum contaminant levels. Primary maximum contaminant levels are the enforceable maximum contaminant levels for various water quality indicators that have been established to protect public health. Secondary maximum contaminant level (SMCL) are a nonenforceable numerical limit set by the US Environmental Protection Agency for a contaminant on the basis of aesthetic effects to prevent an undesirable taste, odor, or appearance that would have an adverse impact on the public welfare.

Maximum Contaminant Level Goal (MCLG). A federally nonenforceable, health-based goal established by the US Environmental Protection Agency under the Safe Drinking Water Act Section 1412(b)(4) for each contaminant regulated by a national primary drinking water regulation. The maximum contaminant level goal is set at the level at which no known or anticipated adverse effects on human health occur and for which an adequate margin of safety exists.

Maximum Day Demand (MDD). The volume of water consumption used on the highest-consumption day in a year.

Maximum Residual Disinfectant Level (MRDL). A level of a disinfectant added for water treatment that may not be exceeded at the consumer's tap without an unacceptable possibility of adverse health effects. For chlorine (HOCl) and chloramines ($NHxCly$, where $x = 0–2$, $y = 1–3$), a public water system will be in compliance with the federally proposed maximum residual disinfectant level when the running annual average of monthly averages of samples taken from the distribution system, computed quarterly, is less than or equal to the maximum residual disinfectant level. (That is, every 3 months, the average of the previous 12 monthly averages is assessed.) For chlorine dioxide (ClO_2), a public water system will be in compliance when no two consecutive daily samples taken at the entrance to the distribution system exceed the maximum residual disinfectant level. Maximum residual disinfectant levels will be enforceable in the same manner as maximum contaminant levels under Section 1412 of the Safe Drinking Water Act. Convincing evidence exists that addition of a disinfectant is necessary to control waterborne microbial contaminants. Notwithstanding the maximum residual disinfectant levels listed in Section 141.65 of the Disinfectant/Disinfection By-Products Rule (*Federal Register,*

July 29, 1994), operators will be able to increase residual disinfectant levels of chlorine or chloramines (but not chlorine dioxide) in the distribution system to a level and for a time necessary to protect public health to address specific microbiological contamination problems caused by such circumstances as distribution system line breaks, storm runoff events, source water contamination, or cross-connections.

Maximum Residual Disinfectant Level Goal (MRDLG). A maximum level of a disinfectant added for water treatment and for which no known or anticipated adverse effect on human health would occur, allowing for an adequate margin of safety. Maximum residual disinfectant level goals are nonenforceable health goals and do not reflect the benefits of adding the chemical for the control of waterborne microbial contaminants.

Membrane Filtration. Filtration using a microfiltration or ultrafiltration membrane.

Membrane Process. A treatment technique using a membrane technology.

Metal Release. Products of corrosion that do not remain attached to pipe walls but are released from the pipe to enter the water contacting the pipe. The forms of released metal may be dissolved (ionic), colloidal, particulate, or combinations of all three.

Microbe. A microscopic organism, either plant or animal, invisible to the naked eye. Examples are algae, bacteria, fungi, protozoa, and viruses.

Microbiological Regrowth. The presence of a persistent population of bacteria in a water distribution system. Such regrowth may lead to problems in the compliance monitoring of total coliform bacteria.

Microbiologically Influenced Corrosion (MIC). Corrosion that is initiated or accelerated by the presence of microbes and their metabolites. Sulfate-reducing bacteria are commonly associated with microbially influenced corrosion.

Monobromoacetic Acid (MBAA) ($CH_2BrCOOH$). A haloacetic acid containing one bromine atom. It is formed during the chlorination of waters containing moderate amounts of bromide (Br^-), and it may also be formed during the ozonation of bromide-containing waters.

Monochloramine (NH_2Cl). A chloramine species produced from the mixing of chlorine (HOCl) and ammonia (NH_3). Typically, monochloramine and a small percentage of dichloramine ($NHCl_2$) are formed. Monochloramine is used as a disinfectant, especially for distribution system residual maintenance.

Monochloroacetic Acid (MCAA) ($CH_2ClCOOH$). A haloacetic acid containing one chlorine atom. Typically, dichlororoacetic acid ($CHCl_2COOH$) and trichloroacetic acid (CCl_3COOH), rather than monochloroacetic acid, are the principal haloacetic acids formed during chlorination of waters containing low amounts of bromide (Br^-).

Monte Carlo Analysis. A method of statistical analysis that involves selecting sample data in proportion to their true or estimated probability of occurrence through a random number generator and using that sample as input to some process (i.e., model) in order to estimate the probability of a particular output from the process. Although Monte Carlo analysis can be applied to a single variable, it is more commonly used to define the attributes of joint probability distributions and multivariate probability distributions. For example, one might wish to define the probability distribution (density function) of total water demand 20 years in the future as it is derived from a functional relationship of population forecasts, current and planned conservation programs, planned changes in water prices, and historical water use per capita. With estimated or known probability distributions for each of these input variables and an equation that links the separate input variables to total water demand, Monte Carlo computer programs are available (for personal computers and mainframes) that will combine the effects of the separate probability distributions into a single probability distribution around the forecast of water demand. This procedure is accomplished by taking hundreds (or thousands) of iterations of sample selections from each of the input variables and then combining them through the functional relationship until a combined distribution takes shape for the projected variable.

Nanofiltration (NF). A pressure-driven membrane separation process that generally removes substances in the nanometer size range. Its separation capability is controlled by the diffusion rate of solutes through a membrane barrier and by sieving and is dependent on the membrane type. In potable water treatment, nanofiltration is typically used to remove nonvolatile organics larger than the 200–500-dalton molecular weight cutoff (e.g., natural and synthetic organics, color, disinfection by-product precursors) and multivalent inorganics (for softening).

National Primary Drinking Water Regulation (NPDWR). A legally enforceable regulation that applies to public water systems as defined in Section 1401 of the Safe Drinking Water Act. It is also referred to as a primary standard. Primary standards protect drinking water quality by limiting the levels of specific contaminants that can adversely affect public health and are known or anticipated to occur in water. The regulations (1) apply to public water systems, (2) specify contaminants that may have any adverse effect on the health of persons, (3) specify for each contaminant a maximum contaminant level or treatment technique, and (4) contain criteria and procedures to ensure compliance.

Natural Organic Matter (NOM). A heterogeneous mixture of organic matter that occurs ubiquitously in both surface water and groundwater, although its magnitude and character differ from source to source. Natural organic matter contributes to the color of a water, and it functions as disinfection by-product precursors in the presence of such disinfectants as chlorine (HOCl). Humic substances (e.g., fulvic acid) represent a significant fraction of natural organic matter in surface water sources.

Nebulized. A liquid reduced to a fine spray. A device that does this is called a nebulizer. For example, aqueous samples can be introduced into the flame of an atomic absorption spectrophotometer via a nebulizer.

Negative Pressure. A gauge pressure less than the existing atmospheric pressure when the latter is taken as a zero reference. Negative pressure can develop in a filter bed caused by high head loss. When this occurs, the gauge pressure in the bed can drop to less than atmospheric, often resulting in the release of dissolved gases from the water and the formation of gas bubbles. Negative pressure is also called negative head.

Nephelometric Turbidity Unit (NTU). A unit for expressing the cloudiness (turbidity) of a sample as measured by a nephelometric turbidimeter. A turbidity of 1 ntu is equivalent to the turbidity created by a 1:4,000 dilution of a stock solution of 5.0 milliliters of a 1.000-gram hydrazine sulfate ($(NH_2)_2 \times H_2SO_4$) in 100 milliliters of distilled water solution plus 5.0 milliliters of a 10.00-gram hexamethylenetetramine ($(CH_2)_6N_4$) in 100 milliliters of distilled water solution that has stood for 24 hours at 25 ± 3° Celsius.

Nitrate (NO_3^-). An oxidized ion of nitrogen. Nitrifying bacteria can convert nitrite (NO_2–) in the nitrogen cycle. Sodium nitrate ($NaNO_3$) and potassium nitrate (KNO_3) are used as fertilizer.

90th Percentile Concentration. A term used in conjunction with water sampling standards as required under the Safe Drinking Water Act (SDWA). It is calculated by first arranging all sample results in order from the lowest concentration of specific contaminant to the highest concentration of specific contaminant. Next, each sample result is assigned a number, starting with the number 1 for the lowest concentration result up to the highest concentration result. The number corresponding to the highest concentration result is equal to the total number of samples collected from a particular water system for a specific contaminant. The total number of samples is then multiplied by 0.9. The sample result with the number corresponding to this calculated value is the 90th percentile concentration.

Nitrite (NO_2^-). An intermediate oxidized ion of nitrogen. Nitrifying bacteria can convert ammonia (NH_3) to nitrite (NO_2^-) to nitrate (NO_3^-) in the nitrogen cycle. Sodium nitrite ($NaNO_2$) is used in curing meats. It has also been associated with adverse health effects.

Nitrite Oxidizing Bacteria (NOB). Gram-negative bacteria that catalyze the second step of aerobic nitrification, i.e., the oxidation of nitrite to nitrate. Nitrobacter and Nitrospira are found in most soil, freshwater, and marine habitats and in biological wastewater treatment plants. They are obligate chemolithoautotrophs and depend on carbon dioxide fixation for biosynthetic carbon.

N-Nitrosodiethylamine (NDEA). A volatile and clear yellow liquid used as an additive to gasoline and lubricants, as a stabilizer in plastics, and as a chemical reagent in research. It has been detected in drinking water at trace levels and is classified by the US Environmental Protection Agency as a probable human carcinogen.

N-nitrosodimethylamine (NDMA) ($(CH_3)_2N_2O$). A synthetic organic chemical that is a known animal carcinogen and is classified as a probable human carcinogen. It has been used in rocket fuels, as a solvent, and as a rubber accelerator. In addition, it may be formed during drinking water treatment by the reaction of secondary amines (R1–NH–R2), nitrite (NO_2^-), and acid.

N-Nitrosodi-n-propylamine (NDPA). A synthetic chemical used in research that appears as a yellow liquid at room temperature. It is a by-product formed during the manufacture of some herbicides and plastics. It also forms in the stomach during the digestion of nitrite-treated foods or food containing certain amines. It has been classified by the US Environmental Protection Agency as a probable human carcinogen.

Non-Potable Water. Water that may contain objectionable pollution, contamination, minerals, or infective agents and is considered unsafe, unpalatable, or both for drinking.

Normal Flow. (1) The flow that prevails for the greatest portion of the time; the mean flow or the average flow. (2) Flow at normal depth.

***Norovirus*.** Genus of viruses in the Caliciviridae family that cause most documented cases of acute non-bacterial gastroenteritis in humans. Formerly called round, structured viruses, human caliciviruses, and Norwalk-like viruses, the genus is subdivided into four genogroups and contains many species and strains, including the Norwalk virus.

Odorant. An odorous substance, especially one that is added to a dangerous odorless substance to warn of its presence.

Operating Pressure. (1) The manufacturer's specified range of pressure—expressed in pounds per square inch (psi), pascals (Pa), or kilopascals (kPa)—within which a water- processing device or water system is designed to function. A range of 30 psi (210 kPa) to 100 psi (690 kPa) is often indicated. Operating pressure is also called working pressure. (2) In membrane systems, the pressure at which feedwater enters the system. It is also called feed pressure.

Operational Evaluation Levels (OELs). As a part of Stage 2 DBPR compliance monitoring, water systems are required to monitor operational evaluation levels (OELs). The OELs are determined by calculating the sum of the two previous quarter's results plus twice the current quarter's results divided by 4. This locational running annual average (LRAA) threshold assists water systems in identifying whether they are likely to exceed the DBP MCL in the next quarter, and gives them a chance to make operational changes. If OELs are higher than the MCL for TTHM or HAA5 at any location in the distribution system, the water system must conduct an operational evaluation, which includes an examination of system source water influence, treatment and distribution operational practices as well as what steps could be considered to minimize future exceedances.

Optimized Corrosion Control Treatment (OCCT). For the purposes of the Lead and Copper Rule, the treatment that minimizes the lead and copper levels at users' taps while ensuring that the treatment does not cause the water system to violate any national drinking water regulation.

Organic Chloramines. Organic compounds formed during the chlorination or chloramination of water containing organic nitrogen (e.g., N-chlorodimethylamine, $(CH_3)_2NCl$). Unlike inorganic chloramines (e.g., monochloramine (NH_2Cl), dichloramine ($NHCl_2$), organic chloramines are not considered to be disinfectants.

Organic Compounds. A carbon-containing compound that is derived from living organisms, e.g., a hydrocarbon, alcohol, ether, aldehyde, ketone, carboxylic acid, or carbohydrate.

Organic Substance. A chemical substance of animal or vegetable origin and having carbon in its molecular structure.

Organoleptic. Affecting or employing one or more of the organs of special sense (e.g., sense of smell or taste).

Orthophosphate. (1) A salt that contains phosphorus as PO_4^{3-}. (2) A product of the hydrolysis of condensed (polymeric) phosphates. (3) A nutrient required for plant and animal growth.

Oxidant. Any oxidizing agent; a substance that readily oxidizes (removes electrons from) something chemically. Common drinking water oxidants are chlorine (Cl_2), chlorine dioxide (ClO_2), ozone (O_3), and potassium permanganate ($KMnO_4$). Some oxidants also act as disinfectants.

Oxidation. (1) A process in which a molecule, atom, or ion loses electrons to an oxidant. The oxidized substance (which lost the electrons) increases in positive valence. Oxidation never occurs alone, but always as part of an oxidation– reduction (redox) reaction. The reduced substance gains electrons and thereby decreases in positive valence. (2) In electrodialysis, a chemical reaction that occurs at an anode and results in the loss of electrons from the anodic material. (3) In ion exchange, a specific attack on the cross-linking of the copolymer of an ion exchange resin by an oxidant (e.g., chlorine, hydrogen peroxide (H_2O_2), or ozone (O_3) leading to degradation (loss of structure of resin beads) and shortening of the resin life.

Oxidation State. (1) For simple monatomic substances, simply the net charge on the atom. Thus, the oxidation states of the ferrous (Fe^{2+}) and ferric (Fe^{3+}) ions are +2 and+3, respectively. (2) For polyatomic species, a numeric value arbitrarily selected for each atom (based on some limited rules) so as to make the algebraic sum of the oxidation states equal to the net charge on the molecule or ion. Thus, in the case of the hypochlorite ion (OCl^-), with an oxidation state of –2 assigned to oxygen (which is done for all oxygen-containing compounds except for peroxides), the oxidation state of chlorine is +1.

Oxidation-Reduction Potential (pE) (ORP). The potential required to transfer electrons from the oxidant to the reductant, used as a qualitative measure of the state of oxidation in treatment systems. It is a dimensionless value. The more positive the value, the more oxidizing the solution; more negative values represent more reducing conditions. Aerobic waters in equilibrium with the atmosphere have a pE value of approximately +13. Anaerobic waters may have a pE of –4.Because electron activity varies with the pH of water, referring to graphs of pE as a function of pH is often useful. In this manner, the range of thermodynamically stable species can be illustrated.

Oxidation-Reduction Reaction. A reaction in which electron transfers occur. Oxidation and reduction always occur simultaneously, and the substance that gains electrons (and is thus reduced) is termed the oxidant.

Oxidize. (1) To combine with oxygen. (2) To change a compound by increasing the proportion of the electronegative part or to change an element or ion from a lower to higher oxidation state. (3) To remove one or more electrons from an atom, ion, or molecule.

Oxygen (O). A nonmetallic gaseous element. Molecular oxygen is O_2 and ozone is O_3. Atmospheric oxygen is the result of photosynthesis.

Ozonation. The process of applying ozone (O_3) to a liquid for disinfection.

Ozone (O_3). An unstable gas that is toxic to humans and has a pungent odor. It is a more active oxidizing agent than oxygen. It is formed locally in air from lightning or in the stratosphere by ultraviolet irradiation; it inhibits penetration of ultraviolet light from the sun to the Earth's surface. It also is produced in automobile engines and contributes to the formation of photochemical smog. For industrial applications, it is usually manufactured at the site of use. It serves as a strong oxidant and disinfectant in the purification of drinking water and as an oxidizing agent in several chemical processes.

Particle Count. The results of a microscopic examination of treated water by a particle counter that classifies suspended particles by number and size.

Particulate. A very small solid suspended in water that can vary widely in size, shape, density, and electrical charge. Colloidal and dispersed particulates are artificially gathered together by the processes of coagulation and flocculation.

Partnership for Safe Water (PSW). A volunteer initiative developed by the US Environmental Protection Agency, the Association of Metropolitan Water Agencies, the Association of State Drinking Water Administrators, the National Association of Water Companies, the American Water Works Association, and the Water Research Foundation to optimize water treatment to better protect the public from Cryptosporidium oocysts, as evidenced by filter water turbidity performance. The program was established in 1995 and commits participating water utilities to a four-phase self-assessment program for optimizing treatment plant operations and performance. Originally developed for surface water treatment plants, the program has since expanded to include groundwater treatment plant optimization. Phase I of the program requires 6 months of compliance with all applicable regulations and consists of the agreement to participate in the program through phase III. Phase II consists of baseline data collection, and Phase III consists of completion of a comprehensive water treatment plant self-assessment and peer-review process. Phase IV consists of demonstrated optimization of treatment plant performance and operations, relative to the PSW's stringent treatment plant optimization goals. Utilities are recognized publicly for their progress through the program's phases.

Partnership for Safe Water Distribution System Optimization Program (PSW-DSOP). A volunteer initiative developed by the US Environmental Protection Agency, the Association of Metropolitan Water Agencies, the Association of State Drinking Water Administrators, the National Association of Water Companies, the American Water Works Association, and the Water Research Foundation to optimize distribution system operations and performance with respect to water quality (maintaining disinfectant residual), hydraulic integrity (pressure management) and physical integrity (reducing the frequency of water main breaks). The program was established in 2011 and was modeled after the treatment program to better ensure that water quality improvements can be measured and validated over time. Similar to the treatment program, participating water utilities commit to a four-phase self-assessment program for optimizing their distribution system operations. Phase I of the program consists of the agreement to participate through at least Phase III of the program. Phase II consists of baseline and annual data collection for distribution system disinfectant residual and disinfection by-product data. Phase III consists of completion of a comprehensive distribution system self-assessment and peer-review process. Phase IV is an achievement of the highest level of demonstrated optimization of distribution system operation and performance, relative to the PSW-DSOP's stringent distribution system optimization goals. Utilities are recognized publicly for their progress through the program's phases.

Passivation. A corrosion control technique that causes the pipe materials to create metal/hydroxide/carbonate compounds that form a film on the pipe wall to protect the pipe.

Pathogen. An organism that can cause disease or infection in a host.

Peak Flow. Maximum flow.

Peak Hour Demand. The greatest volume per hour flowing through a treatment plant for any hour in the year.

Perchlorate (ClO_4^-). A salt or ester of perchloric acid ($HClO_4$), e.g., sodium perchlorate, $NaClO_4$. Sodium perchlorate has been used in explosives, in jet fuel, and as an analytical reagent. Perchlorate has been found in a number of drinking water sources as a result of probable contamination from the solid rocket industry. The US Environmental Protection Agency is currently in the process of determining a national primary drinking water regulation for perchlorate in drinking water and has issued an Interim Lifetime Drinking Water Health Advisory, suggesting that the perchlorate concentration in drinking water not exceed 0.015 milligrams per liter.

Permeability. A measure of the relative ease with which water flows through a porous material. A sponge is very permeable; concrete is much less permeable. Permeability is sometimes called perviousness.

Pesticide. Any substance or chemical applied to kill or control pests.

pH. A measure of the acidity or alkalinity of a solution, such that a value of 7 is neutral; lower numbers represent acidic solutions and higher numbers, alkaline solutions. Strictly speaking, pH is the negative logarithm of the hydrogen ion concentration (in moles per liter). For example, if the concentration of hydrogen ions is 10–7 moles per liter, the pH will be 7.0. As a measure of the intensity of a solution's acidic or basic nature, pH is operationally defined relative to standard conditions that were developed so that most can agree on the meaning of a particular measurement. The pH of an aqueous solution is an important characteristic that affects many features of water treatment and analysis.

Phenol. An aromatic organic compound with the molecular formula C_6H_5OH. It is a white crystalline solid that is volatile. The molecule consists of a phenyl group (C_6H_5) bonded to a hydroxyl group (OH). It is mildly acidic and requires careful handling due to its propensity to cause chemical burns. It is also known carbolic acid.

Phosphate. A salt or ester of phosphoric acid (H_3PO_4).

Phosphorus (P). An essential chemical element and nutrient for all life forms. It occurs in orthophosphate, pyrophosphate, tripolyphosphate, and organic phosphate forms. Each of these forms, as well as their sum (total phosphorus), is expressed in terms of milligrams per liter elemental phosphorus.

Pipe Loop. An experimental apparatus consisting of several feet of pipe complete with joints, elbows, and connections for flow-through testing of corrosion and corrosion control.

Pitting. Highly localized corrosion resulting in deep penetration at only a few spots.

Planktonic. Pertaining to plankton—a diverse group of organisms that live in the water column of large bodies of water and that cannot swim against a current. They provide a crucial source of food to many large aquatic organisms, such as fish and whales.

Plastic Pipe. An identification used for nonmetallic pipe manufactured from different polymers, such as polyvinyl chloride, $(CH_2{:}CHCl)_n$; polyethylene, $(H_2C{:}CH_2)_n$; or polybutylene, $(C_4H_8)_n$.

Plasticizer. An organic compound added to a high-molecular weight polymer to facilitate processing, to increase the flexibility, and to increase toughness of the final product by internal modification of the polymer molecule. Among the more important plasticizers are nonvolatile organic liquids and low-melting solids (e.g., phthalate, adipate).

Plug Flow. A flow regime in which no longitudinal dispersion (mixing in the direction of flow) occurs. The input, or plug, travels through a pipe or unit process in such a fashion that the entire mass or volume is discharged at exactly the theoretical hydraulic retention time of the unit (calculated as volume of the vessel divided by the flow rate).

Plumbing Fixture. A receptacle that receives water and discharges wastewater into a drainage system (e.g., a sink or toilet).

Polyaluminum Chloride (PACl) (Al $(OH_x(Cl)_y(SO_4)_z)$. A hydrolyzed form of aluminum chloride ($AlCl_3$) that is used for coagulation, typically in low-turbidity waters. As a result of its polymeric form, lower dosages can be used compared to metal coagulants. In its molecular formula, y typically ranges from 1.5 to 2.5 and z from 0 to 1.5.

Polyethylene (PE). A tough thermoplastic polymer $(-CH_2CH_2-)_x$ of ethylene $(CH_2{:}CH_2)$ that resists chemicals and absorbs very little moisture. Polyethylene can vary from soft and flexible to hard and rigid depending on the pressures and catalysts used during manufacturing. Low-density polyethylene has a melting point of about 240° Fahrenheit (115° Celsius) and a tensile strength of 1,500 pounds per square inch (10,345 kilopascals); high-density polyethylene melts at 270° Fahrenheit (135° Celsius) and has a tensile strength of 4,000 pounds per square inch (27,580 kilopascals). Polyethylene is commonly used for tubing and piping, food packaging, garment bags, and molded plastic products.

Polymer. A synthetic organic compound with high molecular weight and composed of repeating chemical units (monomers). Polymers may be polyelectrolytes (such as water-soluble flocculants), water-insoluble ion exchange resins, or insoluble uncharged materials (such as those used for plastic or plastic-lined pipe).

Polyphosphate. Any of a broad family of inorganic phosphorus compounds that are commonly referred to as molecularly dehydrated phosphates or condensed phosphates, including hexametaphosphate, tripolyphosphate ($P_3O_{10}^{5-}$), and pyrophosphate ($P_2O_7^{5-}$), among others. Unlike orthophosphate (PO_4^{3-}), for which four oxygen. Atoms always surround a single central phosphate, polyphosphates are arranged as polymeric chains or occasionally rings that vary in their phosphorus-to-oxygen ratio and chain length with different commercial formulations. Inorganic polyphosphates may be protonated like weak acids when they are not coordinated with metal ions, and they tend to break down (hydrolyze) into orthophosphate groups plus molecules of shortened chain length over time and under extreme temperature and pH conditions.

Potable Water. Water that is safe and satisfactory for drinking and cooking.

Potassium Permanganate ($KMnO_4$). A substance in the form of dark purple crystals used as an oxidant in drinking water treatment. Potassium permanganate is used for taste-and-odor control and for iron and manganese removal. Unlike other oxidants, potassium permanganate has not been associated with the production of disinfection by-products.

Pounds Per Square Inch (PSI). A unit of pressure. This is not a Système International unit.

Pourbaix Diagrams. A diagram showing thermodynamically stable species of a metal as a function of potential (relative to the standard hydrogen electrode and pH).

Powdered Activated Carbon (PAC). Activated carbon composed of fine particles and providing a large surface area for adsorption. Powdered activated carbon is typically added as a slurry on an intermittent or continuous basis to remove taste-and odor-causing compounds or trace organic contaminants and is not reused.

Precipitate (ppt). (1) A substance separated from a solution or suspension by a chemical reaction. (2) To form such a substance.

Premise Plumbing. Building plumbing that receives water from the active water distribution system. Premise plumbing within a building provides hot or cold water intended for human consumption (drinking, bathing, showering, cooking, dishwashing, and maintaining oral hygiene).

Pre-oxidant. An oxidant, usually other than chlorine used to treat source water, e.g., ozone (O_3) or potassium permanganate ($KMnO_4$).

Pressure. The force pushing on a unit area. Water pressure is normally measured in pounds per square inch, kilopascals, or feet or meters of head.

Pressure Head. A measurement of the amount of energy in water due to water pressure.

Pressure Management. Methods to optimize pressures in a water distribution network to minimize losses and surge impacts, while maintaining adequate service levels, including firefighting flows.

Pressure Reducer. A control valve that opens to allow flow if the downstream pressure is less than a certain value and that closes when the set pressure is reached. A pressure reducer ensures that the downstream pressure does not become too high. It is used on house services where the distribution pressure is high and in other situations that require reductions from higher pressure planes to lower-pressure planes.

Pressure Regulating Valve (PRV). A valve designed to maintain a set pressure on the downstream side of the valve regardless of pressure changes on the source side.

Pressure Relief Valve. A valve that opens automatically if the water pressure exceeds a certain level.

Pressure Transient. A short-lived pressure wave in a distribution system that can result in hydraulic surges. These waves, having both a positive and negative amplitude, can produce negative pressures that last only seconds and may not be observed by conventional pressure monitoring.

Pressure Zone. An area within a distribution system of a domestic or municipal water supplying which the pressure in the mains is maintained within certain specified limits.

Protozoa. Small, one-celled organisms, including amoebae, ciliates, and flagellates.

Public Water System. As defined in Section 1401(4) of the Safe Drinking Water Act, a system for providing to the public water for human consumption through pipes or other constructed conveyances, if such system has at least 15 service connections or regularly serves at least 25 individuals daily at least 60 days out of the year. Such a system includes (1) any collection, treatment, storage, and distribution facilities under control of the operator of such system and used primarily in connection with such system, and (2) any collection of pretreatment storage facilities not under such control but that are used primarily in connection with such system. A public water system is either a community water system or a noncommunity water system. On Aug. 5, 1998, the US Environmental Protection Agency published guidance regarding interpretation of the definition of *public water system* (*Federal Register,* 63:41939–41946).

Pump. A mechanical device for driving fluid flow, for raising or lifting a fluid, or for applying pressure to a fluid.

Pump Station. A structure containing relatively large pumps and appurtenant piping, valves, and other mechanical and electrical equipment for pumping water. A pump station is also called a lift station.

Quench. To cool a material suddenly or halt a process or reaction abruptly.

Quicklime. Another name for calcium oxide (CaO), which is used in water softening and stabilization.

R2A agar. A medium used for the cultivation and enumeration of heterotrophic bacteria in potable water.

Reclaimed Water. Wastewater that has been treated and recovered for useful purposes, such as landscape irrigation.

Red Water. Rust-colored water. Such color is usually due to the presence of precipitated ferric iron (Fe^{3+}) salts or to dead microorganisms that depended on iron (Fe^{2+}) and manganese (Mn^{2+}).

Reduction. The counterpart of the process of oxidation. Reduction may involve (1) acceptance of one or more electrons by an atom or ion, (2) removal of oxygen from a compound, or (3) addition of hydrogen to a compound.

Reflection. The change in direction of light propagation when deflected by an interface or surface.

Reservoir. An impounded body of water or controlled lake in which water can be collected and stored.

Residence Time. The amount of time that a defined element of fluid or a particle suspended in that fluid spends in a reactor.

Retronasal. The act of perceiving a volatile olfactory stimulus originating in the mouth via transportation of the stimulus molecules up the back of the nasopharynx and into the region of the olfactory receptors from the direction opposite to that aroma. Often mistaken for "taste" by untrained observers.

Reverse Flow. Flow in a direction opposite to the normal flow.

Reverse Osmosis (RO). A pressure-driven membrane separation process that removes ions, salts, and other dissolved solids and nonvolatile organics. The separation capability of the process is controlled by the diffusion rate of solutes through a membrane barrier and by sieving; it is dependent on the membrane type. In potable water treatment, reverse osmosis is typically used for desalting, specific ion removal, and natural and synthetic organics removal.

Revised Total Coliform Rule (RTCR). The US Environmental Protection Agency (EPA) published the Revised Total Coliform Rule (RTCR) in the *Federal Register* (FR) on February 13, 2013 (78 FR 10269) and minor corrections on February 26, 2014 (79 FR 10665). The RTCR is the revision to the 1989 Total Coliform Rule (TCR) and is intended to improve public health protection. All public water systems (PWSs), except aircraft PWSs subject to the Aircraft Drinking Water Rule (ADWR) (40 CFR 141 Subpart X), must comply with the RTCR starting April 1, 2016, or an earlier state effective date. Until then, PWSs must continue complying with the 1989 TCR. Total coliforms are a group of related bacteria that are (with few exceptions) not harmful to humans. A variety of bacteria, parasites, and viruses, known as pathogens, can potentially cause health problems if humans ingest them. EPA considers total coliforms a useful indicator of other pathogens for drinking water. Total coliforms are used to determine the adequacy of water treatment and the integrity of the distribution system. Key provisions of the RTCR include: (1) Setting a maximum contaminant level goal (MCLG) and maximum contaminant level (MCL) for *E. coli* for protection against potential fecal contamination. (2) Setting a total coliform treatment technique (TT) requirement. (3) Requirements for monitoring total coliforms and *E. coli* according to a sample siting plan and schedule specific to the PWS. (4) Provisions allowing PWSs to transition to the RTCR using their existing Total Coliform Rule (TCR) monitoring frequency, including PWSs on reduced monitoring under the existing TCR. (5) Requirements for seasonal systems (such as, Non-Community Water Systems not operated on a year-round basis) to monitor and certify the completion of a state-approved start-up procedures. (6) Requirements for assessments and corrective action when monitoring results show that PWSs may be vulnerable to contamination. (7) Public notification (PN) requirements for violations. (8) Specific language for CWSs to include in their Consumer Confidence Reports (CCRs) when they must conduct an assessment or if they incur an *E. coli* MCL violation.

Risk Assessment. (1) The process of analyzing threats to and vulnerability of a facility, determining the potential for losses, and identifying cost-effective corrective measures and residual risk. (2) A qualitative or quantitative evaluation of the environmental and/or health risk resulting from exposure to a chemical or physical agent (pollutant); combines exposure assessment results with toxicity assessment results to estimate risk.

Risk Characterization. The integration of the following to determine an estimate of risk: information for a chemical, microbe, or process that can produce harm (hazard identification); the expected dose-response relationships; and the extent and distribution of exposure(s) in the population of concern.

Rotavirus. A member of the Reoviridae family associated with acute enteritis in infants and young children. Rotavirus is the most common cause of severe diarrhea among children. The disease is characterized by vomiting and watery diarrhea for 3-8 days, and fever and abdominal pain occur frequently. Immunity after infection is incomplete, but repeat infections tend to be less severe than the original infection virology.

Rust. A reddish corrosion product occasionally found in water. Rust is formed as a result of electrochemical interaction between iron and atmospheric oxygen in the presence of moisture. It is also called ferric oxide or iron (III) oxide (Fe_2O_3).

Ryznar Stability Index (RSI). An index indicating whether a water has a tendency to corrode (dissolve precipitated $CaCO_3$) or to precipitate in the form of calcium carbonate ($CaCO_3$).

$$RSI = 2pH_s - pH_a$$

Where:

pH_s = the pH of the water if it were saturated and at equilibrium with $CaCO_3$, pH_a = the actual pH Ryznar stability index values between 6.0 and 7.0 indicate slight scale formation or corrosion. Values less than 6.0 or greater than 7.0 indicate varying degrees of scale-forming conditions or corrosive ($CaCO_3$ dissolving) conditions, respectively.

Safe Drinking Water Act (SDWA). Public Law 93-523, enacted Dec. 16, 1974, establishing Title XIV of the US Public Health Service Act, codified generally as 42 U.S.C. 300f-300j-11. It required the US Environmental Protection Agency to set national primary (health-related) drinking water regulations that were the first to apply to all public water systems, as defined by the act, in the United States.

Safety Factor. (1) A numerical factor by which a dose or exposure that is known to produce minimal effects or no effect in an experimental situation (usually involving experimental animals or humans) is multiplied to help prevent adverse effects in a more susceptible species or individual. Safety factors are frequently applied to the determination of "safe" doses or exposures to chemicals. In general, however, the term uncertainty factor is favored, because it more precisely indicates why these factors are used. (2) An amount of water added to demand projections to protect against unforeseen changes in water demand and supply.

Sampling. (1) A process of collecting a portion of a substance for analysis. A goal of sampling is to obtain a portion of the substance that is representative of the whole. (2) In statistics, a selection of a subset of the entire population of subjects or products for analysis such that the measured values obtained collectively represent the entire population within certain confidence limits. In general, a method of selecting individual samples in an unbiased way must be found. The most common approach is to randomly select the samples for analysis.

Scale. A coating or precipitate deposited on surfaces. Examples include oxide flakes forming on metal surfaces, carbonate (CO_3^{2-}) precipitates depositing in piping systems or water heaters, sulfate (SO_4^{2-}) scale precipitating in desalting systems, or silica (SiO_2) scale forming in boilers. Waters that contain carbonates or bicarbonates (HCO^-) of calcium or magnesium are especially likely to cause scale when heated. Scale is also called hard water scale.

Scouring Velocity. The minimum velocity necessary to dislodge stranded material from the boundary of a waterway, conduit, or pipeline by a fluid in motion.

Secondary Maximum Contaminant Level (SMCL). A nonenforceable numerical limit set by the US Environmental Protection Agency for a contaminant on the basis of aesthetic effects to prevent an undesirable taste, odor, or appearance that would have an adverse impact on the public welfare.

Sediment. (1) Solid material settled from suspension in a liquid. (2) Mineral or organic solid material that is being transported or has been moved from its site of origin by air, water, or ice and has come to rest on the ground surface either above or below sea level. (3) Inorganic or organic particles originating from weathering, chemical precipitation, or biological activity. (4) Any solid phase settling out of a liquid phase, e.g., deposits in rivers and lakes or sludge in clarifiers.

Sedimentation. A treatment process using gravity to remove suspended particles.

Sequestering. A chemical reaction in which certain chemicals (sequestering or chelating agents) "tie up" other chemicals, particularly metal ions, so that the chemicals no longer react. Sequestering agents are used to prevent the formation of precipitates or other compounds.

Sequestering Agent. (1) A chemical that forms complexes with metallic ions in solution so that the metallic ions may no longer be precipitated. For example, calcium soap precipitates are not produced from hard water treated with sodium hexametaphosphate. (2) Any agent that prevents an ion from exhibiting its usual properties because of close combination with an added material.

Service. The small water line—including such components as the meter, corporation stop, curb stop, and meter couplings—that provides a facility to move water from a water main to or into a building.

Service Line. The pipe and all appurtenances that run between the utility's water main and the customer's place of use, including fire lines.

Set Point. The position at which a process controller is set. The set point is the same as the desired value of the process variable.

Sewage. The used water and water-carried solids from a community (including used water from industrial processes) that flow to a treatment plant. The term sewage usually refers to household wastes, but it has been replaced by the term wastewater which also includes stormwater, surface water, and groundwater infiltration in the wastewater that enters a wastewater treatment plant.

Shewhart Process Control Charts. A process control chart used to illustrate statistical quality control. The concept was developed in the 1930s by Walter A. Shewhart. Although these graphs can be of several varieties, a typical Shewhart chart is a plot of a series of replicated measurements compared with a mean value and control limits.

Short-circuiting. A hydraulic condition in a basin in which the actual flow time of water through the basin is less than the design flow time (i.e., less than the tank volume divided by the flow).

Silicates. A group of compounds containing oxygen and silica (SiO_2). Silicates are considered anodic corrosion inhibitors, combining with the free metal released at the anode site of corrosion activity and forming an insoluble metal–silicate compound. Silicates can also be used to sequester soluble iron and manganese present in source waters to help prevent oxidation and the formation of red and black water.

Siphon. A hydraulic process in a closed system in which a low enough pressure created at the siphon discharge permits a fluid to flow upward and be transferred across a higher elevation as long as the ultimate discharge point is at a lower elevation than the hydraulic grade of the system at the siphon starting point. For example, a siphon will transport water up and over the edge of a container as long as the discharge point is below the water surface in the container and the system is closed (i.e., no air gaps occur in the siphon). To create such a condition.

Six-Year Review. The Safe Drinking Water Act (SDWA) requires the US Environmental Protection Agency (EPA) to review each national primary drinking water regulation at least once every six years and revise them, if appropriate. As part of the "Six-Year Review," EPA evaluates any newly available data, information and technologies to determine if any regulatory revisions are needed. Revisions must maintain or strengthen public health protection. In July 2003, the First Six-Year Review (called the Six-Year Review 1) was conducted and reviewed 69 national primary drinking water regulations (NPDWRs) - 68 chemical NPDWRs and Total Coliform Rule (TCR). After this review, EPA found it was appropriate to revise the TCR. In March 2010, the Second Six-Year Review (called the Six-Year Review 2) reviewed 71 NPDWRs. 67 NPDWRs did not need any revisions, while 4 were candidates for regulatory revision (acrylamide, epichlorohydrin, tetrachloroethylene, and trichloroethylene). Additionally, 14 other NPDWRs were reviewed but not detailed because of recent or ongoing regulatory actions (DBPs, lead & copper, microbials). The Third Six-Year Review (called the Six-Year Review 3) was completed in 2016 and identified 8 NPDWRs as candidates for revision (Cryptosporidium, Giardia, Heterotrophic Plate Count, Legionella, Viruses, Chlorite, Haloacetic Acids, and Trihalomethanes).

Soda Ash (Na_2CO_3). A common name for commercial sodium carbonate.

Sodium Hydroxide (NaOH). A strong caustic chemical used in treatment processes to neutralize acidity, increase alkalinity, or to raise the pH value. It is also known as caustic soda, sodium hydrate, lye, or white caustic.

Soft Water. Water having low hardness or low concentrations of the calcium ion,magnesium ion, or a combination of both ions. Generally, soft water exhibits a hardness within the range of 0 to 75 milligrams per liter (0 to 4.4 grains per gallon) as calcium carbonate ($CaCO_3$).

Solder. A metal or metallic alloy that is melted and used to join metallic surfaces.Many solders contain lead and tin. Historically, 50:50 tin: lead solder was used to join copper piping. The US Environmental Protection Agency now requires that low lead solders be used when the solder is exposed to potable water.

Solubility. The ability or tendency of one substance to blend uniformly with another. Solids vary from 0 to 100 percent in their degree of solubility in liquids, depending on the chemical nature of the substances. To the extent that they are soluble, solids lose their crystalline form and become molecularly or ionically dispersed in the solvent to form a true solution. Liquids and gases are often said to be miscible in other liquids and gases rather than soluble.

Source Water. The supply of water for a water utility. Source water is usually treated before distribution to consumers, but some groundwaters are of such a quality that they can be distributed untreated. This term is preferred over raw water.

Speciation. The chemical makeup of a group of compounds. For example, for disinfection by-products, the speciation may be limited to chlorine-substituted by- products or may alternatively include a mixture of chlorine- and bromine-substituted species.

Specific Conductance. The conductivity of a solution as measured using a standard cell with a 1-centimeter width. Specific conductance is expressed in units of micromhos or microsiemens at 77° Fahrenheit (25° Celsius).

Spot Flushing. The removal of water from a specific location or area by flowing water from a hydrant blowoff or service connection.

Stage 1 Disinfection Byproducts Rule (Stage 1 DBPR). US Environmental Protection Agency regulations for disinfectants and disinfection by-products promulgated Dec. 16, 1998 (63 *Federal Register* 69390-69475), focused on reducing the levels of disinfectants and disinfection by-products in treated drinking water. It set maximum residual disinfectant levels for chlorine, chloramines, and chlorine dioxide and set maximum contaminant levels (MCLs) for two groups of organic disinfectant by-products, trihalomethanes and haloacetic acids, and for two inorganic disinfection by-products, chlorite and bromate. The maximum residual disinfectant levels were set at 4.0 milligrams per liter (as Cl_2) for chlorine and chloramines and 0.8 milligrams per liter (as ClO_2) for chlorine dioxide. Regulated for the first time under this rule, the MCL of the sum of five haloacetic acids (HAA5) was set at 0.060 milligrams per liter as a running annual average. The MCL for the sum of four trihalomethanes (THMs) was set at 0.080 milligrams per liter, for chlorite at 1.0 milligrams per liter, and for bromate at 0.01 milligrams per liter.

Compliance with these MCLs is determined by the running annual average of the monitoring results from all compliance sites within a distribution system. A treatment technique was also mandated for total organic carbon (TOC).

Stage 2 Disinfection Byproducts Rule (Stage 2 DBPR). US Environmental Protection Agency regulations, promulgated January 4, 2006 (71 *Federal Register* 388–493), intended to set limits on peak and average levels of disinfection by-products (DBPs) in drinking water supplies, and so reduce the risk of potential reproductive and developmental health effects and cancer. USEPA's Stage 2 DBPR requires compliance with a total trihalomethane maximum contaminant level (MCL) of 0.08 milligrams per liter and a total haloacetic acid MCL of 0.06 milligrams per liter, both based on a locational running annual average at the new Stage 2 DBPR sampling sites identified by the Initial Distribution System Evaluation (IDSE). This rule maintained the same level of MCLs for these two groups of organic disinfectant by-products but changed how compliance with the rule was calculated. Under Stage 2, monitoring data at each location must meet the MCL, based on the locational running annual average.

Stagnation. A condition that occurs when water is motionless, or nearly motionless, and does not flow in a stream or move in a lake. Similar no-flow or low-flow conditions can occur in a finished water storage reservoir or in a distribution system piping network.

Static Pressure. When water is not moving, the vertical distance from the specific point of interest to the water surface. The static pressure is the static head multiplied by the specific weight of water.

Steady State. A condition in which the input energy equals the output energy. This term may apply to any continuous physical, chemical, or biological process.

Steel Pipe. Pipe manufactured out of rolled steel sheets or plates with welded longitudinal or spiral seams. For water applications the interior can be lined and a coating applied to the exterior for corrosion protection.

Storage Tank. A compartment used to accumulate the product water from a water treatment unit so that sufficient quantity, pressure, or both are available for intermittent periods of higher flow-rate water use. A storage tank is also called a clearwell.

Stratification. (1) The formation of separate mixing layers in a body of water usually characterized by different temperature regimes and usually called thermal stratification. (2) Collectively, the existence of beds or laminae. Such stratification is also called lamination. (3) The layering of granular media on the basis of grain size during backwashing (fluidization). (4) An arrangement into classes, groups, or strata. (5) An arrangement or deposition of sedimentary material in layers; of sedimentary rock in strata marked by change in color, texture, dimension of particles, and composition. (6) A method of analyzing epidemiologic data to assess and control confounding bias. The population sample is separated for analysis into smaller groups based on specified criteria, such as age, gender, or other characteristics. For example, in a study of the possible association of bladder cancer and drinking water exposure, an appropriate analysis would be to stratify on smoking status to determine whether risks differ for smokers and nonsmokers.

Sulfate (SO_4^{2-}). An inorganic ion that is widely distributed in nature. It may be present in natural waters in concentrations ranging from a few to several thousand milligrams per liter.

Sulfate Reducing Bacteria (SRB). Bacteria characterized by their ability to reduce sulfate under anaerobic conditions, producing hydrogen sulfide (H_2S). Sulfate is used as an electron acceptor for the oxidation of organic materials. Sulfate-reducing bacteria are generally found in anaerobic soils and sediments, as well as in areas of drinking water systems where appropriate conditions exist. These organisms are involved in pipe corrosion. The sulfate-reducing bacteria of most importance belong to the genera *Thiobacillus* and *Desulfovibrio*.

Sulfide (S^{2-}). An anion often present in groundwater, especially in hot springs. Its common presence in wastewaters comes partly from the decomposition of organic matter, sometimes from industrial wastes, but mostly from the bacterial reduction of sulfate (SO^{2-}). Acidification of solutions of soluble sulfide4compounds leads to the evolution of hydrogen sulfide (H_2S), a foul-smelling gas that can be poisonous.

Sulfur (S). A nonmetallic element that occurs either free or combined (especially in sulfide (S^{2-}) and sulfate (SO^{2-}) compounds). Sulfur exists in several forms, including as yellow crystals.

Sulfur Dioxide (SO_2). An extremely effective oxidizing and reducing agent. It has many industrial uses, including use as a food additive (for inhibition of browning, of enzyme-catalyzed reactions, or of bacterial growth). Sulfur dioxide is also used as a dechlorination agent in drinking water facilities.

Supervisory Control And Data Acquisition (SCADA). A computer-monitored alarm, response, control, and data acquisition system used by drinking water facilities to monitor their operations.

Surface Water. All water on the surface, as distinguished from subsurface water or groundwater.

Surface Water Treatment Rule (SWTR). The common name for the US Environmental Protection Agency regulation promulgated June 29, 1989, that set maximum contaminant level goals for *Giardia lamblia*, viruses, and *Legionella*, as well as National Primary Drinking Water Regulations for public water systems using surface water sources or groundwater under the direct influence of surface water. These regulations include (1) criteria under which filtration (including coagulation and sedimentation, as appropriate) are required and procedures by which the states are to determine which systems must install filtration, and (2) disinfection requirements.

Surge. A sudden increase in the movement or pressure of water in a pipe or container, usually caused by opening or closing of a valve, starting or stopping of a pump, or trapping of air in a water main.

Surge Tank. A tank used to attenuate the pressure increase or decrease resulting from a sudden change in fluid momentum. A surge tank accumulates water in response to a pressure surge and supplies water in instances of moderate negative gauge pressure, thereby preventing damage to pumping and piping equipment. Surges result when sudden changes in a flow regime occur, such as when pumping equipment is stopped or valves are closed too quickly.

Taste. To ascertain the flavor of a substance by taking a portion into the mouth; to perceive or recognize by the sense of taste. Taste is one of the special senses that perceives and distinguishes the sweet, sour, bitter, or salty quality of a dissolved substance. It is mediated by taste buds on the tongue. Strictly speaking, taste refers to the four basic tastes just discussed, whereas flavor refers to the blend of taste and smell sensations evoked by a substance in the mouth.

Taste and Odor. The combination of sensations perceived by the mouth (i.e., by taste buds on the tongue) and nose.

Telemetry. The use of an electrical link between a transmitter and a receiver. A telephone line can serve as the electrical link, but more frequently, the connection is wireless.

Temperature. (1) The thermal state of a substance with respect to its ability to communicate heat to its environment. (2) A measure of the thermal state on some arbitrarily chosen numerical scale, e.g., Celsius, kelvin, or Fahrenheit.

Ten State Standards. A reference manual providing guidance on the design of water treatment plants, prepared under the supervision of representatives from 10 states, including Illinois, Indiana, Iowa, Michigan, Minnesota, Missouri, New York, Ohio, Pennsylvania, and Wisconsin, as well as the province of Ontario, Canada.

Topography. The configuration of a surface, including its relief and the position of its natural and artificial features.

Throttle. A mechanism by which flow is managed by constriction or obstruction.

Topology. (1) A branch of mathematics that deals with nonmetric spatial relationships, such as connectedness and adjacency. (2) In a geographic information system, the data that deal with topological relationships among point, line, and area objects (e.g., identifying which streets meet at an intersection, which streets bound a block, or which blocks front on a street).

Total Chlorine. The total concentration of the chlorine in a water, including the combined chlorine (i.e., inorganic and organic chloramines) and the free available chlorine.

Total Chlorine Residual. The total amount of chlorine residual present after a given contact time in a water sample, regardless of the type of chlorine.

Total Coliform (TC). The group of bacteria used as warm-blooded animal fecal pollution indicator organisms of drinking water quality.

Total Coliform Rule (TCR). A rulemaking of the US Environmental Protection Agency (EPA) that set National Primary Drinking Water Regulations for total coliforms, fecal coliforms, and *Escherichia coli.* The rule was promulgated June 29, 1989 (54 *Federal Register* 27544-27568) and amended January 15, 1991 (56 *Federal Register* 1556-1557). In September 2008, USEPA signed an Agreement in Principle to revise the Total Coliform Rule.

Total Dissolved Solids (TDS). The weight per unit volume of solids remaining after a sample has been filtered to remove suspended and colloidal solids. The solids passing the filter are evaporated to dryness. The filer pore diameter and evaporation temperature are frequently specified.

Total Kjeldhal Nitrogen (TKN). The total concentration of organic nitrogen and ammonia nitrogen, i.e., nitrogen with a valence of −3 (N^{3-}). The amount of organic nitrogen can be obtained by separately measuring the ammonia nitrogen and subtracting that value from the total Kjeldahl nitrogen.

Total Organic Carbon (TOC). A measure of the concentration of organic carbon in water, determined by oxidation of the organic matter into carbon dioxide (CO_2). Total organic carbon includes all the carbon atoms covalently bonded in organic molecules. Most of the organic carbon in drinking water supplies is dissolved organic carbon, with the remainder referred to as particulate organic carbon. In natural waters, total organic carbon is composed primarily of nonspecific humic materials. Total organic carbon is used as a surrogate measurement for disinfection by-product precursors, although only a small fraction of the organic carbon will react to form these by-products. Quantitatively, total organic carbon is determined by removing interfering inorganic carbon, such as bicarbonate (HCO^-), and oxidizing the organic carbon to carbon dioxide. Typically, the carbon dioxide is then measured with a nondispersive infrared detector.

Total Phosphorus. The sum of all phosphorus forms in a sample.

Trace Metal. A metal that is present in relatively small concentrations in water.

Transient. Passing through from one place to another.

Transmission Main. A water main that transports source water to a water treatment plant or treated water to a distribution grid system.

Treatment Technique (TT) Requirement. A requirement of the National Primary Drinking Water Regulations that specifies, for a particular contaminant, one or more specific treatment techniques that lead to a reduction in the level of the contaminant sufficient to achieve compliance with the requirements of 40 *Code of Federal Regulations (CFR)* Part 141.

Tribromoacetic Acid (TBAA) (CBr_3COOH). The bromine-substituted analog of trichloroacetic acid that may be formed during the chlorination or ozonation of bromide-containing water. Research suggests that this haloacetic acid is relatively unstable in water.

Trichloramine (NCl_3). A compound that is formed by the chlorination of water containing ammonia (NH_3). This compound is also called nitrogen trichloride. Trichloramine is the last in a sequence of compounds formed by the continual addition of ammonia—first monochloramine (NH_2Cl), then dichloramine ($NHCl_2$), and finally trichloramine. Thus, this species is not normally detected in chloraminated waters. Trichloramine has little disinfection capacity and is a source of taste and odors. Upon additional chlorination, trichloramine will be converted to elemental nitrogen gas, and any additional chlorine (HOCl) will be measured as free chlorine.

Trichloroacetic Acid (TCAA) (CCl_3COOH). A haloacetic acid containing three chlorine atoms. Typically, this acid and dichloroacetic acid ($CHCl_2COOH$) are the principal haloacetic acids formed during chlorination, except in water containing moderate amounts of bromide ion (Br^-). Trichloroacetic acid is also commercially prepared for miscellaneous industrial uses, including use as an herbicide.

Trihalomethane (THM) (CHX_3). Any of numerous organic compounds named as derivatives of methane (CH_4) in which three halogen atoms (chlorine, bromine, iodine, singly or in combination) are substituted for three of the hydrogen atoms. Trihalomethanes (THMs) are formed during the disinfection of water with free chlorine. Because of their carcinogenic potential and other possible health effects, the presence of these compounds in drinking water is regulated by the US Environmental Protection Agency at a maximum contaminant level of 0.080 milligrams per liter for the sum of the four regulated THMs, referred to as total trihalomethanes (TTHMs).

Tubercle. (1) A small knob or button of rust formed on the inside of an iron pipe. (2) An attached deposit of metallic salts in pipes, frequently resembling a bunch of grapes.

Tuberculation. (1) The formulation of tubercles in pipe. (2) Localized corrosion at scattered locations resulting in knoblike mounds.

Turbidity. (1) A condition in water caused by the presence of suspended matter, resulting in the scattering and absorption of light. (2) Any suspended solids that impart a visible haze or cloudiness to water and can be removed by treatment. (3) An analytical quantity, usually reported in nephelometric turbidity units, determined by measurements of light scattering.

Turbulence. A pattern of water flow characterized by cross currents and eddies that mix the water (as opposed to a streamlined or laminar flow pattern). Turbulence may be caused by excessive flow rates, by curves or rough surfaces in the flow channel or by the turbulence promoters (such as baffles) purposely created to mix the water. Turbulence significantly increases pressure drop in a system.

Turnover. The phenomenon of vertical circulation that occurs in large bodies of water—also known as overturn. It is caused by the increase in density of water at temperatures greater or less than 39.2° Fahrenheit (4° Celsius), the temperature of maximum density. In the spring, as the surface of the water warms toward 39.2° Fahrenheit (4° Celsius), the water increases in density, becomes heavier, and tends to sink, producing vertical currents; in the fall, as the surface water cools toward 39.2° Fahrenheit (4° Celsius) and therefore becomes heavier, it also tends to sink. Wind also creates such vertical currents.

Total Dynamic Head (TDH). The difference in height between the hydraulic grade line on the discharge side of the pump and the hydraulic grade line on the suction side of the pump. This head is a measure of the total energy that a pump must impart to the water to move it from one point to another.

Tracer Study. A study using a substance that can readily be identified in the water (such as a dye) to allow one to determine the distribution and rate of flow in a basin, pipe, groundwater, or stream channel.

True Color. The color of water from which the turbidity has been removed.

Ultraviolet Light (UV). Radiation having a wavelength shorter than 390 nanometers (the shortest wavelength of visible light) and longer than 10 nanometers (the longest wavelength of X-rays). This wavelength puts ultraviolet light at the invisible violet end of the light spectrum. UV light may be used as a disinfectant or to create hydroxyl radicals (OH).

Unidirectional Flushing. A distribution system flushing method using closed valves and the opening of consecutive fire hydrants to create single-direction water flow pattern, providing a sufficient velocity to cause a scouring action. This method provides an effective flushing action with less water than is used by randomly opening fire hydrants in a conventional method.

Uniform Corrosion. Corrosion that results in an equal amount of material loss over an entire pipe surface.

US Department of Agriculture (USDA). An agency of the US government's executive branch responsible for improving and maintaining farm income and expanding markets abroad for agriculture products. It works to enhance the environment and maintain production capacity by helping landowners protect soil, water, forests, and other natural resources. It administers national programs for rural development, credit, conservation, and research, and it safeguards and ensures standards of quality in the daily food supply through inspection and grading services.

US Department of Homeland Security (DHS). An agency of the US government's executive branch formed by the Homeland Security Act of 2002 and intended to provide a unifying core for the vast national network of organizations and institutions involved in efforts to secure the United States. This department was created primarily from a conglomeration of existing federal agencies in response to the terrorist attacks of September 11, 2001.

United States Environmental Protection Agency (USEPA). An independent agency of the executive branch of the US government, created December 2, 1970, to permit coordinated and effective government action on behalf of the environment. It administers the implementation of environmental laws to abate and control pollution systematically through monitoring, standard setting, enforcement, and research activities.

Uninterruptible Power Supply (UPS). A power supply (usually batter operated) that is used in the event of a commercial power failure. Most businesses isolate computers and other critical equipment by supplying power through the batteries at all times. Other businesses use motor-driven generators to supply power. The motors normally use commercial power, and if they fail they automatically switch over to a battery supply. These systems are extremely reliable and provide good clean power, free from the voltage fluctuations common to commercial power grids.

Unregulated Contaminant Monitoring Rule (UCMR). A program of the US Environmental Protection Agency (US EPA) mandated by the Safe Drinking Water Act (SDWA) that collects data for contaminants suspected to be present in drinking water but that do not have health-based standards. The program was established in 1988, and a number of activities have been part of the program. Every 5 years, US EPA must review the list of contaminants, largely based on the Contaminant Candidate List (CCL). The SDWA Amendments of 1996 created changes in the program, including the formation of the National Contaminant Occurrence Database (NCOD) to store all of the data and to reduce the monitoring requirements for small water systems.

Upstream. In the opposite direction of a flow.

UV-254 Absorption. A measure of water's ultraviolet absorbance at a wavelength of 254 nanometers. This value is an indirect measure of compounds containing double bonds (including, but not limited to, aromatic compounds). Therefore, this measurement has been considered representative of the humic content of natural organic matter, as well as acting as a surrogate for disinfection by- product precursors.

Vacuum. A space in which the pressure is far less than normal atmospheric pressure so that the remaining gases do not affect processes being carried on in the space. In a water system, a vacuum is created when water is discharged from a low point in a closed system without a point for air to enter the system. Such a vacuum can result in backflow from connections to the system.

Vacuum Breaker. A mechanical device that prevents backflow due to a siphoning action created by a partial vacuum. A vacuum breaker allows air into the piping system, breaking the vacuum.

Vacuum Valve. A valve that admits air to relieve vacuum conditions and that permits accumulated air to escape. Such a valve is also called a vacuum relief valve or an air-and-vacuum valve.

Valve Box. A housing that encloses the operating nut of a gate valve and extends to the ground surface, allowing an access opening for an operating or valve key to be inserted and connected to the operating nut so that the valve may be opened and closed. It is also called a buffalo box.

Valve Exercising. The routine and scheduled closing and opening of valves on a specific time period such as once a year or every six months.

Valve. A mechanical device installed in a water distribution system or treatment plant to close off or regulate the flow of water in the system.

Vapor Pressure. The pressure at which a substance (e.g., water) changes from the liquid or solid state to the gaseous state.

Variable Frequency Drive (VFD). A motor speed controller that uses adjustment of the applied power frequency to affect motor speed control. This type of controller is used with induction motors.

Variable Speed Pump. A pump that can be operated at a variable speed to change the flow rate delivered in the process. Variable speed pumps are driven by motors that can draw variable current, thereby modulating the speed and changing the flow rate. Variable speed pumps are often used to reduce pump cycling (i.e., on-off operation) and to allow continuous operation of downstream processes.

Velocity. The time rate of change of position of a body. Velocity represents the rate of movement of water flowing in a pipe, conduit, drainage way, or basin. It is expressed in units of distance per time, such as feet (meters) per second.

Venturi Effect. Phenomenon in a Venturi device whereby the constricted throat cause an increase in flow velocity and a decrease in pressure. The phenomenon can be used to measure flow to draw other fluids into the flow through a small inlet.

Viruses. A minute organism not visible by light microscopy. A virus is an obligate parasite dependent on nutrients inside cells for its metabolic and reproductive needs. It consists of a strand of either deoxyribonucleic acid or ribonucleic acid, but not both, separated by a protein covering called a capsid. A computer program that replicates itself through incorporation into other programs.

Viscosity. A factor that represents a measure of a fluid's internal resistance to flow. The great the resistance, the stickier the fluid and the greater the viscosity. The viscosity of a fluid is an important property in the analysis of liquid behavior and so must be considered when designing water conveyance methods such as pumps.

Volatile. Capable of turning to vapor (evaporating). This term should be associated with a temperature and pressure because they determine whether or not a chemical is volatile. A chemical may be volatile under one set of conditions but not another.

Volatile Organic Compound (VOC). A class of organic compounds that includes gases and volatile liquids. Many volatile organic compounds are used as solvents. A number of these compounds are regulated.

Vulnerability Assessment (VA). A systematic analysis required by the US Environmental Protection Agency of all water systems serving more than 3,300 people that is used to determine the risks posed to the operations of water supply, treatment, and distribution systems. VAs are conducted by the state primacy agency and serves as the basis for the state to allow a water system to lessen monitoring frequencies for certain contaminants. VA is also known as risk assessment or threat assessment.

Vulnerability Self-Assessment Tool (VSAT™). A proprietary software developed to assist water and wastewater utilities to perform vulnerability assessments using a qualitative risk assessment methodology. The software is available free of charge on the US Environmental Protection Agency Website for wastewater utilities, drinking water utilities, and utilities providing both services.

Water Age. The time from when water is treated to when it reaches the customer.

Water Demand. A schedule of the water requirements for a particular purpose, as for irrigation, power, municipal supply, plant transpiration, or storage

Water Hammer. The phenomenon of pressure oscillation that occurs in pipes when a valve is opened or closed very rapidly, creating a sound similar to someone hammering on a pipe. When a valve position is changed quickly, the water pressure in a pipe increases and decreases in a very quick sequence, potentially causing serious damage to the system.

Water Information Sharing And Analysis Center (Water ISAC). A rapid-notification system and information resource that was developed to convey information regarding threats to American, Canadian, and Australian drinking water and wastewater systems through an Internet-based, highly secure subscription service available only to water and wastewater utilities. It provides a means for rapid dissemination of alerts, expert analysis, and other information specific to the water community.

Water Meter. An instrument, mechanical or electronic, used for recording (in cubic feet, gallons, or cubic meters) the quantity of water passing through a particular pipeline or outlet. In water-processing systems, meters may initiate certain functions, such as automatically starting the regeneration cycle in an ion-exchange system.

Water Quality. The chemical, physical, and biological characteristics of water with respect to its suitability for a particular purpose. The same water may be acceptable for one purpose or use but unacceptable for another, depending on its characteristics and the requirements for the particular use.

Water Quality Monitoring (WQM). The process of measuring water quality characteristics, usually at regular intervals over a long period of time, at the same location.

Water Quality Parameter (WQP). A measurable physical, chemical, radiological, or biological characteristic of an aquatic environment. Such a parameter is also called a water quality determinant or water quality variable.

Water Storage. Water that is contained in tanks, reservoirs, clearwells, and the distribution systems, and is readily available for use. This can include water for a specific purpose, such as fire protection.

Water System Operation (WSO). A term usually referring to the operation of water distribution system. The system begins with treatment plant output, and included the transmission and distribution system, clear water reservoirs, pump stations, valves, hydrants, and other ancillary facilities. The water system operator controls flows, pressures, levels, and supply to and within the distribution system. The system is usually operated through a supervisory control and data acquisition (SCADA) interface. The operations center or load control is staffed by certified water system operators.

Waterborne Disease. A disease transmitted through the ingestion of contaminated water. Water acts a passive carrier of the infectious agent or chemical. Water is only one mode of transmission, and epidemiological investigation is necessary to identify water as the source of infection. Chlorea (caused by Vibrio chloerae) and typhoid fever, the classic waterborne diseases, continue to occur but are important primarily in the developing countries of the world. Outbreaks of these diseases are occasionally reported in the United States and other more industrialized countries.

Waterborne Disease Outbreak (WBDO). A significant occurrence of acute infectious illness in at least two persons that is epidemiologically associated with the ingestion of water from a public water system that is deficient in treatment, as determined by the appropriate local or state agency.

Waterborne Pathogen. Any of a group of pathogenic microorganisms that are excreted or shed by infected humans, infected animals, or both in their feces and are then acquired by ingestion. Waterborne organisms may be transmitted directly by ingestion of water or indirectly by, e.g., contamination of food during washing or processing or transmission through the skin or the anal area during recreational use of water (as in hot tubs).

Wholesale Customer. Another water system, usually smaller, that purchases water or has water delivered to it from another water system (public, private, or cooperative), on a regular basis at least 60 days per year and not on an emergency basis.

World Health Organization (WHO). A specialized agency of the United Nations with primary responsibility for international health matters and public health. Created in 1948, the World Health Organization promotes the development of comprehensive health services, the prevention and control of diseases, the improvement of environmental conditions, the development of health-related human resources, the coordination and development of biomedical and health services research, and the planning and implementation of health programs. The organization is headquartered in Geneva, Switzerland.

Yellow Water. Water that is colored as a result of the presence of iron. The iron can originate either from the source water or from corrosion reactions with iron piping. Yellow water is sometimes referred to as red water.

Zinc (Zn). A metallic element that is an essential growth element in terms of nutrition. Zinc is used industrially in alloys, electroplating, batteries, and fungicides.

Zero-Order Kinetics. In kinetics, the rate of a reaction and the concentrations of the reactants maybe expressed as a relationship between the two. A zero-order reaction does not depend on the concentration of the reactant, whereas in a single-order (first-order) reaction, the reaction rate is directly proportional to the concentration of one of the reactants. In a double-order (second-order) reaction, the rate of the reaction is directly proportional to the product of two reactants or the square of the concentration of a single reactant.

This page intentionally blank.

Index

Note: *f.* indicates figure; *t.* indicates table

D

E

Q

R

T

This page intentionally blank.

AWWA Manuals

M1, *Principles of Water Rates, Fees, and Charges*, #30001
M2, *Instrumentation and Control*, #30002
M3, *Safety Management for Water Utilities*, #30003
M4, *Water Fluoridation Principles and Practices*, #30004
M5, *Water Utility Management*, #30005
M6, *Water Meters—Selection, Installation, Testing, and Maintenance*, #30006
M7, *Problem Organisms in Water: Identification and Treatment*, #30007
M9, *Concrete Pressure Pipe*, #30009
M11, *Steel Pipe—A Guide for Design and Installation*, #30011
M12, *Simplified Procedures for Water Examination*, #30012
M14, *Backflow Prevention and Cross-Connection Control: Recommended Practices*, #30014
M17, *Fire Hydrants: Installation, Field Testing, and Maintenance*, #30017
M19, *Emergency Planning for Water Utilities*, #30019
M20, *Water Chlorination/Chloramination Practices and Principles*, #30020
M21, *Groundwater*, #30021
M22, *Sizing Water Service Lines and Meters*, #30022
M23, *PVC Pipe—Design and Installation*, #30023
M24, *Planning for the Distribution of Reclaimed Water*, #30024
M25, *Flexible-Membrane Covers and Linings for Potable-Water Reservoirs*, #30025
M27, *External Corrosion Control for Infrastructure Sustainability*, #30027
M28, *Rehabilitation of Water Mains*, #30028
M29, *Water Utility Capital Financing*, #30029
M30, *Precoat Filtration*, #30030
M31, *Distribution System Requirements for Fire Protection*, #30031
M32, *Computer Modeling of Water Distribution Systems*, #30032
M33, *Flowmeters in Water Supply*, #30033
M36, *Water Audits and Loss Control Programs*, #30036
M37, *Operational Control of Coagulation and Filtration Processes*, #30037
M38, *Electrodialysis and Electrodialysis Reversal*, #30038
M41, *Ductile-Iron Pipe and Fittings*, #30041
M42, *Steel Water-Storage Tanks*, #30042
M44, *Distribution Valves: Selection, Installation, Field Testing, and Maintenance*, #30044
M45, *Fiberglass Pipe Design*, #30045
M46, *Reverse Osmosis and Nanofiltration*, #30046
M47, *Capital Project Delivery*, #30047
M48, *Waterborne Pathogens*, #30048
M49, *Butterfly Valves: Torque, Head Loss, and Cavitation Analysis*, #30049
M50, *Water Resources Planning*, #30050
M51, *Air Valves: Air-Release, Air/Vacuum and Combination*, #30051
M52, *Water Conservation Programs—A Planning Manual*, #30052
M53, *Microfiltration and Ultrafiltration Membranes for Drinking Water*, #30053
M54, *Developing Rates for Small Systems*, #30054
M55, *PE Pipe—Design and Installation*, #30055
M56, *Nitrification Prevention and Control in Drinking Water*, #30056
M57, *Algae: Source to Treatment*, #30057
M58, *Internal Corrosion Control in Water Distribution Systems*, #30058
M60, *Drought Preparedness and Response*, #30060
M61, *Desalination of Seawater*, #30061
M63, *Aquifer Storage and Recovery*, #30063
M65, *On-Site Generation of Hypochlorite*, #30065
M66, *Cylinder and Vane Actuators and Controls—Design and Installation*, #30066
M68, *Water Quality in Distribution Systems*, #30068

This page intentionally blank.

www.ingramcontent.com/pod-product-compliance
Lightning Source LLC
LaVergne TN
LVHW080008210926
848481LV00011B/2974
* 9 7 8 1 6 2 5 7 6 2 2 6 9 *